CP 10ª

AMERICAN GEOGRAPHY
INVENTORY & PROSPECT

Preston E. James & Clarence F. Jones, *Editors*

John K. Wright, *Consulting Editor*

Maps by John C. Sherman

Published for the

Association of American Geographers

by Syracuse University Press – 1954

FOREWORD

American Geography: Inventory and Prospect, the result of a truly cooperative effort, appears in the year that marks the semi-centennial of the Association of American Geographers. While not an official publication of the Association, nor originally intended as an anniversary memorial, the appearance of the volume at this time is indeed most welcome and appropriate. Birthdays are occasions for looking back over the years to note what has been accomplished and for taking stock of the present as a basis for future action. This volume has not the purpose to review the past for its own sake. Rather, as the title indicates, it brings together the experiences of modern American geographic research to provide guideposts for the decades ahead. It considers concepts or basic generalizations that have been used in professional geographic work, both those that have been found wanting and those that today steer our efforts; and the procedures or methods that have been found useful in geographic research. To the questions, so often asked: "What is geography?" and "What do geographers do?" this book gives some of the answers. It is written not only for the trained geographer, but also for the educated layman, for the apprentice geographer and for the worker in another discipline who may want to know what American geographers are thinking and doing and what they hope to accomplish.

These statements need some elaboration and qualification, lest the aim and content of the book be misunderstood. Although this volume is published by the Association, it does not everywhere reflect the thoughts of all members of the Association, nor even of the Editors or of those who were responsible for the final draft of the chapters. The endeavor for sincere cooperation has resulted in some cases in finding a compromise, whereas in others controversies stand revealed. It is likely that each member of the Association will take exception to one or another thought expressed in this volume. Nevertheless, the completion of this project is in itself an indication that there exists a broad basis of substantial agreement among American geographers regarding the aims and methods of their discipline.

A book that deals with the fruits of American learning in the last two generations exposes the reader, especially the apprentice-geographer, to the risk of mistaking a survey of the Here-and-Now for all that there is to know in his world of ideas. Surely nothing was further from the minds

v

of the instigators of this project than such a monument to provincialism. This work must be seen in the perspectives of space and time. American Geography is but one road among the many that together form the international network of Western scientific thought. Richard Hartshorne's *The Nature of Geography*, published in the *Annals of the Association of American Geographers* in 1939 and still widely used in the training of geographers, is evidence of the recognition of this fact.

No less important is the need to see this book in historical perspective. The fifty years of the Association of American Geographers coincide with the normal period of "living memories," that is, of personal experience added to what was learned at first hand from teachers and older associates. This easily results in magnifying the significance of development in the last half century, or at least in over-emphasizing the gap between this last period and the preceding era. While there is no doubt that the founding of the Association signalized a great change and advance in American geography, it did not make a beginning out of nothing. John K. Wright's recent work, *Geography in the Making, The American Geographical Society 1851-1951*, reveals the debt to our predecessors who worked without the benefit of a professional association.

A glance at the year 1904 shows how much it was a link rather than a beginning. The eighth International Geographical Congress convened at Washington and New York and was attended by such masters as William Morris Davis, Paul Vidal de la Blache, and Albrecht Penck. Ellen Churchill Semple's *American History and its Geographic Conditions*, published in the fall of the previous year, was arousing wide interest. Ellsworth Huntington was beginning to develop his ideas on climatic changes. In Germany Friedrich Ratzel died in this year and Alfred Hettner initiated his now classic series of articles on methodology. In England Halford J. Mackinder presented his paper on "The Geographical Pivot of History" before the Royal Geographical Society. All this underlines the truth—perhaps self evident—that the present volume must be viewed as an item in the chorology and history of geographical ideas. It can only be hoped that it will prove to be a significant item.

It was in the nature of the project that geography would be dealt with solely as a scientific discipline. Important as this aspect is for the scholar it by no means characterizes all sectors of the field. There is, for instance, the teaching of geography, a vital activity for most professional geographers. Then too, geography, like history, has its broadly cultural aspects, appealing to the poetic and artistic imagination. The delight one experiences when viewing a beautiful landscape, the insight gained by reading a regional novel, or the esthetic satisfaction derived from seeing a good map, all these and other pleasures are part of geography and, fortunately, not

limited to licensed practitioners. Even in the field of science or its application, geographers hold no monopoly of geographical interests and ideas,
and this is all to the good.

In brief, then, this volume makes no claim to have covered all phases of
American geography. It is perhaps best characterized as a progress report
on the objectives and procedures of geographic research as of January,
1954, rather than as a definitive compendium of the present content of
geography. Even the most complete survey of present factual knowledge
is of far less value than a few constructive ideas that may guide future
research. If the project stimulates the *doing* of geography instead of the
mere talking about it, much will have been accomplished.

A brief review of the history of the project will aid in understanding the
scope of this volume. Ten years ago some geographers, while discussing
their common experiences, agreed that the preceding two decades had
been of great significance for American geography. In the late 1920's and
in the 1930's geography began to play an important practical role in its
application to land-use planning studies. World War II created an unprecedented demand for geographers and required new lines of thought,
new methods, and new devices. It was felt that the experiences of these
decades should not be lost, or retained only among those who directly
participated in these events, but ought to be discussed and presented as a
guide to further professional development. A preliminary step in this direction was taken at a meeting of the Committee on Training and Standards in the Geographic Profession, National Research Council, held at
Hershey, Pennsylvania, October 6-8, 1944, under the chairmanship of
Richard Hartshorne.[A]

In the following three years there was uncertainty regarding the ways
and means of bringing these ideas to fruition. A great step forward was the
action of the National Research Council, through its Division of Geology
and Geography, in setting up a number of committees, each of them to
plan, coordinate, and promote the project in its own field. In the spring of
1949 a meeting was held at Evanston, Illinois, which was attended by the
chairmen of the National Research Council committees and members of
an earlier committee of the Association appointed by Robert S. Platt. At
this conference the plan for a symposium on American geography took
definite shape. Preston E. James and Clarence F. Jones were asked to guide
the general organization of the project. Funds were obtained from the
Social Science Research Council, from the National Research Council,
and through the latter from the Office of Naval Research, to meet the

A. The report, by the chairman, of the findings of this conference, together with the
list of those participating, is presented in the *Annals of the Association of American
Geographers*, 36 (1946): 195-214.

costs of a number of meetings of the committees. These conferences afforded valuable opportunities for threshing out the controversial problems on content which each of the committees had to consider and to resolve. Meetings of the whole central committee were held in October 1952 and in March 1953.

Although the chairmen of the committees and in two cases, individual authors, are responsible for the content of the final versions of their respective chapters, the overall coordination and editing had to be done by a small group. Preston E. James and Clarence F. Jones undertook the task, and John K. Wright acted as consulting editor.

From the start the whole program called for wide discussion, far beyond the circles of the various committees. The authors who initiated each chapter were asked to express the opinions of their committees and not solely their own personal view. In all but two cases, the original draft of each chapter was subjected to critical appraisal. An early version of the chapter on regional study was presented by Derwent S. Whittlesey at the 1950 meeting of the Association and most chapters were widely distributed among colleagues presumed to be interested in the specific topics under discussion. The resulting comments led to numerous revisions, and for several chapters the original form has been almost completely lost. Thus the book represents the combined thoughts of hundreds of professional geographers. One of the most worth-while accomplishments of the program has been the development of such widespread discussion of objectives, methods, and concepts.

At the conclusion of a project of such complexity, to which so many people have devoted their best efforts, a simple expression of gratitude is hard to formulate. Certainly we owe much to the National Research Council, the Social Science Research Council, and the Office of Naval Research. Without their moral and financial support this work could not have been brought to its completion. The committee chairmen and the Editors deserve our deep-felt thanks for the devotion to their difficult and time-consuming tasks. But many other colleagues, whether members or not of a committee, have made significant contributions. The satisfaction of having been a participant in this exciting venture will be their reward. If one name is to be mentioned in this acknowledgment it must be that of Preston E. James. With persistence and good humor he has put through this project in the face of obstacles that would have discouraged almost any other man.

J. Russell Whitaker,   Jan O. M. Broek,
President, Association of   Chairman, Publications
American Geographers   Committee

COMMITTEE ON AMERICAN GEOGRAPHY

THE ASSOCIATION OF AMERICAN GEOGRAPHERS

Meredith F. Burrill
Andrew H. Clark
Charles M. Davis
Richard Hartshorne
G. Donald Hudson
Preston E. James, Chairman
Clarence F. Jones, Co-chairman
Clyde F. Kohn

Hoyt Lemons
Harold M. Mayer
Raymond E. Murphy
Robert S. Platt
Arthur H. Robinson
J. Russell Whitaker
Derwent Whittlesey
John K. Wright

COMMITTEES OF THE NATIONAL RESEARCH COUNCIL

DIVISION OF GEOLOGY AND GEOGRAPHY

Chairmen

Cartography, Arthur H. Robinson
Economic Geography, Raymond E. Murphy
Field Techniques, Charles M. Davis
Population and Rural Settlement, Clyde F. Kohn
Historical Geography, Andrew H. Clark
Physical and Bio-Geography, Hoyt Lemons
Political Geography, Richard Hartshorne
Regional Geography, Derwent Whittlesey
Social Geography, Meredith F. Burrill
Urban Geography, Harold M. Mayer

CONTENTS

CHAPTERS AND PRINCIPAL AUTHORS

FIGURES

INTRODUCTION: THE FIELD OF GEOGRAPHY[A]

GEOGRAPHY AS A PROFESSIONAL FIELD
THE FACE OF THE EARTH
WHO IS A GEOGRAPHER?

THE METHOD OF GEOGRAPHIC RESEARCH
LIKENESSES AND DIFFERENCES AMONG PLACES
THE REGION AS A GEOGRAPHIC GENERALIZATION
THE MAP AS AN ANALYTIC DEVICE
   Points, Lines, and Areas on Maps
   Continuities, Discontinuous Distributions, and Discontinuities
   Accordant and Discordant Relations
THE DEVELOPMENT OF CONCEPTS
TWO CONCLUSIONS REGARDING GEOGRAPHIC METHOD

THE SUBDIVISIONS OF GEOGRAPHY

THE PROSPECT

A. Original draft by Preston E. James. It has been discussed in detail by the Committee on American Geography and has been revised on the basis of many critical comments which emerged from group discussions and from study by individuals. Among the many geographers who have contributed to the ideas herein formulated special mention must be made of Richard Hartshorne, J. Russell Whitaker, and John K. Wright.

2

CHAPTER ONE

INTRODUCTION: THE FIELD OF GEOGRAPHY

GEOGRAPHIC scholarship in the United States has a history which closely parallels that of the republic itself. Jedidiah Morse published his first compendium on the geography of America in 1789, and by 1810 a considerable body of information on the infant country was available to compilers [1].[B] When the American Geographical Society was founded in 1852, the members brought together a common interest in the support of exploration, and in hearing or reading reports on strange lands and peoples [2]. During the second half of the 19th century, American geomorphologists developed advanced concepts regarding the nature and origin of landforms, American climatologists gathered and used recorded data to depict the patterns of weather elements as they had never been depicted before in America, and field men, such as John Wesley Powell, engaged in the inventory of resources and in the classification and mapping of land. Since 1900, however, the tempo of geographic scholarship has greatly increased. This was partly the result of stimulating ideas and procedures brought over from the European geographers, especially from the Germans and French, but it was also a result of the increasing demand for geographic understandings as the economic and political horizons of America widened.

During this time, also, the number of professionally trained geographers greatly increased. In 1904, when William Morris Davis called together a group of scholars to found the Association of American Geographers, not more than a hundred persons in the United States could have been described as professional geographers. As the Association reaches its fiftieth anniversary, the number is nearing 2,000, which is still small compared with other similar professional groups. From these few, however, have come works which demonstrate the value of geographic studies: the value both for the satisfaction of intellectual curiosity about the world we live in, and also for the clearer understanding of the nature of the world's current difficulties and of the issues involved in the decisions that must be made to face them. With the increased demand for American leadership in intellectual affairs as well as in the practical problems of living, all the

B. Numbers in brackets refer to references at the end of each chapter.

fields of scholarship will be called upon for increasingly effective service, geography among the others. Some geographers envision the probability of a rapid expansion of numbers, and are concerned that the geographic profession should merit the wider recognition and greater responsibilities which seem certain to come to it.

What are the concepts and procedures of geographic study that have been found valuable during the past half century? And what are the frontiers of thought which challenge geographic investigation? This seems to be a time for stock-taking, a time to make an inventory of the recent decades of professional work in the United States, and to examine the prospect that lies before us.

GEOGRAPHY AS A PROFESSIONAL FIELD

Geography is a very old field of study. Writers of geography were among the earliest scholars of antiquity, and spoken geography must have been widely practised long before the invention of writing. During the long history of man's effort to gain further understanding of the forces and objects of his environment, the essential nature of geography remained unchanged. Today as in the past, geography is concerned with the arrangement of things on the face of the earth, and with the associations of things that give character to particular places. Those who face problems involving the factor of location, or involving the examination of conditions peculiar to specific locations, are concerned with geography, just as those who must be informed about a sequence of events in the past are concerned with history. Many people, in various professions and in many kinds of business, write history and geography. Professional historians develop methods and set standards for historical writing; professional geographers bear a similar responsibility with respect to the writing of geography.

Geographic writing may be addressed to several kinds of audiences. It may be popular and descriptive, or highly theoretical; it may be practical and directly applicable to current problems of the sort commonly handled also by engineers, economists, public administrators, or business men. But whatever its tone, geography always is concerned with the characteristics of places and with the significance of likenesses and differences among places on the face of the earth.

THE FACE OF THE EARTH

The face of the earth is complex, fashioned and continually changed as it has been by processes of different kinds proceeding at different rates or tempos. The basic conditions in what may be called terrestrial space result from the operation of physical processes. In accordance with biotic processes plants grow in the soil, water, and air, and animals live in intricate relationship to plants and to each other. In the midst of these physical and biotic conditions man himself carries on his economic, social, and political affairs in accordance with processes described as cultural or societal. The interaction of these diverse processes gives character to the face of the earth, and produces the contrasts and similarities of its differentiated parts.

The study of terrestrial space involves knowledge both of processes and of the resulting phenomena. A process is a sequence of change systematically related as in a chain of cause and effect. The phenomena that can be observed at any one moment of time result from the operation of these sequences of change. Because the various processes are irregularly distributed over the earth, so are the phenomena they produce. Things are present in some places, absent from others, and they may vary in intensity or motion from place to place. The interpretation of the arrangement and character of the phenomena of terrestrial space requires an understanding of the processes which have produced them.

The phenomena which are irregularly distributed over the earth are also irregularly associated with other phenomena in particular parts of the earth. Phenomena that are systematically related to each other because they are the product of one kind of process are associated on the face of the earth with other phenomena produced by quite different processes. This is what William Morris Davis meant when he wrote of the unsystematic natural groupings of things. Some of the phenomena associated in a particular place are unsystematically related because they are produced by quite different processes.

In ancient times, when the nature of all these processes was only dimly understood, geographers were mainly concerned with the identification of the phenomena that gave distinctive character to different areas. They described what they could see and they made the first attempts to find ways to measure things and to place them on maps. They speculated, sometimes with keen insight, regarding the sequences of cause and effect that had produced the characteristics of the earth's different areas. During the great age of exploration, which began in the 15th century, geographers engaged in perfecting the methods of observation and of recording on maps the outlines of land and water or the courses of rivers. Little by little the understanding of the processes developed, and the phenomena systematically related to any one process were identified.

Out of an original concern with the characteristics that identified different regions and countries emerged a greater concern with the operation of specific processes wherever they might be located. The various fields that emerged from these systematic studies were defined in terms of the processes they investigated. These fields gradually became distinctive and developed into what are recognized as separate disciplines of learning. Workers in each separate discipline developed their own methods and devised special instruments to measure and describe more precisely the particular sequences of change which interested them. Some processes could be isolated in a laboratory where controlled experiments were devised to eliminate all irregularities or modifications of the sequence imposed by outside influences. Other processes, notably those related to human behavior, could not so easily be isolated in a laboratory; these were isolated symbolically by the use of such phrases as "other things being equal." In the emerging disciplines, including physics, chemistry, the earth sciences, and eventually the social sciences, extraordinary gains were made in the understanding of processes; the methods used brought results of profound importance. Meanwhile geography, from which some of these new disciplines had emerged, continued to focus its attention on the characteristics of countries, and thereby became increasingly more encyclopedic.

Modern geography, in a sense, is very new. Instead of constituting the first step in a process of scholarship, it now constitutes one of the more advanced steps. Instead of fumbling among the phenomena of the face of the earth in the search for system and order, geography now can call upon the understandings provided by the sciences devoted to the study of specific processes, the systematic sciences. Geography is still the field that deals with the associations of phenomena that give character to particular places, and with the likenesses and differences among places. But it plays this role now with the background of systematic knowledge provided by other fields. Geography looks at particular places with a view to understanding how specific processes operate on the face of the earth where other things are not equal.

In this role geography makes three kinds of contribution: 1) it contributes toward a better understanding of the earth as the habitat of man by extending the findings of the other sciences; 2) it provides a means for testing the validity of certain concepts developed by the other sciences by applying them in particular places; and 3) it offers its own peculiar perspective to the clarification of the issues involved in problems of public or private policy.

WHO IS A GEOGRAPHER?

Obviously no one person could become a specialist in all aspects of such a broad field. Geography does, in fact, cover everything on the face of the earth in the sense that any phenomenon, physical, biotic, or cultural, that is irregularly distributed over the earth can be studied by geographic method. This method can be applied to the study of either material or non-material phenomena. Persons who undertake to carry on geographic studies must specialize in order to develop competence in a portion of the field. Nevertheless, whether they specialize on the physical, biotic, or cultural aspects of geography, the analysis of the meaning of likenesses and differences among places involves the use of certain common concepts and methods: basic to the whole field is the regional concept; fundamental to the effective study of geographic phenomena is the method of precise cartographic analysis. In whatever part of the field a geographer specializes he finds common ground with other geographers in two ways: 1) he accepts the fundamental concept that differing patterns and associations of phenomena on the earth produce similarities and contrasts between places, and that these similarities and contrasts are significant in terms of continuing processes of change and therefore worth studying: and 2) in order to understand the significance of area differentiation more fully he defines categories of patterns and associations and studies them in their areal relationships.

Scholars who are identified professionally as geographers are not the only ones who make use of the geographic method or who study and write what is, in fact, geography. Geographic writing is done by anthropologists, economists, sociologists, political scientists, botanists, zoologists, geologists, business men, and many others. Students who are seeking ever more complete understanding of the operation of a specific process will properly pursue the study of this process and its related phenomena to their actual occurrence in particular places. It would be unrealistic to say that in studying places they do work which only professionally trained geographers should do. Workers in the social sciences regularly make use of the so-called historical method, and many of them write history. They would be condemned if they failed to make use of historical concepts or if they failed to use historical method expertly. These same scholars should be encouraged to make use of geographic concepts and geographic methods, and to do so with a similar degree of expertness. Failure to make adequate use of geographic concepts and methods has been recognized, and not only by geographers, as a common weakness in current social science, a result perhaps of the neglect, until recently, of geographic training in the professional preparation for work in these fields. It is equally essential that geographers should learn in greater measure than is today common to understand and use the concepts and methods of the related disciplines.

THE METHOD OF GEOGRAPHIC RESEARCH

To persons who wish to work with geography, the geographic profession offers certain concepts and methods derived from experience in the study of likenesses and differences on the face of the earth. In the several chapters of this volume the basic themes are repeated, with variations developed by writers who possess special competence in one or another of the branches of geography. The regional concept, which constitutes a fundamental part of the geographer's point of view, underlies the discussion of all the various branches of the subject, and therefore requires special treatment in this introduction.

LIKENESSES AND DIFFERENCES AMONG PLACES

No two places on the face of the earth are identical. This is true even if attention is focused on a minute spot. It is possible, however, to identify areas that are more or less homogeneous with respect to this or that phenomenon or combination of phenomena. There is no such thing as a "unit area," for even the smallest of these more or less homogeneous segments of terrestrial space could be further subdivided if it were worth while to do so. There is no such thing as an indivisible area of completely uniform character.

Nevertheless, if the complexity of the face of the earth is to be brought into manageable segments for purposes of examining the causes and consequences of area differentiation, to examine each minute spot separately would defeat the endeavor. Geographers must always seek to generalize, to define categories in terms of selected criteria, to formulate meaningful classifications. This is the method of all science; yet among the sciences there are two kinds of generalization—those which are applied to the classification of phenomena and those which are applied to the description of the ideal operation of isolated processes. Geography deals with the first primarily, and with the second only in so far as certain processes have not yet been examined by one or another of the systematic sciences. The same kind of approach is followed by other disciplines: the sociologist defines general classes into which the diverse individuals of a society are grouped for purposes of study; the geologist defines categories of rock in each of which a certain range of characteristics is permitted. The geographer identifies areas, each with a certain kind of meaningful homogeneity; these areas he calls regions.

THE REGION AS A GEOGRAPHIC GENERALIZATION

During the five years in which the Committee on Regional Study (see chapter two) under the leadership of Derwent Whittlesey examined and discussed the regional concept there inevitably emerged a sharpening of certain definitions. The dictionary defines a region as a large, indefinite area, a major division of the earth's surface; in popular usage the word usually connotes a relatively large, continuous area having some general quality of homogeneity, yet with boundaries not precisely defined; for example, the Middle West. Regional geography in the traditional sense seeks to bring together in an areal setting various matters which are treated separately in topical geography. Whittlesey's committee, attempting to formulate more precise concepts and methods, arrived at a definition of the word region which differs from popular usage in that it implies: 1) an area of any size; 2) an area homogeneous in terms of specific criteria; and 3) distinguished from bordering areas by a particular kind of association of areally related features and therefore possessing some kind of internal cohesion. The word region, as used in this volume, refers to areas of this sort; the regional method refers to procedures by which regions, so defined, are identified and utilized for geographic purposes. The regional method is not to be confused with regional geography in the traditional sense.

The regional concept and the regional method involve the geographic generalization of phenomena associated in area. This is accomplished by defining area categories in terms of selected criteria relevant to a particular problem or purpose, disregarding, as all generalizations must, conditions considered to be irrelevant to the purpose sought. A system of regional divisions is justified if it illuminates the factors or elements of a problem; it is not justified if it obscures these factors or elements. Such a system is based on the selection of parts of a whole designed to clarify the understanding of a situation which otherwise would remain less clearly understood. Whether, or to what degree, it accomplishes this purpose is a matter of judgment or critical appraisal. There can be no such thing as a "correct" system of regions, or a system of "true regions;" no one system of regions is right and all others wrong. There are as many regional systems as there are problems worth studying by geographical method.

THE MAP AS AN ANALYTIC DEVICE

The regions which a geographer defines and observes on the face of the earth are plotted at a reduced scale on maps. The map is the fundamental instrument of geographic research. In the field it provides a precise way of recording observations. It reduces the patterns of area differences on the earth to a size permitting close analysis and comparison. The map is an eloquent form of presentation for those trained to read its symbols, a form of presentation much more concise than can be achieved by the use of word symbols alone. It provides the geographer with a method for reaching a degree of precision in measurement comparable to that achieved by the statistical method, but it is a geometric rather than an arithmetic precision.

Points, Lines, and Areas on Maps

The phenomena which produce likenesses and differences among places on the face of the earth form three different kinds of patterns when they are transferred to maps at reduced scales. Phenomena actually occupy volumes of space: for example, soil types extend below the surface of the ground, air masses and climates extend far above the surface, and political units extend indefinitely in both these directions. But when phenomena are plotted at reduced scale, as they must be for geographical analysis, they appear as patterns of points, patterns of lines, or patterns of areas. Some phenomena, such as houses, mines, or springs, occupying relatively small volumes of space, appear as points marked by out-of-scale symbols on all but very large-scale maps. Other phenomena with linear form, such as rivers, roads, or fences, appear on maps as lines, also with out-of-scale symbols. Such phenomena as landforms, climates, vegetation cover, land use, or political authority, appear on maps as areas.

Continuities, Discontinuous Distributions, and Discontinuities

When areas are plotted on maps for purposes of geographic analysis three contrasted kinds of differentiation may be recognized: 1) differences of degree; 2) differences of enumeration; and 3) differences of kind.

Differences of degree exist when a phenomenon extends continuously over the earth, varying only in intensity from place to place. This is known as a continuity. Examples of continuities are air temperature, degree of slope, or ratios such as cropland to total area. The technique of plotting continuities on maps involves the use of different kinds of isograms, or lines of equal value or ratio. There are two kinds of isograms: isarithms which represent a constant value or intensity pertaining to every point through which they pass, such as isotherms or contours; and isopleths which measure ratios such as cropland per square mile, pertaining to specified areas through which they pass. Any kind of isogram drawn through a continuity indicates the direction of greatest variation in intensity, which is always at right angles to the line. It does not mark a boundary between different kinds of things, for the same kind of thing is present on both sides of the line.

Differences of enumeration exist when a phenomenon arranged in a pattern of points must be plotted on a map at such a reduction of scale that each individual item can no longer be shown with its own symbol, even by an out-of-scale symbol. The treatment of population raises such a problem, for to show the actual position of each individual person on a population map would not be possible. In such a case the pattern is said to be one of discontinuous distribution. To plot discontinuous distributions on a map requires the definition of areas of enumeration, that is, of areas within which the count of individuals is summed up. The enumeration district of the census is such an area. Discontinuous distributions can be shown as continuities when numbers of people are related to area, as people per square mile.

Differences of kind exist when a phenomenon is arranged in discrete (that is, individually distinct) areas, each distinguished from bordering areas by differences of quality or character. The resulting pattern forms a discontinuity. Examples of discontinuities are land and water, forest and grassland, or cropland distinguished from other kinds of land use. Discontinuities are plotted on maps by applying criteria or specific definitions for the separation of different kinds of things. Although their boundaries may be transitional, they nevertheless separate areas which are occupied by different classes of phenomena as defined by the criteria, not areas occupied by the same kind of phenomena with differing intensity. Lines separating discontinuities are not to be confused with isograms even when the matching of a pattern of discontinuities with a pattern of continuities reveals that the two kinds of lines coincide.

Regions must be discontinuities. A region, as the term is used in this book, is an area which is not only homogeneous in terms of specific criteria, but also possesses a unique character in terms of a particular association of features. An objective of regional study, therefore, is to transform continuities and discontinuous distributions into areas that differ in quality or character from bordering areas. The problems involved in these technical procedures of geographic research are discussed in chapter two.

Accordant and Discordant Relations

Cartographic analysis brings to light different kinds and degrees of areal relationship between phenomena. Of course everything on the face of the earth has some relation to everything else, but many relationships are discordant, and many, also are accordant. Accordant relations vary in the degree of correspondence. If that is so complete that two phenomena occupy exactly the same area on the earth, the boundaries outlining the areas would coincide. Most often in the analysis of areal relations geographers deal with discernable similarities, varying in degree, between the areas of occurrence of two phenomena. The identification of an accordant relationship, however, is no proof of causal relationship, although it may offer a strong indication of the existence of some causal connection. Even where two phenomena coincide in area, the proof of causal connection must rest on the identification of the process or processes that have produced the observed results. It is entirely possible that neither of the phenomena observed is the cause of the other.

THE DEVELOPMENT OF CONCEPTS

Out of geographic study two different kinds of concept may be developed: that of regional systems, and that of causal relations.

The first kind of general concept is based on the understanding of the region as a geographic generalization. It has been suggested that because the association of phenomena at any place on the earth is always unique, no generalization is possible, but only descriptions of unique situations. But the uniqueness of individual cases underlies all scientific generalization, even in such an exact science as physics. Every human being in a society is unique. In all science it is necessary to define categories based on criteria considered relevant to the problem and to disregard irrelevant details. In the field of geography, as long as attention is focused exclusively on the differences between places, the unique quality of each place is emphasized, and for some purposes this is exactly what should be done. However, likenesses between places may also be important, and for some purposes more important than the differences. The identification of the similarities between places is a geographic method of arriving at a generalization, a generalization, that is, of the characteristics of area.

The second kind of concept that may be developed from geographic study has to do with causal relations, and requires the time perspective as well as that of space. At the beginning of the present century, and for some time thereafter, geography in America was dominated by the concept of environmental determinism. There are some geographers and many non-geographers who still think of geography as a subject which deals exclusively with the reaction or response of mankind to the physical or biotic environment. Environmental determinism involves a causal relationship between the conditions of the physical earth and the activities of man. From studies guided by this basic idea, and from the many modifications suggested from time to time, came such terms as response, adjustment, influence, geographic factors, and many others. From the French geographers Vidal de la Blache, Brunhes, and their followers, a modified concept regarding the causal relations between man and his environment, known as possibilism, was brought into American geography. The physical and biotic environment is conceived as setting limits, offering opportunities among which the inhabitants have several possible choices. Geographers today recognize that a positive determinism can exist, but only when there is a change in the environment due to the operation of physical or biotic processes. In a relatively stable environment, no example of positive determinism has been demonstrated by acceptable method and consequently the concept is no longer considered useful as a guide to geographic understandings.

Actually the rise of the concept of environmental determinism offers an example of the danger of arriving at conclusions regarding causal relations without the study of process in the time perspective. The idea that men were influenced or even controlled by their physical and biotic environment was given much apparent support from field evidence. Much of the evidence, however, was faulty because of the technique used. Again and again in the writings of the heyday of environmental determinism coincident area relations between natural and cultural phenomena were asserted but not demonstrated on maps. When the area relations are properly analyzed, patterns are often found to correspond but not to coincide. The detailed examination of the places where two corresponding patterns fail to coincide reveals the existence of processes which invalidate the theory of environmental determinism as applied in detail.

The general concept commonly accepted today by professional geographers is based on the idea that the physical character of the earth has different meaning for different people: that the significance to man of the physical environment is a function of the attitudes, objectives, and technical abilities of man himself. With each change in any of the elements of the human culture the resource base provided by the earth must be reevaluated.

Geographical generalizations, like those of all science, must be constantly subject to critical study and review. The regional divisions carefully established by an earlier generation of geographers for the specific purposes of that generation may need to be revised through the application of new criteria in order to define regions more directly relevant to current problems. Concepts regarding causal relations, even if widely accepted at one time, must be scrutinized anew by each succeeding generation of geographers.

Two Conclusions Regarding Geographic Method

Two conclusions may be offered concerning the application of the geographic method to the study of likenesses and differences among places on the earth and their significance in terms of causes and consequences. The first stems from an understanding of the contrasts in the kinds of processes at work on the earth. Considering the various methods of measuring and describing these processes, and the differences in the tempos at which they proceed, ranging from the relatively slow changes of geologic time to the rapid rate of current technological change, it seems that an attempt to define regions based on phenomena produced by too many different kinds of processes is dangerous and could lead to serious errors of interpretation. There are some geographers who prefer to define several parallel systems of regions in the same area, each based on phenomena produced by one process or by a group of closely related processes, and then to analyze the areal relations of the several regional systems. Regions based on anything approaching the total content of area must be regarded with a specially critical eye, unless and until they have been validated by a comparison of the areal relations of the component regional systems.

The second conclusion is that geography cannot be strictly contemporary. If the meaning of area differences is sought in terms of causes and consequences this inevitably involves the time perspective, for processes must operate through time. There can be no full explanation of present geography without use of the methods of historical geography, which deals with the geographies of the past and with the changes in geographic phenomena through time. Because of obvious limitations in a scholar's time and competence, and because of the difficulties involved in finding and interpreting much of the relevant data from the past, the historical approach is usually restricted to studies of those past circumstances most clearly explanatory of present conditions. However, this restriction is justified only by the assumption that by tracing back a single process through time the scholar will be able to link past conditions with present circumstances. This is not entirely true; for at all stages in the past, as in the present, the operation of any one process is modified by the total environment in which it operates. A full understanding of contemporary geographic phenomena requires the full perspective of past geographies; and as a step toward such an ideal understanding, the re-creation of the past, without restriction to those phenomena directly related to present conditions, is greatly needed. The regional concept, then, embraces not only the idea that the patterns and associations of phenomena in particular places give character to those places, but also that the meaning of likenesses and differences among places is to be understood only in the light of complex continuous change growing out of the past, and going on into the future.

There are certain dangers inherent in the application of the second conclusion. It involves proceeding from observations of fact to the application of hypotheses concerning origins and developments. The sequences of change which are called processes are intellectual concepts, tested, to be sure, by the direct observation of what are thought to be the resulting phenomena of area differentiation. But once a sequence of change has been clearly described, it is easy to think evidence has been found to support it, and considerable imagination and independence of mind are required to find conflicting evidence. It is only necessary to recall the years during which innumerable peneplains were identified and described: an example, it would seem of the intellectually stifling effect of a clearly-stated but over-simple theory. Yet in spite of the ever-present and essentially human temptation to find what one is looking for, one cannot well go to the extreme of refusing to look for anything. The deeper understanding of the patterns and associations of phenomena which produce area differentiation involves a search for meanings in terms of causes and consequences; and this search inevitably moves away from the strictly contemporary.[c]

THE SUBDIVISIONS OF GEOGRAPHY

Almost all scholars who have thought deeply about the nature of geography agree on the essential unity of the field. The various kinds of duality which have been popular in the past, such as regional as opposed to topical geography, or physical as opposed to human geography, seem to have obscured rather than illuminated the true nature of the discipline. The latter kind of separation between the physical and human aspects continues to hinder the full and balanced development of geography, for it persists in textbooks, in academic organization, and also in the research agencies of government and in the organization of the research councils. This separation seems to have resulted from the 19th century attempt to divide all knowledge into science, meaning natural science, social studies, and humanities. Such a division is intolerable for geographers, for they must deal with man as well as that which is not man (now commonly defined as nature), and the two are intimately intermixed wherever man has been on the earth. Geography, which has to do with places on the earth, simply cannot be made to fit into so arbitrary a classification of knowledge. Actually, there is just one kind of geography.

c. The discussion of the nature and development of geography on the foregoing pages is based on the views of a large number of geographers of the past as well as the present, in Europe as well as in America, who have examined their field in oral discussions, and in published studies. The more influential of these up to 1939 are compiled and critically reviewed in Hartshorne's The Nature of Geography [3]. Since then Ackerman has contributed a most effective statement of the unity of topical and regional geography [4]. Certain parts of the discussion of the regional concept and the regional method are from James' presidential address before the Association of American Geographers, 1952 [5]. Other references to papers cited in the preparation of this chapter on the field of geography are listed [6; 7].

On the other hand, the limitations of the human mind require that individual scholars specialize in certain parts of the whole field. No one person could become competent in all aspects of geography. Therefore each individual scholar must define a subdivision or subdivisions of the whole for his own purposes. It is necessary to specialize both areally and topically.

The several chapters of this volume are not intended to represent a definitive subdivision of the field of geography. They are the parts of the whole in which geographers in the United States have been actively at work in the past fifty years, or in which the committee believes it is most essential that geographers should work. Such a popular subdivision as economic geography is found to include so many different kinds of processes that it is no longer possible to specialize in it as a whole. Economic geography, therefore, is represented by some five topical fields, each of which offers opportunities for true specialization. Yet these five fields might be differently defined, and several other possible kinds of specialties are suggested in the chapter dealing with the other economic fields. There are possible fields of specialization that are not mentioned. Each individual who seeks to develop the professional competence that marks him as a geographer must define his own fields of specialization; he must select the part or parts of the earth in which he wishes to work, and he must select the processes or topics that he wishes to master. The kind of competence achieved determines the kinds of problems that can be attacked. The committee hopes that this volume will stimulate new thinking, new approaches to old problems, rather than lead to any crystallization of thought along the lines here presented.

THE PROSPECT

Professional geography in America is both old and new. It is old because it represents a traditional field of interest of long heritage. It is new because of its role in bringing the findings of the systematic sciences back to the realities of particular places on the earth. Geography is a single discipline because: 1) it has a distinctive point of view and a distinctive methodology; 2) it has a history or tradition through which geographic thought has grown and geographic concepts have been formulated, tested, and revised; and 3) it plays a part in the contemporary world not played by any other field or by all other fields together. It is new in the significance of the role it plays; but it is old in terms of its traditional point of view.

In the long run, professional geography as a distinctive field will grow or decline in proportion to the contribution it makes to understanding.

Such a contribution may be purely to the satisfaction of a deep-seated desire to know and understand the nature of the world and man's part in it. But geography can also be a practical subject. Confronted with changing economic, social, and political values as a result of the radical changes in the technology of living, the persons charged with responsibility for decisions of policy, as well as the thinking citizens who in a democracy share that responsibility, must seek clarification of the issues involved in their decisions. Geography can contribute additional depth of perspective.

Professional geographers are well aware that many of the possibilities of their field of study are only beginning to be developed. But how are these potentialities to be realized? There is need for more geographic research, and more publication of the results; there is need for more production per capita; and there is need for more effective criticism of geographic work, whether that work is done by persons listed professionally as geographers or by persons in other disciplines who undertake studies of geographic quality. There is need for more cooperative work by teams of scholars with different areal and topical specialties. Geographers need to know more about the processes as they are measured and described in the cognate systematic fields; but if the workers who achieve competence in these special aspects of geography and in the cognate fields to which they are closely related, are to make their most effective contribution to geography, they must adhere to the geographic point of view and practise incontrovertible geographic methods. They are charged with the study of the operation of processes in the total environments of particular places, and in this role they must bring to the systematic studies the touch of reality that springs from the use of the regional concept.

This volume seeks to reflect as faithfully as possible the conditions of a field of study in the early stages of its development. Two world wars and a depression have created a demand for the kinds of understandings that geography can contribute; and as the demand and the contribution continue, expansion of the field is inevitable. This inventory of the experience of the past few decades is undertaken for the purpose of setting up guideposts to point toward the frontiers. There is need that the advance toward these frontiers be based on sound theory and acceptable practise.

REFERENCES

1. BROWN, R. H. *Mirror for Americans, Likeness of the Eastern Seaboard, 1810,* American Geographical Society Special Publication No. 27, New York, 1943.
2. WRIGHT, J. K. *Geography in the Making: The American Geographical Society, 1851-1951,* American Geographical Society, New York, 1952.
3. HARTSHORNE, R. "The Nature of Geography: A Critical Survey of Current Thought in the Light of the Past," *Annals of the Association of American Geographers,* 29 (1939): 171-645; Reprinted by the Association and available in the Central Office, Library of Congress, Washington, D. C.
4. ACKERMAN, E. A. "Geographic Training, Wartime Research, and Immediate Professional Objectives," *Annals of the Association of American Geographers,* 35 (1945):121-143.
5. JAMES, P. E. "Toward a Further Understanding of the Regional Concept," *Annals of the Association of American Geographers,* 42 (1952):195-222.
6. DAVIS, W. M. "An Inductive Study of the Content of Geography," *Bulletin of the American Geographical Society,* 38 (1906):67-84.
7. ——— "Principles of Geographic Description," *Annals of the Association of American Geographers,* 5 (1915): 61-105.

THE REGIONAL CONCEPT AND THE REGIONAL METHOD[A]

ANTECEDENTS OF THIS REPORT
THE RISING TIDE OF INTEREST
DISCUSSION AND PUBLICATION
POLITICAL SECTIONALISM AND REGIONAL PLANNING
HUMAN ECOLOGY
DUALISM IN REGIONAL GEOGRAPHY

THE REGION: THEORY AND PROCEDURE
A MONISTIC GOAL FOR REGIONAL STUDY
THE PROCEDURE OF REGIONAL STUDY

ESSENTIAL CONSIDERATIONS IN REGIONAL STUDY
CRITERIA
CATEGORIES
Single, Multiple, and "Total" Regions
Uniform and Nodal Regions
CHARACTERISTICS
CORES AND BOUNDARIES
Cores of Regions
Boundaries of Regions
COMPAGES
A SUGGESTED HIERARCHY FOR COMPAGES
REGIONAL CONSCIOUSNESS
The Outlook for Regional Consciousness

(*Continued on next page*)

A. Original draft by Derwent Whittlesey. For the procedures followed in the preparation of this chapter, and the names of other persons who worked closely with the committee, see footnote B.

19

THE METHOD OF REGIONAL STUDY
APPROACHES
ATTITUDES
TOOLS
   Sources of Information
   Techniques of Analysis
POINTS OF REFERENCE

FRONTIERS OF REGIONAL RESEARCH
REGIONAL ELEMENTS
TOOLS AND TECHNIQUES OF INVESTIGATION
MODES OF PRESENTATION

CHAPTER TWO

THE REGIONAL CONCEPT AND THE REGIONAL METHOD

GEOGRAPHERS ARE in general agreement that regional study is an essential part of their craft. However, they do not claim for their discipline exclusive rights to the regional concept. The recognition of regional distinctions figures in all disciplines that deal with features which vary from place to place on the earth. History, while concerned primarily with tracing human events through time, finds those events occurring in particular areas. Each of the disciplines that treats of one kind of process or group of processes gives consideration to the resulting phenomena as they are associated in particular places. This aspect of each such discipline is generally known as ecology.

In geography, the subject of investigation and presentation is the areal differentiation of the face of the earth. Geography focuses on the similarities and differences among areas, on the interconnections and movements between areas, and on the order found in the space at or near the earth's surface. It utilizes the ecological contributions of other disciplines in so far as they aid in interpreting spatial distributions. Likewise geography reaches back through time for pertinent space-order of the past, and sometimes it finds in the historical perspective a background for forecasting trends of change in space-order.

This chapter examines the regional concept and the method of regional study from the geographic point of view. The word "region" has been traditionally used, and remains widely current, as meaning an uninterrupted area possessing some kind of homogeneity in its core, but lacking clearly defined limits. Its unifying traits may or may not be explicitly stated. More often than not the term refers to an area smaller than a subcontinent, but too large and varied to be readily identified as uniform throughout.

For many years geographers have been trying to shape and sharpen the technical meaning of the term "region" into an instrument more powerful than the non-technical usage provides. No other word exists to convey the idea embodied in the procedure of regional study, and a word to express the regional concept is found in all the principal languages. In this volume the word is employed to mean an area of any size throughout which accordant areal relationship between phenomena exists. The area is singled out by applying specific criteria to earth-space, and it is homogeneous in terms of the criteria by which it is defined. But a region, as the word is employed here, is more than homogeneous; it possesses also a quality of cohesion that is derived from the accordant relationship of associated features. The observation and measurement of the phenomena brought to the fore, by specific criteria, from the diversified background, and the search for accordant areal relationships among these phenomena, constitute the regional method or the procedure for discovering order in earth-space. The order is expressed in the form of regional patterns made up of specifically defined characteristics and distributed within clearly outlined borders.

The term "area" is almost universally used to mean a geometric portion of earth-space, with no implication of homogeneity or cohesion.

ANTECEDENTS OF THIS REPORT

The following presentation is intended to strike a balance between general American practice in regional study, and the goals and procedures envisioned as attainable by students who have deeply probed the subject.[B] In this balance is combined the inventory and prospect called for by the title of the volume.

Regional study is in active evolution. Hence the committee submits its findings as a report on current status, and assumes that these findings will continuously be modified and elaborated.

Meanwhile, the statement of present views and judgments may be serviceable as a frame of existing ideas and procedures, and a springboard for

B. Early in 1948 a statement of purpose asking for comments, criticisms, and contributions was circulated widely among American geographers. A committee selected from among those who responded prepared a detailed outline of the proposed chapter which was sent to all who had expressed an interest. Six revisions of the original outline were successively circulated between 1948 and 1950, followed by two full texts in 1951 and 1952. In these revisions were incorporated the suggestions received from nearly 150 professional geographers. The committee held two formal, all-day sessions, and at every opportunity its members joined with others for impromptu but intensive discussion. During the 1950 meeting of the Association of American Geographers the then current report was the subject of a well-attended open session lasting more than two hours. In one or another version the matter has been thrashed out in seminars at half a dozen universities. Finally the report was read critically three separate times by the Committee on American Geography. Among the many who have contributed to this chapter, there are several who have repeatedly and discerningly given criticism and aid: Charles C. Colby, Clark N. Crain, Samuel N. Dicken, Eric Fischer, John H. Garland, Howard L. Green, Richard Hartshorne, Preston E. James, Stephen B. Jones, Wellington D. Jones, Lester E. Klimm, James C. Malin, Harold H. McCarty, Raymond E. Murphy, Allen K. Philbrick, Robert S. Platt, Malcolm J. Proudfoot, Kirk H. Stone, Benjamin E. Thomas, Edward L. Ullman, Stephen S. Visher, J. Russell Whitaker, and John K. Wright.

further refinement and new conceptions. The pool of American geographic thinking on regions brims with the inflow of comments and ideas from many sources. Indeed, so large a part of the report springs from these currents and sources that published papers are cited only to trace some specific origin or direction of the flow. Besides those few publications, account must be rendered of the great volume of books and papers that have served as a background of this report.

First in number and variety are the thousands of studies of individual regions of many types. Curiosity about regions has always been part of geographers' thinking. No studies of individual regions are listed here, because complete coverage is impossible and any selection could hardly amount to more than random choice. All the same, a wide variety of such regional presentations has been constantly in the minds of all who have contributed to this report.

Many an author, in presenting a particular region, has incorporated an explicit statement of his philosophy about regions. No attempt has been made to garner that scattered harvest, rich though it might be. Instead, the titles cited are confined to history, definition, classification, and technique; in broad terms, to papers and books dealing with the nature of the region and of regional study. Moreover, they are drawn mainly from American participation in the progress of regional thinking. Care has been taken to cite works that incorporate footnotes and lists which extend the bibliography to reasonably complete coverage.

THE RISING TIDE OF INTEREST

The beginnings of organized regional study were made in northwestern Europe after 1750. They have been traced by Richard Hartshorne [1:35-93]. In the United States of America, geography was popularized a little later as compendious descriptions of the earth and of its parts arranged by political divisions. Critiques by Ralph H. Brown disclose the character and range of regional material on the United States published around 1800 [2; 3]. An account of the regional concept in the United States, with appended bibliographies, has recently been written by Fulmer Mood and Vernon Carstensen in a book on regionalism in America [4:5-98, 99-118].

At the level of university scholarship, the study of the earth by Americans was imported piecemeal from various European sources between 1860 and 1925, but not always from exponents who were most representative or influential in their homelands [1:96-129]. The study of regions as a useful means of investigating the earth entered the universities only sporadically before 1900. In subsequent years, as geography took its place in university curricula, the several continents were customarily studied as regions. Courses so organized were likely to begin with aspects of the natural environment, such as climate and terrain, and to conclude with aspects of human geography. The center of interest was the regional variety within the continent under study, and not the systematic exemplification of geographic topics. Topical study was confined to geomorphology and climatology, and was closely associated with geology. In view of this activity and concentration, it is surprising that no reasoned statement of regional study appears to have been published by any American geographer before World War I.

In 1907, fifty-five members of the Association of American Geographers reported the branches of geography in which they were most interested. Only one, Isaiah Bowman, specified regional geography [5]. In 1910, Walter S. Tower, at the time an incisive teacher of regional geography, published an outline critique of the geographic discipline. Among his ten subdivisions of the subject matter [6] neither regional geography nor the region appears. Four years later, he and eight others, of twenty-five American geographers, rated regional studies among the three most needed lines of investigation in the geographic discipline. Of the nine, however, several were thinking exclusively of regional aspects of the natural environment [7]. The emphasis laid upon regions of the natural environment before World War I has been summarized by W. L. G. Joerg [8]. In 1916, Mark Jefferson called attention to a rising interest in human geography, mentioned three aspects of the subject, and stressed the distribution of population densities [9: 3-4].

The dawning American interest in the region, whether for its own sake or to serve as a tool, reflected European movements at the turn of the century. In all three principal Old World centers of geographic thought, the region was conceived as a key subject for both technical and empirical exemplification. Paul Vidal de la Blache set the pace by encouraging students under his supervision to write monographs on the regions of France. Several of these became classics [10; 11]. Alfred Hettner led the way for German geographers by stating, first in 1905 and more fully in 1927, the place of regional study in geography [12: 217-218, 293-317, 398-404]. In Britain, the interest inspired by A. J. Herbertson [13] remained focused on natural regions, both for the purpose of defining the subdivisions of the earth and of providing a base for further regional study. Among others, Percy M. Roxby and J. F. Unstead continued this emphasis [14; 15].

The concept of synthetic natural regions found little favor in America, where investigation focused on component elements of the natural environment. Study of geomorphic regions, initiated by J. W. Powell [16: 65-100] was continued under the leadership of Nevin M. Fenneman for the United States [17]. It was extended by W. N. Thayer to cover Canada [18]. A parallel study of climatic regions took the form of constructive criticisms of Wladimir Köppen's classification (adumbrated in 1900 and elaborated in 1931). These have been reviewed by F. Kenneth Hare [19:111-134]. Essays into regional subdivision of other elements of the natural environment have likewise been greeted attentively by geographers. Most of them have originated with ecological specialists in the natural sciences, but they are found especially useful for geographic purposes if a geographer has had a hand in their production, as in a paper by C. C. Adams [20]. Similar studies undertaking to recognize regions of the several cultural elements are fewer and apparently more controversial than those dealing with the natural elements.

The germinating preoccupation with regions found fertile ground at the end of World War I. The then young geographers comprised the first considerable group trained in American universities as geographers rather than geologists. Their wartime service turned them still further toward aspects of geography other than geomorphology. Thus their attention was directed toward regions by a combination of unplanned circumstances.

DISCUSSION AND PUBLICATION

Within half a dozen years, translations or interpretations in English spread before the Americans commentaries out of continental Europe: from Vidal de la Blache and Jean Brunhes in France; from Sten De Geer in Sweden, and from several German sources through Carl O. Sauer [21; 22; 23; 24]. Study of these methodologies broadened the American horizon of geography.

During the same years the younger men turned eagerly to the field as their laboratory, where they could study the landscape more narrowly than had been the practice in geographic description, even when based on direct observation. In 1923 a dozen enthusiasts began to devote a long weekend each year to some aspect of geography observable in the field, giving special attention to techniques for studying it. Limitation of time and place resulted in an unintentional accent on localities (microgeography). The participants and their advanced students quickly disseminated the fruits of the conferences among the major centers of research and graduate training in geography. The study of small areas became popular and the new field techniques were widely adopted.

The interplay between ideas freshly imported from Europe and experience in the field fused in discussions of overall aspects of regional study, especially of problems of definition and presentation. By 1930 American students had begun to move in an orbit of their own, although they remained cognizant of European advances, such as the summary of classifications of regions by a committee of the (British) Geographical Association [25]. At about the same time, through statements by Rupert B. Vance [26] and others, it became evident that the kind of work done by geographers and by ecologists in the natural sciences chimed with objectives of students of the social sciences, both academicians and regional planners.

The width of the lens focused on the nature of the region is measured by the contributors to four symposia published during the first half of the 1930's. Three were reports of conferences: a conference on regional phenomena held by the Division of Anthropology and Psychology of the National Research Council and the Social Science Research Council [27]; and two sessions of the Association of American Geographers [28; 29]. The fourth symposium was the work of the National Resources Committee [30: 138-154].

These joint inquiries and exchanges of opinion as to the scope and treatment of regions were paralleled by publications that showed a maturing of thought on the subject of regions. Statements by thirty geographers, collected in 1930 and 1932 by A. E. Parkins, were paraphrased as twelve views of "geography in America, its content, its philosophy" [31]. Only two of those views were indubitably regional, although two others combined regional study with the search for interrelationships between man and his natural environment. Bowman surveyed the classifications of regions and relations between natural and cultural phenomena within the region, in one of seven chapters dealing with geography in relation to the social sciences [32: 144-199]. Charles C. Colby, in tracing the history of American geographic science, without giving a separate heading to regional geography strongly advanced "the vital quality of the regional idea," as permeating the thought of American geographers, whatever their specialties [33]. V. C. Finch, after noting that a reaction had set in against the method of regional geography as the study of chorology, argued the case for that kind of regional study [34].

The reaction alluded to by Finch was an attack on the value of descriptive regional studies, and especially studies of micro-regions, a spark that touched off a vigorous controversy and eventuated in Hartshorne's historical and philosophical treatise, where the underlying discussion may be followed in two chapters and a major section of a third. [1:250-365; 436-486; reference numbers 220-224].

POLITICAL SECTIONALISM AND REGIONAL PLANNING

The stream of regional thought has been ruffled by cross-currents from other sources; two of them are sectionalism and regional planning. In the large and varied United States, areas exhibiting marked individuality have long been known as "sections." The inhabitants of a section tend to develop group solidarity in which a salient element is regional consciousness. The give-and-take between sections has always been a matter of practical politics in the federal government; under the name of sectionalism, it was a subject frequently discussed by Frederick J. Turner [35: 193-206, 287-314].

A desire to avoid haphazard use of the land in rapidly growing communities led first to city planning and then to planning for larger regions. Many regional plans are confined to one or more adjacent administrative units. At the same time, planners with broader interests have adopted the section as an appropriate unit for overall planning [36: 187-193]. So treated, sections assume the guise of administrative entities with political potential [37: 296-308; 38: 1-16; 39].

Both sectionalism and regional planning impinge upon the geographer's interest in regions. He may take a hand in the effort to guide land-occupance. Even if he does not, he is bound to come upon dissonances between his objectives in regional study and the objectives of planners, particularly those imbued with sectional fervor. It seems possible to resolve these dissonances, so that the geographer may work effectively as a planner. He can soften strident political overtones of sectionalism by calling attention to the mutual interdependence of the parts of a region separated by administrative, sectional, or national boundaries [40; 41: 49-50, 57-79; 37: 309-314]. He can be on his guard against the common tendency of geographers and planners to assume cause-and-effect before it has been proved, and as an alternative to present the desirability of "strict adherence to inductive reasoning from circles of facts, rather than from assumptions of causal . . . relationship among facts," to use the words of Hugh M. Raup [42:345; 43].

Political aspects of regions and regional planning in Germany, France, and Britain, discussed by Robert E. Dickinson [37: 245-249, 260, 273-292], show that the issues are not confined to the United States.

HUMAN ECOLOGY

Another side of regional thought which has been disturbed by crosscurrents is that which borders on ecology. The extension of the ecological viewpoint from the biological to the physical and social sciences has led to the employment of some terms in more than one sense. The most troublesome among these is "human ecology," which seems to have been used orally by both geographers and sociologists before being pinned down in print by exponents of each subject. In 1922 Harlan H. Barrows entitled his presidential address before the Association of American Geographers "Geography as Human Ecology" [44]. A year earlier the sociologist Robert E. Park had tentatively suggested modelling the study of human societies on plant communities and calling it human ecology. [45:555].

As defined by Barrows, human ecology covers the relationship between people and place, "a field cultivated but little" in 1922. Subsequently that field has been intensively tilled, but geographers have rarely labelled their work human ecology, in the sense defined by Barrows. Sociologists have adopted the term to designate the symbiosis in human communities analogous to that in plant or animal communities. A paper first published in 1915 by Park [46] was reprinted in 1925 with the added words: "the city has been studied in recent times from the point of view of its geography and still more recently from the point of view of its ecology" [47:1]. Since then the term human ecology has been used increasingly by sociologists who are at pains to point out its debt to and differences from human geography [48; 49]. As sociologists now use the term it is restricted to associations of people with people. The origins of this usage have been traced, and its current standing is covered in two recent publications [50: 1-18; 51]

Those geographers and sociologists who have hoped that human ecology might be extended to cover a fusion of those two sets of relationships (people to place, and people to people) have omitted or disregarded a third set of relationships. These are the ecological aspects of the natural sciences (the relationships of place to place) which were early formulated and long since firmly established. Only a very slight difference of viewpoint distinguishes an ecological study from a geographic study in a topical field. But since that difference concerns space, the basic organizing concept of the geographer, it is to him all-important [52].

DUALISM IN REGIONAL GEOGRAPHY

Regional thought has also been burdened by two forms of dualism. For more than a century a division has been made between natural and societal phenomena[c], or in more familiar words between physical and human geography. This duality evolved as a phase of the history of natural sciences. It comes down as an inheritance of the period when the long tradition of geography as a humanistic subject was buried under the weight of 19th century geology, and obscured by overemphasis on physiography and climatology. For two generations before 1900 physical geography eclipsed human geography on both sides of the Atlantic [1:86-90,

c. Phenomena resulting from the presence and acts of man are sometimes described as cultural phenomena, using culture in the anthropological sense. The chairman of this committee has suggested the term societal to stand for manmade phenomena. The latter term used in this chapter implies the same meaning as cultural in the anthropological sense.

102-126], and the resultant blurring of the geographic objective and method fogged the American scene for a full generation after 1900.

The humanistic tradition in geography, after being reasserted in the late 19th century, has occasionally led its followers to the extreme of claiming only the human half of geography, relinquishing physical geography to the several natural sciences. This principle would leave out of geography the vital elements of the natural environment, except as these elements might be borrowed in unassimilated form from the ecological aspects of the natural sciences. A logical but senseless corollary to this position would be the abandonment of human geography to the various social sciences, thereby discarding geography altogether as an intellectual discipline. In practice most geographers retain physical geography as natural environment, but not as geophysics. This logic permits the retention of human geography as societal environment, but not as sociology.

Geographers are now tending to avoid this duality of natural and human geography, feeling that it is an over-simplification that carries seeds of error. The unwarranted assumption of natural cause and societal effect has often been decried; equally deplorable, but less clearly recognized, is the obscuring of cause-and-effect on the same side of the cleavage line. Many students prefer a three-fold division: physical, biotic, and societal. This grows out of the fact that the physical, biological, and societal disciplines examine three quite different sets of processes, not two; and that these disciplines follow three quite different sets of procedures, not two.

Dualism of a different sort appears in the custom of distinguishing between topical and regional geography. Such a distinction was made by the classical Greek scholars; it was reformulated by Varenius; and today it underlies the choice of fields for specialization made by most professional geographers. Those in academic posts almost invariably divide their time between the two aspects in both research and teaching. Their two-fold pursuit generally takes the form of: 1) coverage of the entire earth or large parts of it with regard to one feature or group of closely related features, such as climate, vegetation, agriculture, or manufacturing; and 2) coverage of most if not all features as they are associated in one of the larger divisions of the earth, such as Latin America, Europe, or Southeast Asia. Such dualistic practice fits a widely adopted prescription for the organization of a graduate school of geography formulated by William Morris Davis [53: 122-123].

This form of dualism, like the separation of physical from human geography, has been treated historically by Hartshorne [1: 414-426], and its place in the work of contemporary geographic study has been analyzed by Edward A. Ackerman [54: 122-124, 129]. Both reach the conclusion that these two aspects of geographic study are not separable and that any attempt to insist on the possibility of separating them leads only to confusion of ideas and terms. Doubts of the adequacy of regional geography, voiced by a few who advance exclusive claims for topical geography, vanish in the light of those conclusions.

The antecedents of regional study reach back to several sources, as the foregoing recapitulation bears witness. Published commentaries on the region by both geographers and others, have grown out of vigorous oral discussion, mainly since the end of World War I. Regional studies of individual areas have appeared in unbroken lineage ever since the renaissance. The 20th century has produced innumerable examples in Germany and other European countries, and in France nearly the whole of geography is regional. In the United States regional geography came into high favor after 1920, supported by the new field techniques. Clearly, it is a perennial and an irrepressible interest.

THE REGION: THEORY AND PROCEDURE

Against the background sketched previously, the committee, pondering the history and philosophy of regional study in geography, decided to undertake a fresh inquiry into the nature of regions. As a result of this quest, the committee came to see the region as a device for selecting and studying areal groupings of the complex phenomena found on the earth. Any segment or portion of the earth surface is a region if it is homogeneous in terms of such an areal grouping. Its homogeneity is determined by criteria formulated for the purpose of sorting from the whole range of earth phenomena the items required to express or illuminate a particular grouping, areally cohesive. So defined, a region is not an object, either self-determined or nature-given. It is an intellectual concept, an entity for the purposes of thought, created by the selection of certain features that are relevant to an areal interest or problem and by the disregard of all features that are considered to be irrelevant.

A Monistic Goal for Regional Study

The regional method conceived in these terms is a method common to all phases of geographic study. Yet in the entire body of publications on the region that were examined by the committee no comprehensive analysis of the elements and characteristics of the region was discovered. The prime objective in regional study in North America has been the presentation of the characteristics of particular regions, a natural tendency in a world and continent where thousands of presumptive areal homogeneities challenge inquiry and interpretation. The studies produced have treated different categories and ranks of regions with little or no thought of their relations to each other and with no attempt to formulate a systematic method of handling them.

Search for a frame to support a comprehensive analysis led to the view that geography may properly be considered a monistic discipline studied by two approaches, rather than a dualistic study falling into two discrete parts [54: 133-134; 55]. The study of a topical field in geography involves the identification of areas of homogeneity, which is the regional approach; the study of regions that are homogeneous in terms of specific criteria makes use of the topical approach, because the defining criteria are topical.

A key to a workable monistic frame of reference for regional study was discovered in identifying the characteristics whereby areas are differentiated and regions recognized. These are listed on pages 37-40. The prime factors pertaining to regional study marshalled themselves in this checklist. Regions, when tested by these indices, were seen to range in one comprehensive series from those defined in terms of single features to those representing highly complex associations of features.

THE PROCEDURE OF REGIONAL STUDY

How shall areas be differentiated and regions recognized? It seems that among the numerous geographers who have contributed to this report there is a considerable difference of opinion regarding the procedure of regional study. The underlying purpose in all cases is the same: to reach a fuller comprehension of the order of earth-space. Some feel that this purpose has been served when regions are identified, their internal arrangement has been brought out, and their external relations to other regions have been measured: in short, that these things constitute the order of earth-space. Others emphasize the need for seeking the significance of observed order in terms of causes and consequences. The first group finds its challenge in the areal homogeneities that are apparent on the face of the earth and is concerned to examine these areas more closely with regard to the quality of homogeneity, the internal connections, and the external relations. The second group finds its challenge in unsolved questions of the relations between processes and phenomena, of the modifications of process in particular places, and of the areal relations of phenomena. Of the two approaches to regional study, the regional and the topical, the first group of geographers emphasizes the regional approach, the second group the topical approach.

The committee was not disposed to judge between these two approaches to regional study, because it recognized that differences of emphasis are essential for the full development of the subject. Every homogeneous area can be analyzed into its topical elements, a procedure likely to sharpen the investigator's perception of its make-up. Conversely, attention to the regional setting of a topic broadens the student's understanding of its connections. Indeed, the regional and topical points of view are not really separable. The experienced student of regions will make use of both approaches, yet the degree to which he emphasizes one or the other may turn out quite different products.

The regional approach to regional study starts with the homogeneous area, which is accepted as a hypothesis. The area is then examined with a view to discovering its components and connections [56: 545-547; 57: 355]. The region is analyzed with respect to the various elements which in association give it character, and is interpreted sagely against the investigator's background and grasp of topical geography. The region, seen as a complex association of features, guides the procedure.

The topical approach to regional study starts with a problem. There is a question of cause-and-effect to be answered or a question of policy to be clarified. The topics or features relevant to the problem are defined and their regional patterns brought out separately and compared. Accordant areal relations are identified by cartographic analysis. The complex association of features seems less important than do the component regional systems that make it up.

For both of these approaches the regional concept is fundamental, and for both of them some parts of the regional method are employed. Yet the results can be strikingly different. The one seeks the greatest possible synthesis, the other the most complete analysis. The one has an appeal for scholars with a certain turn of mind, the other has an appeal for scholars of a somewhat different turn of mind. The largest measure of progress in regional study seems likely to be achieved by the successful merging of the two approaches.

ESSENTIAL CONSIDERATIONS IN REGIONAL STUDY

Various attributes of regions have been mentioned or implied in some of the technical discussions already cited. No over-all statement regarding these attributes has been found, nor any systematic analysis of how these attributes might be used in regional study. There is no generally accepted terminology. In an attempt to fill the gaps, certain essential considerations regarding the attributes of regions will be discussed under six headings: 1) criteria; 2) categories; 3) characteristics; 4) cores and boundaries; 5) compages; and 6) regional consciousness.

CRITERIA

The region, in the technical sense proposed in this chapter, is an area in which accordant areal relations produce some form of cohesion. It is defined by specific criteria and is homogeneous only in terms of these criteria. The face of the earth with its complex associations of phenomena could theoretically yield an infinite variety of regional patterns, each brought forth by the application of different criteria.

Obviously not all patterns of homogeneous areas can have equal significance. Simply drawing lines on maps based on random criteria may be good fun and might even be justified by the hope that hitherto unknown patterns and relationships might be discovered. But an investigation which is successful in identifying and presenting regions must seek meaningful patterns, and should contain the demonstration that the patterns presented are, in fact, significant.

What are meaningful patterns? In geographic study a homogeneous area has meaning when it can be shown to correspond or coincide in its position on the earth with other homogeneous areas. But the identification of an accordant relationship, as stated in the introduction to this volume, does not prove a causal relationship. The regional pattern has both meaning and significance when it can be interpreted in terms of systematically related processes, operating through time.

These general principles can best be clarified by an example. For the sake of simplicity the treatment of a single feature, slope, will be considered. Slope is a continuity, and isograms of various kinds can be applied to it as criteria for the purpose of bringing out the variations of degree. Contour lines (20 feet, 40 feet, 60 feet and so on) indicate the direction of greatest degree of slope, which is at right angles to the lines. But the area between the 20-foot contour and the 40-foot contour, although it is homogeneous with respect to the criteria defining it, is not a region unless it can be shown to have accordant relations with other phenomena. In parts of the United States where detailed studies of land quality and land use have been made, it was found that slopes ranging from 0° to 3° could be plowed and cultivated without serious loss from accelerated erosion. The same observations elicited the criterion of 8° as the maximum slope suitable for tillage by machinery. The slope region 3°-8° suffers soil erosion when so handled, but not enough to offset the advantage of using machines. The areas homogeneous in terms of these criteria are meaningful areas because of the accordant relationship between slope and erosion; they are slope regions. They are shown to be further significant in that the process of accelerated erosion is causally related to the degree of slope.

Effective regional study is founded on the selection of meaningful criteria. The purpose of the study having been stated, criteria intended to bring out relevant distinctions can sometimes be formulated in advance of actual field observation. Such formulations are hypothetical and subject to considerable revision in the field. Even when the criteria for identifying regions have been adjusted to the area under study, they need to be continuously scrutinized and evaluated as the field work progresses. An experienced field observer notes certain apparently recurring associations, such as relatively level land and land used for crops without serious soil erosion. He attempts to identify and state the criteria that will bring forth regional patterns having the closest degree of accordance.

Criteria previously used in other areas or pertinent to other problems should be introduced only after testing. The slope regions mentioned above exemplify this warning. The slope class defined as 0°-3°, appropriate in the area where it was devised, may prove to be too inclusive in some areas. Testing in Puerto Rico led to reducing the criterion to slopes of 0°-2°. The slope class defined as 3°-8° is found to be inapplicable to regions of the humid tropics where the hoe is the common agricultural implement and machinery is not used.

The criteria by which regions are identified not only determine the outlines of homogeneous areas; they also determine the amount of variation or range in character permissable within a region. If this variation is so great that it obscures instead of illuminates areal relations with other phenomena a new set of criteria must be formulated.

CATEGORIES

The regional method fits a frame of study widely used in non-laboratory subjects, namely, the sorting and grouping of data according to specified criteria. It is a method of examining areal differentiation on the face of the earth, of finding similarities between areas, and of revealing the patterns of interconnection between areas. Ideally the regional pattern that results would be the product of mechanical sorting of the data with a minimum of value judgments. Actually, the study of geography unavoidably and frequently entails the making of value judgments of several kinds. Among them, generalization takes first place because geographers always seek to reduce earth-patterns to a scale, less than outdoor size, that permits analysis and comparison.

This inherent quality of geographic study can be stated as a ratio: neither the earth nor any part of it can be reproduced on a scale as large as 1/1; at any scale smaller than 1/1 some generalization of the phenomena that occupy area is inevitable. One test of a competent geographic product is the skill evinced in making the value judgments on which illuminating generalizations are built. Such is the essential nature of the regional method, and of geography.

There are many different categories of regions. The committee once listed more than fifty pertaining to nearly every aspect of the physical, biotic, and societal environments, defined in terms relevant to a great variety of purposes. Regions range from relatively simple delineations of single features, such as slope categories, to highly complex areas embracing the entire content of the human occupance of earth-space. Regions also differ according to the nature of their internal cohesion and structure.

Single, Multiple, and "Total" Regions

In considering the classes of regions it is useful to think of them as ranged in three basic types: 1) those defined in terms of single features; 2) those defined in terms of multiple features; and 3) those defined in terms which approach the totality of the human occupance of area. These types may be outlined as follows:

1. Single-feature regions, such as the slope categories in the example previously given, in each case delineate an individual phenomenon that is examined in relation to other phenomena in the search for accordant relationships. The geographers who prefer the topical emphasis in their approach to regional study favor the construction of regions of this type and the identification of recurrent associations of phenomena through the technique of matching maps of two or more single-feature systems. Such regions must not be thought of as "unit areas" in the sense that they are not further divisible; for within the limits set by the criteria they include a certain amount of variation or range of character, as within the limits of 3° of slope, or even within the area of a single field of grain.

2. Multiple-feature regions are differentiated on the basis of combinations or associations of features. Sometimes they may be constructed through matching single-feature regions; or else they may be sufficiently distinctive and cohesive to be observed and mapped directly in the field. Such regions fall into three subtypes:

a. Associations of intimately connected features which are highly cohesive because they have been produced by one kind of process. Examples are climates (defined as an interplay of temperature and moisture), soil types (defined in terms of slope, soil, and drainage), or types of agricultural land use (defined by the mode of handling a particular association of crops and livestock). Terms designating many different varieties of regions of this subtype may be found in the dictionary.

b. Associations of features less intimately connected than those of the preceding subtype because they have been produced by different kinds of processes, as, for example, an economic region (defined in terms of a resource base together with its associated use). Regions of this subtype lack dictionary names, but are widely employed in geographic study.

c. Associations of features only very loosely connected. One of these is the traditional natural region (theoretically defined in terms of climate, terrain, soil, vegetation, animal life, water and minerals). Its counterpart, the total cultural region (presumably compounded of economic, social, and political distributions), has not been effectively defined.

3. The third major type of region is differentiated in terms of the entire content of human occupance of area. Such a region is an association of

inter-related natural and societal features chosen from a still more complex totality because they are believed to be relevant to geographic study.

A good many geographers studying regions of type 3 have assumed that they imply an obligation to sort regionally the entire content of earth-space. All who have commented on this point during the preparation of this report have agreed that so comprehensive an undertaking lies beyond the competence of geography, because in every such region items are found that have no meaning when referred to the features, processes, and sequences being investigated. This intermixture of relevant and irrelevant is demonstrated in the field where the student is confronted with the whole mass of phenomena, and the impropriety of including everything in his interpretation becomes glaringly apparent. When the reach of such regional studies is restricted to the human occupance of area, the concept promises to be useful.

Some are critical of this kind of regional study, even when thus limited, because they believe that it generalizes so many diverse elements that it may be untruthful. In this view, type 3 appears as a special sort of multiple-feature region, and would be listed in the foregoing classification as type 2d. Others are convinced of the necessity for making a distinction between types 2 and 3. They hold that a separate frame is needed for encompassing that part of the entire content of earth-space which they believe to lie within the field of geography and also within the grasp of geographers.

To find a term without burden of other connotations, and to avoid misunderstanding arising from words now in use, it is here proposed to adopt for the type of region differentiated as to human occupance of area the term *compage*.[D]

Uniform and Nodal Regions

No matter what criteria are invoked in defining them, geographic regions of all kinds may also be grouped under two heads according to whether they are uniform or nodal.

Uniform regions are so throughout. The uniformity is not complete, for there is always a certain range of characteristics permitted by the criteria, and there are irrelevant differences which are disregarded. But within the

D. Compage (kŏm-pāj (e)'), singular noun. Used 1550-1694; now obsolete. An adaptation from the Latin *compages*, a joining together, structure; from *com-*, together, plus *pag-*, root of *pangere*, to fasten or fix.

Compages (kŏm-pā-jĭz, or kom-pa-jez), both singular and plural. Used since 1638. A system or structure of many parts united (Webster). A whole formed by the compaction or juncture of parts; a framework or system of conjoined parts; a complex structure; a solid or firm structure (Oxford).

(It is recommended that compage be used in the singular, and compages in the plural only.)

limits set by the criteria, regions of this kind are uniform. A climatic region is an example. If it is a multiple-feature region its uniformity is defined in terms of the association of features.

Nodal regions are homogeneous with respect to internal structure or organization. This structure includes a focus, or foci, and a surrounding area tied to the focus by lines of circulation. For example, an area of newspaper circulation is a single-feature nodal region, the trade area of a town a multiple-feature nodal region. Nodal regions of like character may lie adjacent to each other, or one such region may be surrounded by nodal regions of different character. A nodal region may coincide with other nodal regions of different character, selected by the application of different criteria. Internally nodal regions are marked by a diversity of function that goes far beyond the range of minor variation permitted in uniform regions. Circulation, including the movement of people and goods, communications, and other aspects of movement, is a primary attribute. Hence the nodal region is bounded by the disappearance or the differential weakening of the tie to its own focus in favor of some other focus. Its boundary lines tend to run at right angles to the lines that tie it together.

CHARACTERISTICS

Once criteria have been set up for defining regions of any category, the regional pattern is disclosed by applying the criteria to facts concerning the area, facts obtained by observation and inquiry. The task is to discover existing features, processes, and sequences, and to generalize the connections between them in order to illuminate the regional qualities and eliminate distracting details.

To test the soundness of the criteria chosen, it is useful to check the areas thereby differentiated against certain norms that are generally accepted as regional traits or characteristics. Some of them are almost too obvious and basic to require statement; others are not widely utilized, perhaps because they have not been clearly recognized. A check-list of characteristics that serve the regional geographer is given here.

Several characteristics pertain to both uniform and nodal regions. Hence *every kind of region* differentiated may properly be checked against these items.

A1. The region is unique, in that it differs in location from all other regions of the same category.

A2. The region enfolds a three-dimensional segment of earth-space. Viewed broadly regions are intimately associated with the thin water-envelope that limits human life. This hydrosphere, including water-vapor and ice, lies between the atmosphere and the lithosphere and penetrates both. Regions may extend indefinitely above and below the hydrosphere. In practice, however, most regional patterns are shown as possessing length and breadth on the earth's surface, and only occasionally, as in a block diagram, is the third dimension above or below the surface shown.

A3. The region incorporates an association of coherent features. This characteristic exists in all kinds of regions. Areas homogeneous in terms of single features are to be considered as regions only when they are shown to possess areal qualities accordant with one or more other regional systems.

A4. The present character of the region is partly derived from conditions that existed and events that occurred in past times—historical, archeological, and geological. Changes in physical, biotic, and societal features rarely occur without leaving traces significant for the study of succeeding regions. Thus, Roman roads and Roman law are imprinted on the landscape of contemporary Western Europe. The pace and velocity of change may also need to be taken into account, as well as the mere persistence of the evidence of change. The present is merely the latest moment available for observation. These matters are discussed in chapter three.

A5. The region is defined by criteria inherent in the category to which it belongs, not by traits that pertain to other categories of regions. For example, a climatic region is not defined by vegetation. By means of separate analyses, different kinds of regions are kept distinct from each other, even where they occupy the same, or nearly the same area, as do Mediterranean climate and Mediterranean vegetation. A degree of correspondence in area between regions belonging to different categories may occur; sometimes it amounts to coincidence. This correspondence may be accidental, or it may betoken some degree of cause and effect, often in association with still other features. To attribute the traits of any one category of region to any other often leads to confusion, if not also to false conclusions. Nevertheless, where data covering inherent traits are lacking or insufficient, tentative regional divisions can sometimes be deduced from related known aspects of the geography of the area under study. For instance, a polar climate might be areally mapped as coincidental with tundra, a vegetational distribution. Such regions should be presented as hypothetical only, accompanied by a warning to the reader.

A6. The region occupies a fixed position in a hierarchy of regions of the same category, in which those of each successively higher rank consist of aggregations of regions of the next lower rank (for example, minor civil divisions, counties, States, the United States). Conversely a given region may be one subdivision of a region of higher rank. Regions that form subdivisions of another region are, of course, smaller than the latter, but otherwise the ranks have no connotation of size. No region can belong to more than one rank.

Hereafter, in order to keep in view these two equally useful approaches to the construction of regions, this concept of hierarchies will be designated by the term "aggregation-subdivision." Every category of regions, from those of the single-feature group to compages, has its own ranks of aggregation-subdivision, designated by terms appropriate to the regional character.

A *uniform region* needs to be checked for two distinctive characteristics in addition to those applicable to uniform and nodal regions alike.

B1. The uniform region is homogeneous because all parts of its area contain the feature or features by which it is defined. No region is uniform in the absolute sense, for all regions are generalizations based on selected items. It follows that the single-feature uniform region drawn at a very large scale (that is, in the lowest rank of the hierarchy of regions) contains the fewest irrelevant features and so approaches most closely to uniformity. The compage drawn on a very small scale includes the greatest number of irrelevant features, and so is the least strictly uniform. (It should be noted that in a compage either uniform or nodal characteristics may dominate.)

B2. The uniform region includes among the features by which it is defined a certain range of intensity or character permitted by the criteria. The range is least at the bottom of the hierarchy and greatest at the top. Because of these permitted variations within the region, it is not without textural variety or grain; for example, the crests and troughs of a region of ridges and valleys, or the checkering of crop and fallow in a region of extensive wheat growing.

While the uniform region has two characteristic attributes of its own, the *nodal region* has no less than four, in addition to those applicable to all regions.

C1. The nodal region is homogeneous because the whole of its area coincides with an integrated design of internal circulation. This unity of organization, and not the spread of specific features throughout its whole area, differentiates it from other regions.

C2. The nodal region contains a focus (occasionally more than one) that serves as a node of organization. The focus is likely to be a center of communication and is most often urban. It may lie outside the region in exceptional cases, but it must be closely connected with the region by one or more lines of communication; for example, the port of Oporto is connected with the upriver Port Wine District. The same place may serve as the focus of two or more nodal regions. These may be in different ranks of a single category, as where a city is a county seat and a State capital, both functions being administrative. Or they may belong to different categories, whether in the same or different ranks; for example, Salt Lake City, which is a State capital, the headquarters of the Mormon Church, and the transport center of the Great Basin.

C3. The nodal region is enmeshed by a pattern of circulation. This circulation may be an expression of mobility or communication only, such as exchange of people, goods, ideas, or telephone calls. It may involve force, as in government exercised from a focal political center. Lines of force in this sense should be clearly distinguished from mere movement or lines of flow. The circulation need not be equipollent in every direction; indeed, it rarely is. Its reach may vary, as when a focus lies at one side of its region. Its intensity may vary, as in the volume of trade between a commercial center and a thinly peopled rural area on one side, and on the other a more densely populated and urbanized area.

C4. The focus of a nodal region is linked to the remainder of the region by ties of different intensity and different character. Commonly the focus lies within the core area, beyond which spreads the marginal area of the region. Distance itself tends to weaken the ties to the focus, as the perimeter of the region is approached. Nearly always other conditions of the area (it may be the terrain, a language boundary, or a pattern of trade restrictions) so modify and contort the simple effect of distance that an ideal concentric design is not developed. Aside from varying intensities of connection with the focus, ties differ in character. In a succession of areas tributary to a city the following distinct patterns may appear: a close net of roads and rails, affording commuter travel; prolongation of main routes with regular but less frequent traffic; areas with no rails and few poor roads. All these areas lie within the same nodal region because their relations to other nodes are unimportant in comparison.

CORES AND BOUNDARIES

By definition, regions are differentiated segments of earth-space. When portrayed on a map the several segments must be separated, either by lines or zones. Where regions are sorted out of a continuity (see chapter one, p. 10), the boundaries are given precise definition. They are troublesome mainly because of the difficulty of deciding upon the appropriate isograms, although making the physical measurements may be taxing. Regions of this kind have no cores, since the transition from one limit to the other is continuous (as 0° to 3° of slope). Where regions are defined by discontinuities, or occupy an area of discontinuous distribution, regional peripheries are likely to be acutely troublesome because they are transitional, or zonal, and at the same time critical.

The attention of geographers has perennially been drawn to boundaries, because of the need for regional demarcation in a discipline that centers on the variation or variety of associated phenomena in earth-space. Preoccupation with peripheries has diverted emphasis from areas where the criteria are met more closely. This is most serious in the case of regions defined by differences of kind and differences of enumeration.

Cores of Regions

A region marked off by differences of kind in a discontinuity includes an area where the characteristics of the region find their most intense expression and their clearest manifestation. This may conveniently be called the core of the region. Although the whole region is homogeneous in terms of the criteria by which it is defined, the peripheral parts of it are distinguished from the core by an increased intermingling of extraneous characteristics. Hence the selection and presentation of the regional character is best accomplished by exposition of the core area.

In uniform regions the core comes closest to the ideal expression of the criteria whereby the region is selected. In a single-feature language region, for example, the core includes that part of the whole area where a single mother tongue is found. The peripheral areas show an increasing proportion of other languages until the limits of the region, as defined by the criteria, are reached. In a multiple-feature region such as the Corn Belt, a certain range of characteristics is tolerated by the criteria of its definition; within the core of the Corn Belt, however, are found all the characteristics commonly used to describe the whole region. The core of a uniform region possesses two qualities which may be blurred in the peripheral zone: it exists as a recognizable and coherent segment of earth-space, defined by the criteria of its selection, and it differs notably from neighboring core areas.

In nodal regions, the core is the most representative portion of the entire area and the part most closely tied to the focus. For example, in a region of newspaper circulation, the core is the area approaching complete coverage. It should be noted that the core and the focus are not synonymous, even though the focus ordinarily lies within the core. The focus is one of the salient features of the nodal region's structure; the core is the epitome of the region's character, whether uniform or nodal. The cores of adjacent nodal regions may closely resemble one another, because the distinguishing feature is nodality and not uniformity.

Within any category, an occasional region may have so little internal variety that the core embraces the whole of it, but more commonly the periphery is transitional, partaking of the character of two or more cores. It follows that comparison of different cores produces a sharper areal distinction between regions than can be found in the fuzzy interpenetration of peripheral areas. Hence rough delimitation of cores is commonly the easiest and surest way of undertaking areal differentation and comparison.

Boundaries of Regions

Boundary lines used to separate regions are of three sorts, depending on whether they are drawn in discontinuities, continuities, or areas of discontinuous distribution.

Boundary lines in a discontinuity separate regions which differ in kind. Sometimes such boundaries are sharp and easily observed in the field. For example, the edge of an ore body where it is cut off by a fault is clear even when examined in minute detail. There are places where climatic boundaries between humid and arid conditions are abrupt, as along the crest of the Sierra Nevada. Boundaries separating regions created by the activities of man are usually sharper than the naturally marked boundaries they replace, as where open grassland and deep forest stand side by side as a result of repeated fires, or where the line between desert and oasis is drawn along an irrigation ditch. Discontinuities contrived by man may be essentially linear: only the width of a street often separates blighted and redeveloped areas in a city. Sharpest of all, since they are geometric lines with length but no breadth, are the boundaries that men survey, such as political boundaries or the limits of land property.

But most boundary lines in discontinuities are not abrupt. The peripheral zones are marked by a varied intermixture or interfingering of component features. Vegetation boundaries, especially if they have not been modified by human occupance, are often notably transitional, as where the boreal forest borders the tundra through a wide zone of mixed tundra and stunted trees. In parts of the low latitudes the zone of mixture between forest and savanna is so wide that further study may bring out an additional category of forest-grassland mixture. The humid-semiarid boundary in North America is commonly shown as a single line based on an average of many years, yet its zonal character is revealed by plotting the position of the boundary year by year. Examined in detail the boundaries between soil types often exhibit such a mixture of characteristics on their margins that the beginner is bewildered in his attempt to map them. Man-made boundaries are also transitional in many places, as where the commercial core of a city is being enlarged at the expense of a bordering blighted area, or where the city is expanding into the rural countryside through what has come to be known as a rurban zone. A geometric boundary usually evolves into a zone, minor as where a hedge marks a property line, or major along national boundaries, especially in densely settled areas. Along some national boundaries workers pass without interruption back and forth, as between Canada and the United States at Detroit, or between Belgium and France; along other boundaries there are elaborate fortifications, customs barriers, and other structures scattered over a wide zone on either side of the line.

The demarcation of boundaries in discontinuities may give trouble in actual practice, although the rules of procedure are simple in abstract statement. In uniform regions the boundary is drawn where the distinctive characteristics of adjacent core areas are least discernable or fade into each other. In nodal regions the boundary is drawn where attraction to adjacent foci is equal. If there is no overlap, the boundary is drawn where the attraction is zero.

In some cases the boundary of a multiple-feature region in a discontinuity can be most effectively located by matching the maps of several component single-feature regions. These boundaries may all fall within a boundary-girdle where they correspond or perhaps even coincide. The meaning of the boundary-girdle depends on the meaning of the several regional categories superimposed to make it. Mere piling of lines does not prove the existence of an important regional division unless the significance of the associated regions is determined; but such a piling of lines points to the probability that a significant boundary can be derived.

Quite apart from the quality of the boundary as a sharp line or a wide zone of transition, and independent of its demarcation, is the geometry of its depiction on a map. If the map-scale is very small, the width of a pen line may spread a sharp linear boundary over more than its fair share of space, or may occupy the whole breadth of a zonal boundary. A zonal boundary around a single-feature region comes closest to exact map representation when it is drawn at a scale which permits the pen line to coincide with the zone of transition. A boundary representing correspondence of more than one feature is precise only where the separate lines coincide.

Boundary lines drawn in a continuity are of very different character. They do not represent divisions between differences of kind, for the same kind of feature exists on both sides of the line. The differences are those of degree or intensity. The lines reveal the pattern of the phenomenon in question: the direction of greatest variation in intensity is always at right angles to the bounding line. In this book, the kind of isogram that connects points of equal value is called an isarithm; the kind that passes through areas of equal ratio is called an isopleth. In either case, isograms are, at least theoretically, precise [58].

Because of this quality of isograms drawn athwart a continuity, they are useful in determining the position of boundaries in pertinent discontinuities. The boundaries of agricultural regions, for example, may be given strong support through the conversion of discontinuities into continuities and the plotting of isopleths, as when areas of crops or pasture are converted to ratios of cropland or pastureland to total area [59]. The quantitative character of isograms as computed lines on a map does not confer special validity on them as boundaries of discontinuities, because the criteria for determining them are based on value judgments in the first place. Yet isograms do provide for a uniform application of the selected criteria, and therefore may serve as guides to the demarcation of the boundary of a discontinuity.

The boundaries used to outline areas of discontinuous distribution, such as those within which population is enumerated, present special problems of interpretation and procedure. These are discussed in chapter four.

Whatever the kind of boundary, on a map the line depicting it always strikes the eye. It therefore tends to appear more real than the zone it symbolizes, and to divert the attention from the cores it separates.

COMPAGES

Geographers and other workers who make use of regions investigate and present single- and multiple-feature regions in much the same way. These two categories of regions are generally recognized as segments of earth-space produced by subjecting the complex associations of phenomena on the face of the earth to scrutiny according to criteria set up for a specific and stated purpose.

There appears not to be the same unity of view among those who are concerned with the still more inclusive kind of region. The great majority of historians, anthropologists, and sociologists seem not to have questioned the nature of regions as objects or given segments of spatial totality. In a recent symposium on regionalism, Vance invokes the term "true regions" as if they existed as objective facts rather than as intellectual concepts [4: 119-140]. Four of the fifteen chapters in that volume are restricted by title to single-feature regions. In all the other chapters the authors appear to assume that the kind of region they are discussing includes spatial totality except where they specify otherwise. Like workers in other disciplines, numerous geographers postulate the existence of a region which is undefined but which is assumed to approach totality of the combined physical, biotic, and societal content of area. This is the meaning most commonly attached to the word region when it is used without qualification. Acceptance of the region as objective reality has been increasingly criticized by geographers, and it is flatly rejected in this book as being incompatible with the position that the region is a device for segregating areal features. (See, however, Chapter Seven, page 179.)

Superficially, the idea of a "total" region appears simple, but once its surface has been probed, its underlying, disorderly complexity is brought to light and its unfitness as a guide to regional order is demonstrated. The earlier view appears to have been given currency through failure to examine the concept of totality. That word may mean the mere sum total of the contents of space, as in a trash can where the miscellaneous matter has some degree of unity by virtue of its existence within the confines of a certain container. If this meaning is transferred to the study of a region, literally everything in the designated segment of earth-space becomes grist for the regionalist's mill. Geographers and others have attacked the omnivorous study of spatial totality as being undiscriminating, futile, even dangerous.

Proving a particular meaning of totality worthless as a guide to regional study does not brush aside the fact that studies simulating spatial totality have long been made and continue to appear. They fill a need and may be unimpeachable. When analyzed, the acceptable studies are found to exemplify a second meaning of totality, namely, the sum total of items linked together by ties of functional association, as are the parts of a motor or the chapters of a textbook.

Among the possible ways of approaching total functional associations the one suggested here makes use of the concept of the community of features that depict the human occupance of area. The fusion of these elements gives a direction to study that is essentially geographic. To express this concept, the committee proposes the adoption of the obsolete term compage as explained previously (footnote D, p. 36). The compage is, by definition, something less than spatial totality; but it does include all of the features of the physical, biotic, and societal environments that are functionally associated with man's occupance of the earth.

A very large proportion of all geographers have interested themselves in the study of compages or of areas akin to compages. Even those who prefer the topical approach to regional study are guided by the ultimate purpose of comprehending as much of the totality of the occupance as is relevant to a problem. Those who feel that the compage is beyond the grasp of a single scholar except as an ideal, may nevertheless accept it as a practical possibility for a team of scholars. Therefore, it may not be claiming too much to assert that such study comes close to the eternal function of geography. It draws upon the ecological segment of every related subject in physical, biotic, and social science. In turn it may be useful to related disciplines. Thus, social scientists may use compages to illuminate their own work, and sometimes as a foundation for it. Because it is the keystone of the geographic arch, the compage is here singled out for treatment at greater length than the regions of other categories.

Like all regions of any one category, the compage is uniquely located. It embraces a combination of elements found nowhere else. Strictly construed, therefore, it has no counterparts, and this has baffled some students who desire to make regional comparisons. Construed more loosely, however, compages are likely to possess many features in common that will repay comparative treatment.

Both the regional and the topical approach to regional study can be applied to the study of compages. In fact, the regional approach usually starts with the acceptance of hypothecated compages based on impression. The experienced field observer of geographic phenomena out-of-doors senses the existence of a compage before he has taken the steps required by unprejudiced scholarship to prove the accuracy of his impressions. On the other hand, there are others, perhaps less experienced, who do not trust their impressions and who prefer to build up the compage from a wide variety of more or less experimental single- or multiple-feature regions. As already suggested, the greatest progress is likely to come from some merging of these two approaches.

There is some danger in undertaking a study of a compage with any standard list of criteria. A review of the practice of regional study suggests that too often geographers have entered upon a field investigation with a check-list of items to observe, and have proceeded to fill in the list with unrelated and uncoordinated information. Yet the criteria for studying a compage need not comprise the entire gamut of physical, biotic, and societal features. Rather the selection of phenomena to be observed, classified, and mapped should be made on the basis of relevance to the particular compage under investigation, and should subsequently be arranged for presentation in the order of their significance to it. The common method, used so often in German regional studies, of starting with the geologic past and progressing through the physical and biotic features to the societal aspects of the area does provide closely comparable treatment, yet one which sometimes fails to illuminate the unique character of the region. Alternatively to a rigid formulation, the items applicable to the compage under study may be listed in a rough approximation to their significance. One case may start with mining and minerals; in another case neither of these features needs to be mentioned; in a third, mining and minerals may appear on the list of elements, but far below the treatment of agriculture, soil, and climate. Once the items relevant to the unique character of a particular compage have been ascertained, they are closely observed and thoughtfully analyzed in order to bring out the distinctive characteristics of the region. In this procedure, no standard list of criteria or of items to be observed is adopted as universally applicable for the purpose of differentiating compages.

On the other hand, the student of compages must not overlook the possibility that some of the items he has discarded may be highly significant. Those who set up the criteria and the items to observe as they proceed in the field study of a compage, run three risks: some items may appear to be irrelevant and so be omitted from consideration, yet may later prove to be critical; other items may seem to have so little weight that they are given only cursory and possibly insufficient examination; and still others, through oversight or preconception, may escape notice altogether. Those who work with a prearranged check-list of all the conceivable elements of a compage are at least aware of the elements they are disregarding.

Because of the importance of wide training and experience in the field as a background for the study of compages, and because few scholars possess the entire range needed, at least while they are still physically able to undertake field study with enthusiasm, some geographers urge that compages can best be attacked cooperatively. This is a practical problem of procedure as yet unsolved. It involves the use of teams of scholars, each member responsible for the treatment of a rather narrow range of topics. Such area-research teams commonly include workers in the geographic aspects of other disciplines. Limited experience with such cooperation suggests the need for developing special procedures for bringing the diverse members of a team together so that they do, in fact, work jointly and not separately. Geographers who have specialized in the study of particular areas would seem to be well qualified to lead such teams of scholars in parts of the earth where they possess special competence.

A Suggested Hierarchy for Compages

The general neglect of the meaning of differences of scale or degree of generalization in studies of areal differentiation is a lacuna in geographic thinking which the committee believes should be filled as soon as possible. Every kind of region, whether based on one feature or more than one, whether uniform or nodal, can be organized in a hierarchy of ranks ranging from those with the least degree of generalization to those with the greatest degree of generalization. Those of less generalization can be combined into aggregates at a higher rank; and those of higher ranks can be subdivided.

No close accord could be found among geographers regarding the number of ranks that would most effectively present the several categories of regions, nor regarding the distinguishing qualities of different ranks. Preston E. James has suggested that the adjective topographic should be applied to those studies carried out on scales large enough to permit the mapping of the specific features of human occupance (such as single fields and farms); and that the adjective chorographic should refer to studies at a greater degree of generalization where the specific features are merged or aggregated in regions of higher rank [55]. Others have urged employment of as many as six ranks.

With particular reference to compages the committee came to the conclusion that four ranks are needed. To establish a hierarchy so widely acceptable as to lead to its adoption would, it is felt, greatly advance the study of compages. Three requirements should be met by any suggested system of ranks of aggregation-subdivision: it should make clear the salient differences between the ranks, find terms for naming them, and suggest ranges of map scale suited to both investigation and presentation. In the following proposed hierarchy for compages, illustrations are offered in the interest of clarity despite the risk that specialists in the areas named may consider them inappropriate examples. The four ranks are entitled localities, districts, provinces, and realms.

1. *Locality*. The lowest rank of the compage has been widely studied and with more precise measurement than any other. It has attracted the attention of American geographers over a period of thirty years. Its procedures are fixed in American usage. Examples (arbitrarily cited from Europe and Asia) are the market area of any one village of the Beauce in France or of the Chengtu Plain in China. There is general acceptance of the word locality as the designation of this rank of compage.

The locality comprises the daily orbit of a community where place has a maximum reality and meaning for its inhabitants. It may be rural or urban. In rural areas the locality usually comprises the area focusing on a social center, a village, or a ranch-house. In urban areas the locality is a neighborhood similarly focused on a social and economic center. The locality, besides being small in compass, epitomizes the human occupance of area in its simplest form. The specific elements of the occupance (the fields and farms, the buildings, the trade centers, and the rest) are examined on scales large enough to permit the minimum degree of generalization. Although it is recognized that a single field of grain could be further subdivided if the objectives of geographic study demanded it, the study of compages usually need not go into the detailed differences within a single field. The geographer may, however, wish to observe and map the layout of the ingredients of a farmstead, or compare the varying productivity of individual fields with the underlying soil types, or record the use of buildings in the commercial center of a town. These details can be examined in locality studies, where the most direct relations between man and the land are clearly visible.

The scale of the maps on which field observations are plotted must be large enough to permit the examination of relevant detail. Where fields and farms are very small, map scales as large as 1/10,000 may be needed to permit plotting areal differences as small as a fraction of an acre. Where the specific features of the occupance are somewhat larger, a scale of 1/20,000 may be useful; but scales smaller than that can scarcely be used in locality studies. Scales larger than 1/10,000 may be required in detailed urban studies, or in the studies of individual farm units. Once examined and interpreted on large scales, the results can be presented on scales of 1/50,000 more or less.

2. *District.* Localities may be grouped into ensembles, each with a character so distinct that it is accepted as a unit by laymen as well as by students of regions. Often they have names familiar to those living in them or nearby. Examples, built upon those cited for the locality, are the Beauce and the Chengtu Plain, with Chartres and Chengtu as their respective foci. The term district is suggested as being suitable for this rank of compage.

The outstanding quality of the district is its unified nature, usually arising from an association of features having close functional cohesion. The Beauce is a district of wheat farms, thanks to well-drained limestone soil in a country short of favorable conditions for its traditional staple food. The Chengtu Plain is an irrigated rice area, thanks to the presence of abundant water, flat land, and impermeable soil, all favoring the special conditions needed for the staple food most prized in southeast Asia.

Districts, like localities, are also studied on maps of large enough scale, so that the specific impress of the human occupance is not lost to view. For the study of the miniature fields of Chengtu a larger scale is needed than for the study of Beauce. But districts cannot well be examined on maps much smaller than 1/50,000 unless the texture of the features is unusually coarse. The characteristics of a district can be suitably presented on scales not smaller than 1/250,000.

3. *Province.* Neighboring districts, each possessing a high degree of unity within its own bounds, may be grouped to form a compage of higher rank. Such aggregation is accomplished by identifying certain features, common to all, that permit fitting the individual districts into a more general region. A compage of this rank has striking internal variety in contrast to the strongly unified district and the highly cohesive locality. Although its internal contrasts may be considerable, it nevertheless possesses considerable homogeneity, as defined by the criteria of its selection, and often has a name in common parlance. Many regions of this rank have been made the subject of geographical study because they lend themselves to unified treatment and at the same time permit coverage of large expanses of territory fraught with considerable complexity. Examples are the Paris Basin of France and the Red Basin of China. A compage of this rank may be called a province.

If non-political proper names are applied to particular provinces of this type there should be no confusion with the names of political subdivisions of countries called provinces. Compages of this rank have often been called "major geographic regions," but the arguments against that designation are impressive. The province is no more geographic than are compages of other ranks, nor than any other kinds of regions being studied for geographic ends. The word region, being the generic term, cannot be preserved in the title of any one of the scores of kinds of regions without confusion. The adjective major is unnecessary if a term without an adjective is found adequate. Province has the further advantage of equating this rank of compage with the same rank in a generally used hierarchy of geomorphic regions [17].

The study of a province requires techniques quite different from those used for studies of districts and localities. No one has ever scrutinized a province from any one viewpoint. Whereas localities and even districts can be examined with respect to the specific details of the occupance, provinces must be treated in terms of more general concepts. Specific kinds of land use or specific types of soil must be generalized into crop associations or soil associations. Mapping areal varieties of the sorts distinguished in locality or district studies by direct field observations drops out. It is replaced in the field by piecemeal observations and mapping. This may be sampling, or it may consist of generalizing from selected observations that leave to one side the mass of data presumed to be irrelevant. The highly generalized elements of a province cannot be observed and noted *in toto.*

Provinces are studied on scales too small to show specific items of occupance, but presumably not much less than $1/1,000,000$. On that scale the smallest areal division that can be mapped occupies a space at least four miles on a side. Scales in the suggested range are large enough to permit cartographic analysis, because the data are not too generalized to be used to identify accordant relationships. Yet it requires considerable skill and experience, usually derived from long apprenticeship in making studies of locality and district, to map and interpret reliably the features of a province. The characteristics of a province cannot be clearly presented on scales smaller than $1/5,000,000$.

4. *Realm.* Several provinces in any one area may possess enough similarity or mutual ties to be profitably treated as a unit. Sometimes a compage of this rank has a proper name of its own, but more often it is given a title by writer or speaker, afterwards adopted by others who sense its utility. Examples are Northwest Europe, and Green (Central and South) China. Certain judgments concerning the human occupance of area can be applied to these large portions of the earth, but the degree of generalization is necessarily so great that many important details are disregarded.

The compages eventuating from such study are called realms. They appear to be useful in teaching and in presenting a broad regional sweep to the general public.

Investigation of the realm by the regional method is limited to a few combinations and contrasts of features, connections, and sequences. These alone can be shown on the small scale to which the portrayal of realms is confined, and they can usually be presented on scales smaller than 1/5,000,000.

REGIONAL CONSCIOUSNESS

Regional consciousness is a form of group consciousness that derives from a sense of homogeneity of area. It usually applies to uninterrupted space, whether a continuous extent of land or the opposite coasts of a unifying body of water. Regionalism, although a word much used, is inadmissible in geographic study because it is overlaid with special meanings and generally serves non-geographic interests. A recent critique of regionalism points out that the word represents not only a state of mind (regional consciousness), but also a frame for collecting information about areas, a hypothesis to account for the interrelation between areas, a tool of administration and planning, and a cult [4: 381-393]. These four connotations do not include the dictionary definition, sectionalism. Those who feel that the term regional consciousness is clumsy, may prefer to substitute the single word, regionality, which lacks dictionary definition, and is therefore available as a synonym.

The most obvious occurrence of regional consciousness is in political regions. When identified with sovereign states it is rampant under the name of nationality or patriotism. It may appear in particular areas within a state, strongly if the state is composed of sharply differentiated regions, as in the Soviet Union, weakly or limited to certain phases of life when contacts with other regions are intimate, as in northern interior United States. Internal barriers may heighten and perhaps occasion regional awareness, as in Canada. Where a sovereignty is divided along lines that mark sub-nationalities, regional consciousness may quietly persist without rancor, as in cantons of Switzerland; or it may be invoked as an aid in a struggle for autonomy, as in Scotland, or for independence, as in Poland. A feeling of insecurity or inferiority may promote regional consciousness, whereas the portions of sovereign states seated firmly in the political saddle, such as the Humid Pampa of Argentina, are unlikely to feel the need of it. Constituent members of a sovereign state may join forces to gain regional advantages, as the states of Northeast Brazil have done. Conversely, regional consciousness may provide foundation stones for a superstructure jointly raised, as Benelux.

Regional awareness appears in conurbations, not because they may have a legal status different from the surrounding countryside, but because they contrast with it in appearance, comprise social units with distinct economic functions and common problems of servicing, and possess both focus and nodal pattern.

In regions of the higher ranks, self-awareness may be furthered by the leadership of a focal city, as in the case of Spokane in the Inland Empire. In rare and disputed instances, two or more foci may simultaneously be centers of a single regional outlook, as San Francisco and Los Angeles in California. Landscape in the sense of scenery, whether beautiful or not, is often a factor in creating regional awareness.

It seems to be true that the only areas, other than political regions, in which a sense of regional consciousness appears, are compages. If names belonging to regions of other categories have a connotation of regional consciousness they are being used in a dual sense. The following samples of the dual usage are taken at random from several categories. Locational regions: East End (London), an urban neighborhood or locality. Climatic regions: Thermal Belt of North Carolina, a resort district. Landform regions: Southern Appalachians, a province of distinctive occupance. Soil regions: Black Belt of Alabama, a district having an economic and social character of its own. Vegetation regions: Landes, a French *pays* of district rank. Crop regions: Bordeaux Wine area, a district set apart by the appearance and mode of rural life. Mining regions: the Iron Range northwest of Lake Superior, a collection of isolated localities. Language or religious regions: French Canada, a province sharply differentiated by its way of social life.

From the foregoing it will be observed that compages of all ranks may have overtones of regional consciousness. Self-awareness appears reliably in the locality. Where people are in close touch with each other, only a hermit can avoid a sense of community with his neighbors. Its existence in the district stems from a common way of life, as in the *pays* of France. On analysis, this seems to grow out of contacts between small groups of people, repeated and varied, and continued long enough to establish a sense of shared existence. It may be accentuated where there is also isolation, either natural or societal. Central Alaskan settlements in the pioneering stage exhibit vigorous regional consciousness, as does likewise the relict Wendish (Slavic) settlement in the Spree near Berlin. Provinces have often been differentiated in part by the criterion of regional consciousness. Brazil's South may be cited in point. Some realms have only vague suggestions of regional consciousness, but in others it clearly appears, perhaps only where there is a firm connective element, as sea-trade in Atlantic Europe, or Arab culture in the Near East.

Compages of any rank may lack regional consciousness. In such cases, their reality as regions may exist only in the minds of their authors, and not in the hearts of their residents. Thus, West Africa and the Pocahontas Coal Region have utility as geographic tools, but they appear to lack regional consciousness.

Where it does exist, regional consciousness appears to deepen human solidarity within a region, thus contributing to its stability. A region clearly aware of itself is likely to appear to its inhabitants and to outsiders, to have an independent existence. This psychology is an element of the regional complex. It helps to stamp the area with tangible features, such as a distinctive architecture and a tempo of human movement. It is an expression of innate loyalties that reinforce the external evidences of its regional individuality.

The intensity of attachment to a region varies with individual inhabitants, with the character of the region, and with the impact of history. Switzerland and Nebraska may serve as contrasting illustrations of extremes. Likewise, any one person may feel attachment to more than one region, say to neighborhood, city, province, and nation.

Section is a term with a connotation of regional consciousness. It is widely used in the United States by historians and sociologists, and therefore needs to be understood by geographers who may wish to equate it with their concept of the region. It has been defined as the historical doublet for the geographers' region. It has been utilized as an areal fulcrum for political leverage, with the object of achieving autonomy and independence. In both senses it has a political character and is dynamic. As currently used in geographic writings, it appears to add nothing but confusion to the geographer's study of areal differentiation.

The Outlook for Regional Consciousness

Regional consciousness appears in a good many areas. Where it is coextensive with political units, its boundaries are clear. To assess its weight in the balance with the other features that make up a region is baffling, because it is intangible and therefore not readily measured by instruments familiar to geographers. Yet it cannot be ruled out in the qualitative sense, even though no means have been devised to measure its quantitative value for recognizing and interpreting regional character. Regional consciousness occurs more often than not in compages. This has led some to think of them as analogous to biological creatures, a dangerous habit of mind because it leads to untenable conclusions.

That regional awareness will continue indefinitely to hold its place among regional phenomena is doubted by some, presumed by others. It is possible to demonstrate current waxing or waning in one or another particular place. Where the regions of several categories tend to coincide, regional awareness is usually present. A notable case is California; there climate, terrain, water supply, agriculture, and cultural heritage have for a century contributed to a regional differentiation which has evolved out of the socio-political unity of the United States. Conversely, when isolation in a generally uniform area gives way to economic and social intercourse, regional feeling is likely to weaken. Several instances can be found in Africa, where enemy tribes have expanded across the no-man's-land that formerly separated them, and now associate peaceably.

Certain trends in contemporary life tend to undermine regional consciousness, particularly in its more bigoted expression, usually called provincialism. With technological advance, increased ease of communication and movement smooths away differences rooted only in tradition, and blurs and widens the peripheral zones between regions. The current tendency to substitute the mobile individual for the earthbound social group as the unit of society, loosens or breaks the hold of group-solidarity and weakens regional cohesion. North Americans need to remember that this shift is farther advanced in the United States and Canada than it is in any other part of the world.

In spite of fraying in many places, regional consciousness, once firmly established, maintains a tenacious hold. This is notably true so long as natural and societal environment together provide a distinguishing set of conditions for regional individuality. The distinctive quality of Tehuantepec within Mexico illustrates this tendency. It may be equally true even though one or more contributing aspects of the natural environment is obliterated. Thus drainage of marsh barriers and increase in waterborne traffic between Netherlands and Germany leave the two nationalities still confronting each other along their historical delta-margin boundary. A recognizable pattern of regional consciousness may persist in the face of altered conditions in the societal environment. Today, in the Solid South, two parties contest local elections, each deriving its strength from the same districts as did their ante-bellum forebears. In France the map of party locale has changed little through four republics, two monarchies, and two empires.

On the whole, regional awareness seems to be a concomitant of areal diversity, and promises to persist as an element in regional differentiation. It appears least subject to change where the natural environment is harsh and isolating to an extreme degree, and the societal environment is sharply or traditionally distinctive. Where conditions of nature or culture are markedly changing, the pattern of regional consciousness may be altered without destroying its essence. New ways of using space, and faster communication may result in changes in shape or size of regions, or in their disappearance, without obliterating regional consciousness. It may persist as an element in the new pattern of regions, as does the sense of nationality in conquered states. Or it may be transferred from one set of phenomena to another. This seems to have occurred in New England, where factory cities have largely replaced rocky farms as the focus of regional awareness. Some have assumed that a single political frame for the entire earth would obliterate regional consciousness. This seems unlikely, because, as political loyalties come to cover larger areas, governmental policy will have to take account of the diversity of pattern in regions of other categories.

THE METHOD OF REGIONAL STUDY

To pursue an investigation that will reach the goal of regional presentation as formulated in the foregoing pages requires a method. The method will compel the student to approach his investigation with a clearly defined purpose or problem, to recognize the attitude and preconceptions he brings to the chosen area, and to possess skill in the use of the tools with which he attacks it. The procedure may be described as a succession of stages, although in practice the steps may be taken in almost any order and sometimes combined.

APPROACHES

The initial step in any regional study is the clear statement of its basic purpose. Many objectives lead geographers and others to study regions, or to apply the regional method to the study of problems. Since the purpose of the study determines the kind of region sought and the kind of treatment given it, a clear statement at the start of an inquiry is essential for the investigator, and a clear statement in the final report is essential for the reader. The purpose may be to satisfy curiosity about a particular region, to differentiate two or more unlike regions, or to compare similar regions. A different purpose may be to clarify the way in which a region or its components function in a more inclusive area, or to answer a question of cause-and-effect. A still different purpose may be to lay a foundation for private or public policy, usually in setting up operational, administrative, or military regions. An element of geography is inherent, perhaps basic, in every economic, social, political, or military situation, for human affairs always occur in a place and are in some measure conditioned by that place. The regional approach can, therefore, make some contribution to every interest or problem of mankind on the earth.

The objective determines the choices which follow. If he has not already done so, the investigator must decide between what have here been called the regional approach to regional study and the topical approach to regional study. In the first instance he accepts a hypothetical region and proceeds to uncover its components: its forms, its functions, and the sequences of change that have produced it. In the second instance, he selects criteria relevant to his problem, applies them to the earth in the search for accordant relations among regional systems, and builds up his larger regions on the basis of the functional association of their parts. As pointed out earlier in this chapter, the two approaches are not mutually exclusive, for in practice the student may make use of both.

Whether the approach is mainly topical or regional, the degree of generalization appropriate for the purpose of the study must be established by placing the region in its rank in the hierarchy of aggregation-subdivision. This may be done either by analysis, which starts with the region of the highest rank and works down to the regional rank desired, or it may be done by synthesis, working up from the lowest rank. Breakdown and build-up can be used to check each other, thus confirming the authenticity of the chosen region. Initial rapid analysis has the advantage over synthesis of providing a useful orientation for the investigator.

Once the region is in view it becomes the subject of internal examination, whether by the regional or the topical approach. Form, function, and sequence are arrayed as the three aspects of regional character.

The forms are mapped as component regional systems having accordant areal relations. To this end criteria are selected, experimentally at first, with the object of bringing out the maximum measure of valid accordance. The regional forms so arrived at have meaning in picturing the regional structure.

The forms are expressions of the functions; hence, a study that disregards the functional associations within a region is lifeless. For example, in the regional study of a city it is not enough to differentiate the commercial core on a map; it is necessary also to identify the specific commercial functions for the performance of which the core exists. Further, its relations to other functional forms, such as manufactural, residential, and recreational, should be established. Both the arrangement of the patterns and the interaction of the operations within a region are basic in regional geography. From the discovery of recurring patterns and associations in nature's workshop, the student can attain an understanding of process just as valid as that derived from repeated experiment under the controlled conditions of a laboratory.

A high percentage of regional studies are reports on matters current at the time the study was made. But not even the geographer most concerned with contemporary conditions dares lose sight of the time sequence in which his region belongs, because forms once created tend to remain etched on the landscape. So long as these forms are there, they exert the force of inertia upon changing functions. Thus, in a city the street pattern laid down for residences nearly always remains unchanged when business encroaches upon it.

The region having been examined internally, must then be considered as a part of a larger whole. Goods, people, ideas, and administrative authority may all move from region to region, and each region needs to be examined with reference to its part in this pattern of spatial interchange.

The study of regions of the lowest rank, except for the practical purpose of solving a particular economic problem, can lack the breadth necessary for a thorough grasp of their significance. Some geographers have seen a danger in endless depictions of localities, set in no meaningful matrix. A different danger, lack of balance, may threaten the treatment of a region of higher rank, when studied alone. The risks can be avoided through the comparative study of regions in different areas. Comparisons are valid between analogous regions provided the categories to which they belong and their status in the hierarchy of aggregation-subdivision are clearly recognized.

ATTITUDES

The general purpose of regional study as the recognition, delimitation, and interpretation of segments of earth-space permits a considerable range of attitude on the part of the student as to the immediate objective sought. All regional geography appears to reach its objectives through one or another of three modes of operation.

1. Academic study, for the sake of understanding the present state of a region and so much of its past as contributes to that understanding. Here the discovery of generalizations is a usual, but not an inevitable concomitant. Scientific curiosity prompts such a study, and it may be formalized in terms of the elements associated within the region as follows: where A is found, x, y, and z are also present. The conclusions permit or lead to diagnosis of the meaning and import of the discovered relationships. The investigation and presentation of the findings serve to increase knowledge of the earth for the general public, for teaching, and for transfer to cognate disciplines.

2. Practical study of the present state of a region, with a view to prospects of impending or current change, and possibilities for alteration. Curiosity regarding the future motivates this kind of study, with useful applications of the findings in mind. The elements are associated in the region according to the formula: if A is wanted, full account must be taken of x, y, and z; change made in A may alter also x, y, and z. The conclusions result in prescription in anticipation of alteration (either prevention or improvement). The findings constitute materials

helpful to the engineer and planner in government or business, while substance for students and teachers is a likely by-product. Generalizations may be discovered incidentally.

3. Prejudged study of the present state of a region with the avowed intention of altering it along preconceived lines. Reform is the underlying motive of the study, and commitment to a cause tends to turn the student into an advocate, blind to findings that do not support his views and unwilling to consider multiple working hypotheses. The elements formulated run as follows: A ought to be wanted, therefore x, y, and z must be bent to achieve A. The conclusion is bound to be a nostrum, deleterious to all who come in contact with it. *Geopolitik* is the best-known example of this attitude. Its generalizations have no reliable relation to truth.

TOOLS

The tools of research in regional investigation are the same whatever the approach or the attitude of the student. Furthermore, since the regional method is applied to all aspects of geographic research, and in fact to all observational (as opposed to laboratory) disciplines that deal with phenomena on the earth, the instruments are those common to a wide variety of problems and objectives. Since these tools and techniques are discussed at length in certain of the later chapters of this volume, they need only be outlined here. It is convenient to consider them under two headings: sources of information, and techniques of analysis.

Sources of Information

Documentary sources include published and unpublished descriptions, reports, statistics, photographs, and maps. The student of regions must develop specialized knowledge of the documentary sources, and the geographic and other bibliographies in which sources are listed. The handbook, "Aids to Geographical Research," by J. K. Wright and E. T. Platt is the outstanding bibliography of bibliographies [60]. The relatively new field of map intelligence or map analysis, which involves both the knowledge of map series and the methods of evaluating maps, is discussed in chapter twenty-six.

An aspect of documentary study in regional, as distinct from historical, geography may escape notice unless mentioned. This is the use of all pertinent sources, including air photographs, to shed light on the previous stages in the sequence of occupance. To handle this safely and successfully, geographers should have some familiarity with sequence as studied in other disciplines: to wit, the written record of history; the archaeological evidence of prehistoric times; the cultural stages studied in anthropology; the ecological concept of succession; climatic change; and geomorphology, the surface manifestations of earth history.

Direct observation in the field and the interview of informants, two other techniques for gathering information, are treated at length in chapter twenty-four.

Air photo interpretation is still another tool of regional research which, with respect to the visible elements of the landscape, can supplement direct observation in a most effective way. This new field is discussed in chapter twenty-five.

Techniques of Analysis

Information gathered from the above sources is subject to analysis by the methods of regional study. Since regional analysis is always undertaken in terms of symbols, it is convenient to recognize:

Analysis by expository methods, using word symbols;

Analysis by statistical methods, using mathematical symbols;

Analysis by cartographic methods, using map symbols;

Analysis by photo-interpretation methods, using photo-interpretation keys.

Points of Reference

The reader has a right to expect certain explicit statements in any presentation of the results of regional study. These are diverse in kind, but all are important, because taken together they enable the prospective reader to estimate the value of the study to him.

1. The purpose of making the study should preface the report. The information conveyed and the conclusions reached are bounded by the intent of the undertaking.

2. The scope of the study should be indicated, including its degree of generalization or scale, and the intensity of the coverage.

3. The criteria used for identifying the regional pattern should be listed.

4. The reliability of the information presented should be evaluated, including a critique of the major sources used in the report.

5. The period at which the study was made should be stated, including dates of field investigations and of principal collections of other data used. Lacking this information the reader can only assume the date of publication, which may belie the facts.

FRONTIERS OF REGIONAL RESEARCH

The preceding pages point to certain conclusions regarding directions in which regional research can profitably be advanced. The evaluation of the concepts and the testing of the methods of regional study constitute the frontiers of this field. Accepted practice usually lies somewhat behind the frontier. So important is the regional method for all phases of geographic study that attention to the development of its procedures along sound lines is urgently needed. The frontier may be surveyed under three major headings: regional elements, the tools of investigation, and the modes of presenting the findings.

REGIONAL ELEMENTS

Both of the approaches to regional study, the regional and the topical, need to be developed by their proponents. Both methods would be advanced by examples of regional study in which the procedure as well as the results are discussed. Evaluation of different methods is not simplified by the tendency to omit technical treatment of procedures, as, for example, the explanation of the reasons for choosing particular criteria in the definition of regions.

The criteria set up for selecting regions cannot be standardized. The very term criteria implies value judgments unavoidable in the development of meaningful generalizations. The chosen criteria can be evaluated only in terms of the purpose for which the study is undertaken. Nevertheless, some groupings presumably are used more than others. It would be helpful to make available sample sets of criteria that have been used in successful regional study. Such samples might well clarify the procedures that enter into regional differentiation and comparison, and might sort these procedures into physical, biotic, and societal groups. In like fashion, sample regional type-studies made according to clearly formulated criteria would help to clarify the basis and range of regional differentiation.

The arrangement of regions by categories in a single series, from single-feature regions to compages, appears not to have been suggested elsewhere. Nor has there been a clear recognition of the distinction between uniform and nodal regions, although the idea has been germinating for some time [61: 171]. Much remains to be done in the evaluation of these several kinds of regions, and in experimentation with the use of each kind for particular purposes.

The concept of a hierarchy of rank within each category of region has not been uniformly recognized. In certain categories, notably physiographic regions, there is a generally accepted hierarchy, while in others the concept of hierarchy has not been applied. Until each kind of region has been examined in that light, the number of ranks needed for each of the several categories will remain unknown. The proposal put forth in this chapter regarding the ranks of compages is an effort to find common ground; but it cannot be assumed that four ranks, which were found desirable for compages, will necessarily be the preferred number for other kinds of regions. Furthermore, the names thus far chosen for the ranks of different categories of regions are unrelated. Unequated ranks and unrelated nomenclature make it difficult to compare regions belonging to different categories. Considerable error has resulted from incautious comparison of different ranks, as when a high-rank climatic region, such as Köppen's rainy climate with mild winters, is compared with a lower-rank agricultural region, say the United States Corn Belt. An ordered set of terms would not only facilitate comparison but would also sharpen the procedures of inter-regional examination.

The correlation of hierarchical rank with map scale and degree of textural generalization is beginning to interest students of regions. The suggestion that the adjective topographic be used to apply to those scales at which it is possible to map the specific features of the human occupance, and that the adjective chorographic apply to scales on which a somewhat greater degree of generalization beyond the specific features is required, while not widely accepted, is another effort to provide for the comparative treatment of ranks and to make students of regions more conscious than they have been of the importance of the concept of hierarchy. This whole frontier of investigation is challenging: not only do the ranks themselves remain to be defined, but also it seems possible that the criteria should be selected on somewhat different principles in each rank of a hierarchy.

A number of characteristics that serve as brands in the round-up of regional features have been used, but the whole check list that appears on pages 37-40 has not heretofore been set down for reference. Additions to the list or modifications of it will presumably be made on the basis of experience in using it.

Boundaries are among the most thoroughly discussed aspects of regions. Without minimizing their importance, the major emphasis may properly be shifted to regional cores, as a means of focusing on the essence of regional character rather than on special manifestations of it.

In this report, the compage, alone among categories of regions, has been considered in detail, in the hope of bringing study of it to a fresh frontier. Its broad sweep across the whole of geography has left a confusing trail of misunderstanding that urgently needs to be cleared up. The numerous careful studies of compages of the lowest rank point out a path to equally painstaking studies of higher ranks.

The validity of regional consciousness is much debated. Regardless of the furor, it enters into regional study at a number of points and refuses to be denied. Analysis of its character and estimates of its proper place now stop at a frontier far short of the ideal.

TOOLS AND TECHNIQUES OF INVESTIGATION

All the tools and techniques of regional study outlined previously, and many of which are discussed in other chapters, await sharpening by continued experimentation. There are perhaps two frontiers which need special emphasis here.

One has to do with the procedure for the selection of sample studies. American geographers during the past few decades have made notable advances in the methods of relatively large-scale studies of small areas, especially where such studies were being undertaken for application to current economic problems. Other chapters discuss such investigations as those of the Michigan Land Economic Survey, the Tennessee Valley Authority, or the survey of Puerto Rico. But there are not enough trained geographers to secure reasonably prompt coverage by such detailed studies even if selections are limited to highly critical areas. The rapid extension of studies of all ranks would greatly widen the contribution of professional geography to the study of problems of land classification and land use. Since the regions defined at some median scale are recognized as made up of associations of regions of the next lower rank, a method is needed to reveal the component parts of the association. Students of statistics have devised sampling procedures with a high degree of reliability; students of regions are challenged to propose procedures for sampling areal rather than numerical groupings. Some steps already taken toward this end are discussed in chapter twenty-four.

Another frontier in the techniques of regional research involves the direction and operation of teams of scholars. Individual effort sometimes appears too puny to cope with the complexity of regions, especially those of the higher ranks. Teams are composed of persons identified professionally with geography and with other disciplines, but all must be eager to concentrate on a particular part of the earth. The reason for teamwork is to broaden the base of competence by arranging for the joint study of particular places by scholars with different and complementary backgrounds of training and experience. The idea of making a combined attack had its beginning before World War II, as a device for furthering scholarly research, notably among the specialists in the Latin American field. During the war, much of the work of information-gathering agencies, such as the Research and Analysis Branch of the Office of Strategic Services, was organized as interdisciplinary study, with practical goals of regional analysis in view (see chapter twenty-three).

Since the war a good many area programs have been instituted in American universities. The programs differ in the emphasis on research versus training, and the teams rarely include representatives of all the disciplines which should be making contributions. As the cooperative spirit of the war effort waned, the several disciplines moved back to their separate tracks. As that has occurred, the joint approach to area studies has tended to emphasize one or another aspect of the region, most likely the one in which the director of the program is expert. The need for organized search for the most effective procedures in these programs remains acute. Practical goals are not the same as those in wartime, but they still challenge. Regional planning is probably the most provocative and promising of them.

Even if no trained geographer is present on an area team, geography of a sort emerges from the joint effort; but the time has passed when the regional method, as developed by geographers, can be overlooked or when irresponsible and sporadic geography can be reckoned as acceptable. Area research, which in actuality is joint regional study, must be developed on the basis of sound geographic method [62: 636-646, 655-664].

MODES OF PRESENTATION

The presentation of the results of regional study has its frontier no less than the methods of investigation. An understanding of the elementary principles of English composition, of the sort so often neglected in our schools and colleges, is fundamental. This involves the construction of a properly balanced outline of major and minor headings, an appreciation of the function of the paragraph, the role of the topic sentence, and a clear view of the content of an effective introduction and conclusion. Training is needed here, not experimentation for the basic principles have long been known.

There are certain other aspects of presentation, however, in which experimentation is needed. One has to do with the order or succession of topics, especially when dealing with compages. In presenting compages, a procedure inherited from the history of the geographic discipline is still widely followed. The items of the physical and biotic environment are presented first, and those of the societal environment later. The physical items may conform to a rigid succession: climate, landforms, soil, and the rest. Alternatively, the scale of the paper may determine the initial topic: in the realm so treated, climate comes first; in the province, landforms; in the district, soils. The other physical items are fitted in next, to be followed by plants and animals. The elements of the societal environment are brought in separately and afterward, but in no set order. Either of these arrangements of topics unrealistically separates closely related features, such as climate, natural vegetation, and crops. It also lays an arbitrary stress on the natural environment as a whole, and especially upon the first topic taken up.

Instead of any fixed order of topics in presenting compages, a flexible arrangement of the subject matter appears to be a desirable substitute, and is being increasingly employed. The order is determined by the findings, and so is neither arbitrary nor haphazard.

If the purpose of the study is the expository and interpretative description of a region, first place is given to the most significant or critical element or elements, followed by the others in diminishing order of importance in that particular region. Elements inextricably intertwined are thus treated in blocks, giving the writer the advantage of an unforced succession of topics. By this arrangement, the reader is left with a sharp impression of the appearance and essential character of the place, partly through the order of topics and partly through the emphasis which first place confers.

If the purpose of a regional study is to contribute to the solution of economic, social, or political problems, the presentation may well begin with a treatment of those elements on which the proposed application rests, followed by those less relevant. Skill and care are required to follow this sequence without leaving the reader with a distorted concept of regional actualities.

Of equal importance to expository writing in the presentation of regional studies, are the perennial problems of cartographic procedure and method in preparing maps, graphs, and diagrams. A parallel and still different problem is the handling of photographs, sketches, and similar illustrative materials.

Geographers have long been attracted by the notion that regional studies could be presented almost wholly in map form with a bare minimum of text, somewhat as in an atlas. The possibilities of an advance along these lines within the limitations of cost and within the competence of the average reader, who is unlikely to have much experience in reading and interpreting maps, invites experimentation.

To advance the entire field of regional study, a series of world maps might be prepared. Each of these maps would present one kind of region, say climatic or agricultural, and would present existing knowledge regarding this kind of region at the highest rank, comparable to the realm among compages. For example, climatic regions of this rank have already been presented in accordance with several systems of world-wide application, such as those of Köppen or Thornthwaite (see chapter fourteen). Similarly vegetation regions have been presented according to various systems (see chapter nineteen). Some preliminary essays in agricultural regions and manufactural regions at this same rank have appeared (chapters ten and twelve), but there is ample room for further work. Not only are there other categories of regions which have never been examined with this degree of generalization in mind, but there is also a need for critical appraisal and revision of those systems already propounded. A comprehensive series of such maps on a uniform scale would serve as frames of reference for studies of the lower ranks of aggregation-subdivision, and would permit individuals concerned with any rank to coordinate their studies and present them in terms understandable to regional students everywhere.

Regional study, which underlies and is applicable to all aspects of geography, is by no means fixed or standardized. A student of regions is still forced to think for himself, to develop his own competence in making value judgments, and to bring to bear upon both accepted routines and untried possibilities the freshness of his own imagination. The only requirements are fundamental intellectual honesty and an aptitude for geographic endeavor. But geography offers little comfort to those persons who feel happiest when they can proceed along standardized lines, with a set outline of items to be filled in or steps to be taken. Furthermore, every move along the frontier of regional research is certain to be subjected to critical evaluation. The student who offers tentatively a new regional system or a new procedure in regional study need feel no sense of frustration if his system or procedure is not accepted or is so thoroughly modified that it bears little resemblance to his original concept.

A few recent writers have attacked the regional concept itself, and voice despair as to its prospects. They point to the miscellaneous assortment of descriptive reports, each dealing with a small, specific area, few of them comparable enough to be built into larger regions. They point to the absence of an accepted system of regional hierarchies, and to the conflicting definitions of the region and of regional geography. They recommend devoting more effort to topical studies, to the relation of phenomena to process. Yet such studies, in so far as they are focused on the processes as modified in particular places, inevitably make use of the regional method. The topical approach to regional study is a positive effort to give regional study added significance; it shows promise of providing a sound basis for the regional approach to regional study.

There appears to be no slackening of devotion to regional study among geographers and among geographically-minded workers in other disciplines. The regional concept has proved its vitality over the centuries; but it has also eluded the grasp of those who pursue it from a point of view too narrowly systematic. Now with great advances in knowledge of process in cognate fields, and with many successful applications of this knowledge to the ecological aspects of closely allied disciplines, the need for regional study is more urgent than ever and its prospects are brighter. An ever-welcoming public awaits the results.

66   THE REGIONAL CONCEPT AND THE REGIONAL METHOD

REFERENCES

1. HARTSHORNE, R. "The Nature of Geography," *loc. cit.* reference 3 chapter one.
2. BROWN, R. H. "The American Geographies of Jedidiah Morse," *Annals of the Association of American Geographers*, 31 (1941): 145-217.
3. ————. "Materials Bearing Upon the Geography of the Atlantic Seaboard, 1790 to 1810," *Annals of the Association of American Geographers,* 28 (1938): 201-231.
4. JENSEN, M. ed. *Regionalism in America,* Madison (Wisconsin), 1951.
5. BRIGHAM, A. P. Unpublished answers to a printed questionnaire circulated by the secretary of the Association of American Geographers, *Archives of the Association.*
6. TOWER, W. S. "Scientific Geography: the Relation of its Content to its Subdivisions," *Bulletin of the American Geographical Society,* 42 (1910): 801-825.
7. ROORBACH, G. B. "The Trend of Modern Geography," *Bulletin of the American Geographical Society,* 46 (1914): 801-816.
8. JOERG, W. L. G. "The Subdivision of North America into Natural Regions: A Preliminary Inquiry," *Annals of the Association of American Geographers,* 4 (1914): 55-83.
9. JEFFERSON, M. "Some Considerations on the Geographical Provinces of the United States," *Annals of the Association of American Geographers,* 7 (1917): 3-15.
10. DEMANGEON, A. *La Picardie et les régions voisines Artois-Cambrésis-Beauvaisis,* Paris, 1905.
11. BLANCHARD, R. *La Flandre,* Paris, 1906.
12. HETTNER, A. *Die Geographie: ihre Geschichte, ihr Wesen und ihre Methoden,* Breslau, 1927.
13. HERBERTSON, A. J. "The Major Natural Regions: An Essay in Systematic Geography," *Geographical Journal,* 25 (1905): 300-312.
14. ROXBY, P. M. "The Theory of Natural Regions," *Geography,* 13 (1925-1926): 276-382.
15. UNSTEAD, J. F. "A System of Regional Geography," *Geography,* 18 (1933): 175-187; ROBINSON, G. W. S. "The Geographical Region: Form and Function," *Scottish Geographical Magazine,* 69 (1953): 49-58.
16. POWELL, J. W. "Physiographic Regions of the United States," in *Physiography of the United States,* National Geographic Society Monograph No. 1, Washington, 1895: 65-100.
17. FENNEMAN, N. M. "Physiographic Divisions of the United States," *Annals of the Association of American Geographers,* 18 (1928): 261-353.
18. THAYER, W. N. "The Northward Extension of the Physiographic Divisions of the United States," *Journal of Geology,* 26 (1918): 161-185, 237-254.
19. STAMP, L. D., and WOOLDRIDGE, S. W. eds. *London Essays in Geography,* Cambridge (Massachusetts), 1951.
20. ADAMS, C. C. "The Relation of General Ecology to Human Ecology," *Ecology,* 16 (1935): 316-335.
21. VIDAL DE LA BLACHE, P. *Principles of Human Geography,* (DE MARTONNE, E. ed.), New York, 1926.
22. BRUNHES, J. *Human Geography,* Chicago, 1920; republished abridged, London, 1952.
23. DE GEER, S. "On the Definition, Method and Classification of Geography," *Geografiska Annaler,* 5 (1923): 1-37.
24. SAUER, C. O. "The Morphology of Landscape," *University of California Publications in Geography,* 2 (1925): 19-53.
25. UNSTEAD, J. F., and three other authors. "Classifications of Regions of the World," *Geography,* 22 (1937): 253-282.
26. VANCE, R. B. "The Concept of the Region," *Social Forces,* 8 (1929): 208-218. Reprinted with negligible changes in *Human Geography of the South,* Chapel Hill (North Carolina), 1932: 3-19.

27. National Research Council, Division of Anthropology and Psychology, and Social Science Research Council. *Conference on Regional Phenomena, April 11-12, 1930*, Washington (National Research Council), 1930.
28. "Conventionalizing Geographic Investigation and Presentation," *Annals of the Association of American Geographers*, 24 (1934): 77-122.
29. "A Conference on Regions," *Annals of the Association of American Geographers*, 25 (1935): 121-174.
30. National Resources Committee. *Regional Factors in National Planning*, Washington, 1935.
31. PARKINS, A. E. "The Geography of American Geographers," *Journal of Geography*, 33 (1934): 221-230.
32. BOWMAN, I. *Geography in Relation to the Social Sciences*, New York, 1934.
33. COLBY, C. C. "Changing Currents of Geographic Thought in America," *Annals of the Association of American Geographers*, 26 (1936): 1-37.
34. FINCH, V. C. "Geographical Science and Social Philosophy," *Annals of the Association of American Geographers*, 29 (1939): 1-28.
35. TURNER, F. J. *The Significance of Sections in American History*, New York, 1932.
36. ODUM, H. W. *Southern Regions of the United States*, Chapel Hill (North Carolina), 1936.
37. DICKINSON, R. E. *City, Region and Regionalism: A Geographical Contribution to Human Ecology*, London and New York, 1947.
38. MOORE, H. E. "What is Regionalism?" *Southern Policy Papers No. 10*, Chapel Hill (North Carolina), 1937.
39. KOLLMORGEN, W. M. "Crucial Deficiencies of Regionalism," *American Economic Review: Papers and Proceedings*, 35 (1945): 377-389.
40. WHITAKER, J. R. "Regional Interdependence," *Journal of Geography*, 31 (1932): 164-165.
41. Conference at Chapel Hill, North Carolina, April 21, 1945. "Regionalism in World Economics," *World Economics: Bulletin of the Institute of World Economics*, 3 (1945): 49-132.
42. RAUP, H. M. "Trends in the Development of Geographic Botany," *Annals of the Association of American Geographers*, 32 (1942): 319-354.
43. MALIN, J. C. "Ecology and History," *Scientific Monthly*, 70 (1950): 295-298.
44. BARROWS, H. H. "Geography as Human Ecology," *Annals of the Association of American Geographers*, 13 (1923): 1-14.
45. PARK, R. E., and BURGESS, E. W. *Introduction to the Science of Sociology*, Chicago, 1921.
46. PARK, R. E. "The City: Suggestions for the Investigation of Human Behavior in the City Environment," *American Journal of Sociology*, 20 (1915): 577-612.
47. PARK, R. E., BURGESS, E. W., and McKENZIE, R. D. *The City*, Chicago, 1925.
48. McKENZIE, R. D. "The Ecological Approach to the Study of the Human Community," *American Journal of Sociology*, 30 (1924-1925): 287-301.
49. ———. "Demography, Human Geography, and Human Ecology," in *The Fields and Methods of Sociology*, (BERNARD, L. L. ed.), New York, 1934: 52-56.
50. ALIHAN, M. A. *Social Ecology*, New York, 1938.
51. HAWLEY, A. H. *Human Ecology*, New York, 1950.
52. PARK, R. E. *Human Communities*, volume 2 of the *Collected Papers of Robert E. Park*, Glencoe (Illinois), 1932.
53. DAVIS, W. M. "A Graduate School of Geography," *Science*, 56, (1922): 121-134.
54. ACKERMAN, E. A. "Geographic Training, Wartime Research, and Immediate Professional Objectives," *Annals of the Association of American Geographers*, 35 (1945): 121-143.
55. JAMES, P. E. "Toward a Further Understanding of the Regional Concept," *Annals of the Association of American Geographers*, 42 (1952): 195-222.
56. JONES, S. B. "The Enjoyment of Geography, *Geographical Review*, 42 (1952): 543-550.

57. PLATT, R. S. "Environmentalism Versus Geography," *American Journal of Sociology*, 53 (1948): 351-358.
58. JONES, W. D. "Isopleth as a Generic Term," *Geographical Review*, 20 (1930): 341.
59. ———. "Ratios and Isopleth Maps in Regional Investigation of Agricultural Land Occupance," *Annals of the Association of American Geographers*, 20 (1930): 17-195.
60. WRIGHT, J. K., and PLATT, E. T. *Aids to Geographical Research*, American Geographical Society Research Series, No. 22, New York, 1947.
61. PLATT, R. S. "Field Approach to Regions," *Annals of the Association of American Geographers*, 25 (1935): 153-174.
62. "Area Studies," *International Social Science Bulletin*, 4 (1952): 633-699.

HISTORICAL GEOGRAPHY [A]

THE SCOPE OF HISTORICAL GEOGRAPHY
THE BEGINNING AND ENDING OF STUDIES IN HISTORICAL GEOGRAPHY
GEOGRAPHY AND HISTORY
CONFUSIONS IN TITLE

OLD WORLD TRADITION OF HISTORICAL GEOGRAPHY
CLASSICAL TRADITIONS
MEDIEVAL INTERESTS
MEDIEVAL-MODERN TRANSITION
MORE RECENT ANTECEDENTS

HISTORICAL GEOGRAPHY IN NORTH AMERICA
CONTEMPORARY INVENTORY
THE FUTURE OF AMERICAN HISTORICAL GEOGRAPHY
Research in Historical Geography
Training for Research in Historical Geography
Teaching of Historical Geography

CONCLUSION

A. Successive drafts of this chapter were submitted by the chairman of the committee responsible for its writing to other committee members, including Herman Friis, Dan Stanislawski, and J. Russell Whitaker, who contributed liberally in both substantive and editorial criticism. Still others, notably Richard Hartshorne, made important suggestions. The basic approach, organization, and final decisions on wording are, however, the responsibility of the chairman of the committee, Andrew H. Clark.

CHAPTER THREE

HISTORICAL GEOGRAPHY

JUST AS THE USE of the regional concept is implied, if not explicit, in most kinds of geographical analysis, so also must the student of geography turn to a consideration of the past circumstances of the things that interest him and of the processes by which these things have changed. Some geographers have concentrated their attention on such past geography, especially upon what may be termed geographical change through time. Such a focus of interest is called historical geography.

The genetic approach to geographical study inevitably leads to an examination of the past. This does not mean that one is to seek simple causes in the past to account for contemporary conditions, but rather that the conditions observed at any period of time are to be understood as momentary states in continuing and complex processes of change. Simple cause and effect relations are elusive, for no matter how far back a scholar may penetrate there is always a more distant past calling for further investigation. The genetic approach focuses attention on processes, for whatever interests us in the contemporary scene is to be understood only in terms of the processes at work to produce it. It is not, therefore, a search for origins in any ultimate sense, but rather views the present, or any particular time, as a point in a long continuum.

THE SCOPE OF HISTORICAL GEOGRAPHY

Any study of past geography or of geographical change through time is historical geography, whether the study be involved with cultural, physical, or biotic phenomena and however limited it may be in topic or area. Historical geography, therefore, is like regional geography in that its concepts and methods are applicable to all branches of the subject. It is not a field of topical specialization, for any topic can be examined in terms of its past circumstances or of its changes. Studies in historical geography, however, do require the use of particular procedures, are based on certain kinds of data in considerable part derived from documentary sources, and demand a special background of training. An individual scholar thus may well become a specialist in historical geography as it is applied to the study of a specific group of processes in a particular part of the earth.

The Beginning and Ending of Studies in Historical Geography

The problem of the limits in time within which studies of past geography may fall has greatly concerned many who have thought or written of historical geography. "When may one begin?" and "What distinguishes the geography of the recent past from that of the present?" are the forms the questions usually take.

To insist that historical geography begins where history, as opposed to prehistory, begins would assume some inherent necessity for written records in studying the past geography of an area. Archaeological reconstructions alone have sufficiently demonstrated that no such necessity exists. The reasons for denying the validity of such a division apply with almost equal force to any other. There is, indeed, no logical date or period in time when such studies may properly be said to begin. If physical geography is something more than a summation of geological, climatological, ecological, and similar evidence, then a physical historical geography must exist, which utilizes the kind of evidence that is also studied, often in arbitrarily restricted categories, by the historical geologists, paleontologists, and paleoclimatologists. It is true that for periods before the Pleistocene and for much of that epoch, such studies either do not exist, or have been attempted only by scholars from one of these systematic fields. In practice, "dawn" for the historical geographer rarely antedates the late Pleistocene; he has shown little interest in ages devoid of human culture. In logic, however, his license as a scholar allows him to go back in time as far as he has interest and competence.

The division which any student makes between past and present is also a highly subjective one. Geographers of the contemporary scene more often than not use data from some time back in the past in analyzing what they call present circumstances. Historical geographers often project the significance of their conclusions forward in time. The only proper criterion to apply, in attempting to assess the contribution to historical geography in any study, is to judge its relevance to the reconstruction of past geographies and to the interpretation of geographical changes through time.

Geography and History

Historical geography is distinct from history in both viewpoint and methodology. It is, however, a study which gives that attention to differentiation through time necessary to an adequate understanding of circumstances at a point in time. In this respect it is parallel in method to history and the distinction between the two needs to be clearly stated.

History is here taken to mean that discipline which is concerned primarily with human society in its various facets, the character of changes in these through time, and the ideas or other circumstances, including natural conditions, that have influenced those changes. The emphasis in geography, in contrast, has been chiefly directed to the surface of the earth, in whole or in part, and to areal associations and differentiations thereon. This is a clear difference, derived from what have been the core accomplishments of the two fields, and based solidly on the training, experience, and competence of scholars in the respective disciplines.

If this distinction be accepted, historical geography can be defined as the study of the past circumstances of, or of changes in, phenomena of concern to geography. Its relationship with history as a genetic study and its place within geography as a whole can then be recognized with a minimum of confusion, and the basic history-geography contrast between relative emphases on social change through time and area differentiation over the earth's surface is in no degree compromised.

CONFUSIONS IN TITLE

The words history and geography can be otherwise combined to designate a particular field of study and this has sometimes led to misunderstandings. Geographical history (or the geography of history as it has less happily often been called), which is concerned with the significance of areal differentiation and natural circumstance to changes in society and culture, is one such field, but it would seem to lie well on the history side of the history-geography borderlands. Even more clearly distinct is the history of geography.[B] Unfortunately some of our major bibliographical listings still persist in using either historical geography or the history of geography for an omnibus classificatory title which includes all of these distinct categories.

In making such distinctions, where somewhat similar titles have created confusion, there is of course no intention to value one of the fields distinguished above another. But this chapter is about historical geography and not geographical history nor any other field with which it has been, or may be confused, and there is a danger that the exposition and argument that follow will lose much of their point if the distinctions are not clearly drawn at once. The study of the geography of the past, seen as an entity, or entities, changing through time, is what is meant by the descriptive title, historical geography. However prominently changes in time figure in such a study, however genetic its emphasis, historical geography is not history in the traditional sense; it is an approach to the study of any and all geography as understood by geographers and as described in other chapters in this volume.

B. See note x, page 94.

OLD WORLD TRADITION OF HISTORICAL GEOGRAPHY

Space does not permit an elaborate discussion of the long evolution of historical geography. However, some brief references to the history of its exposition in the western cultural tradition are appropriate.

CLASSICAL TRADITIONS[c]

In the legends and mythology from which the Homeric poems grew, one can identify some fact mixed with the fancy which attempted to re-create the past of the peoples and lands of the Aegean area. Later, Herodotus in his famous prologue to the narrative of Greek resistance to Persian invasions, is clearly a geographer and historian at once [11]. His background studies included a critical examination of earlier manuscripts along with arduous field work.

The philosophers and scientists who contributed so much in the classical Greek and Hellenistic literature to the background of mathematical and physical geography were little concerned with past geography; but the expansion of Greek culture, by the colonizing activities of the Greek cities throughout the Mediterranean-Black Sea lands, and through the later conquests by Alexander the Great, assured an audience for descriptive accounts of newly occupied areas, and these descriptions were forthcoming in some quantity. In them one finds a kind of historical regional geography in which past and present characteristics of areas were intimately interwoven and in which changes in character were at least described, if little analyzed.

Of the later writers Strabo in particular is supposed to reflect a change in geographical interest from the more rigorous systematic approach of the Greek philosophers to a descriptive emphasis typical of the spirit of Roman expansion [12]. He shared the combined historico-geographic

c. There are, fortunately, many studies of classical geography and geographers published in English from which the position of historical geography at the very beginning of our traditions may be inferred. Besides the early chapters of such relatively recent surveys as that of R. E. Dickinson and O. J. R. Howarth [1] we have Sir Edward H. Bunbury's classic study [2], unfortunately long out of print and hard to come by, and H. F. Tozer's work [3] which was reissued in 1935 with notes by M. Cary. The latter's still more recent *Geographic Background of Greek and Roman History* [4] proved most revealing for the purpose of this analysis, as did J. O. Thomson's somewhat mistitled *History of Ancient Geography* [5]. A needed check on the generalizations of these historians was provided by E. H. Warmington's translated excerpts from the writings of Greek geographers [6]. Other helpful sources were the studies of ancient exploration and discovery, particularly that of H. E. Burton [7] and the joint work of Warmington and Cary [8], both of which concentrate on the ancient world, and the early chapters of J. N. L. Baker's more comprehensive study [9]. The sources in languages other than English, especially French and German, are much wider in scope and more varied in emphasis; one of the best examples is Hugo Berger's definitive work on Greek geography [10].

outlook of Herodotus, yet whereas the chief emphasis in Herodotus was on history, in Strabo it was on geography. In the seventeen books of Strabo's *Geography*, the author's interest in the different areas is never narrowly limited to a strictly contemporary account. Among writers with greater or less interest in geography who followed, however, this tradition of Herodotus and Strabo was not maintained. There is, for example, very little of it from the imaginative encyclopedist, Pliny, or the mathematical astronomer, Ptolemy.

MEDIEVAL INTERESTS[D]

Historical geography had little place in the geographical interests of the scholars of early and medieval Christendom in Europe. It is true that lack of information and inhibition of critical faculty led to constant repetition of inaccurate description and dubious interpretation;[E] but that much of both was drawn from very old sources does not make their use an exercise in historical geography.

Our present tendency to classify scholars of the past in terms of our own disciplinary subdivisions has led to the conclusion that most Moslem geography,[F] contemporary with that of medieval Christendom, had little historical focus. Al-Biruni's *Kitab-al-Hind* [20], or "Great Book of India," is an arguable exception. Yet if one examines the writings of many Moslem scholars now classed as historians one finds a great deal of substantial work ranging from geographical history to historical geography. As early as the 9th century al-Baladhuri [21] was writing history with a marked geographical emphasis. In the 14th century Abulfida [22] was including similar material in his encyclopedia. Most important of all Moslem scholars in this regard, however, was Ibn Khaldun, one of the greatest historians of any age or culture. From the easily available Prologomena to his *Kitab-al-Ibar* or "Universal History," [23] it is evident that he was

D. The standard source in English for the medieval period, Sir Charles Beazley's *The Dawn of Modern Geography*. . . . , [13] has recently been reissued. In addition Lloyd A. Brown's *The Story of Maps* [14], George H. T. Kimble's *Geography in the Middle Ages* [15], and John K. Wright's *The Geographical Lore of the Time of the Crusades* [16] were all consulted with profit. The last three have extremely useful bibliographies.

E. Thus Roger Bacon, the Franciscan, whose relative independence of mind has led many to rank him as the originator of modern scientific method, described Ethiopia in terms which Claudius Ptolemy would have considered naïve and careless, and which depended on unverified information up to 1500 years old.

F. Kimble [15] and Wright [16] have discussions as well as bibliographies of many of the Moslem writers and Carl Schoy [17] has a brief treatment of the theme. Recently two modern Moslem students, Nafis Ahmad [18] and S. M. Z. Alavi [19], have made more extensive studies of the men concerned and their times. See also the article on geography in the *Encyclopedia of Islam*.

vitally concerned with the significance of different geographical situations, and different man-land relationships in agricultural or pastoral economies, in relation to the developing characters of peoples, civilizations, and regions. He presents a wealth of exceedingly useful references to geographical circumstance in time and place.

MEDIEVAL-MODERN TRANSITION[G]

William Camden's *Britannia*, first appearing in 1586, has been rather widely accepted as the first modern example of our genre. Whatever its intent, it was far from satisfactory as historical geography. In fact "to Camden and his contemporaries, as to many historians to this day, the simple elucidation, identification, and mapping of the place names occurring in historical records deserved the title of 'restoration of ancient geography' " [25: 10]. Yet in Camden's younger continental contemporary Philipp Clüwer (Cluverius) one does first indisputably recognize deliberate and fundamentally successful attempts at historical geography. Clüwer died the year after the birth of Bernhard Varen (Varenius), author of the celebrated *Geographia Generalis* which appeared in Amsterdam in 1650. To our great misfortune both died young, but the writings of Clüwer, notably his volumes on Germany and Italy [26], had perhaps even more influence on geographical thought in the late 16th and early 17th centuries than Varen's organization of systematic geography.

MORE RECENT ANTECEDENTS[H]

Neither of the 19th century figures whom historians have accorded a joint divinity in our professional Olympus, can be said to have contributed much directly to historical geography. Alexander von Humboldt's *Essai.....de la Nouvelle Espagne* [29] has sometimes been cited, and it is certainly a comprehensive regional study with an historical bias in its concern for genesis, but in none of his epochal work did he concentrate his attention on geographical changes as such. Karl Ritter, while often perhaps more of an historian than a geographer, never achieved a methodologic approach which would have allowed him to exploit fully his vast

G. Eva G. R. Taylor's meticulously scholarly volumes on the geography of English writers of the period from the late 15th to the early 17th centuries [24; 25] are the basic references here, especially with regard to Camden.
H. The history of European geography in the 17th, 18th, and 19th centuries is mostly written in German and French. Alfred Hettner's frequent contributions in *Geographische Zeitschrift*, which he edited, and his comprehensive methodological monograph [27] are basic references here. Richard Hartshorne has critically reviewed the most important of the German material [28] including a quite full exposition of Hettner's ideas, and his study provides much the best available bibliography in any work written in English. Hartshorne has a particularly valuable section on Ritter with a generous bibliographical listing.

store of knowledge and almost uniquely suitable training in the direct service of historical geography. However, here and there in his writings are extremely useful bits which arouse in us regret that more was not forthcoming. These are particularly evident in the two dozen monographs that are interspersed through thirteen volumes of Die Erdkunde [30], and deal chiefly with the historical background and distribution of certain plants and animals, and with agricultural, pastoral, or mineral products.[1] These studies are important not only for the themes and methods for research in historical geography which they indicated, but also because they stimulated such later writers as Alphonse de Candolle on domesticated plants [31] and Eduard Hahn, who developed the first coherent theory of the domestication of animals [32].

Meanwhile, a different sort of historico-geographical interest was developing among geographers, stemming from such varied sources as the environmental determinism of Charles L. de S. Montesquieu [33], Johann G. von Herder [34], and Henry T. Buckle [35], and the evolutionary hypotheses of Sir Charles Lyell [36], and Charles Darwin [37]. This approach presupposed a necessary developmental sequence of culture which was in accord with simply conceived physical principles and, therefore, closely tied to the physical environment. Among geographers, Friedrich Ratzel's [38] influence in this direction was great, although careful study of his writings shows that he held diverse views on this question. Reaction, however, was swift, and although some geographers were diverted to a kind of deterministic geographical history and others withdrew in protest to the refuges of purely physical geography, a more balanced and objective approach to the study of past geography developed, perhaps most notably among the school of French regionalists led by Paul M. J. Vidal de la Blache.[J] Perhaps the most effective attack on the environmentalists, however, was to be that of the French historian Lucien Febvre [42].

The purpose of this chapter allows no extensive review of European contributions of the past few decades to the literature of historical geography. They have been many and their scope has been broad, as a very small selection of titles indicates. A few examples of works in German are: Alfred Klotz's analysis of the geography of Caesar's writings [43], Konrad Kretschmer's reconstruction of past Central European geography [44], F. Oelmann's Haus und Hof im Altertum [45], Siegfried Passarge's

1. The essay on tea (Vol. 3: 229-256 of Die Erdkunde) indicates the general approach by its subtitle: "Historisch-geographische und ethnographische Verbreitung der Thee-Kultur, des Thee-Verbrauchs und Thee-Verkehrs, zumal auf dem Landwege, aus dem Süden China's durch Tübet, die Mongolei, nach Westasien und Europa, über die Urga und Kjachta."

J. Vidal's deep interests in historical geography are apparent in many periodical articles [39], in the introduction to his celebrated atlas [40], and in his collaboration in the great historical series on France edited by Ernest Lavisse [41].

interpretation of the pre-Magellan geography of the Pacific islands [46], Karl Sapper's investigation of the pre-Columbian economic geography of animals in the new world [47], and Carl Schott's *Landnahme und Kolon-isation in Canada* [48]. To such more explicitly historical studies might well be added many economic and regional studies in German with a contemporary emphasis which include much historical background. Robert Gradmann's *Süddeutschland* [49] is one of the clearest examples. Moreover this sampling ignores much work in Pleistocene geomorphology, and extensive studies of changing climates, vegetation associations, and soils.

The Vidal school and other geographers writing in French have continued their earlier emphases. Although German and British interest in the geography of prehistoric times has also been keen, French writers, like Pierre Deffontaines [50], have contributed perhaps the most interesting studies methodologically in this particular subject. Demangeon's studies of geography and history in the period of the French revolution [51], his joint study with Febvre on the Rhine [52], and Gallois' interest in the changing northern and eastern frontiers of France [53] are indicative of what has been done. There are also many pertinent studies by non-geographers. No student of European historical geography, for example, can fail to profit from Henri Pirenne's medieval urban studies [54], A. Jardé's report on cereals in ancient Greece [55], or E. Dienne's somewhat earlier history of the drying up of lakes and marshes in France before the Revolution [56]. Because of the tradition of joint training in history and geography in French universities, it is unusual to find a geographer who has not a keen appreciation of historical geography or an historian without concern for geographical history.

This random sampling of a few works in French and German ignores not only the bulk of material in those tongues, but a substantial catalogue of material in other languages. If an attempt were made to choose from this mass some group of studies which has been particularly contributory to later developments in historical geography, August Meitzen's revolutionary work on folk and cultural movements in Europe [57] and Eduard Hahn's investigations of the development and spread of plant and animal husbandry might be selected [32].

One reason often advanced for the relative neglect of historical geography on this side of the Atlantic is the most unfortunate unwillingness, or inability, of American geographers by and large, whether journeymen or apprentices, to read languages other than their own. Yet there now exists, in English, coming from Britain in the past few decades in particular, a rich literature in historical geography in which American students can easily find examples and stimulation.

The historians of British historical geography have tended to be catholic in their selections of examples from the broad joint historico-geographic field and have included perhaps more geographical history than historical geography in their discussions, partly because, in the 19th century, the major interest in whatever might have been called historical geography existed chiefly among recognized historians. J. N. L. Baker [58] derives the geographical interests of early 19th century British historians[K] to a marked degree from B. G. Niebuhr's *History of Rome* [59], although he acknowledges the influence of some early geologists. This interest among historians in studies that are also of great interest to geographers has persisted into later decades.[L] Of rather special importance was the work of the economic historians deriving especially from William Cunningham, whose *Growth of English Industry and Commerce* [83] appeared in 1882, and extending through Sir John Clapham, his associates and students [84]. Clifford Darby [85] among historical geographers acknowledges an especial debt to the encouragement of Clapham.

British geography through most of the 19th century had many sources of support; none, it must be observed, were very substantial until late in the century. There were the historians already mentioned, some geologists, and an assortment of explorers, surveyors, military men, and government officials; of these the historians were certainly among the most important and an early and abiding interest in historical geography was established in their profession. Others, however, did help: although her contributions have rarely been so regarded, it now seems possible to rank Mary Somerville as one who encouraged the historical view. In her famous *Physical Geography* [86], which first appeared in 1842, she gave attention to some of the same themes in historical geography that were afterward expounded by the American, George Perkins Marsh (e.g. the

K. Important among nineteenth century British historians who had such interests were Thomas Arnold [60], T. B. Macaulay [61], A. P. Stanley [62], H. T. Buckle [35]. Charles Kingsley [63], H. F. Tozer [3 and 64], J. R. Green [65], E. A. Freeman [66], Sir Charles Lucas [67], and Sir George Adam Smith [68]. Tozer's *Geography of Greece*, and Smith's *Historical Geography of the Holy Land* deserve special mention although Macaulay's famous third chapter on the geography of Restoration England is certainly much more widely known.

L. A highly selective list of more recent British historians who have contributed substantially to historical geography might include: Sir William Ashley [69], M. Cary [4], R. G. Collingwood [70], Lord Ernle (or Reginald Prothero) [71], H. Reginald H. Hall [73], F. J. Haverfield [74], D. G. Hogarth [75], G. C. Home [76], J. N. L. Myres [70; 77], D. Randall-McIver [78], J. M. Thompson [79]. G. M. Trevelyan [80], Warmington [8; 81], and A. E. Zimmern [82]. Special attention should be called to Trevelyan's early chapters of the volume on *Blenheim* of his three-volume *England under Queen Anne* which contain an excellent social geography of early 18th-century England and, incidentally, compare most favorably with Macaulay's effort (see note K) on the 17th century.

influence of changing culture on changing nature), and Marsh has many references to her writings [87].

With the full flowering of geography in the universities the interest in historical geography among its leaders at the turn of the century and after was made clear by Sir Halford Mackinder [88], and Marion Newbigin [89], in particular. This interest and the encouragement of scholarship in the field is clearly apparent in the work of their students and successors, Britain's contemporary geographers. Many historians of geography, notably Eva G. R. Taylor [24; 25] and E. W. Gilbert [90], have extended their researches into the distinct but adjacent field of historical geography. Darby [85], and Gordon East [91] have been, with little question, in the van of the productive scholars in historical geography, although East has probably always been as much or more interested in what might better be called geographical history. While it is true that few others have specialized strongly in historical studies, perhaps the majority of leading British scholars in the profession have been moved to make one or more contributions, however involuntary, to historical geography.[M]

In concluding this skeletonized appreciation of British accomplishment in historical geography one should not ignore the geography of prehistoric times, to which "joint scholars" like H. J. Fleure [102] and Darryl Forde [103] and geographers like Kenneth Cumberland [104] have indeed made contributions. This field is especially well developed, however incidentally or without consciousness of geographical purpose, by many of Britain's leading archaeologists.[N]

HISTORICAL GEOGRAPHY IN NORTH AMERICA

In North America few studies at any time have been made avowedly as historical geography. What has been done has often been incidental to other purposes. It is therefore not surprising that one must go far beyond the thin ranks of recognized geographers in search of examples. Before 1925 there were few scholars in the United States trained to the doctorate specifically as geographers, and the early band of enthusiasts who formed the Association of American Geographers (1904) was drawn from many fields.

M. Another of our eclectic listings might include at least: S. H. Beaver [92], E. G. Bowen [93], P. W. Bryan [94], Vaughn Cornish [95], O. J. R. Howarth [1; 96], Rodwell Jones [97], L. W. Lyde [98], Hilda Ormsby [99], Norman Pounds [100], and O. H. K. Spate [101].

N. A few of these are: M. C. Burkitt [105], V. Gordon Childe [106], Grahame Clark [107], O. G. S. Crawford [108], Sir Cyril Fox [109], C. F. C. Hawkes [110], H. J. E. Peake [102], Stuart Piggott [111], W. J. Sollas [112], and R. E. M. Wheeler [113].

All too few modern geographers will think of George Perkins Marsh [87] as one of their own, but in his concern with the changes that man had directly effected in the non-cultural complexes of the earth's surface, he was among the first, and was one of the greatest, of our historical geographers. At the turn of the new century Nathaniel Shaler's [114] writings reflected similar interests. Moreover, the contributions of the geologist-physiographers, from whose ranks came many of the founders of North American geography, were substantial in their interpretations of past physical geography. Basic to William Morris Davis' [115] theory of the development of landforms, for example, was a supposed succession of stages through which the surface of an area would pass as a result of the erosion process. So simple and persuasive was the statement of this succession that it stimulated the formulation of analogous concepts in other fields (see chapters seven and sixteen).º Its application to historical-cultural geography, however, was largely left to the incidental consideration of the historians of the expanding United States and Canada, who wrote their histories and developed their theories with a fine regard for regional character and the areal differentiation of both nature and culture.ᴾ In a time when the conception of a cultural geography of North America was barely formed by indigenous scholars, the nominal historians plowed ahead. Although, as historical geography, their efforts often suffer from the incidental nature of their geographical interest, and from their lack of appropriate training, their writings compare more than favorably in this respect with those of the great majority of recent American historians. Some of course, held the seductive hope of interpreting culture and its development in terms of simple physical laws and wrote in the spirit of what can only be called a deterministic geographical history.

In such a vein, indeed, was much of the work of two writers, both early members of the Association of American Geographers, who have long been recognized as our first professional colleagues avowedly in the historico-geographical field. Some antecedents of environmental determinism have been mentioned before; its concepts proved particularly attractive to many biologists and geologists steeped in theory derived

o. Carl Sauer has taken an essentially contrary view in a recent publication [147:2]: "Among geographers, William Morris Davis delayed somewhat our learning about the physical earth by his systems of attractive but unreal cycles of erosion. . . . Such concepts are sometimes, but improperly, called 'evolutionistic.' Actually evolution operates by continuing variation and divergence. It does not return to a previous condition and rarely rests."

p. Among the most prominent on whom the historical geographer must lean heavily are: H. H. Bancroft [116], H. E. Bolton [117], A. G. Doughty [118], Francis Parkman [119], F. L. Paxson [120], Adam Shortt [118], Reuben Gold Thwaites [121], and Frederick Jackson Turner [122], but they had scores of associates and contemporaries who were only less important for this purpose [123; 124; 125].

from Darwin and Lyell. It is, therefore, less surprising than might at first appear to find the geologist, Albert Perry Brigham [126], as one of the chief exponents of this interpretation of human geography. Better known to contemporary American geographers, many of whose leaders she taught at one time or another, is Ellen Churchill Semple [127], who interpreted Friedrich Ratzel's theses to her American colleagues and students. Perhaps in part because their theory of predominant environmental influence could only be illustrated historically, both Brigham and Semple found their first interests in the North American historical record; their earlier work carefully scans that record for examples of such influence.

It is interesting to read the several contributions of each with an eye to segregating the historical geography in them from that which might better be regarded as geographical history. However subjective such a division must be, it reveals how much solid historical geography both actually succeeded in presenting, an appreciation which has often been denied them because of their influence in establishing a deterministic bias in the exposition of geography in our primary and secondary schools. Although much of Ellen Semple's extensive production must be classed as geographical history, the individual parts of the great series ultimately collected in *The Geography of the Mediterranean Region* [128] are important documents in any inventory of American historical geography. Similarily Brigham's studies of geographical influence are paralleled and succeeded by his stimulating descriptions of changing transportation routes through the Appalachians and the extension of wheat growing in North America [126].

During this same period other joint interests in geography and history were becoming apparent among geographers. H. H. Barrows taught a course in the geographical history of the United States for many years at Chicago, and although its lasting influence in terms of subsequent publication may have been greater among those of his students who became historians than among students in his own field, it undoubtedly had a strong influence on many of the latter. The study of the Middle Illinois Valley [129], his only important publication in this field, illustrates his approach. Carl Sauer perhaps adumbrated his later interests in his frequent references to past conditions in his studies of Illinois in 1916 and of the Ozarks in 1920 [130]. In 1918, A. E. Parkins published his historical geography of Detroit [131]. Anthropological interest was evident in H. J. Spinden's studies of agricultural origins and distributions in America published in 1915 [132], and in A. L. Kroeber's work on California place names of Indian origin published in the same year [133]. The varied sources from which the interpretation of past geography appeared are illustrated by H. A. Gleason's work on the vegetational history of the

Middle West [134], and M. Rostovtzev's report on a large Egyptian estate in the Third Century B. C. [135].

CONTEMPORARY INVENTORY

During the past two decades or so, writers who are not professional geographers have continued to do important work in historical geography. Notable among these are Bernard DeVoto [136], Harold A. Innis [137], James C. Malin [138], and Walter P. Webb [139]. Stimulating, also, are the works dealing intimately with relatively small areas, such as the fine series by Joseph Schafer that have appeared under the surprising title of *Wisconsin Domesday Book* [140], and R. C. Buley's Pulitzer-Prize-winning *The Old Northwest* [141]. There is plenty of material in such books to stimulate local studies of historical geography in almost all parts of the country.

The concern of this volume, however, is mainly with the work of professional geographers. During the past few decades publications in historical geography by geographers have increased considerably, although perhaps the rate of increase has not been so great as that in some other branches of geography. Two names dominate the contemporary record: Ralph H. Brown [142; 143; 144; 145] and Carl O. Sauer [146; 147]. The former was an independent, active, non-controversial scholar whose interest in historical geography developed, without any clearly defined antecedents, from field research in the western mountain country and archival research in the history of American geography. Rarely were his methodological conceptions explicitly stated, but he did make some brief comments in 1938 [143: 201-204] which reveal the "geography of the past" viewpoint in unmistakable fashion. His two outstanding contributions are *Mirror for Americans, Likeness of the Eastern Seaboard, 1810,* and *Historical Geography of the United States* [144]. The former is a reconstruction of past geography for a particular period; the latter exemplifies the themes of geographical change [145]. Together they may well constitute a monument not only to Brown himself, whose untimely death followed close on the publication of the second volume, but to the coming of age of American historical geography. Brown's developmental treatment in his second work was strongly influenced by the regional emphasis so strong in American geography in the late twenties and early thirties.

If interest in past geography was rarely explicit, genetic interpretations of regional character have had a large place in a score or more studies examined for this critique. With landscape a fashion in terminology, "changing landscapes" became a manner of expressing interest in historical geography. A new approach was suggested by Derwent Whittlesey in 1929 with the expression "sequent occupance" [148], which has remained a popular and useful descriptive term. In developing this concept, Whittlesey suggested that stages could be recognized "during which human occupation of an area remains constant in its fundamental aspects" [149: 451]. In other words he saw great variations in the tempo of change in the way people live. Rapid and profound changes in the occupance of areas, for example, were introduced by the coming of the railroads or of the motor vehicles and paved highways. By identifying periods of relatively slow change the historical geographer would then select those past geographies the reconstruction of which was especially significant for his purposes.

Three outstanding publications of this period which are clearly historical geography in the sense of this chapter appear to illustrate the point that American geographers generally, to the degree that they have been interested in past geography, have usually been concerned with geographical change through time. These are James W. Goldthwait's "A Town That Has Gone Downhill" [150], Preston E. James' "Blackstone Valley," [151], and Robert B. Hall's "Tokaido: Road and Region" [152]. While each focuses on some periods more than on others, attention to change is in the forefront of each. In some other studies one notes an unfortunate tendency to use the present as the major criterion of what, in past geography, may be reasonably considered, an unnecessary and illogical restriction on the student of historical geography.

A few other examples of recent interest in past geography deserve mention here. Glenn Trewartha's studies of the Driftless Hill Lands [153], and those undertaken by his students in the same area, are of varied emphasis but fall in general into such a classification. His later papers on settlement types and forms [154] are successful essays into the historical aspects of cultural geography. Parkins followed his pioneer study of the historical geography of Detroit [131] with his historical studies of the South [155], although whether these are more in the field of geographical history or historical geography may be hard to assess. J. Russell Whitaker is known to us chiefly for his work in conservation, but it is clear that this interest in resources has grown along with an appreciation for historical geography [156]. Stanley D. Dodge has persisted with his careful historical studies of settlement in the United States [157]. Although Derwent Whittlesey specialized in political geography and in the study of African settlement, he has nevertheless drawn from his keen historical interests stimulating contributions in past political geography and geographical history [158].

A study of the dissertation topics accepted for degrees in geography at the Universities of the United States and Canada up to 1945 makes it clear that some of the more important departments have accepted, and in some cases actually stimulated, research in historical geography whether or not identified by that name.[Q] Yet remarkably few of the students who cut their research teeth on such topics have given substantial evidence of continuing interest.

With all of this widespread, but often ephemeral, interest in historical geography elsewhere in the United States and Canada, Carl Sauer and those who have studied with him at the University of California at Berkeley have been widely recognized as the major contributors, apart from Brown, to a modern American literature in this field of specialization. Two important contrasts between their writings and the work of Brown should be recognized at once. The latter conceived of the study of past geography as being something different from the study of the antecedents of the geography of the present. The members of the Berkeley school have appeared to feel that such a distinction is illusory. They have attacked their problems historically because they have believed that approach fundamental to their explanatory purpose. For example, though much of Sauer's teaching in lectures and seminar has been concerned with the past, he has not used the expression "historical geography" as a course title, and seems not to have thought of what he has taught as essentially distinct from an interpretation of, or a contribution to the understanding of, present geography.

The second contrast may perhaps follow from the first: Brown tended to embrace the cross-sectional approach, whereas the Berkeley group has shown little interest in it, even in the modified form implied in the concept of sequent occupance. The Berkeley emphasis has been rather upon cultural processes and what is here termed geographical change through time, and there has been little concern with making nice distinctions between geography and history or with dodging the appellation "culture historians." But, whatever their avowed or implicit purpose, the work of Sauer and his students has to be classed as historical geography as that expression is here defined.

Q. The influence of Dodge, Hall and James at Michigan is reflected in a long list of dissertations and seminar papers in which the historical approach is clearly evident: Charles M. Davis [159], W. Bruce Dick [160], Eric H. Faigle [161], Robert M. Glendinning [162], Otto E. Guthe [163], Henry M. Kendall [164], Alfred H. Meyer [165], Leonard S. Wilson [166]. Quite a number of the dissertations at Chicago, showing at different times influences of both Barrows and Whittlesey, including those of Parkins [131], Dodge [157], E. N. Torbert [167], and Whitaker [156], many of those at Clark, e.g. those of James [168] and Merle Prunty [169], and some at Wisconsin, e.g. those of Guy-Harold Smith [170] and John Weaver [171] but, curiously, less clearly that of Brown, are to some degree historical. Lester Klimm's dissertation at Pennsylvania [172], Walter Kollmorgen's at Columbia [173], Stephen B. Jones' at Harvard [174], and J. Wreford Watson's at Toronto [175] also had at least something of this emphasis.

Although Sauer himself is perhaps best known to the profession for his methodological writings, his substantive contributions have been forthcoming continuously and with evidence of great breadth of interest [130; 146; 147; 176; 179; and 200]. In area he has reported especially on different parts of the American Southwest and on many sections of western and central Mexico. In time his interests have extended back to the beginnings of culture in the Western Hemisphere and to the Mesolithic of the old world. Although the themes have been varied too, there has been an overall interest in the changing relationship of man with plants, animals, and soils. Perhaps this is the theme in historical geography that has most stimulated those of his students who have worked in the geography of the past.

Whereas interest in historical geography as a primary research field has only occasionally been sustained beyond the doctoral dissertation by geographers trained in other centers, most of the Berkeley group who have completed degree requirements with such research have maintained that interest with active investigation and publication since. Because Latin America offered a field for research where the significance of culture history to contemporary cultural geography was especially clear, many of this group besides Sauer have chosen that area for much of their work,[R] but more have given their attention to the United States or to other parts of the world.[S] Students in other departments at Berkeley, like Richard J. Russell [192], were strongly influenced; visitors for short periods had their interests stimulated, as did Herman Friis (see ref. 33 in chapter four), and John Weaver [171]; and others with similar interests also worked there for a time, as did Jan Broek [193]. Many of the degree students, of course, did not elect historical geography as a research field and some who did a dissertation and a little further work with such an emphasis, like John Leighly [194] and Warren Thornthwaite [195], later moved to distinctly different lines of study.[T]

An explanation of the leading role of the Berkeley group within recent American historical geography should be a part of this inventory, but it is not easily made. Sauer, of course, set an example in his own work; but rarely, if ever, did he deliberately channel the interests of his students.

R. The Latin American historical group includes Donald Brand [176], Henry Bruman [177], Webster McBryde [178], Peveril Meigs [179], James Parsons [180], Dan Stanislawski [181], and Robert West [182].

S. These include especially George Carter [183], Andrew Clark [184], Leslie Hewes [185], Fred Kniffen [186], Edward Price [187], Halleck Raup [188], Erhard Rostlund [189], Joseph Spencer [190], and Wilbur Zelinsky [191].

T. Notably Robert Bowman, David Blumenstock, Samuel Dicken, John Kesseli, and Robert Richardson. Thornthwaite's dissertation was in part an historical geography of Louisville, Kentucky.

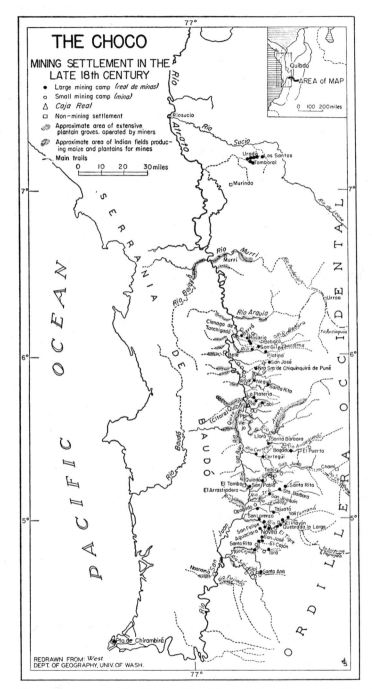

Fig. 1. Mining Settlement in the Choco, Colombia, late 18th Century, by West [182].

87

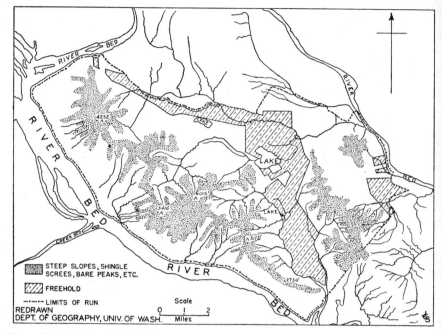

Fig. 2. Freehold Land in Relation to Terrain in a Part of New Zealand, by Clark [184].

His strong emphasis on physical and biotic geography and on field observation supplies no apparent clue. An approach to cultural geography
through the study of the materials and conclusions of anthropology,
archaeology, and economic history unquestionably was influential, as was
a disdain for formal disciplinary boundaries. Perhaps the key to understanding the situation lies in his untiring insistence upon what he called
"the genetic approach." The application of this principle of a search for
origins through study of development by both himself and his students
may explain why the rest of the profession has considered them essentially historical geographers, rather than students of the contemporary.

In a retrospect of American historical geography from Marsh to Sauer
it is evident that most investigation of past geography has been undertaken merely as one of the adjuncts to an understanding of the general
problems of the geographer. Viewed in this way the examples of American historical geography given in the foregoing analysis could be greatly
multiplied. Hartshorne's studies of European boundaries and boundary
zones [196] and Kirk Stone's investigations of the background of Alaskan
settlement are clear examples [197]. The former's interest lay in an understanding of contemporary problems of political geography, the latter's in
planning for new settlement, yet each made an important contribution to

historical geography. To extend this list is an insistent temptation; one should include many more studies like Hildegard Johnson's fine pieces on the settlement of Germans and Scandinavians in the American midwest [198] and Loyal Durand's historical studies of the American cheese industry [199]. Indeed Brown and Semple have been among the few who have made a clear distinction between the geography of past and present. This analysis, however, is one of accomplishment, rather than one of intention.

Perhaps the largest single theme, equally clear in the writings of Marsh and the Berkeley group and recurring insistently in other studies, has been the emphasis on the historical record of man's use, alteration, and rearrangement of his only potentially permanent resources: water, soil, vegetation, and animal life. Certainly the enthusiasts for this theme have had a telling argument to support its study, for, unless geographers can somehow learn what has been done in the past with resources necessary to our future, our land and resource planning will lack one of its potentially most useful bases.

THE FUTURE OF AMERICAN HISTORICAL GEOGRAPHY

In practice the study of historical geography in America in the past few decades has fallen far short of the theoretical scope outlined at the beginning of this chapter. There has been almost no conscious historical research in the physical branches by geographers, and relatively little research in the topical cultural fields has been focused particularly upon the past. Interest is, however, increasing and in the concluding sections of this report some suggestions will be made for the future, partly to encourage the wider teaching of historical geography and the training of students for its study, but chiefly to indicate promising avenues for research investigation.

Research in Historical Geography

Nothing said here should be construed as a recommendation that the aspects of historical study which are directly explanatory of contemporary situations and problems should be neglected; rather such research should be intensified and extended. But there is much of importance to be learned from considering past geographies and their changing characters as worthy of study in themselves. From whichever point of view, the historical approach must be maintained and strengthened for the enrichment of geography as a whole.

A large part of research in historical geography might be called pure research in that it is not directly focused on contemporary applications. One should, however, neither take pride in that out of some twisted sense of intellectual snobbery, nor be ashamed of it in an age when pure research has so often proved more productive and practical than its applied counterpart. The answers one finds may have more or less relevance to contemporary geography, but they will always aid an understanding of the world as it is through a study of what it has been. Through such answers one gains substance and authority for qualified forecasts of the future.

A promising theme[v] is that of the effect on changing culture of certain processes of physical geography: climatic change, natural alteration of vegetation, reorientation of drainage, or advance and recession of coastal waters. This suggests a further need for the devotion of time and effort to purely physical and biotic historical geography. Certainly some of the more extreme theses of the determinists in the field of geography would thereby receive the testing they so richly deserve.

G. P. Marsh [87], R. L. Sherlock [201], and others have studied and speculated on man's role as an agent in changing the "natural" environment, but they have barely scratched the surface of what should be known. Does forest clearing really affect the climate of an area, and if so, in what degree? Has culturally accelerated soil erosion been a problem, not just of yesterday, but of thousands of years? What evidence supports the belief that soil erosion in the Mediterranean lands was directly related to the clearing of the vegetation cover? How extensive and significant have been the alterations of faunal associations by selective fishing or hunting, or in other ways? What has been the ultimate hydrologic effect of drainage here and irrigation there for thousands of years, and how has it altered local soils? Are the Argentine Humid Pampa, the American prairie, the Sudanese savanna, the New Zealand tussock, to any substantial degree culturally induced grasslands, and if so, how were the changes effected? Such are a few of the more obvious problems in this category, for which definitive answers are yet lacking.

Geographers have long been interested in the patterns of population and settlement, the arrangement of cities and towns, of industrial and commercial establishments, and of areas of specialized economic production. These interests are presented in the chapters of this book from four through thirteen. In many cases geographers have not only mapped and described the contemporary arrangement of the various aspects of human occupance, but they have also offered explanations of the observed patterns, and many of these explanations have been convincingly presented. Yet without the historical perspective through which the varying significance of the physical land and changing cultures might be discerned, geo-

v. Some of these themes were first suggested by Carl Sauer in his presidential address to the Association of American Geographers in 1941 [200: 17-23].

graphers have lacked the evidence to support the conclusions so confidently reached. The explanation of the observed patterns of dispersed and agglomerated settlements, for example, can only be reached after a study of the past and of changes in pattern through time. The record of man's experience with the land in a particular area is fundamental to any attempt to plan for the better use of land in that area. Planning efforts, of the kind described in chapters six or twelve, are most effective when illuminated by insights drawn from the only equivalent to a laboratory a geographer possesses—the record of past geography.

Of special interest to geographers are the problems of resource utilization, including the use of the land for agriculture. Various kinds of public agencies take steps to modify or even make over the land-use practices of large areas. But how much is known about previous experience either in the same or other areas regarding the effect of such steps? If a new crop, a new animal, or a new kind of machine is introduced into the economy of a rural area what results may be expected? The immediate and direct effect is usually foreseen; but a direct effect has its impact on some less direct phenomenon until the change has been felt throughout the whole of the area, and some of the results that are not foreseen may prove disastrous. The efforts of the rural sociologists, agricultural economists, and others would be greatly aided by reports from historical geographers regarding what has happened in similar situations in the past.

At present the maladjustment of population and resources in different parts of the world, and in the world as a whole for that matter, is receiving great public attention. Optimists or pessimists, all are anxious to try to raise the level of living of peoples throughout the world. But what policies will lead most surely to these obvious ends? And how must general policies be modified to apply to particular places? The answers usually require more knowledge of the cultural geography of the areas concerned than is available. Because the record of past changes has not been made clear, no one knows what elements in a specific area are in balance and what ones are not.

Geographers do not make the absurd claim that research in historical geography can reach definitive, all-inclusive answers to such problems, but they do suggest that such research can make a contribution and can throw light that is not gained from any other perspective. Imbalance of population and resources, cultural receptivity and cultural lag, rapid changes in techniques and population, indeed most of the themes arising in the present problems have recurred again and again in the past. Today we are faced with a world that is changing rapidly because of new technical devices. Geographers may wish with great fervor to give direction to the change, or, if that is impossible, to make some intelligent guesses

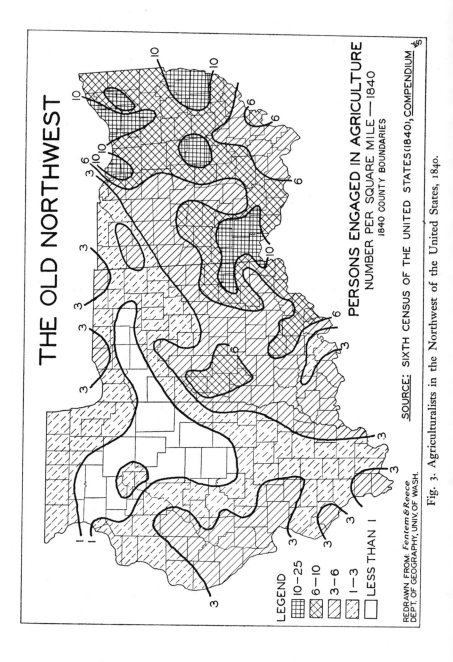

Fig. 3. Agriculturalists in the Northwest of the United States, 1840.

as to where the changes may lead. Many of the desired insights are to be gained from past experience. There is no dearth of problems; only of curiosity to see them and of industry to solve them.

Training for Research in Historical Geography

The professed viewpoint of this chapter, which stresses the theoretical inclusiveness of historical geography, would perhaps suggest that a well-trained historical geographer was simply a properly-prepared geographer. Instead of a theoretical discussion of this point, a few brief suggestions based largely on the record of the training and teaching of Brown and Sauer, as to what has gone into, or what they have felt should go into, such a training, will be made.

A rigorous preparation in the usual physical branches of geography would seem to be prerequisite and more training than has been usual in the theory and practice of geographic cartography and in the interpretation of air photographs is certainly to be recommended.[v] Some neglected fields, like plant and animal geography, might well be more vigorously tilled. Sound training in field observation, including plenty of actual practice, cannot be bypassed. These statements will seem truisms to the geographer; they are emphasized here as a friendly warning to students who, with a background of training in history or one of the other social sciences, may wish to undertake research in historical geography. They need no union cards, but they do need appropriate tools.

The effective use made by many historical geographers of the ideas and materials of anthropology and archaeology suggests the utility of more attention to the literature and methodology of these sciences than apprentice geographers have ordinarily given. Field experience in either will usually be beyond the reach of the student, but if he becomes a good field man as a geographer he will learn to recognize, explore, and sometimes interpret the kind of evidence geographers commonly leave to sister disciplines. This again emphasizes the necessity for careful field training for the potential historical geographer. Field work in his specialty may be even more difficult than in some other kinds of geographical work, but it may be, at the same time, more rewarding. It is, for example, often mandatory that maps and documents should be taken into the field and the findings of archival and library research must be constantly checked in the areas to which they have reference. Above all, the historical geographer must learn to read the record of the past from a wide variety of kinds of evidence to be found only in the field.

The converse of this proposition is also true; field observation must be continually checked against the documentary record and the documents, verbal, statistical, and cartographic, often provide the direction for

v. Note, for example, figures 26 to 33 in chapter twenty-five.

effective organization of field observation. Brown was, and Sauer is, a superb field observer, but each had also to make himself a master of the techniques of documentary research[w] The warning to those who might underestimate the importance of training in physical geography or field work in historical geography must thus be supplemented by a stricture on the geography student to master the basic tools in the use of documents. The acolyte must, moreover, read all the history he can. A wide reading is necessary, not only for facts and interpretations but for the development of a critical attitude toward interpretation.

In the field of history there is nothing more important to one who would work in historical geography, or in any branch of geography for that matter, than the history of geography itself. The story of the development of geographic concepts and procedures is both fascinating and important. It is important because of the perspective it throws on the research undertakings of the moment, and on the origin of professional traditions and concepts. It is important because it points to the blind alleys already explored in previous generations that, for this reason, need not be explored again. Yet in many centers of geographical training not enough attention is paid to this.[x] For students who wish to recreate the geography of a past period, as Brown has shown so conclusively, the history of geography as it was practiced in that period may provide an essential approach to historical geography.

Yet much as all this asks of the would-be historical geographer, there remain skills and areas of learning which would be of great utility but which few geographers can hope to master. In the investigation of the historical geography of prehistoric and ancient times, for example, no man could hope to be, at once, an archaeologist, epigraphist, physical anthropologist, glacial geologist, or, indeed, the score of other things he might have to be if he undertook to "go it alone." Teamwork is increasingly a hallmark of research in the social as well as in the natural sciences, and the geographer of the past must depend upon it heavily. He may find it necessary or even prefer to make his investigations and do his writing alone, but what he does must yet rest heavily upon the findings of other scholars and he must   usually consult frequently with them. This is,

w. Students should early become acquainted with our major sources of documentary material, particularly with the Library of Congress and the National Archives where geographers such as Arch Gerlach and Herman Friis, respectively, are strongly encouraging the use of the extensive records of the American past, in both map and text.

x. John K. Wright has dealt with these matters in a number of articles seeking to develop an appreciation of the need for attention to the history of geography [202]; his book on the development of geography in the United States commemorating the hundredth anniversary of the American Geographical Society in 1952 is a fine example of such work [203].

however, the normal situation of the geographer and is not unique to the historical approach.

Teaching of Historical Geography

The steadily increasing number of courses in the historical geography of the United States and Canada, listed by our universities and colleges, is one of the many evidences of growing interest in the geography of the past which the preparatory study for this inventory has revealed. Apart from growing interest in geography in general, and historical geography in particular, this increase may be in part attributed to a realization that few other kinds of study are better calculated to stimulate the student's curiosity about his own land and culture. The active development of similar courses concerned with other regions, or on various topical themes may, likewise, stimulate a wider interest in the rest of the world.

Whether as historical geography in itself, or as an historical introduction to more standard courses, such an approach should have an even greater place in our educational program. The experience of many college teachers suggests that there are few better vehicles for introducing the geographer's viewpoints to the general student body, and a broader participation in our offerings must be considered important to all professional and scholarly preparation, as well as to general education for citizenship, if our convictions about the importance of our own field are soundly based. The historical approach not only leads more non-geographers to an interest in geography, but in turn has the important function of leading geographers to wider interests outside their traditional lines of training. Overspecialization within geography is a denial of one of geography's fundamental virtues.

CONCLUSION

This discussion has considered a rational and workable conception of what historical geography can be, its role in the development of geographical study in the Western cultural tradition, its place in the short history of American geography, and some suggestions for its future in terms of research, training, and teaching.

The historical approach has been part of much geographical study of all kinds and for all purposes. The historical view of any kind of geographical problem should be developed and encouraged. At the same time more attention should be directed toward the geography of the past and geographical change through time as subjects in themselves. The rationale of historical geography is that through its study we may be able to find more complete and better answers to the problems of interpretation of the world both as it is now and as it has been at different times in the past.

The opening chapter of this volume discussed the general problems and points of view of the geographer, whatever his special interest or bias. If any geographer, in his analyses, concentrates upon too limited a period of time he may deny himself a great body of data helpful to sound conclusions. Historical investigation vastly expands the potential data-base, a particularly noteworthy service where the dangers of generalization from scanty evidence are most obvious. The geographer has no opportunity to set up controlled experiments in which he can watch nature and culture combining to create the character of regions and the vast variety of inter- and intra-regional patterns of natural and cultural forms. He can, however, look back through the record in an attempt to see how these combinations have worked. This is the nature of, and a major justification for, historical geography.

Increased interest in historical geography in research, in teaching, and in the training of scholars for its effective study is an urgent need. A strengthening of the position of the historical approach within our field will help to guarantee the continued vitality of geography.

REFERENCES

1. DICKINSON, R. E., and HOWARTH, O. J. R. *The Making of Geography*, Oxford, 1933.
2. BUNBURY, E. H. *A History of Ancient Geography*, (2nd ed.), 2 vols., London, 1883.
3. TOZER, H. F. *A History of Ancient Geography*, Cambridge, 1897; reissued, Cambridge, 1935, with notes by M. CARY.
4. CARY, M. *The Geographic Background of Greek & Roman History*, Oxford, 1949.
5. THOMSON, J. O. *History of Ancient Geography*, Cambridge, 1948.
6. WARMINGTON, E. H. *Greek Geography*, Library of Greek Thought, London, 1934.
7. BURTON, H. E. *The Discovery of the Ancient World*, Cambridge, (Mass.), 1932.
8. WARMINGTON, E. H., and CARY, M. *The Ancient Explorers*, London, 1929.
9. BAKER, J. N. L. *A History of Geographical Discovery and Exploration*, London, 1931; new edition, revised, London, 1937.
10. BERGER, E. H. *Geschichte der wissenschaftlichen Erdkunde der Griechen* (2nd ed.), Leipzig, 1903.
11. *Herodotus* . . . English translation by A. D. GODLEY, 4 vols., London and New York, 1924; WELLS, J. *Studies in Herodotus*, Oxford, 1923; RENNELL, J. *The Geographical System of Herodotus* . . . , 2 vols., London, 1830; BOBRIK, H. *Geographie des Herodot* . . . , with atlas of 10 maps, Königsberg, 1838.
12. *The Geography of Strabo, with an English Translation by Horace L. Jones*, 8 vols., London and New York, 1917-1933.
13. BEAZLEY, C. R. *The Dawn of Modern Geography* . . . , 3 vols., London, 1897-1906; reissued, by lithograph, New York, 1949.
14. BROWN, L. A. *The Story of Maps*, Boston, 1949.
15. KIMBLE, G. H. T. *Geography in the Middle Ages*, London, 1938.
16. WRIGHT, J. K. *The Geographical Lore of the Time of the Crusades*, New York, 1925.
17. SCHOY, C. "The Geography of the Moslems of the Middle Ages," *Geographical Review*, 14 (1924): 257-269.

18. AHMAD, N. *Muslim Contributions to Geography*, Lahore, 1947.
19. ALAVI, S. M. Z. "The Development of Arab Geography," *Indian Geographical Review*, 22, no. 3 (1947), and 23, no. 2 (1948).
20. SACHAU, E. L. *Albiruni's India: An Account of the Religion, . . . Geography . . .*, London, 1914.
21. HITTI, P. K. *The Origins of the Islamic State . . .*, 2 vols., New York, 1916-24, contains a translation of al-Balādhurī's major account of the Moslem conquests in the early centuries of expansion.
22. ABŪ-L-FIDĀ, *Géographie d'Aboulféda*, translated from the Arabic into French with commentary and notes, vols. 1 and 2, part 1, by J. T. REINAUD, Paris, 1848; vol. 2 part 2 by S. GUYARD, Paris, 1883.
23. IBN KHALDŪN, *Les prolégomènes d'Ibn Khaldoun . . .*, translated from Arabic to French with commentary by M. DE SLANE, 3 vols., Paris, 1863-1868; reissued, Paris, 1934-1938.
24. TAYLOR, E. G. R. *Tudor Geography, 1485-1593*, London, 1930.
25. ————, *Late Tudor and Early Stuart Geography*, London, 1934.
26. CLÜWER (CLUVERIUS), P. *Germaniae Antiquae*, Leyden, 1616; *idem. Italia Antiqua*, Leyden, 1624.
27. HETTNER, A. *Die Geographie: Ihre Geschichte, Ihr Wesen und Ihre Methoden*, Breslau, 1927.
28. HARTSHORNE, R. "The Nature of Geography," *loc. cit.* reference 3, chapter one.
29. HUMBOLDT, A. VON. *Essai politique sur le royaume de la Nouvelle Espagne*, 5 vols., Paris, 1811-1812; other major works include: *Ansichten der Natur . . .*, Tübingen, 1808, Stuttgart and Tübingen, 1849; *Relation . . . du Voyage aux Régions Équinoxiales du Nouveau Continent*, 3 vols., Paris, 1814-1825; and *Kosmos . . .*, 5 vols., Stuttgart, 1845-1862.
30. RITTER, K. *Die Erdkunde, im Verhältniss zur Natur und zur Geschichte des Menschen, oder allegemeine, vergleichende Geographie als sichere Grundlage des Studiums und Unterrichts in physikalischen und historischen Wissenschaften*, 19 parts in 21 vols., Berlin, 1822-1859.
31. CANDOLLE, A. L. P. P. DE. *Origin of Cultivated Plants*, 3rd English edition, London, 1939.
32. HAHN, E. *Die Haustiere und ihre Beziehungen zur Wirtschaft des Menschen; Eine geographische Studie . . .*, Leipzig, 1896; *idem. Die Entstehung der Pflugkultur, (unseres Ackerbau)*, Heidelberg, 1909; and *idem. Von der Hacke zum Pflug . . .*, Leipzig, 1914.
33. MONTESQUIEU, C. L. DE S. *De l'esprit des lois . . .*, first published, Geneva, 1748, more generally available in his collected works, edited by Laboulaye, 7 vols., Paris, 1875-1879, and published in English translation as *The Spirit of Laws*, London, 1902.
34. HERDER, J. G. VON. *Ideen zur Philosophie der Geschichte der Menscheit*, vols. 4-7 of Section 3 of his . . . *Sämmtliche Werke*, Stuttgart, 1827-1830; first published in 4 vols., 1784-1791. Excerpts with notes by KUNO FRANCKE are available in English in *The Warner Library of the World's Best Literature*, vol. 13, pp. 7259-7276.
35. BUCKLE, H. T. *History of Civilization in England*, 2 vols., London, 1857-1861. A revised edition, 2 vols., London, 1925-1931, is widely available.
36. LYELL, C. *Principles of Geology . . .*, 2 vols., London, 1830-1832, has run through many English and American editions as have his *Elements of Geology*, London, 1838, and . . . *Antiquity of Man*, London, 1863.
37. DARWIN, C. R. *On the Origin of Species . . .*, London, 1859, had many later editions, e.g., Modern Library, New York, 1936, as did his *The Variation of Animals and Plants under Domestication*, London, 1868, and *The Descent of Man . . .*, London, 1871.
38. RATZEL, F. *Anthropogeographie*, Stuttgart, vol. 1, 1882 and 1889, vol. 2, 1891 and 1912.

39. VIDAL DE LA BLACHE, P. M. J. "Les voies de commerce dans la géographie de Ptolémée," *Comptes rendus Academie des Inscriptions et Belles-Lettres*, 4e series, 24 (1896): 456-483; *idem*. "Note sur l'origine du commerce de la soie par voie de mer", *ibid*. 25 (1897): 520-527; and, *idem*. "La géographie de l'Odyssée d'après l'ouvrage de M. Bérard," *Annales de Géographie*, 13 (1904): 21-28.

40. ———. *Histoire et Géographie. Atlas Général Vidal-Lablache*, Paris, 1894; revised, Paris 1905, 1909, 1951.

41. ———. "Tableau de la géographie de la France," vol. 1, part 1, of Ernest Lavisse, ed. *Histoire de France* . . . , Paris, 1903.

42. FEBVRE, L. P. V., and BATAILLON, L. *A Geographical Introduction to History*, New York, 1925, a translation of . . . *La terre et l'évolution humaine* . . . , Paris, 1924.

43. KLOTZ, A. "Geographie und Ethnographie in Caesar's *Bellum Gallicum*," *Rheinisches Museum*, 83 (1934): 66-96.

44. KRETSCHMER, K. *Historische Geographie von Mitteleuropa*, vol. 4 of *Handbuch der mittelalterlichen und neuren Geschichte*, München and Berlin, 1904.

45. OELMANN, F. *Haus und Hof im Altertum*, vol. 1 of *Die Grundformen des Hausbaus*, Berlin, 1927.

46. PASSARGE, S. "Die Politisch-geographischen Grundlagen des Südseegebiets vor dem Eintreffen der Europäer," *Petermann's Mitteilungen*, 72, (1926): 209-212.

47. SAPPER, K. "Jagdwirtschaft, Tierhaltung, und Tierzuchtung der Indianer in vorkolumbischen Zeit," *Mitteilungsblatt der deutschen Gesellschaft für Völkerkunde*, Hamburg, 9, (1939): 41-56.

48. SCHOTT, C. *Landnahme und Kolonisation in Canada: am Beispiel Südontarios*, Schriften des Geographisches Institut des Univeristät Kiel, Band 6, Kiel, 1936.

49. GRADMANN, R. *Süddeutschland*, Stuttgart, 1931.

50. DEFFONTAINES, P. "Essai de géographie préhistorique du Limousin et son pourtour sédimentaire," *Annales de Géographie*, 42 (1933): 461-476; *idem*. "Essai de géographie préhistorique de la Tchécoslovaquie," *L'Anthropologie*, 40 (1930): 275-283.

51. DEMANGEON, A. "Géographie et histoire: les paysans du Nord de la France pendant la révolution française," *Annales de Géographie*, 34 (1925): 62-66.

52. DEMANGEON, A., and FEBVRE, L. *Le Rhin: problèmes d'histoire et d'économie*, Paris, 1935.

53. GALLOIS, L. L. J. . . . *Les variations de la frontière française du Nord et du Nord-Est*, Paris, 1918.

54. PIRENNE, H. *Medieval Cities: Their Origins and the Revival of Trade*, translated from the French by F. D. HALSEY, Princeton, 1925; *idem*. . . . *Les villes et les institutions urbaines*, 2 vols., Paris, 1939.

55. JARDÉ, A. . . . *Les céréales dans l'antiquité grecque*, vol. 1, "La production," *Bibliothèque des Écoles françaises d'Athènes et de Rome* . . . , fasc. 130, Paris, 1925.

56. DIENNE, É. *Histoire du desséchement des lacs et marais en France avant 1789*, Paris, 1891.

57. MEITZEN, A. *Siedlung und Agrarwesen der Westgermanen und Ostgermanen, der Kelten, Römer, Finnen und Slawen*, 3 vols. and an atlas, Berlin, 1895, esp. vol. 1, *Wanderungen, Anbau, und Agrarrecht der Völker Europas nördlich der Alpen*.

58. BAKER, J. N. L. "The Last Hundred Years of Historical Geography," *History*, New Series, 21 (1936): 193-207; and *idem*. "The Development of Historical Geography in Britain during the Last Hundred Years," *The Advancement of Science*, 8, 32 (March 1952): 406-412.

59. NIEBUHR, B. G. *Römische Geschichte*, 3 vols., Berlin, 1828-1832; translated as *The History of Rome*, 3 vols., London, 1851, is the work referred to, but his *Lectures on Ancient Enthnography and Geography* . . . , 2 vols., translated by LEONARD SCHMITZ, Boston, 1854, is also relevant.

60. ARNOLD, T. *History of Rome*, 3 vols., (5th ed.), London, 1848, contains chapters on "The City of Rome, its Territory and Scenery" and on "Physical History."
61. MACAULAY, T. B. . . . *History of England* . . . 2 vols., (5th ed.), London, 1849, especially the third chapter of vol. 1 describing the condition of England in 1685.
62. STANLEY, A. P. *Sinai and Palestine* . . . , London, 1856, which is something of a precursor of SIR GEORGE ADAM SMITH's more famous work of 1894.
63. KINGSLEY, C. *The Roman and the Teuton*, London, 1864.
64. TOZER, H. F. *Lectures on the Geography of Greece*, London, 1873.
65. GREEN, J. R. *A Short History of the English People*, London, 1875, which was followed by his joint work with GREEN, A. S. A. (S.) *A Short Geography of the British Islands*, London, 1879, and *The Making of England*, London, 1885.
66. FREEMAN, E. A. *Historical Geography of Europe*, London and New York, 1881; the 3rd edition, edited by J. B. BURY, appeared in 1903.
67. LUCAS, C. *Introduction to a Historical Geography of the British Colonies*, Oxford, 1887; *idem. Historical Geography of the British Colonies*, 7 vols., Oxford, 1888-1920.
68. SMITH, G. A. *The Historical Geography of the Holy Land* . . . , London and New York, 1894, and *idem. Atlas of the Historical Geography of the Holy Land* . . . , London, 1915.
69. ASHLEY, W. *The Bread of our Forefathers* . . . , Oxford, 1928.
70. COLLINGWOOD, R. G., and MYRES, J. N. L. *Roman Britain and the English Settlements* (The Oxford History of England), Oxford, 1936.
71. ERNLE, LORD (PROTHERO, R. E.). *English Farming Past and Present*, (6th ed.), edited by SIR (ALFRED) DANIEL HALL, London, 1941.
72. GRAS, N. S. B. *The Economic and Social History of an English Village (Crawley, Hampshire)*, A. D. 909-1928, Harvard Economic Studies . . . , vol. 34, Cambridge (Mass.), 1930.
73. HALL, H. R. H. *The Ancient History of the Near East, from the Earliest Times to the Battle of Salamis*, London, 1913.
74. HAVERFIELD, F. J. *The Romanization of Roman Britain*, (4th ed.), Oxford, 1923.
75. HOGARTH, D. G. *The Ancient East*, New York, 1915.
76. HOME, G. C. . . . *Roman London, A. D. 43-457*, London, 1948.
77. MYRES, J. N. L. "The Marmara Region: a Study in Historical Geography," *Scottish Geographical Magazine*, 40 (1924): 129-150.
78. RANDALL-MacIVER, D. *The Iron Age in Italy* . . . , and *The Etruscans*, both Oxford, 1927; and *idem. Italy before the Romans*, Oxford, 1928.
79. THOMPSON, J. M. *An Historical Geography of Europe*, (800-1789), Oxford, 1929.
80. TREVELYAN, G. M. *England Under Queen Anne*, 3 vols., London, 1930-34.
81. WARMINGTON, E. *The Commerce between the Roman Empire and India*, Cambridge, 1928.
82. ZIMMERN, A. E. *The Greek Commonwealth*, (5th ed., rev.), Oxford, 1931.
83. CUNNINGHAM, W. *The Growth of English Industry and Commerce*, Cambridge, 1882, and 1927-38.
84. CLAPHAM, J. H. *The Woollen and Worsted Industries* . . . , London, 1907; *idem. The Early Railway Age, 1820-1850*, vol. 1 of *An Economic History of Modern Britain*, 3 vols., Cambridge 1926-38.
85. DARBY, H. C., ed. *An Historical Geography of England before A. D. 1800*, Cambridge, 1936; *idem. The Medieval Fenland*, and *The Draining of the Fens*, both Cambridge, 1940; *idem.* "The Changing English Landscape," *Geographical Journal*, 117 (1951): 377-398; *idem. The Domesday Geography of Eastern England*, Cambridge, 1952, the first multi-volume series.
86. SOMERVILLE, M. *Physical Geography*, London, 1948 (the first of many editions). J. N. L. BAKER has a study, "Mary Somerville and Geography in England" in *Geographical Journal*, 111 (1948): 207-222.

87. Marsh, G. P. *The Earth as Modified by Human Action*, New York, 1874 (really a later edition of *Man and Nature, or Physical Geography as Modified by Human Action*, New York, 1864).

88. Mackinder, H. *Britain and the British Seas*, New York, 1902.

89. Newbigin, M. *Canada, The Great River* . . . , New York, 1926.

90. Gilbert, E. W. *The Exploration of Western America, 1800-1850, an Historical Geography*, Cambridge, 1933.

91. East, W. G. *An Historical Geography of Europe*, (4th ed. rev.), London, 1950; *idem.* "Historical Geography," from *The Spirit and Purpose of Geography*, with S. W. Wooldridge, London, 1951: 80-102; and *idem. The Geography Behind History* . . . , London and New York, 1938.

92. Beaver, S. H. "The Development of the Northamptonshire Iron Industry, 1851-1930," in *London Essays in Geography*, L. D. Stamp and S. W. Wooldridge, eds., Cambridge (Mass.), 1951: 33-58.

93. Bowen, E. G. *Wales; A Study in Geography and History*, Cardiff, 1941.

94. Bryan, P. W. *Man's Adaptation of Nature: Studies of the Cultural Landscape*, London, 1933.

95. Cornish, V. *The Great Capitals, An Historical Geography*, London, 1923.

96. Howarth, O. J. R. . . . *A Geography of Ireland*, Oxford, 1911.

97. Jones, L. R. *The Geography of London River*, London, 1931.

98. Lyde, L. W. *Peninsular Europe* . . . , London, 1931.

99. Ormsby, H. R. *London on the Thames* . . . , London, 1924.

100. Pounds, N. G. J. *An Historical and Political Geography of Europe*, London, 1947.

101. Spate, O. H. K. "Toynbee and Huntington: A Study in Determinism," *Geographical Journal*, 118 (1952): 406-424.

102. Peake, H. J. E., and Fleure, H. J. *The Corridors of Time*, 9 vols., London and New Haven, 1927-1936.

103. Forde, C. D. *Habitat, Economy and Society: a Geographical Introduction to Ethnology*, London, 1934.

104. Cumberland, K. B. "Aotearoa Maori: New Zealand about 1780," *Geographical Review*, 39 (1949): 401-424.

105. Burkitt, M. C. *Prehistory* . . . , Cambridge, 1925.

106. Childe, V. G. *Man Makes Himself*, London, 1941; and *idem.* "Adaptation to the Post-Glacial Forest of the North Eurasiatic Plain," in MacCurdy, G. G. *Early Man* . . . , Philadelphia and New York, 1937: 233-242.

107. Clark, J. G. D. *Prehistoric Europe: The Economic Basis*, New York, 1952.

108. Crawford, O. G. S. *Topography of Roman Scotland, North of the Antonine Wall*, Cambridge, 1949.

109. Fox, C. F. *The Personality of Britain*, Cardiff, 1943.

110. Hawkes, C. F. C. *The Prehistoric Foundations of Europe to the Mycenean Age*, London, 1940.

111. Piggott, S. *Prehistoric India* . . . (Pelican Books), Harmondsworth, 1950.

112. Sollas, W. J. *Ancient Hunters and Their Modern Representatives*, (3rd ed. rev.), New York, 1924.

113. Wheeler, R. E. M. *Prehistoric & Roman Wales*, Oxford, 1925.

114. Shaler, N. S. *Nature and Man in America*, New York, 1891; *idem. Man and the Earth*, New York, 1905.

115. Daly, R. A. "Biographical Memoir of William Morris Davis, 1850-1934," *National Academy of Science, Biographical Memoirs*, 23 (1945): 263-303.

116. Bancroft, H. H. *California Pastoral* . . . *1769-1848*, San Francisco, 1888.

117. Bolton, H. E. *The Spanish Borderlands* . . . , New Haven, 1921.

118. Shortt, A., and Doughty, A. G. *Canada and Its Provinces* . . . , 23 vols., Toronto, 1914-17.

119. Parkman, F. *The Oregon Trail*, Boston, 1925.

120. PAXSON, F. L. *History of the American Frontier, 1763-1893*, Boston, 1924.
121. THWAITES, R. G., ed. *Early Western Travels, 1748-1846* . . . , 32 vols., Cleveland, 1904-07; and *idem. The Jesuit Relations* . . . , 73 vols., Cleveland, 1896-1901.
122. TURNER, F. J. *The Frontier in American History* (1st ed., 1920), New York, 1948.
123. HULBERT, A. B. *Historic Highways of America*, 16 vols., Cleveland, 1902-1905.
124. KELLOGG, L. P. *The French Regime in Wisconsin and the Northwest*, Madison, 1925.
125. WINSOR, J. *Cartier to Frontenac: Geographical Discovery in the Interior of North America in its Historical Relations, 1534-1700* . . . , Boston, 1894.
126. BRIGHAM, A. P. *Geographic Influences in American History*, Boston, 1903; *idem. From Trail to Railway Through the Appalachians*, Boston, 1907; *idem.* "The Development of Wheat Culture in North America," *Geographical Journal*, 35 (1910): 42-56; *idem.* "Problems of Geographic Influence," *Annals of the Association of American Geographers*, 5 (1915): 3-25.
127. SEMPLE, E. C. *American History and its Geographic Conditions*, (1st ed.) 1903, revised in collaboration with C. F. JONES, Boston, 1933; *idem. Influences of Geographical Environment* . . . , New York, 1911.
128. ———. *The Geography of the Mediterranean Region: its Relation to Ancient History*, New York, 1931.
129. BARROWS, H. H. *Geography of the Middle Illinois Valley*, Illinois State Geological Survey Bulletin No. 15, Urbana, 1910.
130. SAUER, C. O. *Geography of the Upper Illinois Valley and History of Development*, Illinois State Geological Survey Bulletin No. 27, Urbana, 1916; *idem. The Geography of the Ozark Highland of Missouri*, Chicago, 1920.
131. PARKINS, A. E. *The Historical Geography of Detroit*, Michigan Historical Commission Publication, University Series No. 3, Lansing, 1918.
132. SPINDEN, H. J. "The Origin and Distributions of Agriculture in America," *Proceedings of the 19th International Congress of Americanists of 1915*, Washington, 1917: 269-276.
133. KROEBER, A. L. "California Place Names of Indian Origin," *University of California Publications in American Ethnography and Archeology*, 12 (1916): 31-69.
134. GLEASON, H. A. "The Vegetational History of the Middle West," *Annals of the Association of American Geographers*, 12 (1922): 39-85.
135. ROSTOVTZEV, M. (also ROSTOVTZEFF). *A Large Estate in Egypt in the Third Century, B. C.; A Study in Economic History*, University of Wisconsin Studies in the Social Sciences and History, No. 6, 1922.
136. DEVOTO, B. *Across the Wide Missouri*, Boston, 1947; and *idem. The Course of Empire*, Boston, 1950.
137. INNIS, H. A., ed. *Select Documents in Canadian Economic History, 1497-1783*, Toronto, 1929; and *idem. The Fur Trade in Canada*, New Haven, 1930.
138. MALIN, J. C. "Grassland, 'Treeless', and 'Subhumid' . . . ," *Geographical Review*, 37 (1947): 241-250; and *idem. The Grassland of North America: Prolegomena to its History*, Lawrence (Kansas), 1947.
139. WEBB, W. P. *The Great Plains*, New York, 1927, Boston, 1931.
140. SCHAFER, J. *Wisconsin Domesday Book*, 4 vols., Madison, 1922, 1927, 1932, 1937.
141. BULEY, R. C. *The Old Northwest*, 2 vols., Bloomington, 1951.
142. DODGE, S. D. "Ralph Hall Brown, 1898-1948," *Annals of the Association of American Geographers*, 38 (1948): 305-309, gives a nearly complete bibliography for Brown.
143. BROWN, R. H. "Materials Bearing upon the Geography of the Atlantic Seaboard, 1790-1810," *Annals of the Association of American Geographers*, 28 (1938): 201-231.
144. ———. *Mirror for Americans: Likeness of the Eastern Seaboard, 1810*, New York, 1943; and *idem. Historical Geography of the United States*, New York, 1948.

145. CLARK, A. H. "Ralph Hall Brown's Beitrag zur amerikanischen Historischen Geographie," *Die Erde,* 5 (1953): 148-152.
146. SAUER, C. O. *Geography of the Pennyroyal,* Kentucky Geological Survey, Series 6, Vol. 25, Frankfort (Kentucky), 1927; "The Road to Cibola," *Ibero-Americana,* 3 (1932): 1-58; "The Distribution of Aboriginal Tribes and Languages in Northwestern Mexico," *ibid.,* 5 (1934): 1-94; "Aboriginal Population of Northwestern Mexico," *ibid.,* 10 (1935): 1-33; "American Agriculaural Origins: A Consideration of Nature and Culture," in R. B. LOWIE, ed., *Essays in Anthropology in Honor of A. L.* Kroeber, Berkeley, 1936: 279-297; "Destructive Exploitation in Modern Colonial Expansion," *Proceedings of the International Geographical Congress,* Amsterdam, 2, 3C, 1938; "Theme of Plant and Animal Destruction in Economic History," *Journal of Farm Economics,* 20 (1938): 765-775; "The Settlement of the Humid East," in *Climate and Man,* United States Department of Agriculture Yearbook, 1941: 157-166; "The Personality of Mexico," *Geographical Review,* 31 (1941): 353-364; "A Geographic Sketch of Early Man in America," *ibid.,* 34 (1944): 529-573; "The Relation of Man to Nature in the Southwest. . . ." *The Huntington Library Quarterly,* 8 (1945): 115-151; "Early Relations of Man to Plants," *Geographical Review,* 37 (1947): 1-25; "Environment and Culture During the Last Deglaciation," *Proceedings of the American Philosophical Society,* 92 (1948): 65-77; "Colima of New Spain in the Sixteenth Century," *Ibero-Americana,* 29 (1948): 1-104; "Grassland Climax, Fire and Man," *Journal of Range Management,* 3 (1950): 16-21. (See also references 130, 147, 176, 179, and 200.)
147. ———. *Agricultural Origins and Dispersals,* Bowman Memorial Lectures, Series 2, American Geographical Society, New York, 1952.
148. WHITTLESEY, D. "Sequent Occupance," *Annals of the Association of American Geographers,* 19 (1929): 162-165.
149. ———. "Coastland and Interior Mountain Valley: A Geographical Study of two typical localities in Northern New England," in JOHN K. WRIGHT, ed. *New England's Prospect,* New York, 1933: 446-458.
150. GOLDTHWAIT, J. W. "A Town That Has Gone Downhill," *Geographical Review,* 17 (1927): 527-552.
151. JAMES, P. E. "The Blackstone Valley. . . . ," *Annals of the Association of American Geographers,* 19 (1929): 67-109.
152. HALL, R. B. "Tokaido: Road and Region," *Geographical Review,* 27 (1937): 353-377.
153. TREWARTHA, G. T. "The Prairie du Chien Terrace: Geography of a Confluence Site," *Annals of the Association of American Geographers,* 22 (1932): 119-158; "French Settlement in the Driftless Hill Land," *ibid.,* 28 (1938): 179-200; "A Second Epoch of Destructive Occupance in the Driftless Hill Land, 1760-1832," *ibid.,* 30 (1940): 109-142; "Population and Settlements in the Upper Mississippi Hill Land During the Period of Destructive Exploitation (1760-1832)," in *Proceedings of the Eighth Scientific Congress,* Washington, May 8-10, 1940 (Washington, 1943): 183-196.
154. ———. "The Unincorporated Hamlet: One Element of the American Settlement Fabric," *Annals of the Association of American Geographers,* 33 (1943): 32-81; "Types of Rural Settlement in Colonial America," *Geographical Review,* 36 (1946): 568-596.
155. PARKINS, A. E. *The South: Its Economic-Geographic Development,* New York, 1938.
156. WHITAKER, J. R. "Negaunee, Michigan: An Urban Center Dominated by Iron Mining," *Bulletin of the Geographical Society of Philadelphia,* 22, 1931: 137-174, 215-240, and 306-339; "The Development of the Tobacco Industry in Kentucky: A Geographical Interpretation," *ibid.,* 27 (1929): 15-22; "Regional Contrasts in the Growth of Canadian Cities," *Scottish Geographical Magazine,* 53 (1937): 373-379; "Our Forests, Past and Present," in A. E. PARKINS

and J. R. WHITAKER, *Our Natural Resources and Their Conservation*, New York, 1939: 229-250; "Tennessee—Earth Factors in Settlement and Land Use," *Tennessee Historical Quarterly*, 5 (1946): 195-211; and "Historical Geography in School and College," *Peabody Journal of Education*, 27 (1949): 3-15.

157. DODGE, S. D. "Sequent Occupance of an Illinois Prairie," *Bulletin of the Geographical Society of Philadelphia*, 29 (1931): 205-209; "A Study of Population in Vermont and New Hampshire," *Papers of the Michigan Academy of Science, Arts and Letters*, 18 (1932): 131-136; "The Frontier of New England in the Seventeenth and Eighteenth Centuries and its Significance in American History," *ibid.*, 28 (1942): 435-439; and "Periods in the Population History of the United States," *ibid.*, 32 (1946, published 1948): 253-260.

158. WHITTLESEY, D. *Environmental Foundations of European History*, New York, 1949; *The Earth and the State* (2nd ed.), New York, 1944.

159. DAVIS, C. M. "The High Plains of Michigan," *Papers of the Michigan Academy of Science, Arts and Letters*, 21 (1935): 303-341.

160. DICK, W. B. "A Study of the Original Vegetation of Wayne County, Michigan," *ibid.*, 22 (1936): 329-334.

161. FAIGLE, E. H. "Some Aspects of the Urban Geography of Syracuse," *ibid.*, 23 (1937): 349-359.

162. GLENDINNING, R. M. "The Lake St. Jean Lowland, Province of Quebec," *ibid.*, 20 (1934): 313-341.

163. GUTHE, O. E. "The Black Hills of South Dakota and Wyoming," *ibid.*, 20 (1934): 343-376.

164. KENDALL, H. M. "The Central Pyrenean Piedmont of France," *ibid.*, 20 (1934): 377-414; "The Occupance of the Lower Vézère Valley," *ibid.*, 16 (1931): 299-312.

165. MEYER, A. H. "The Kankakee 'Marsh' of Northern Indiana and Illinois," *ibid.*, 21 (1935): 359-396.

166. WILSON, L. S. "The Functional Areas of Detroit, 1890-1933," *ibid.*, 22 (1936): 397-409.

167. TORBERT, E. N. "The Evolution of Land Utilization in Lebanon, New Hampshire," *Geographical Review*, 25 (1935): 209-230.

168. JAMES, P. E. "Geographic Factors in the Development of Transportation in South America," (1923), appearing under same title in *Economic Geography*, 1 (1925): 247-261.

169. PRUNTY, M. C. JR. "Evolution of the Agricultural Geography of Dyer County, Tennessee" abridged in "Physical Bases of Agriculture in Dyer County, Tennessee," *Journal of the Tennessee Academy of Science*, 23 (1948): 215-235.

170. SMITH, G.-H. "The Settlement and Distribution of Population in Wisconsin: A Geographical Interpretation" (1927), abridged in *Geographical Review*, 18 (1928): 402-421.

171. WEAVER, J. C. "Barley in the United States: A Historical Sketch," *Geographical Review*, 33 (1943): 56-73.

172. KLIMM, L. B. *The Relation Between Certain Population Changes and the Physical Environment in Hampden, Hampshire, and Franklin Counties, Massachusetts, 1790-1925*, Philadelphia, 1933.

173. KOLLMORGEN, W. M. *The German-Swiss in Franklin County, Tennessee: A Study of the Significance of Cultural Considerations in Farming Enterprises*, U. S. Department of Agriculture, Bureau of Agricultural Economics, Washington, 1940.

174. JONES, S. B. "Human Occupance of the Bow-Kicking Horse Region, Canadian Rocky Mountains," dissertation, 1934; "The Forty-Ninth Parallel in the Great Plains: The Historical Geography of a Boundary," *Journal of Geography*, 31 (1932): 357-368.

175. Watson, J. W. "The Geography of the Niagara Peninsula," dissertation, 1945; "The Changing Industrial Pattern of the Niagara Peninsula: A Study in Historical Geography," *Ontario Historical Society Papers and Records*, 37 (1945): 49-58; "Mapping a Hundred Years of Change in the Niagara Peninsula," *Canadian Geographical Journal*, 32 (1946): 266-283; "The Influence of the Frontier on Niagara Settlements," *Geographical Review*, 38 (1948): 113-119.

176. Sauer, C. O., and Brand, D. "Prehistoric Settlements of Sonora with Special Reference to Cenos de Trincheras," *University of California Publications in Geography* 5 (1931): 67-148; *idem.* "Aztatlan: Prehistoric American Frontier on the Pacific Coast," *Ibero-Americana*, 1 (1932): 1-92.

177. Bruman, H. J. "Some Observations on the Early History of the Coconut in the New World," *Acta Americana*, 2 (1944): 220-243.

178. McBryde, F. W. *Cultural and Historical Geography of Southwest Guatemala*, Smithsonian Institution, Institute of Social Anthropology, Publication No. 4, Washington, 1925.

179. Meigs, P. "The Dominican Mission Frontier of Lower California," *University of California Publications in Geography*, 7 (1935): 1-192. Meigs also published, jointly with C. O. Sauer, "Lower California Studies I: Site and Culture at San Fernando de Velicata," *University of California Publications in Geography*, 2 (1927): 271-302.

180. Parsons, J. J. "Antioqueño Colonization in Western Colombia; an Historical Geography," *Ibero-Americana*, 32 (1949): 1-225.

181. Stanislawski, D. "The Origin and Spread of the Grid-Pattern Town," *Geographical Review*, 36 (1946): 105-120; "The Political Rivalry of Patzcuaro and Morelia, an Item in the Sixteenth Century Geography of Mexico," *Annals of the Association of American Geographers*, 37 (1947): 135-144; "Early Spanish Town Planning in the New World," *Geographical Review*, 37 (1947): 94-105; *The Anatomy of Eleven Towns in Michoacan*, University of Texas Institute of Latin American Studies, Latin American Studies No. 10, Austin, 1950.

182. West, R. C. "The Economic Structure of the Mining Community in Northern New Spain; The Parral Mining District," *Ibero-Americana*, 30 (1949): 1-169; *Colonial Placer Mining in Colombia*, Baton Rouge, 1952.

183. Carter, G. F. *Plant Geography and Culture History in the American Southwest*, Viking Fund Publications in Anthropology, No. 5, New York 1945; "Origins of American Indian Agriculture," *American Anthropologist*, 48 (1946): 1-21; "Evidence for Pleistocene Man in Southern California," *Geographical Review*, 40 (1950): 84-102.

184. Clark, A. H. "Historical Explanation of Land Use in New Zealand," *Journal of Economic History*, 5 (1945): 215-230; "Field Research in Historical Geography," *The Professional Geographer* (old series) 4 (1946): 13-22; *The Invasion of New Zealand by People, Plants and Animals: The South Island*, New Brunswick, 1949.

185. Hewes, L. "The Geography of the Cherokee Country of Oklahoma" (dissertation, 1940) of which various parts have been published with additions and revisions in: *Economic Geography*, 18 (1942): 401-412; *ibid.*, 19 (1943): 136-142; *Geographical Review*, 32 (1942): 269-281; and *Chronicles of Oklahoma*, 22 (1944): 324-337; *idem.* "Some Features of Early Woodland and Prairie Settlement in a Central Iowa County," *Annals of the Association of American Geographers*, 40 (1950): 40-57; "The Northern Wet Prairie of the United States: Nature, Sources of Information, and Extent," *ibid.*, 41 (1951): 307-323.

186. Kniffen, F. B. "Lower California Studies III: The Primitive Cultural Landscape of the Colorado Delta," *University of California Publications in Geography*, 5 (1931): 43-66; "The American Agricultural Fair: Time and Place," *Annals of the Association of American Geographers*, 41 (1951): 42-57; "The American Covered Bridge," *Geographical Review*, 41 (1951): 114-123.

REFERENCES   105

187. PRICE, E. T. "The Melungeons: A Mixed-Blood Strain of the Southern Appala-
     chians," *Geographical Review*, 41 (1951): 256-271.
188. RAUP, H. F. "The German Colonization of Anaheim, California," *University of
     California Publications in Geography*, 6 (1932): 123-146; "The Susquehanna
     Corridor: A Neglected Trans-Appalachian Route," *Geographical Review*, 30
     (1940): 439-450; "Place Names of the California Gold Rush," *ibid.*, 35 (1945):
     653-658.
189. ROSTLUND, E. "Freshwater Fish and Fishing in Native North America," *Univer-
     sity of California Publications in Geography*, 9 (1951): 1-314.
190. SPENCER, J. E. "Changing Chungking: The Rebuilding of an Old Chinese City,"
     Geographical Review, 29 (1939): 46-60; "The Development of Agricultural
     Villages in Southern Utah," *Agricultural History*, 14 (1940): 181-189.
191. ZELINSKY, W. "The Population Geography of the Free Negro in Ante-Bellum
     America," *Population Studies*, 3 (1950): 386-401; "The Log House in Georg-
     ia," *Geographical Review*, 43 (1953): 173-193; "The Historical Geography
     of the Negro Population in Latin America," *Journal of Negro History*, 34
     (1949): 153-221.
192. RUSSELL, R. J. "Climatic Change Through the Ages," in *Climate and Man*,
     United States Department of Agriculture Yearbook, 1941: 67-97.
193. BROEK, J. O. M. *The Santa Clara Valley, California: a Study in Landscape
     Change*, Utrecht, 1932; "The Relations Between History and Geography,"
     Pacific Historical Review, 10 (1943): 321-325.
194. LEIGHLY, J. B. "The Towns of Mälardalen in Sweden: A Study in Urban Morph-
     ology," *University of California Publications in Geography*, 3 (1928): 1-134,
     from dissertation, 1927; "The Towns of Medieval Livonia," *ibid.*, 6 (1939):
     235-314; "Settlement and Cultivation in the Summer-Dry Climates," in *Cli-
     mate and Man*, United States Department of Agriculture Yearbook, 1941:
     197-204.
195. THORNTHWAITE, C. W. "Climate and Settlement in the Great Plains," in *Climate
     and Man*, United States Department of Agriculture Yearbook, 1941: 177-187.
196. HARTSHORNE, R. "The Franco-German Boundary of 1871," *World Politics*, 2
     (1950): 209-250.
197. STONE, K. H. "Some Geographic Bases for Planning New Alaskan Settlement,"
     Alaskan Science Volume, issued by The Arctic Institute of North America,
     1952; "Populating Alaska: The United States Phase," *Geographical Review*,
     42 (1952): 384-404.
198. JOHNSON, H. "The Distribution of the German Pioneer Population in Minne-
     sota," *Rural Sociology*, 6 (1941): 16-34; "The Location of German Immi-
     grants in the Middle West," *Annals of the Association of American Geo-
     graphers*, 41 (1951): 1-41.
199. DURAND, L. JR. "The Migration of Cheese Manufacture in the United States,"
     Annals of the Association of American Geographers, 42 (1952): 263-282.
200. SAUER, C. O. "Foreword to Historical Geography," *Annals of the Association of
     American Geographers*, 31 (1941): 1-24.
201. SHERLOCK, R. L. *Man as a Geological Agent*, London, 1922.
202. WRIGHT, J. K. "A Plea for the History of Geography," *Isis*, 8 (1926): 477-491;
     "The History of Geography: A Point of View," *Annals of the Association of
     American Geographers*, 15 (1925): 192-201; "Where History and Geography
     Meet: Recent American Studies in the History of Exploration," *Proceedings
     of the Eighth American Scientific Congress . . . 1940*, vol. 9, *History and
     Geography*, Washington (1943): 17-23; "Terrae Incognitae: The Place of
     Imagination in Geography," *Annals of the Association of American Geo-
     graphers*, 37 (1947): 1-15.
203. ———. *Geography in the Making: The American Geographical Society, 1851-
     1951*, New York 1952.

THE GEOGRAPHIC STUDY OF POPULATION [A]

THE RANGE OF POPULATION GEOGRAPHY

AMERICAN CONTRIBUTIONS TO POPULATION GEOGRAPHY

THEORY AND METHODS

THE MAPPING OF POPULATION

THE RECONSTRUCTION OF PRE-CENSUS POPULATION PATTERNS

THE ANALYSIS OF POPULATION TRENDS

MIGRATION

CULTURAL CHARACTERISTICS

FRONTIERS FOR RESEARCH

PROBLEMS OF MAPPING POPULATION

OTHER FRONTIERS AND TRAINING REQUIREMENTS

[A]. Original draft prepared by Preston E. James.

CHAPTER FOUR

THE GEOGRAPHIC STUDY OF POPULATION

THE GEOGRAPHIC STUDY of population is an underdeveloped topical field. The irregularity of the distribution of mankind over the earth and the differences from place to place in the racial and societal character of the population are facts which underlie all studies in social science, including those of human geography. Yet not only have few American geographers pursued population studies as a topical specialty, but also in economic and political geography, where population is certainly involved, this aspect of the geographic picture has seldom been given adequate treatment. Glenn Trewartha in his presidential address to the Association of American Geographers in 1953 presented the case for population geography and urged greater attention to it [1].

The geography of population is concerned with area differentiation of a peculiar kind. People scattered over the earth do not form a continuity, nor can it be said that they form discrete units except when viewed as individuals. Actually population studies make use of differences of enumeration within specific areas, with what are described in chapters one and two of this book as discontinuous distributions. The fundamental problem of population geography is the search for a systematic method of outlining enumeration areas that are meaningful in terms of the question being investigated.

Geographers always deal with phenomena which are so distributed over the curved surface of the earth that the patterns of distribution cannot be seen in one glimpse. Geographers always seek to reduce these large areas to maps of a size permitting analysis. Thus they are concerned with the methods of generalizing from specific detail.

In the field of population geography the need is to find a way to generalize the distribution of people without obscuring the relationships of man to the other phenomena with which he is areally associated. Most census data are gathered and summed up within enumeration districts arbitrarily outlined, often districts which are politically defined. Almost always these districts are irrelevant to the problems the geographer is investigating. Population density derived as a ratio between total numbers in an enumeration district and the total area of the district is a crude figure which at best reveals only dimly the main outlines of population patterns, and which in many cases gives wholly unreal patterns.

Geographers are not the only ones concerned with the search for better methods of generalizing the patterns of population. It is of fundamental importance also to demographers, sociologists, and social anthropologists. Workers in these other disciplines, whose training is not geographical, need the assistance of professional geographers in facing this challenge.

THE RANGE OF POPULATION GEOGRAPHY

To recognize population geography as a distinct topical specialty is not to think of it as separable from the whole field of geography. Like other topical specialties this one offers a core of related problems, a central theme around which to organize geographic investigations. It is justified as a topical field because it deals with a certain group of phenomena and the systematically related processes, and because the study of these matters involves training in a certain methodology, in this case the statistical method. The objective of population geography is to define and to bring forth the significance of differences from place to place in the number and kind of human inhabitants. Starting with number and kind, however, areal relationships are sought with all the other significant phenomena of the human habitat, and meanings can be found in a wide variety of causal connections with the total physical, biotic, and cultural environment.

The professional geographer is not the only scholar who works in this topical field. In fact, most of the studies in population geography have been undertaken by people who do not think of themselves professionally as geographers. It is the special responsibility of the population geographer to propose methods, set standards, and provide for the critical appraisal of geographic work whether done by geographers or by people in other disciplines. Also he may well point out to the demographers ways in which the application of the regional concept and the regional method may bring new illumination to the study of population problems.

Persons who deal with enumerations of people are faced with certain common difficulties. Chief among these is the illusion of numbers, the false idea that because one is supplied with numerical data he is dealing with something precise. Unfortunately population data gathered by reliable methods are available for few parts of the world. The methods of gathering data are so varied that the figures for different countries are not directly comparable. For many areas the currently accepted figures of total population are only guesses; informed estimates of the population of China, for example, range from 300 to 600 millions [2]. Figures for birth rates and death rates based on actual enumeration are available for only 30 percent of the world's population. The indispensible information concerning age and sex structure exists in reliable form for only 43 percent of the total world population, exists in vague form for 23 percent, and is entirely lacking for the remaining 34 percent [3:17-18]. The demographers have refined their statistical method far in advance of the data on which they must depend.

The study of a population goes beyond the count of total numbers. Demographic studies require other kinds of information, pertaining, for example, to: 1) age and sex structure, portrayed by the familiar pyramidal diagram; 2) ethnic composition; 3) economic occupations; 4) urban and rural groups; 5) religious groups; 6) political and social groups. Countries can be classified according to the proportions of the total population in each category, as when the age-sex diagram is used to distinguish an old population from a young one, or one with a lack of balance between the sexes; or when agricultural countries are distinguished from industrial ones on the basis of the proportions of the population employed in farming, manufacturing, or other activities.

Population is never static. Changes take place through periods of time and these are perhaps poorly revealed by periodic censuses, at time intervals which are often out of phase with the periods of history. During the rise of the Industrial Society in the last two centuries one country after another has entered and some have passed through the so-called demographic cycle. A Pre-Industrial Society is characterized by high birth rates and high death rates. With the beginnings of industrialization and the increase of urban living, the death rate drops while the birth rate remains high. The result is a sharp up-turn in the net rate of increase. This is followed, it seems, by a gradual decline in the birth rate, and the achievement of a relatively stable population at a higher density, with longer life span and higher average age. Demographers are concerned to observe the operation of this cycle, this process, and to note the departures from the normal or theoretical curve in specific countries or other political units.

Population also changes because of migration. Notable migrations of mankind have taken place in the past, most of them before the days of census-taking. Modern migrations of Chinese into Manchuria or Germans into West Germany are also poorly recorded. The modern migration from rural areas to the cities, which is appearing in all parts of the world, can, in many countries, be studied by actual count.

All these enumerations can be examined in different ways. The demographer seeks to formulate the theoretical curve of population change, and to refine it on the basis of more and more precise statistical procedures; he examines the departures from the theory in particular places; and he seeks to predict the population of decades to come. A part of what he does is in the field of population geography. When countries or areas are classified on the basis of the characteristics of the population and when these are examined for areal relations with other phenomena, this also is a part of population geography, no matter who does it.

Fundamental to all these analyses, however, is the basic map of population, on which the total number of people counted in a specific area is shown, or the degree of density computed. To this aspect of population study geographers can make a great contribution. There is need for a refinement of cartographic methods comparable to the refinement already achieved by the demographers in their statistical procedures. The fact that this must be done in the face of inaccurate and generally inadequate data should be no more of a deterrent than was this same fact for the scholars who developed the statistical method. Censuses are costly, and students of population are fortunate to have even such data as are available.

AMERICAN CONTRIBUTIONS TO
POPULATION GEOGRAPHY

Although the geographic study of population may properly be considered as underdeveloped, several Americans have made contributions in this field. In many cases population studies are incidental to the investigation of economic questions and these papers are more properly discussed in subsequent chapters on the various aspects of economic geography. Among such papers are those of O. E. Baker on population and food supply [4], and H. H. McCarty on the economic factors involved in the interpretation of population problems [5]. A few papers stand out as major contributions all the more conspicuously because of the small number.

In this inventory of American population geography, studies will be discussed under the following six headings: 1) theory and method; 2) the mapping of population; 3) the reconstruction of pre-census population patterns; 4) the analysis of population trends; 5) migrations; and 6) cultural characteristics.

THEORY AND METHOD

The more important ideas and methods used by American geographers in studies of population geography were imported from Europe. There is a vast European literature in this field, as indicated by the bibliographical work of Dörries who lists nearly 5,000 titles between 1908 and 1938 [6]. For the past three decades the methods used in the mapping of population in America have been strongly influenced by Sten de Geer, whose discussion of the procedures and categories he used in mapping the population of Sweden (1919) appeared in the *Geographical Review* in 1922 [7]. Sten de Geer's map, on a scale of 1/500,000 carries an amazing amount of detail regarding population and settlement patterns. He made use of dots for rural people, and globes of proportional volumes for urban concentrations.

At about this same time M. Aurousseau published several papers on population geography, emphasizing the need for more attention to this field and suggesting some of the basic concepts and methods [8;9]. In these papers he outlined the concept of the expansion ratio, a measure of the potential increase of population in an area; and he examined the method of defining the ultimate limits of population density beyond which further increase could not take place without either a lowering of the standard of living or a change in the technology. He was concerned to find a workable formula for distinguishing between urban and rural people and offered a definition of the former based on economic occupation. He suggested a classification of urban functions. Aurousseau, who derived most of his ideas from European experience, exerted a great influence on American thinking in this field.

Another geographer in the United States whose influence was great in both the theoretical and applied aspects of population studies was Mark Jefferson. This productive American geographer published a large number of papers on population subjects. He was among the first to express dissatisfaction with enumerations within arbitrary census areas; early in the present century Jefferson was distinguishing between the geographical city and the political city, and pointing out the fact that statistics by political cities were not at all comparable [10]. He also studied the statistics of urban populations the world over [11;12], and from these studies he formulated what he called the Law of the Primate City, the generalization that in most countries one city grows to a size almost double that of the next in order [13].

Other studies of method have been devoted to the dot map and the isopleth map as devices for showing relative densities. Again, as in the case of cities, to space the dots on a map uniformly over an enumeration district produces only a crude approximation of the real pattern of population, and one which can be totally misleading. And there is the apparently insoluble problem of showing urban and rural densities on the same scales. A clear-cut demonstration of the very different map patterns that are brought out when the total number of people is mapped in relation to different areas of enumeration was given by J. K. Wright in a paper on Cape Cod [14]. The method he presented in that paper offered a mathematical basis for the arrangement of areal density patterns within census districts; in a subsequent paper he dealt with various measures of distribution [15]. J. W. Alexander and G. A. Zahorchak and others have examined the relative advantages of different techniques for using dots and isopleths [16; 17; 18].

In this category of theoretical studies should be included the challenging paper of J. Q. Stewart [19]. This author, a physicist by training, made use of a mathematical formula to define the grouping of people around urban centers and to predict the movement of people into these centers. As a device for uncovering some kind of theoretical order in the processes leading to the distribution of people this paper may be of great importance, although the geographer is more especially concerned with the modifications of the process in particular places.

THE MAPPING OF POPULATION

In addition to the theoretical and methodological discussion of the problems of population geography, several have attempted to map population. Usually the maps are accompanied by little or no discussion of method. Yet the basic problem, whether recognized explicitly or not, is always the selection of meaningful enumeration areas.

One of the earliest contributions was made by Henry Gannett in 1881. Gannett, who was then the Geographer of the Census, drew up a plan for the subdivision of the national territory for purposes of enumeration, grouping together States which he conceived to possess some degree of homogeneity [20]. Gannett did not try to depart from State areas. Even in 1916 when A. P. Brigham mapped the changes of population density in different parts of New York State he made use of whole counties for his enumeration areas [21]. On the other hand, S. W. Cushing in 1921 departed from the political units for which statistics were available in the construction of an isopleth map of population density for Mexico [22].

Dot maps have been made to show population patterns in several of the States. One of the earliest of these was by C. E. Batschelet, the present Geographer of the Census, in a paper on the State of Pennsylvania [23]. A. B. Wolfe showed in greater detail than the other maps of this period some especially sharp contrasts in population density, as along the front of the Balcones Escarpment of Texas [24]. If the dots were arranged uniformly over the areas of the statistical divisions, such sharp population gradients would not be visible at all. Guy-Harold Smith's map of the population of Ohio made use of globes for the urban concentrations, and in rural areas used a scale of one dot for 25 people [25]. C. J. Bollinger's map of a part of Central Oklahoma [26] shows the density of rural population by isopleths and the size of cities by globes. F. A. Stilgenbauer's map of Michigan [27] uses the dot method, and shows the cities by circles rather than globes.

In addition to these maps of States, there were several attempts to portray population patterns of larger areas on very small-scale maps. Among these may be mentioned Stilgenbauer's maps of North America [28],

P. E. James' map of South America [29], and J. W. Coulter's map of Japan [30].

The Reconstruction of Pre-Census Population Patterns

Geographers have been interested not only in the static picture of population in the current period, but also in the reconstruction of population patterns for periods before there were any censuses. The anthropologist, A. L. Kroeber, made a study of the Indian distribution in North America before the arrival of white men on the continent [31], and W. B. Hinsdale attempted a more detailed picture of aboriginal Michigan [32].

In the United States there are scattered and more or less comparable data for population before the first census. These data have been used for the construction of a series of dot maps which present a picture at least of the areas of relative concentration at different dates; this series, prepared by H. R. Friis, runs from 1625 to 1790 [33]. Other population studies of this pre-census era include those of E. B. Green and V. D. Harrington [34], A. B. Wolfe [35], and S. H. Sutherland [36].

THE ANALYSIS OF POPULATION TRENDS

A publication which bridges the zone of uncertainty between the pre-census population studies, which are frankly based on estimates, and the studies of more recent decades in which census data are more or less comparable was issued by the Bureau of the Census in 1909. In this report, which covers the first twelve censuses from 1790 to 1900, special attention was given to the data for 1790, and to irregular enumerations before the first census [37]. Estimates of population running back to 1610 are included.

Several studies of population trends by American geographers before 1930 dealt more or less directly with the immigration problem. As early as 1894, J. D. Whitney examined the density of population and the rates of population increase as revealed by the censuses of 1870, 1880, and 1890 [38]. Whitney also showed that the great majority of the immigrants of this period had come from Germany, Ireland, England, and Sweden. After the census of 1910 it became apparent that the rate of increase was slowing down, and also that the immigrants were coming chiefly from southern and eastern Europe. There was a considerable effort to point out the dangers of continued unrestricted immigration and of approaching over-population. Yet Jefferson showed in 1925 that Malthus had underestimated the number of years required to double a population [39].

More recently (1946) S. D. Dodge has analysed the population trends of the United States [40]. He showed that the Malthusian rate had obtained from 1660 to 1860, and that the subsequent rates of increase tended more and more closely to approximate those that the first English census showed to be actual for England at the time Malthus wrote. Furthermore, in the first decades of the American censuses the rate for the States of the Atlantic seaboard was already much lower than that of the nation as a whole, the latter being about 33 percent or approximately the Malthusian rate. Dodge included a logarithmic graph of the whole population of the United States, both before and after 1790, and showed that both 1660 and 1860 were points of marked lowering of the rate of increase. By locating the frontiers of settlement at those dates, he demonstrated a connection between the change of the rate of increase and the position of the frontier at places where, for one reason or another, the westward movement was temporarily halted.

Dodge has gone farther into the analysis of curves of growth and decline than has any other American geographer. He approaches the geographical study of differences in growth rates by accepting the underlying concept that all communities in the long run reach a peak of population increase and thereafter decline. On the theoretical curve there are two significant points: first the date of first settlement, and second the date of the peak or the beginning of decline. If there are two or more peaks, the dates of the low points between are also significant. The significant dates for specific communities, such as counties, can be recorded on a map, and lines connecting the same dates can be drawn. The result is an isochronic map [41; 42; 43; 44]. Using somewhat similar techniques, James studied the changing patterns of population in São Paulo State, Brazil [45], N. Mirkowich studied population trends in California [46], and C. F. Kohn examined the trends in the United States as a whole since 1940 [47].

Another method of analysing the trends in population and the movement of the frontier of new settlement involves the preparation of a series of dot maps, each for a different census. This requires a careful study of enumeration districts for each census in order to make certain that the series is in fact comparable. Examples of such studies include the paper by Guy-Harold Smith on Wisconsin [48], and the paper by L. S. Wilson on Minnesota [49].

MIGRATION

Although the internal migration of people in the United States has been going on since the period of first settlement, it has received little attention from geographers. There is not only the movement of people generally toward the west, but also from rural areas into the cities, and from city centers to the suburban fringe. Such internal rearrangements have become increasingly important during the 19th century owing not only to the development of large manufacturing industries, but also to the increasing number of job opportunities in the service businesses and professions. The changes in transportation from horse to railroad, to interurban, and to the automobile on paved highways has further accented the internal migrations. Furthermore these movements can be studied quantitatively because since 1850 the census has reported data on the State and even the county of birth of all the people. The results of these movements have been studied by the rural-settlement geographers and by the urban geographers, but little by population geographers. Most studies of internal migration have been done by sociologists who treat the subject statistically and usually with relatively coarse cartographic techniques.

Two geographical studies in this field, however, should be mentioned. In 1929, Guy-Harold Smith studied and presented on maps the movement of people into and out of Ohio [50]. In 1934, C. W. Thornthwaite's extensive study of internal migration throughout the United States was published [51]. His report contains colored maps which show in a detail not available elsewhere the direction of movement of the American people.

CULTURAL CHARACTERISTICS

The study of the cultural characteristics of a population occupies a border zone between geography and anthropology [52:65-67]. The word culture is used here in the anthropological sense to indicate a way of living, an association of material and non-material traits characterizing a population. Individual traits or groups of traits can be identified and examined geographically, that is for the purpose of determining how such cultural characteristics have combined to produce similarities and differences on the earth, and what kinds of accordant relationships can be found between these characteristics and other phenomena. Such a border zone field should be a stimulating one, receiving ideas from both the overlapping disciplines; yet the fact is that the geographers have contributed so little that, generally speaking, "the cultural anthropologist has to be his own geographer. . . . " [53:227].

There was a period during which American geographers did a considerable amount of work in the study of language, race, and nationality. In 1917 and 1918 the so-called "Inquiry" under the direction of Isaiah Bowman at the American Geographical Society compiled and analyzed a vast number of maps of Europe. The purpose was to provide a basis of geographic understanding on which to carry out the negotiations of the peace conference after World War I. Many of these maps showed racial distributions and proportions, based on language. Numerous other ethnic studies came either directly or indirectly from the example of the Inquiry [54;55].

Although the volume of works of this sort has diminished greatly since 1930, there are two outstanding examples of the study of population in terms of its cultural characteristics. A paper by Richard Hartshorne in

1938 was accompanied by a series of maps showing racial minorities in the United States, based on a careful examination of the census data [56]. And a paper by J. O. M. Broek, published in 1944, analyzes the possibilities for unity or disunity among the emerging nationalities of Southeast Asia, based on a series of four maps showing density of population, average annual increase of population, languages, and religions [57].

Meanwhile there have been some notable examples of cultural studies done by scholars in other fields. The student of law, J. H. Wigmore, published a map of the world's legal systems [58]. And the student of linguistics, Hans Kurath, brought together in one monograph the results of his studies of word usage and pronunciation throughout eastern United States [59]. On 163 maps he presents the arrangement and the accordant relationships of popular language, a document of fundamental importance to studies in historical geography.

FRONTIERS FOR RESEARCH

Perhaps no other field of geography offers such challenging opportunities for basic research studies as are to be met in population geography. All the different aspects of the study of population need attention. There are at least four frontiers of research: 1) the development of more satisfactory methods of mapping the distribution of population; 2) the reconstruction of past population patterns; 3) the search for methods of revealing more clearly the dynamics of population change; and 4) the correlation of studies of the cultural characteristics of population with other geographic phenomena.

PROBLEMS OF MAPPING POPULATION

Population mapping involves two distinct kinds of problems. There are the cartographic problems which have to do with map symbols, visual quality, and other matters of this sort which are discussed in the concluding chapter of this book. And there are the problems of procedure involved in reducing the phenomena of the face of the earth to smaller-scale representation on flat pieces of paper. The geographer, like the astronomer, is always concerned with the reduction of patterns too large to analyse rather than with the enlargement of patterns that are too small; he is especially concerned therefore, with the concepts and methods of generalizing area.

In the field of population geography the question that is basic to all the others is the question of the enumeration area. Within what areas are the enumerations of individuals to be summed up? How are the areas to be outlined? Can a systematic procedure be devised so that the selection of enumeration areas is not left entirely to the intuition of the worker and so that the results can be evaluated? Here is one of the most alluring of the frontiers of geographic research.

The inadequacy of enumeration areas based on political boundaries is widely recognized, yet all too often is forgotten. The story of the popular writer or lecturer who points to the low population density of Australia or of Canada and urges the movement of pioneer settlers into the empty areas is well-known. Obviously figures of density based on the ratio of the number of people to the total area of a country, whether occupied or not, are meaningless. Yet a distinguished student of demography who need not be named, not so long ago pointed out that of all the countries of South America, Uruguay had the greatest density of population. Little Uruguay is, of course, wholly occupied, but nowhere in Uruguay are densities of population comparable with those of parts of Middle Chile, or of Northeast Brazil. Where total numbers of people are enumerated within a large expanse of national territory much of which is unoccupied, the resulting figure is not only meaningless but is sometimes dangerously misleading as to the true situation.

In somewhat greater detail consider the case of New England. Statistics of population enumerated by counties or even by towns do not bring out the real patterns of density. This is because the counties and towns are irrelevant to the population patterns; the patterns are the current result of processes that are unrelated to the political subdivisions.

Yet the fact remains that such census data as are available are summed up by census districts. They are available in manuscript form for certain quite small areas in cities. They are published by townships for the nation as a whole. In most cases the geographer has neither the time nor the staff of assistants necessary to make a new count of his own. Therefore the geographer may disregard the collection of new data, enumerated in newly defined areas, as a practicable way toward the solution of the problem.

There are two ways in which available statistical data can be more effectively related to maps. First is the method of scale reduction. Data for townships, for example, can be plotted on a map of 1/3,000,000, the dots for population being uniformly spaced throughout the township area. When such a map is reproduced at 1/6,000,000 or smaller a degree of refinement has been achieved suitable to the scale. If the nature of the problem being investigated permits the use of maps at such small scales, this procedure is satisfactory. But if population is to be mapped at 1/500,000 and reproduced at 1/1,000,000, the even spacing of dots over township areas would give only a crude approximation. On maps of larger scale, the census enumeration areas, the townships, are too large.

The second method is to define areas within the townships, to subdivide the townships or other statistical areas, and to find a method for rearranging the census data in relation to the new enumeration areas [60:235-240]. The purpose is to find areas that are meaningful because they are relevant to the problem, so that population density can be measured in relation to some systematically defined area rather than to an arbitrary and generally irrelevant area [14]. In the field of political action this kind of a search for politically effective areas is called "gerrymandering;" where the motives are those of understanding rather than action, experimentation with differently defined areas in the search for meaningful area relations is the essence of the regional method. Unless geographers undertake this kind of search, the greatest potential contribution of geographic study to demography and related aspects of sociology and social anthropology will not be made.

When the nature of the problem and the scale of the maps demands greater refinement of geographic procedure, the geographer turns naturally to direct field observation, or to the examination of air photographs. On map scales which permit a direct view of the area being studied it becomes apparent that there is a considerable part of the land that is not occupied by people. In textbooks, a distinction is sometimes made between total national territory and the effective national territory, namely that part of the total territory that is actually used. On small-scale maps the unused parts of the national territory, the empty areas, must be defined in quite general terms, in terms too general to permit precise analysis. But on large-scale maps, at scales that permit the plotting of specific fields and farms (topographic scales), the occupied and unoccupied areas can be given precise definition.

What is an empty area? The census, as has often been remarked, counts people where they sleep, not where they work. But on a farm it would not be useful to insist that the farmstead was the only occupied part of the total area just because the people sleep there. The work area, the fields, are also occupied. Can the concept of the work area be used as a basis for defining enumeration areas? In rural areas possibly; in urban areas the problem is not so simple.

Lester E. Klimm has attempted to delimit such unoccupied areas [61]. The work has been done on a scale of 1/500,000 for the northeastern part of the United States. On this scale large areas may be mapped in which there are no man-made structures within a quarter of a mile. Such areas are said to be empty, although they may be entered temporarily from time to time by wood choppers, hunters, fishermen, or just campers and hikers. About 20 percent of the area studied was found to be empty in terms of this definition, a definition which is based on direct observation of settlement forms in the field.

Within the areas where man-made structures currently in use indicate that the land is occupied the data enumerated by the census must be further localized or re-arranged. Here also, on large-scale maps the distribution of houses seems to offer the best guide to the distribution of population. On chorographic scale maps, where specific fields and farms are too small to show (a map of 1/500,000 or 1/1,000,000, for example), the built-up areas of towns or cities can be distinguished from rural areas, and within the latter the places where people are concentrated and where they are widely scattered can be distinguished. By observation of the settlement pattern, a systematic procedure can be formulated by which the people enumerated in relatively large statistical areas can be arranged in more meaningful areas. If a dot map is made, the dots can be concentrated in those places where field observation or photo-interpretation shows them to be concentrated. If an isopleth map is to be made, some calculation of the enumeration areas geographically defined must be made. An interesting new technique is offered by the "densitometer," an electronic device which can be used to measure the density or spacing of dots. If the dot map is made with adequate precision, this precision can be transferred to an isopleth map through the use of such a tool.

Even where the census enumeration districts in cities are very small, the geographer may find that field observation shows the statistical areas to be less meaningful than others that could be defined. For example, where statistics are summed up by blocks in a city, the results are of little relevance to the study of such problems as purchasing power, marketing arrangements, or traffic flows. The street, including the people living on both sides of it, is an enumeration area of greater relevance than is the block.

Other Frontiers and Training Requirements

The reconstruction of population patterns of the past offers many opportunities for population geographers who are ready to work in the field of historical geography. For the reconstruction of past populations, in addition to the training requirements for historical studies (chapter three), the worker must understand the basic problems of population mapping.

Since maps of present and past populations can only give a static picture, the dynamics of population change must be examined geographically. Dodge's method of relating the growth and decline of communities to a theoretical curve on which there are significant points that can be dated and mapped, should be tested and evaluated. Perhaps instead of a theoretical curve, actual curves can be examined and classified in types each with a definite character. New and significant relationships and understandings may well be discovered when such curves are located on maps.

The study of the cultural characteristics of a population should not be left to the anthropologists. The geographic view point is needed here as elsewhere. Furthermore, such studies offer important results in terms of modern concepts of geography. The concept which, for many geographers, has replaced that of environmental determinism insists that the significance to man of the physical and biotic conditions of an area is a function of the attitudes, objectives, and technical abilities of the people. These are cultural characteristics. To define meaningful categories of culture, and to seek accordant relationships between them and other geographic phenomena offers opportunity to make new contributions to the basic theory of geography.

A person who wishes to specialize in population geography must have a certain rather specific kind of training. He must master the usual geographical techniques: the techniques of direct field observation and photointerpretation, of interviewing informants, and of searching documentary sources, in this case, published and unpublished enumerations. He must also be ready to work closely with cartographers, for many of his problems involve the search for more effective map procedures and methods of graphic presentation. And, in addition, he must be competent in the use of statistical procedures, and be well acquainted with the work of demographers, sociologists, and social anthropologists.

REFERENCES

1. TREWARTHA, G. T. "The Case for Population Geography," *Annals of the Association of American Geographers*, 43 (1953): 71-97.
2. KIRK, D. *Problems of Collection and Comparability of International Population Statistics*, Milbank Memorial Fund, New York, 1949.
3. GEORGE, P. *Introduction a l'étude géographique de la population du monde*, Institut national d'études démographiques, no. 14, Paris, 1951.
4. BAKER, O. E. "Population, Food Supply, and American Agriculture," *Geographical Review*, 18 (1928): 353-373.
5. McCARTY, H. H. "A Functional Analysis of Population Distribution," *Geographical Review*, 32 (1942): 282-293.
6. DÖRRIES, H. "Siedlungs und Bevölkerungsgeographie (1908-1938)," *Geographisches Jahrbuch*, 55 (1940): 3-380.
7. DE GEER, STEN. "A Map of the Distribution of Population in Sweden: Method of Preparation and General Results," *Geographical Review*, 12 (1922): 72-83.
8. AUROUSSEAU, M. "The Distribution of Population: A Constructive Problem," *Geographical Review*, 11 (1921): 563-592.
9. ———— "The Geographic Study of Population Groups," *Geographical Review*, 13 (1923): 266-282.
10. JEFFERSON, M. "The Anthropogeography of Some Great Cities," *Bulletin of the American Geographical Society*, 41 (1909): 537-566; *Idem*. "The Real New York," *ibid.*, 43 (1911): 737-740.
11. ———— "Some Considerations on the Geographical Provinces of the United States," *Annals of the Association of American Geographers*, 7 (1917): 3-15.
12. ———— "Distribution of the World's City Folks: A Study in Comparative Civilization," *Geographical Review*, 21 (1931): 446-465.
13. ———— "The Law of the Primate City," *Geographical Review*, 29 (1939): 226-232.
14. WRIGHT, J. K. "A Method of Mapping Densities of Population: with Cape Cod As an Example," *Geographical Review*, 26 (1936): 103-110.

15. ———— "Some Measures of Distributions," *Annals of the Association of American Geographers*, 27 (1937): 177-211.

16. ALEXANDER, J. W. "An Isarithmic-Dot Population Map," *Economic Geography*, 19 (1943): 431-432.

17. ALEXANDER, J. W., and ZAHORCHAK, G. A. "Population-Density Maps of the United States: Techniques and Patterns," *Geographical Review*, 33 (1943): 457-466.

18. MACKAY, J. R. "Some Problems and Techniques in Isopleth Mapping," *Economic Geography*, 27 (1951): 1-9.

19. STEWART, J. Q. "Empirical Mathematical Rules Concerning the Distribution and Equilibrium of Population," *Geographical Review*, 37 (1947): 461-485.

20. GANNETT, H. *Tenth Census*, U. S. Bureau of the Census, Bulletin 27, Washington, 1881.

21. BRIGHAM, A. P. "The Population of New York State," *Geographical Review*, 2 (1916): 206-217.

22. CUSHING, S. W. "The Distribution of Population in Mexico," *Geographical Review*, 11 (1921): 227-242.

23. BATSCHELET, C. E. "A Picture of the Distribution of Population in Pennsylvania," *Geographical Review*, 17 (1927): 429-433.

24. WOLFE, A. B. "Some Population Gradients in the United States," *Geographical Review*, 18 (1928): 291-301.

25. SMITH, G.-H. "A Population Map of Ohio for 1920," *Geographical Review*, 18 (1928): 422-427.

26. BOLLINGER, C. J. "A Population Map of Central Oklahoma for 1920," *Geographical Review*, 20 (1930): 283-287.

27. STILGENBAUER, F. A., assisted by G. J. HONZATKA, *Population of Michigan*, Detroit, ca. 1933.

28. STILGENBAUER, F. A., assisted by J. WETZEL. *A New Population Map of Middle America*, College of the City of Detroit, 1933; idem. assisted by H. E. JINKS. *A New Map of the Population of Canada with Alaska and Newfoundland*, Detroit City University, 1934; idem. *A New Population Map of the United States*, Wayne University, 1944.

29. JAMES, P. E. "The Distribution of People in South America," in *Geographic Aspects of International Relations*, Lectures of the Harris Foundation, 1937, Chicago, 1938: 217-240.

30. COULTER, J. W. "A Dot Map of the Distribution of Population in Japan," *Geographical Review*, 16 (1926): 283-284.

31. KROEBER, A. L. "Map of Distribution of Ancient Indians on Continent of North America," *American Anthropologist*, 36 (1934): 1-25.

32. HINSDALE, W. B. "Distribution of Aboriginal Population of Michigan," *Occasional Contributions from the Museum of Anthropology of the University of Michigan*, No. 2, Ann Arbor 1932.

33. FRIIS, H. R. "A Series of Population Maps of the Colonies and the United States, 1625-1790," *Geographical Review*, 30 (1940): 463-470.

34. GREEN, E. B., and HARRINGTON, V. D. *American Population before the Federal Census of 1790*, New York, 1932.

35. WOLFE, A. B. "Population Censuses before 1790," *Journal of the American Statistical Association*, 27 (1932): 357-370.

36. SUTHERLAND, S. H. *Population Distribution in Colonial America*, New York, 1936.

37. ROSSITER, W. S. *A Century of Population Growth. . . . 1790-1900*, U. S. Bureau of the Census, Washington, 1909.

38. WHITNEY, J. D. *The United States, Supplement I, Population: Immigration: Irrigation*, Boston, 1894.

39. JEFFERSON, M. "Looking Back at Malthus," *Geographical Review*, 15 (1925): 177-189.

40. DODGE, S. D. "Periods in the Population History of the United States," *Papers of the Michigan Academy of Science, Arts and Letters*, 32 (1946): 253-260.

41. ———— "A Study of Population in Vermont and New Hampshire," *Papers of the Michigan Academy of Science, Arts and Letters*, 18 (1932): 131-136.

42. ———— "Population Regions of the United States," *Papers of the Michigan Academy of Science, Arts and Letters*, 21 (1935): 343-353.

43. ZELINSKY, W. "An Isochronic Map of Georgia Settlement, 1750-1850," *Georgia Historical Quarterly*, 35 (1951): 191-195.

44. DODGE, S. D. "The Frontier of New England in the Seventeenth and Eighteenth Centuries and its Significance in American History," *Papers of the Michigan Academy of Science, Arts and Letters*, 28 (1943): 435-439.

45. JAMES, P. E. "The Changing Patterns of Population in São Paulo State, Brazil," *Geographical Review*, 28 (1938): 353-362.

46. MIRKOWICH, N. "Recent Trends in Population Distribution in California," *Geographical Review*, 31 (1941): 300-307.

47. KOHN, C. F. "Population Trends in the United States Since 1940," *Geographical Review*, 35 (1945): 98-106.

48. SMITH, G.-H. "The Populating of Wisconsin," *Geographical Review*, 18 (1928): 402-421.

49. WILSON, L. S. "Some Notes on the Growth of Population in Minnesota," *Geographical Review*, 30 (1940): 660-664.

50. SMITH, G.-H. "Interstate Migration as Illustrated by Ohio," *Bulletin of the Geographical Society of Philadelphia*, 27 (1929): 301-312.

51. THORNTHWAITE, C. W., assisted by H. J. SLENTZ. "Internal Migration in the United States," *Study of Population Redistribution*, Bulletin 1, Philadelphia (University of Pennsylvania), 1934.

52. KROEBER, A. L. *Anthropology Today, An Encyclopedic Inventory*, Chicago, 1953.

53. TAX, S., EISELEY, L. C., ROUSE, I., VOEGELIN, C. F., eds. *An Appraisal of Anthropology Today*, Chicago, 1953.

54. DOMINIAN, L. *The Frontiers of Language and Nationality in Europe*, New York, 1917.

55. BEYNON, E. D. "Isolated Racial Groups of Hungary," *Geographical Review*, 17 (1927): 586-604.

56. HARTSHORNE, R. "Racial Maps of the United States," *Geographical Review*, 28 (1938): 276-288.

57. BROEK, J. O. M. "Diversity and Unity in Southeast Asia," *Geographical Review*, 34 (1944): 175-195.

58. WIGMORE, J. H. "A Map of the World's Law," *Geographical Review*, 19 (1929): 114-120.

59. KURATH, H. *A Word Geography of the Eastern United States*, Ann Arbor (Michigan), 1949.

60. MONKHOUSE, F. J., and WILKINSON, H. R. *Maps and Diagrams, Their Compilation and Construction*, London, 1952.

61. KLIMM, L. E. "Empty Areas in the Old Northeast: With Examples from New Jersey," *Annals of the Association of American Geographers*, 43 (1953): 178-179.

SETTLEMENT GEOGRAPHY [A]

THE SUBSTANCE OF SETTLEMENT GEOGRAPHY
STUDIES OF PIONEER SETTLEMENT
STUDIES OF SPECIFIC FACILITIES AND THEIR GROUPING
Studies of Architectural Style
Studies of Roads and Properties
Studies of Settlement Ensembles
The Compact Farm Village
The Individual Farmstead
Hamlets and Villages

THE PURPOSES OF SETTLEMENT GEOGRAPHY
STUDIES OF ORIGIN
STUDIES OF FUNCTION

THE PROSPECT FOR SETTLEMENT GEOGRAPHY
SPECIAL PROBLEMS TO BE SOLVED
PRACTICAL IMPLICATIONS OF SETTLEMENT STUDIES
RELATION OF SETTLEMENT STUDIES TO OTHER BRANCHES OF GEOGRAPHY

[A] Original draft by Clyde F. Kohn, with the cooperation of Robert E. Dickinson, Robert B. Hall, and Fred B. Kniffen.

CHAPTER FIVE

SETTLEMENT GEOGRAPHY

IN AMERICAN GEOGRAPHY, the study of settlements is both old and new. It is old in the sense that consideration of settlement runs like a thread through almost the whole fabric of geographic thought, giving unity and coherence to investigations of a wide variety of problems in many different areas. Discussions of the patterns and processes of settlement can be found in regional studies and in the literature of many of the topical fields developed by American scholars. On the other hand, the study of settlements is new in the sense that, as yet, no analytical framework has been developed for settlement geography comparable to the principles of location in industrial geography or of nationalism in political geography. This is an objective yet to be reached.

In general, settlement geography has to do with the facilities men build in the process of occupying an area. These facilities are designed and grouped to serve specific purposes, and so carry functional meanings. Their exterior forms reflect architectural styles of the time and culture from which they spring. Their distribution produces discernible patterns in the landscape. Once created, they are apt to outlast both the function for which they were originally designed and the architectural fashions of their time. For these reasons they reflect changes in man's occupance of an area and are often the only existing landscape expressions of the past.

THE SUBSTANCE OF SETTLEMENT GEOGRAPHY

Settlement studies differ in two fundamental ways. At first, the predominant interest in settlement geography in the United States and in other American countries was to examine the process of occupying pioneer areas. These studies were designed to direct the colonizing of unoccupied farming areas. More recently, there has been a growing interest in settlement features, *per se,* and in their groupings. In Europe where there is a rich variety of architectural styles and of spatial patterns reflecting a long and complex culture history, this aspect of settlement has long been the center of attention. In recent years European concepts and procedures have been applied to the American scene. Both of these interests in settlement geography are continuing. It is with the second interest, however, that much of this chapter deals.

STUDIES OF PIONEER SETTLEMENT

The examination of pioneer areas as places of potential colonization was promoted and directed in America by Isaiah Bowman and by those who worked with him [1; 2; 3]. These scholars identified and described pioneer zones that might be open to new settlement. Specific processes of settlement were analyzed and evaluated for a number of areas that had been occupied during the late 19th and early 20th centuries, especially the semi-arid margins of the middle-latitude grasslands, and the cold margins of the boreal forests. Bowman turned his attention to economic, social, and political processes as they were related to physical and biotic conditions in areas of new or potential settlement. The scholars who worked with him on the pioneer-belt project, many of whom were economists or sociologists, did not always treat these processes with the geographic insight of their leader. Moreover, it now seems correct to say that when the pioneer studies were initiated in the 1920's the settlement of new and marginal lands by small, independent farmers was already outmoded as an economic process of world-wide importance. Yet interest in the settlement of new areas continues [4] and is actively promoted in several American countries. One of the projects of the Commission on Geography of the Pan American Institute of Geography and History has to do with the study of areas of potential colonization. The purpose is practical: that of providing geographic knowledge concerning the qualities of the land as a resource base, and the opportunities it offers or the limitations it imposes with respect to new settlement. Bowman proposed to guide pioneer settlement scientifically, and to avoid the economic and social waste of unregulated and uninformed movement into unfamiliar areas. Those objectives are still valid and might even be broadened to include the promotion of resettlement in areas long occupied.

STUDIES OF SPECIFIC FACILITIES AND THEIR GROUPING

Studies of the character and distribution of specific facilities are beginning to receive more and more attention in the United States. This does not imply that they have been entirely neglected in the past. Specific settlement features, such as house types, the shapes of barns, and the distinctive structures associated with different kinds of agriculture or industry have been examined in detail. Some geographers have also expressed special interest in the size, alignment, and spatial arrangement of roads, fields, and properties. Others have, and are, contributing to a more complete understanding of settlement ensembles: the individual farmstead, the compact farm village, hamlets, and rural villages.

Studies of Architectural Style

European geographers have given much attention to studies of individual settlement features, especially to the classification and mapping of different kinds of rural habitation [5]. This interest was stimulated, no doubt, by the existence in the European landscape of sharp differences within small areas. In the United States, notably in recent decades, there is a uniformity of house types which extends from Maine to California; the architectural styles reflect the dates of construction rather than the areas. On the other hand, this continent is not without its areal contrasts.

Several American geographers have expressed a special interest in the shapes of buildings and in the materials out of which they are built. Fred B. Kniffen drew attention to the purposes and procedures of such studies in 1936 with a report on the house types of Louisiana [6]. His objective was to delineate the several cultural regions of Louisiana, each occupied by a distinctive societal group. To reach this goal he proposed to classify and map several specific cultural features which, taken together, would serve to outline the areas occupied by different kinds of people. His map of house types brought to light important areal differences that, without mapping, could be perceived only dimly. Kniffen also studied the American agricultural fair and associated buildings in different places [7].

Another American geographer, Loyal Durand, Jr., has investigated the shapes of barns as part of a long-range program of research on the dairy regions of North America (chapter ten). In this connection he undertook to describe and map areal differences in the shapes of the barns both in southeastern Wisconsin [8] and in southwestern Michigan [9], and to relate these differences to the distribution of settlements of different kinds, as, for example, those of Yankees, of Germans, and of Scandinavians.

There have been several other publications in which specific types of rural settlement facilities were classified and mapped. These include a profile of dispersed dwellings running from the Great Lakes to the Gulf by R. Finley and E. M. Scott [10]; a study of sod houses by D. S. Gates [11]; a report on certain Mexican house types by A. B. Cozzens [12]; and two papers on Szechwan villages by J. E. Spencer [13]. In a few cases, also, American geographers have studied houses in relation to the climatic conditions in which they are built and the artificial climates produced in them for human comfort. Notable among such studies are those of D. H. K. Lee (chapter twenty-two) [14].

Studies of Roads and Properties

Another aspect of settlement geography of perennial interest is the geometric pattern of roads and properties and the arrangement of buildings with reference to them. The basic ideas and procedures for this kind of work have been derived from the German and French geographers. In 1895, August Meitzen established a point of view in the study of settlements which has been followed by the German workers in this field [15]. Categories of geometric patterns illustrated by specific examples of each were described by R. Martiny [16]. The American geographers have not devoted so much attention to studies of geometric patterns as have their European colleagues. Nevertheless, there are contrasts in settlement patterns in North America that would repay careful investigation.

The parts of eastern United States that were settled before 1785 were surveyed by a system known as "metes and bounds." The pattern of settlement was articulated to the main roads, which in many cases followed old Indian trails. These roads were winding so as to take advantage of river crossings or grades that were easiest for men on foot. The same advantages were important for men on horseback or for horse-drawn wagons. The properties were attached to these main roads, and had irregular outlines. After the survey of properties, the road pattern was elaborated along the property lines. The result is a pattern of irregularly winding roads among fields of odd shapes and sizes.

The Ordinance of 1785 outlined a system of rectangular survey to be applied to all unsettled parts of the new United States. The details of this pattern need not be explained here: essentially the land was marked off into sections, each a square mile in area, and each bounded by north-south and east-west roads. Thirty-six sections made up a township, in which section 16 was set aside for the support of the schools.

In many parts of the Middle West, this rectangular survey was superimposed on a pre-existing pattern of main routes which followed old Indian trails, as in the East. The radial patterns of these older routes spreading from the river crossing at Detroit throughout southern Michigan, and spreading southwest, west, and northwest from Chicago, are still visible [17: 238-245, 294-299]. The main routes west of Lake Michigan trend in a northwest-southeast direction all the way to central North Dakota; around the smaller settlements there are radiating patterns of smaller areal spread. Where these radial roads were developed in a forested country which had to be laboriously cleared, they survived the rectangular survey; but in open grass country, where roads could easily be shifted, as soon as the square sections were surveyed and square properties were fenced off, the roads conformed to the north-south or east-west trend, and in such areas travel in a northeasterly direction, for example, follows a succession of right-angle turns. The result of the rectangular survey is a checker-board arrangement of fields and farms [17: 294-295]; properties of 160 acres, a quarter section, are divided into square 40-acre fields, with

square patches of woods occupying the center of each section in previously forested areas.

In other parts of the United States and Canada there are other distinctive road and field patterns. In areas of French settlement, for example,
the properties are commonly laid out in long, narrow strips running at
right angles to the main roads. This long-lot system, strikingly developed
along the lower St. Lawrence Valley of Canada, can also be observed
along the lower Mississippi in Louisiana and along the Detroit River in
southeastern Michigan. Road and property patterns reflecting the Spanish
land divisions are found in California, New Mexico, and Texas [18].

In 1935, Carleton P. Barnes examined the relative merits of the rectangular and linear patterns of rural settlement [19]. He found that the standard rectangular pattern characteristic of large parts of the United States
was actually very costly in terms of public services. In any block of 16
sections, for example, 32 miles of road must be built, repaired, and kept
open in winter. If this same block were re-surveyed with properties a
mile long and a quarter mile wide, instead of a square quarter mile, and
if roads were built between every two sections, instead of along every
section boundary, there would have to be only 16 miles of road to serve
the same area. This would release a total of 128 acres for agricultural use.
The gain per farm in the area studied in Iowa was estimated at $530. And
there would be additional savings in electric power lines, telephone lines,
mail routes, school bus routes, and other kinds of service [20: 247-248].

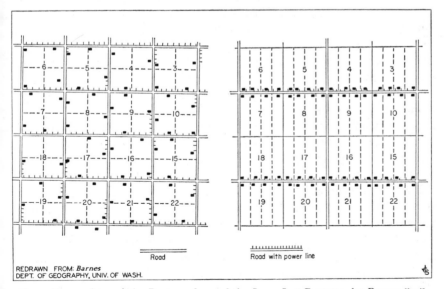

Fig. 4. Comparison of the Rectangular and the Long-Lot Patterns, by Barnes [19].

The fact is that in some areas the rectangular system is in process of changing to a linear system. W. H. Kollmorgen and G. F. Jenks have found a line pattern of rural settlements developing in an area originally laid out on a rectangular pattern in the wheatlands of western Kansas [21]. In this area main east-west roads have developed from three to eight miles apart and there are few north-south roads to connect them. The average size of the farms has increased from 207 acres in 1890 to 1190 acres in 1950 by a process of sale and amalgamation. The first farms sold were those not located on the developed roads. The present occupied farmsteads are arranged along the east-west routes in a linear pattern.

Linear patterns have also appeared elsewhere. They are notable in the Connecticut Valley of southern New England, and in some of the narrow incised valleys of the Appalachian Plateau. Linear patterns have been recommended for the subdivision of areas of new settlement where water from the Columbia River will be provided for irrigation [22].

Studies of Settlement Ensembles

The problems of settlement that have attracted the most attention in the United States are those having to do with the settlement ensemble, with special reference to its functional relationships. Commonly individual buildings and other facilities are grouped together in clusters located at focal points in the patterns of roads. These clusters range through a hierarchy from simple to complex, starting with a cluster of farm buildings around a dwelling. This cluster is known as a farmstead and is made up of various buildings and other facilities arranged in characteristic fashion. This is the most elementary of the settlement ensembles. Clusters of farms and other kinds of buildings make up hamlets or rural villages. Distinctively urban are the more complex clusters that make up towns, cities, and metropolises.

Conceived in these terms, settlement geography covers the whole range from rural to urban settlements, from nomadic to sedentary. It is concerned not only with the buildings grouped around the permanent farm dwelling, but also with the temporary camp of the hunter or herder, or with settlement clusters or agglomerations, running the scale from hamlet, to village, to town, to city, and to metropolis.

But the problems and procedures of settlement geography in rural areas are quite distinct from those pertaining to urban areas. There has appeared among geographers, therefore, a group of specialists who have devoted themselves to the study of urban areas and have developed concepts and techniques peculiar to the urban scene. Other geographers, fewer in number, have devoted themselves to the rural scene. Because of this division of labor, settlement geography as discussed in this chapter deals only with the rural settlement ensembles; chapter six discusses urban geography. Throughout the world, two basic kinds of ensemble have appeared in rural areas. In some places people live in compact rural villages, leaving the farms quite free from buildings; in other places families live on individual farmsteads located on farms. There are many variations and combinations of these two basic patterns.

The Compact Farm Village. The compact village type of settlement occurs in different parts of the world, for a variety of reasons. Robert E. Dickinson has traced some of the complex backgrounds of such villages in Germany [23]. This kind of rural settlement has been studied in Japan by Glenn T. Trewartha [24] and Robert B. Hall [25]; in Mexico by Dan

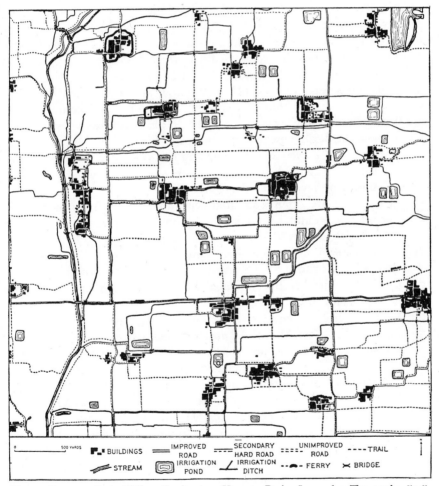

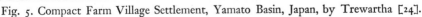

Fig. 5. Compact Farm Village Settlement, Yamato Basin, Japan, by Trewartha [24].

Stanislawski [26]; in Puerto Rico by John P. Augelli [27]; and in various parts of Latin America by Robert S. Platt [28]. In some cases it can be shown that the compact village was adopted as a settlement pattern for purposes of defense; in other cases it was adopted to take advantage of limited supplies of water or to avoid trespassing on a meager acreage of arable land; in many cases the preference for the compact village is rooted in cultural tradition and is not necessarily related to local conditions.

In the United States the compact rural village reflects colonization by homogeneous ethnic or religious groups. It was common in early New England [29; 30] and in eastern New Jersey. In a sense, the clustering of buildings around the plantation center in the South belongs in this category of settlement. Later the Mormons formed compact villages in Utah [31; 32; 33], and there are isolated examples of such communities occupied by Mennonites, Rappists, and Dukhobors. The Amana community in Iowa is a striking example of the survival of a similar type of settlement and of the institutions that supported it [34; 35].

The advantages and disadvantages of the compact farm village over individual farmsteads dispersed throughout an area have often been discussed. In a land only thinly populated, education and community life are fostered by grouping people together. In places where the land is held in common, the flexible use and assignment of fields is aided by placing all the dwellings and other fixed structures in one place. On the other hand, where property is individually owned there is an advantage, especially in rainy country, of having the farmer live on the land he cultivates. Compact settlements are often associated with an extreme fragmentation of farm properties; an individual farmer may own and operate small patches of land in widely-scattered locations. The keeping of livestock in compact villages also creates many undesirable problems.

The Individual Farmstead. The early settlement of the Middle Atlantic colonies was carried out by individual freeholders rather than by organized groups. The large proprietors, unlike their counterparts in most of the Latin American countries, made small farm properties available at nominal cost. From the Middle Atlantic colonies, the westward movement pushed back the frontier of new settlement. The plantation type of settlement to the south and the compact-village type of settlement to the north were transformed by the rapid spread of individual farmsteads, until the latter type became characteristic of most of the Middle West and beyond.

But the individual farmstead is not everywhere the same. Trewartha has described the areal differences in the types of buildings and the layout of the American one-family farmstead [36]. He distinguished the farmstead ensembles that are characteristic of the several major agricultural regions. He found the most elaborate farmsteads in the dairy region of western Wisconsin; those of the New York-New England dairy region he found to be almost as elaborate. The simplest ones are in the Cotton Belt and the Range Livestock Region.

Clyde F. Kohn made use of the associations of farmstead ensembles with agricultural systems to develop keys for the identification of types of farming on air photographs [37]. Based on the size and shape of farms and fields, on the assortment and layout of buildings, and on the arrangement of farmsteads, these keys make possible the more effective use of air photography as a tool of geographic research (chapters twenty-four and twenty-five).

Hamlets and Villages. In areas occupied by occidental people, among whom trade is a basic feature of economic life, a pattern of small semi-rural trade centers makes its appearance. The hamlets and villages are clusters of buildings somewhat more complex than the farmstead, but much less complex than the truly urban centers, which are generally referred to as towns or cities. The semi-rural village, in which certain urban functions are centered, is not to be confused with the farm village described above, for the latter is completely rural and it lacks internal differentiation.

Trewartha has examined the hamlet as an element in the fabric of American settlement [38]. He defines it as consisting of not fewer than five buildings grouped within a linear distance of one quarter mile; of the five buildings four are used as residences, and there is at least one retail store or service establishment. In Wisconsin, where these rural-settlement ensembles have been studied in detail, only grocery stores and elementary schools are to be found in most hamlets, but taverns, filling stations, and churches are not uncommon.

The village is larger and performs more functions. According to the criteria adopted by Trewartha, a village must have at least ten retail or service establishments. In addition to the services found in hamlets, the village generally has at least four other retail businesses selling such things as automobiles, farm implements, appliances, lumber, hardware, or feed. The village may also provide repair shops, and banking facilities, and may have a telephone exchange. Physicians and dentists are generally found only in large villages, with more than 600 inhabitants. The village commonly has no motion-picture theaters or recreational facilities.

Much has been written in Europe and America about the arrangement, spacing, and size of hamlets, villages, and towns. As early as 1915 the American rural sociologist, C. J. Galpin, published a study of the rural communities of Wisconsin [39]. In 1918, Galpin offered some suggestions regarding the distribution of central trading places which would serve most efficiently the needs of the dispersed inhabitants [40]. Walter

Christaller, a German geographer, published his now famous discussion of central places in 1933 [41]. Various writings on the actual and theoretical arrangement of central places were summarized by Dickinson in 1947 [42], and by John E. Brush in 1953 [43]. Brush also tested various concepts in his study of 234 agglomerated settlements in parts of some eight counties in southwestern Wisconsin [44].

These studies constitute steps in the formulation of concepts and procedures in settlement geography. Several writers have drawn up geometric figures to show the theoretical spacing of trade centers of different ranks in an area of evenly dispersed rural population. The radial movement of people and traffic around a central place tends to create circular trade areas and to cause the spacing of trade centers with equivalent functions at approximately equal distances from one another. Within the larger trade areas of the towns, smaller circular trade areas develop around the villages and within these still smaller ones develop around the hamlets. Christaller recognizes that if the whole of an area occupied by a dispersed rural population is to be served by more or less evenly spaced central places each with a circular trade area around it, there must be an overlap of the circles. The only perfect geometric figure to resolve this overlap of adjacent circles is the hexagon, as illustrated in Fig. 6. According to his theory hexagonal trade areas of different sizes surround the towns (C), the villages (B), and the hamlets (A). The roads which serve the hexagonal trade area converge on the central place from six directions.

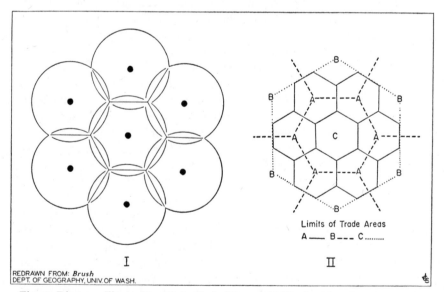

Fig. 6. Diagram Illustrating Christaller's Hexagonal Trade Area Concept [44].

Many students in Europe and America have criticized the hexagonal concept. The American sociologists J. H. Kolb and E. deS. Brunner [45] have observed that the smaller centers (A) do not develop as close to the larger centers as they do to one another. The pull exerted by a trade center varies directly with the size of the center and indirectly with the distance.

Brush, in his study of Wisconsin, finds that the smaller agglomerated settlements developed in the period of transportation by horse and wagon before 1880. When railroads were built, the larger centers developed along rail lines, in some cases at the sites of the earlier hamlets or villages.

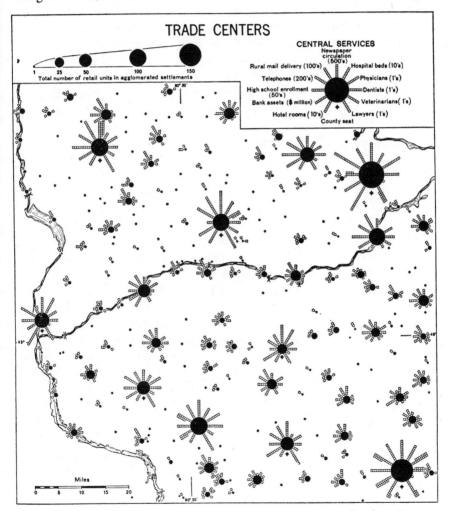

Fig. 7. Central Services of Trade Centers in Wisconsin, by Brush [44]

The smaller places, not on railroads, declined but did not disappear; the hamlets lost many of their earlier commercial functions but remained as clusters of residences. Thus the functional interpretation of the settlement pattern must be a dynamic one, taking account of economics and technological changes. The current pattern includes relict features from earlier periods which are seldom obliterated entirely. The theoretical approach by way of geometric figures obscures the dynamic aspects of functional change.

THE PURPOSES OF SETTLEMENT GEOGRAPHY

Geographers do not examine architectural styles, roads and properties, and settlement ensembles only for the sake of identifying new categories or developing new classifications. To be sure, there was a time when geographers, newly aware of the possibilities of detailed field mapping, limited themselves to the study of shapes [46: 229-236]. But in recent decades studies of the facilities which men build have been undertaken for one of two purposes. One is for the light they throw on historical sequence—studies of origin; the other is for the light they throw on functional relations—studies of functions.

STUDIES OF ORIGIN

The facilities men build in the process of occupying an area, and the grouping of these facilities in the landscape, reflect the way of living of the people who build them. In any one place at any one time, men build houses, lay out roads or property lines, and arrange the structures they use for business, for worship, or for pleasure according to tradition, need, and technical knowledge. But from time to time there are changes, especially in technology. A distinctive set of settlement features and patterns is associated with transportation by horse; a quite different set of facilities and patterns came with the railroad; and still different structures and patterns are associated with the automobile, the paved highway, and the airplane. But once established, settlement ensembles and their distribution are more or less fixed. They remain as relict features when the way of living that they represent is changed. New settlements are superimposed on or intermingled with the older ones. In many parts of the United States the basic patterns were laid out in the period when horses furnished the chief transport power. The barns of that period can be made over into garages, although the outward appearance is not what one would adopt today if a new garage were being built. But the narrow roads with sharp grades and curves are not so easily changed to meet the needs of an automobile age.

Geographers may study settlement ensembles and their distribution for the light they throw on these sequences of change. A map showing the areas occupied by different styles of architecture may reveal forgotten details of the migration of certain ethnic groups. A complex pattern of relict and modern features may be taken apart to show the unfolding of the stages of occupance, each reflecting a period characterized by a certain technology of transport. These are studies in sequent occupance discussed in chapter three [47]. From such studies in the historical geography of settlement it becomes clear that the significance to man of certain physical conditions changes: a slope that is not too difficult for horses becomes a barrier to a railroad; a river that is deep enough for a sailing ship is not nearly deep enough for a steamship. Studies of sequent occupance in terms of settlement provide laboratory reports on the experience of man with his changing problems in specific environments [48; 49].

STUDIES OF FUNCTION

Settlement studies also yield new insights into the arrangement of current economic, social, and political institutions. Every feature of the settlement ensemble takes on its own distinctive shape and position because of the function it performs. Certain structures are built to serve as habitations. In different farming regions, different buildings are grouped around the farm dwelling, according to the needs of the area. The regions of dairying are known by their distinctive barns and silos; the wheat regions by their grain elevators; the sugar-beet or potato regions by their loading ramps; the cotton areas by their gins. Furthermore, each community has its churches, its schools, its playgrounds, and other features as demanded by the needs and traditions of the people. The study of function extends also to the pattern of land use, the nature of land tenure, and other matters relating to the institutional framework in which the settlement is embedded.

THE PROSPECT FOR SETTLEMENT GEOGRAPHY

A review of the work in settlement geography serves to underline again the essential unity of the field of geography as a whole. For, although settlement studies can be pursued as a topical specialty, the treatment of settlement, as pointed out at the beginning of this chapter, also runs as a unifying thread through the whole range of geographic research in which the activities of man and the facilities he builds as a result of these activities are considered.

SPECIAL PROBLEMS TO BE SOLVED

Scholars who wish to devote themselves to the study of settlements will find many important problems to claim their attention. In the first place, such specialists are needed to develop concepts and formulate procedures for the guidance of workers in other branches of geography. Special sessions at the annual meetings of the Association of American Geographers might well be devoted to the discussion of settlement problems as these problems are posed and solved in different areas.

The comparative study of settlements in different cultural areas is needed to broaden the concepts regarding the origin and functional relations of different patterns. Almost nothing has been done in such distinctive areas as the Moslem world, or in parts of the Orient. The world of the Soviet Society offers sharp contrasts in settlement features, ensembles, and in their distribution as compared with those of the capitalist countries. There is need for the compilation of world maps showing the areas of individual farmsteads and the areas of compact farm villages, and of various combinations of these basic types. The comparative study of areas of pioneer colonization, interpreted in the light of contemporary economic institutions and modern technology, should prove to be of great importance to the formulation of policy with respect to the development of underdeveloped areas.

PRACTICAL IMPLICATIONS OF SETTLEMENT STUDIES

Settlement geography can be both theoretical and practical in its objectives. Scarcely any problem in the theoretical aspects of settlement is without practical implications [50]. In the planning for new settlements, whether they are to be in pioneer zones or in areas long occupied, geometric schemes for their arrangement should be drawn up by persons who are fully aware of the many analyses of the advantages and disadvantages of different patterns that have been made. Geographers and rural sociologists can cooperate in seeking the answers to many questions in advance of settlement. How should farmsteads be distributed? What is the best layout of roads and properties to serve a specific area? What are the needs of the new community in the way of rural services, and where can these services be located most effectively? Where an old area of settlement is undergoing change, as when a cotton-growing community is transformed into a cattle and general-farming community with additional employment from new manufacturing industries, the settlement geographer can suggest the alterations in the pattern of roads and the distribution of services that can make the transformation least painful.

RELATION OF SETTLEMENT STUDIES TO OTHER BRANCHES OF GEOGRAPHY

Settlement studies appear as essential elements of research undertakings in many other branches of geography. Students of population geography are generally agreed that the most effective way to ascertain the probable arrangement of people inside enumeration areas is to make use of the pattern of settlement as a guide (chapter four). Settlement patterns are so closely interrelated to the patterns of agricultural land use that any attempt to insist on the study of these two phenomena in separate topical fields would be artificial and would retard the development of both (chapter ten). The treatment of settlement patterns, too, constitutes an integral part of the study of transportation geography (chapter thirteen). Settlement features are included in military geography, not only in tactical studies where the things men build are always recognized as an essential element of the terrain, but also in broader strategic studies (chapter twenty-three). In modern mechanized warfare the pattern of roads and of focal points where roads come together in villages and towns is a major consideration.

Settlement studies bridge the more or less arbitrary line between the rural and the urban (chapter six). The distribution of central places cannot be analyzed as a purely rural phenomenon without reference to the spacing and functions of towns and cities. The fabric of settlement must be looked at from an over-all viewpoint. The separation of studies of rural settlements from the studies of urban settlements, so often necessary or expedient in geographical work, should not be permitted to erect barriers or impede communications between workers in these two fields. For urban geography is in part merely a special phase of settlement study applied to very complex areas possessing sharp internal differentiation, although, as the next chapter points out, it deals also with many other aspects of cities besides their buildings and street patterns. In making this distinction between rural and urban, as in making other distinctions between topical specialties in geography, one should never forget the common concepts and procedures that tie them all together in one single discipline.

REFERENCES

1. Bowman, I. "The Scientific Study of Settlement," *Geographical Review*, 16 (1926): 647-653.
2. ———. *The Pioneer Fringe*, American Geographical Society, Special Publication no. 13, New York, 1931.
3. ———, (and nine other authors). *Limits of Land Settlement*, New York, 1937.
4. Stone, K. H. *Alaskan Group Settlement: The Matanuska Valley Colony*, United States Department of the Interior, Bureau of Land Management, Washington, 1950; idem. "Populating Alaska: The United States Phase," *Geographical Review*, 42 (1952): 384-404.
5. Demangeon, A. "L'habitation rurale en France: Essai de classification des principaux types," *Annales de Géographie*, 24 (1920): 352-375.
6. Kniffen, F. B. "Louisiana House Types," *Annals of the Association of American Geographers*, 26 (1936): 179-193.
7. ———. "The American Agricultural Fair: The Pattern," *Annals of the Association of American Geographers*, 39 (1949): 264-282.

8. DURAND, L. JR. "Dairy Barns of Southeastern Wisconsin," *Economic Geography*, 19 (1943): 37-44.
9. ———. "The Lower Peninsula of Michigan and the Western Michigan Dairy Region: A Segment of the American Dairy Region," *Economic Geography*, 27 (1951): 163-183.
10. FINLEY, R., and SCOTT, E. M. "A Great Lakes-to-Gulf Profile of Dispersed Dwelling Types," *Geographical Review*, 30 (1940): 412-419.
11. GATES, D. S. "The Sod House," *Journal of Geography*, 32 (1933): 353-359.
12. COZZENS, A. B. "The Geographic Background of some Mexican House Types," *Annals of the Association of American Geographers*, 28 (1938): 43-44.
13. SPENCER, J. E. "The Szechwan Village Fair," *Economic Geography*, 16 (1940): 48-58; idem. "The Szechwan Village Tea House," *Journal of Geography*, 41 (1942): 52-58.
14. LEE, D. H. K. "Thoughts on Housing for the Humid Tropics," *Geographical Review*, 41 (1951): 124-147; idem. *Physiological Objectives in Hot Weather Housing*, United States Housing and Home Finance Agency, Washington, 1953.
15. MEITZEN, A. *Siedelung und Agrarwesen der Ostgermanen, der Kelten, Römer, Finnen, und Slawen*, Berlin, 1895.
16. MARTINY, R. *Die Grundrissgestaltung der deutschen Siedlungen*, Petermann's Mitteilungen, Ergänzungsheft 197, Gotha, 1928.
17. JAMES, P. E. *A Geography of Man*, Boston, 1949.
18. BROEK, J. O. M. *The Santa Clara Valley, California: A Study in Landscape Changes*, Utrecht, 1932.
19. BARNES, C. P. "Economies of the Long-Lot Farm," *Geographical Review*, 25 (1935): 298-301.
20. LOOMIS, C. P., and BEEGLE, J. A. *Rural Social Systems*, New York, 1950.
21. KOLLMORGEN, W. M., and JENKS, G. F. "A Geographic Study of Population and Settlement Changes in Sherman County, Kansas," *Transactions of the Kansas Academy of Science*, 54 (1951): 449-494.
22. Columbia Basin Joint Investigations. *Patterns of Rural Settlement in the Columbia Basin*, Washington, 1947.
23. DICKINSON, R. E. "Rural Settlements in the German Lands," *Annals of the Association of American Geographers*, 39 (1949): 239-263; idem. *Germany, A General and Regional Geography*, London, 1953.
24. TREWARTHA, G. T. *Japan: A Physical, Cultural, and Regional Geography*, Madison (Wisconsin), 1945.
25. HALL, R. B. "Some Rural Settlement Forms in Japan," *Geographical Review*, 21 (1931): 93-123.
26. STANISLAWSKI, D. *The Anatomy of Eleven Towns in Michoacán*, Latin American Studies 10, Austin (Texas), 1950.
27. AUGELLI, J. P. "Rural Settlement Types of Interior Puerto Rico: Sample Studies from the Upper Loiza Basin," *Journal of Geography*, 51 (1952): 1-13.
28. PLATT, R. S. *Latin America: Countrysides and United Regions*, New York, 1942.
29. SCOFIELD, E. "The Origin of Settlement Patterns in Rural New England," *Geographical Review*, 28 (1938): 652-663.
30. TREWARTHA, G. T. "Types of Rural Settlement in Colonial America," *Geographical Review*, 36 (1946): 568-596.
31. SPENCER, J. E. "The Development of Agricultural Villages in Southern Utah," *Agricultural History*, 16 (1940): 181-189.
32. NELSON, L. *The Mormon Village: A Study in Social Origin*, Brigham Young University, Study no. 3, Provo (Utah), 1930.
33. ———. *Utah Farm Village of Ephriam*, Brigham Young University, Study No. 2, Provo (Utah), 1928.

34. SHAMBOUGH, B. M. H. *Amana: The Community of True Inspiration,* Iowa State Historical Society, Des Moines (Iowa), 1908.
35. DAVIS, D. H. "Amana: A Study of Occupance," *Economic Geography,* 12 (1936): 217-230.
36. TREWARTHA, G. T. "Some Regional Characteristics of American Farmsteads," *Annals of the Association of American Geographers,* 38 (1948): 169-225.
37. KOHN, C. F. "The Use of Aerial Photographs in the Geographical Analysis of Rural Settlements," *Photogrammetric Engineering,* 17 (1951): 759-771.
38. TREWARTHA, G. T. "The Unincorporated Hamlet: One Element of the American Settlement Fabric," *Annals of the Association of American Geographers,* 33 (1943): 32-81.
39. GALPIN, C. J. *The Social Anatomy of an Agricultural Community,* University of Wisconsin Agricultural Experiment Station, Bulletin 34, 1915.
40. ———. *Rural Life,* New York, 1918.
41. CHRISTALLER, W. *Die zentralen Örte in Süddeutschland,* Jena, 1933.
42. DICKINSON, R. E. *City, Region and Regionalism,* London, 1947.
43. BRUSH, J. E. "The Urban Hierarchy in Europe," *Geographical Review,* 43 (1953): 414-416.
44. ———. "The Hierarchy of Central Places in Southwestern Wisconsin," *Geographical Review,* 43 (1953): 380-402.
45. KOLB, J. H., and BRUNNER, E. DE S. *A Study of Rural Society,* Boston, 1946.
46. HARTSHORNE, R. "The Nature of Geography," *loc. cit.* ref. 3, chapter one.
47. WHITTLESEY, D. "Sequence Occupance," *Annals of the Association of American Geographers,* 19 (1929): 162-165.
48. ACKERMAN, E. A. "Sequent Occupance of a Boston Suburban Community," *Economic Geography,* 17 (1941): 61-74.
49. THOMAS, L. F. "The Sequence of Areal Occupance in a Section of St. Louis, Missouri," *Annals of the Association of American Geographers,* 21 (1931): 75-90.
50. HITCHCOCK, C. B. "Westchester-Fairfield: Proposed Site of the Permanent Seat of the United Nations," *Geographical Review,* 36 (1946): 351-397.

URBAN GEOGRAPHY[A]

EUROPEAN ROOTS OF AMERICAN URBAN GEOGRAPHY

EARLY GROWTH OF AMERICAN URBAN GEOGRAPHY

MAJOR CONCEPTS IN THE FIELD OF URBAN GEOGRAPHY
URBAN FUNCTIONS
SUPPORTING AND TRIBUTARY AREAS
SITE AND SITUATION
DISTRIBUTION OF CITIES
THE ECONOMIC BASE OF CITIES
INTERNAL FORMS AND PATTERNS
URBAN EXPANSION AND THE RURAL-URBAN FRINGE
LAND USES
SIZE OF CITIES

THE PROSPECT
COMPARATIVE STUDIES OF CITIES
HISTORICAL EVOLUTION OF CITY FORMS AND FUNCTIONS
SUBURBAN SETTLEMENTS
URBAN RIBBONS
OPTIMUM CITY SIZE, AND NEW TOWNS
INTERNAL RELATIONS WITHIN CITIES
THE THIRD DIMENSION IN CITIES
FUNCTIONAL AND LOCATIONAL INTER-RELATIONS
APPLICATIONS

[A]. Original draft by Harold M. Mayer with the collaboration of Edward L. Ullman, and the cooperation of Robert E. Dickinson, Chauncy D. Harris, Clyde F. Kohn, Raymond E. Murphy, and Victor Roterus.

CHAPTER SIX

URBAN GEOGRAPHY

GEOGRAPHERS ARE concerned with the study of cities because urban centers constitute distinctive areas. They are the foci of the general patterns of settlement; they are populated to a density rarely encountered in rural areas; they are the portals through which the spatial interchange of goods and ideas connects region with region; they dominate the patterns of economic life. Cities are economically, socially, and politically important out of all proportion to the areas they occupy.

Urban geographers approach the study of cities in different ways. They may be chiefly concerned with the city as a part of the fabric of settlement as defined in the preceding chapter. In this case they are interested in the forms of buildings, the arrangement of streets and railroads, or the relations between these things that men build and the functions they are or were intended to serve. They may examine the forms and patterns of settlement as of today, trace back the evolution of the phenomena of settlement to their origins, forecast the changes to come. Or they may approach the city as an economic phenomenon with associated social and political attributes, by seeking to identify the function or functions underlying city growth or decline, or the role of the city in the economy of the larger area it serves. Actually most urban geographers combine these approaches. Furthermore, they may study cities for the purpose of formulating basic concepts of city growth, city location, or city character; or they may study cities in order to contribute to the solution of practical problems of urban planning.

Many urban studies of each of these kinds are made not only by professional geographers, but also by scholars in other disciplines, notably by sociologists, land economists, and city planners. Some of the basic concepts of urban geography have been developed partly within the framework of other disciplines; yet the particular orientation of the geographer toward the complex of associations within an area has enabled him to develop a distinctive viewpoint within the expanding field of urbanism.

EUROPEAN ROOTS OF AMERICAN URBAN GEOGRAPHY

The analysis and interpretation of cities by both European and American geographers are largely a product of the 20th century. Several noteworthy studies of cities by European geographers greatly influenced the beginnings of the field in the United States. At the turn of the century, works on the location and distribution of urban settlements were published in Germany by Ratzel, Richthofen, and Hettner, and in Britain by Chisholm. Methods of studying the internal pattern, or layout, of cities, and the treatment of cities as part of the landscape were developed in 1899 by Schlüter. A few years later, Hettner posed the problem of the economic support of cities, particularly of ports. The first general urban geography was published by Hassert in 1907, and subsequent papers did much to define the scope and methodology of urban geography in Germany. Many studies of individual cities have since appeared on the continent, among them outstanding papers on Stockholm by Sten de Geer, on Grenoble by Raoul Blanchard, and on Innsbruck, by Hans Bobek. Particularly noteworthy are the contributions of Patrick Geddes [1] whose formulations of principles of urban growth and structure were popularized in Britain and brought to the attention of students of urbanism in the United States by Lewis Mumford [2].

EARLY GROWTH OF AMERICAN URBAN GEOGRAPHY

Urban geography in the United States began about the middle of the 19th century. In 1855, a monograph on cities by H. P. Tappan was published by the American Geographical and Statistical Society [3]. It was not until after 1900, however, that urban studies began to appear in considerable number.

In most of the early writings on urban geography in the United States, the physical sites of individual cities received particular attention [4; 5]. Analyses were made of the relations between the underlying rocks, the soils, and the drainage conditions on the one hand, and building construction, underground installations, canals, cemeteries, and other urban facilities on the other. The emphasis in urban geography, as in other branches of the discipline, was on the relations between man and the land.

Mark Jefferson was a pioneer in the field of urban studies, devoting his attention chiefly to the distribution of cities and of urban population throughout the world [6]. His "law of the primate city" anticipates some of the later work of the so-called social physicists. This generalization states that, with few exceptions, each large nation has one primate city, much larger than any other, which best typifies the character and culture of its region or nation, and which, once having attained primacy, tends to maintain that position by attracting the greatest enterprises and talents from the whole of the supporting area.

More recently the kinetic relations of cities and tributary areas have been examined, in many cases by persons not identified professionally as geographers. K. K. Liepmann defined the commuting patterns around certain English cities [7]; and G. W. Breese completed a similar study of Chicago [8]. Robert E. Dickinson has done much to advance the concept of the relations of the city to its tributary area [9].

Several geographers and others have attempted to formulate empirically derived principles relative to the general distribution of cities, particularly in relation to their hinterlands or service areas. Walter Christaller's notable work on central places in Germany, discussed in the preceding chapter [10], stimulated a further examination of this problem in the United States by Edward L. Ullman and others [11]. G. K. Zipf [12] and J. Q. Stewart [13] have worked out mathematical rules pertaining to the distribution and support of population which have attracted wide attention in view of their possible application in the prediction of urban growth. Their conclusions, however, have aroused considerable controversy [14]. Victor Roterus has initiated work on empirical principles with respect to changes in intra-urban phenomena attendant upon population growth or lack of growth [15].

Studies of American cities generally place less emphasis on the contrasts in street layout and in architectural types than do studies of European cities. This may be due to the fact that most cities in the United States lack long histories involving changes through a succession of strongly differentiated cultures. American urban studies have to deal with another kind of complexity, however. That is the rapidity of the changes taking place because of the development of the technology of transport.

MAJOR CONCEPTS IN THE FIELD OF URBAN GEOGRAPHY

Many criteria have been devised for defining and delimiting urban settlements. Definition in terms of political organization and jurisdictional boundaries is useful for many purposes. However, except for some aspects of political geography, such definitions are generally unsatisfactory from the geographic point of view because the municipal boundaries and the functional boundaries of a city rarely coincide. The delimitation of cities in terms of population also presents disadvantages to the geographer. The United States Bureau of the Census in 1947 re-defined its standard metropolitan areas to include the central city together with whole counties (towns in New England) which are urban in function to a significant degree and which contain dense agglomerations of population [16]. In the 1950 Census urbanized areas were delimited so as to include the central city and contiguous and surrounding areas having a population density of over a fixed limit [17]. Whereas the standard metropolitan area for any given central city or cities has a fixed boundary and includes non-urban areas within counties that are partially urbanized, the urbanized area expands from one census to the next in accordance with the expansion of the area within which high population density and essentially urban conditions exist.

URBAN FUNCTIONS

For the geographer, a most significant criterion for the definition of an urban as opposed to a non-urban agglomeration is the function that the agglomeration performs for the areas outside of its own limits.

Several classifications of cities in terms of functions have been developed within the past decade by American geographers. Chauncy D. Harris and Edward L. Ullman [18], for example, have classified cities as: 1) central places performing comprehensive services for areas outside of the central city; 2) transport cities, which perform break-of-bulk and allied services along transport routes, and are supported by areas that may be remote in distance but close in connection because of the city's strategic location on transport routes; and 3) specialized-function cities, each performing one service such as mining, manufacturing, recreation, education, or administration. Obviously these categories are not mutually exclusive. Manufacturing cities, for example, may owe their importance to a strategic location on transport routes which facilitate the assembly of raw materials and labor and the marketing of products. All functions, moreover, are likely to be present in some degree in every city.

Using the statistics of the United States Census, which gives the occupations of all gainfully employed persons, Harris[19] developed a classification of American cities based upon the relative proportions of employed workers in metropolitan districts engaged in the respective activities. His classes of cities are: 1) manufacturing (two sub-classes depending upon degree to which manufacturing predominates in the economic base); 2) retail; 3) diversified; 4) wholesale; 5) transportation; 6) mining; 7) university; and 8) resort and retirement cities. This classification, being based upon the entire metropolitan area, eliminates discrepancies due to the reporting of occupations at the place of residence of the workers rather than at their place of work, since there is relatively little commuting across metropolitan area boundaries. The classification has the further advantage of permitting the use of a single source of statistics with uniform definitions of occupation; but it has the disadvantage of not using uniform limits in terms of number of persons engaged in a given occupation. Although Harris's application of his classification was to metropolitan areas, it can be used for any kind of area.

SUPPORTING AND TRIBUTARY AREAS

Fundamental to the understanding of city location, or of the spacing of cities, is the definition and delimitation of supporting and tributary areas.

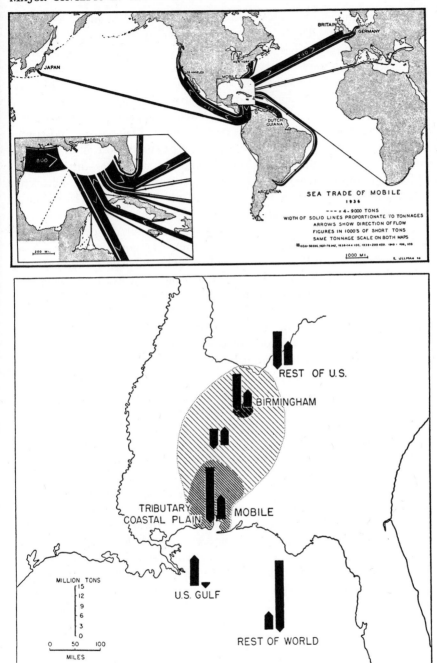

Fig. 8. The Trade Area and the Sea Trade of Mobile, Alabama, by Ullman, [Chap. 13, Ref. 20].

The supporting and tributary area of a city is a kind of multiple-feature nodal region of which the city is the focus (chapter two). Actually the multiple-feature region is made up of a great variety of single-feature regions, for each kind of service and each kind of commercial exchange occupies its own area. When the criteria are applied and the results are plotted on a map each separate boundary may correspond more or less to the others, but they seldom coincide. Where a number of boundaries fall within a narrow zone the boundary of the general supporting and tributary area is drawn. Close to the city, it may be that almost all of the connections are with the one urban center; but at greater distances the increasing competition with other urban centers results in an increasing amount of overlap. Where the pull of one city is about equal to that of a competing city the boundary of the nodal region is drawn.

A great variety of criteria have been suggested for the delimitation of the single-feature components of the general supporting and tributary area. Wholesale connections are usually broken down into such separate items as groceries, dairy products, meat, drugs, hardware, gasoline and oil, and others that may be locally important. A map can be made of the places from which the customers come to the city retail stores, to the banks, or to the offices of real estate companies, physicians, dentists, or lawyers. A map might be made of the places of residence of the people who enjoy the city's recreational facilities. Newspaper circulation and telephone connections provide other components. The area from which a city derives its water supply, or its fuel or electric power may be considered. If the city also has manufacturing industries, each industry has its source of raw materials or semi-manufactured goods, and each industry, also, has its markets.

To delimit such a variety of component single-feature regions for the purpose of defining the general supporting and tributary area of a city requires the employment of a number of different techniques. Some data are available in quantitative terms in published statistical sources. Other data must be gathered from interviews with informed representatives of each business or profession. The boundary of a wholesale area can often be identified most easily by interviews with retail merchants near the borders of the trade area. The plotting of information regarding the flow of traffic on the highways, or the movement of goods in and out by rail, ship, or airplane reveals the volume and direction of external connections (chapter thirteen).

SITE AND SITUATION

In urban studies the areal relations of a city are commonly examined from two different points of view, each representing a different degree of generalization. On maps of large scale the relation between the internal pattern of the city (the arrangement of its functional areas, the layout of its streets and railroads) and the features of the terrain are analyzed: on maps of smaller scale covering a much larger area the external relations of the city as a whole are analyzed. The close-up view reveals the characteristics of the site; the broader view reveals the characteristics of the situation. The significance of both site and situation must be examined historically for they may change radically, even in a short time, as a result of changes in the technology of transportation, changes in the general economic development of the country, or changes in the functions performed in the city.

DISTRIBUTION OF CITIES

From the analysis of the location of individual cities and of city patterns in larger areas have emerged some generalizations, empirical formulae, and deductive theories on the distribution and size of cities in general.

Notable among these is the work of Christaller (discussed also in chapter five, pp. 134-135) [10]. His theory regarding the hexagonal arrangement of central-place cities was based on work in south Germany, and has been evaluated by application elsewhere (see the references to John E. Brush, chapter five, refs. 49, 50). Christaller saw that the simplest geometric figure that would resolve the overlap of three or more tangent circles was the hexagon. He assumed, therefore, that the hexagon was the ideal shape toward which actual supporting and tributary areas of specific central places would tend. He also classified central-place agglomerations by size in categories in which the larger agglomerations perform a greater variety of services than the smaller ones and are supported by larger hexagonal areas (see Fig. 6, p. 134). On this basis, the distance between central places of any one category would be greater by the square root of three than the distance between places of the next smaller category. Christaller found that in South Germany the distance between the smallest central places (hamlets) averages about seven kilometers, or not more than one hour's walking distance from any point to the nearest center. A corollary of the hexagonal trade-area concept is the principle that a town is most efficiently connected with its tributary area by six radiating highways. Larger urban centers may be approached by highways numbering a multiple of six [20].

Several other theories relating to the spacing and size of settlements have been formulated. The so-called rank-size rule states that the population of the largest city in a big country, such as the United States, is essentially equal to the population of any other city in the country multiplied by its rank. Thus the second ranking city would have half the population of the first one. It is recognized that exceptional situations exist in some countries, such as Canada, Australia, or Brazil.

Mathematical formulae for measuring the effect of size and distance in the gravitational pull of a city have been presented by a number of persons, notably Zipf [12], and Stewart [13]. The formula $\dfrac{P_1 P_2}{d}$ states that the relations between one city (P_1) and another city (P_2) are directly proportional to their size and inversely proportional to the distance which separates them. The validity of the formula has been demonstrated by the measurement of the interchange of such items as railway express, or long distance telephone calls (chapter thirteen). Yet it is over-simplified in at least one way: it will fail to measure the interchanges between two cities when a larger city lies between them. Geographers are more interested in the deviations from a general formula as they develop under the circumstances of particular places.

Nevertheless, there is need for further research to test and perhaps revise such formulae. If geometric and mathematical regularities actually exist in the patterns of settlement, and if the intensity of interchange between one place and another can be predicted, such a theoretical norm may be used to measure the degree of departure in particular places, and so as to define the importance of the other things that are not regular. Perhaps more satisfactory results could be obtained by seeking formulae to apply to specific culture areas, not to the world as a whole. And what do these geometric and mathematical regularities mean? Do these abstractions replace the concept of environmental determinism which was popular a generation ago? If so, how are the cause and effect relations to be interpreted?

THE ECONOMIC BASE OF CITIES

The functions which are performed in cities, and which lead people of certain culture areas to group themselves close together, include the production, processing, exchange, and distribution of goods, and the provision of a variety of services for the people of a wide area outside of each city. The measurement of a city's economic base is an important phase of urban geography, although, until recently, most of the attempts to measure it were made by economists who had become interested in geographic problems. A few studies by geographers, including those of St. Louis by Lewis F. Thomas [21] and of Minneapolis-St. Paul by Richard Hartshorne [22], although primarily concerned with locations and patterns of functional areas within cities, foreshadowed the later work of geographers on the economic base of cities.

Homer Hoyt, a land economist, developed a two-fold classification of urban functions which he distinguished as basic and non-basic [23]. The basic functions serve the areas outside of the respective cities or metropolitan areas; the non-basic functions serve the inhabitants of the respective cities or metropolitan areas. The extent to which a given activity is basic may be measured by finding the ratio of persons engaged in that activity to the total population of the city, and then comparing this with a similar ratio for the country as a whole. Thus, for example, if a given activity employs 15 percent of the population of a metropolitan area, as compared to 5 percent of the population of the nation engaged in the same activity, the activity is basic to the economy of the city to the extent of two out of every three persons who are engaged in it. The third person engaged in this activity is presumably supplying the needs of the inhabitants of the city itself.

The same technique can be used to determine the extent to which various activities are under-represented in proportion to the population and labor force and in relation to those of other comparable cities. If available resources be taken into consideration, this information can be used as a guide to the promotion of economic expansion in an urban area.

Economic base studies, using these and other techniques, have been made for a number of the cities of the United States. Such studies were made of New York, Cincinnati [24], and Philadelphia [25], as well as of several smaller cities including Orlando in Florida, Brockton in Massachusetts [26], Oshkosh in Wisconsin, and Des Moines in Iowa [27]. Roterus used this approach in the study of seasonal variations in employment [28].

INTERNAL FORMS AND PATTERNS

As the increasingly complex sources of support for cities have been reflected in the expansion of urban areas, the internal forms and patterns have also become increasingly complex. Until the Industrial Revolution, cities were small and compact. Within their boundaries, structures and land uses for the most part were functionally undifferentiated. Businesses and residences were usually in the same structure, and except for the market places or public squares and a few focal points represented by cathedrals, churches, and administrative buildings, there was little to distinguish one section of a city from another, although in one sense these few focal points might be considered analogous to present-day central business districts. With industrialization and increased specialization, it no longer was possible to carry on the urban functions without separation of places of work from places of residence; and as cities expanded the separation tended to become more marked.

Several generalized concepts of city structure have been developed in recent decades. The three principal generalizations are: 1) the concentric circle or zonal theory; 2) the wedge or sector theory; and 3) the multiple nuclei theory.

The concentric circle or zonal theory of urban growth and structure, developed largely by University of Chicago sociologists, notably E. W. Burgess [29], was an adaptation of the earlier generalized and theoretical descriptions of Von Thunen, who, a century ago, described what he recognized as roughly concentric land-use zones around certain German cities.

According to the Burgess concentric zonal theory, the central business district is the core of the city, around which the respective zones may be found in the following order: 1) a zone of warehousing and light manufacturing; 2) a zone in transition, within which are the blighted areas where commercial and industrial uses encroach upon the residential districts; 3) a zone of workingmens' middle-grade homes; 4) a zone of better residences; and 5) a commuters' zone, beyond which lies the non-urban zone. The development of each zone is characterized by a succession of land uses expanding from each successive inner zone as the city grows. Thus the high-grade residential areas initially developed on the urban periphery gradually become middle-grade, and eventually deteriorate to become blighted areas, perhaps later to be replaced by the expansion of the inner warehousing and light manufacturing zone, and in some cases even by the expanding central business district.

The concentric zone theory of urban structure and growth has been criticised because it was developed largely in one city, Chicago, with inadequate comparative studies of other cities, because its proponents were primarily concerned with the recognition of patterns of pathological conditions, and because it did not sufficiently consider the role of the site, including slope, drainage, and soil conditions. Furthermore, it neglected the relative pulls of the hinterland in various directions, and of lines of transport, which disrupt the idealized patterns. Studies of the detailed land uses and functional areas of various cities have raised questions relative to the validity of the concentric zonal theory [30].

Studies of the internal pattern of cities were greatly facilitated in the 1930's by the information made available through numerous real property inventories, conducted in many cities as work-relief projects. Such studies became increasingly quantitative, and, after the censuses of 1940 and 1950 made data available for population and housing in small areas, such as census tracts and blocks, detailed investigations of a large number of cities became possible.

Among the concepts developed by the use of detailed census and real property inventory data for many cities is the wedge or sector theory. This theory, anticipated by R. M. Hurd [31] and developed by Hoyt with the use of data from about 200 cities [32], describes a city as a series of sectors or wedges extending outward from its core, each wedge being characterized by one type of land use, or by residential development with one outstanding characteristic, such as high rent. According to the theory, growth tends to proceed from any point of origin along transportation routes toward high ground which is free from floods or atmospheric pollution and has scenic interest, and along waterfronts not used for commerce or industry, and tends to avoid "dead-end" directions in which growth is limited by natural or artificial barriers. The highest density of development is along the wedges that are served by the best transportation routes, while the highest-grade residential areas usually occupy those wedges favored by water frontage or other unusual features that place a premium value upon the land. Once started in a particular direction, urban growth tends to retain similar characteristics as it proceeds outward along the wedge. The resultant urban form is often that of a star or diamond, depending upon the number of major radiating routes. Protuberances and higher density development take place along the main transportation lines.

The multiple nuclei theory describes a city as a series of nuclei, each characterized by an activity or group of related activities requiring specialized facilities best provided by the sites and locations that it occupies [18]. Land economists regard the development of urban patterns as the result of competition for sites. The uses best adapted to particular sites can outbid other uses [33]. Hence a stratification of land uses results, with those uses best able to afford a particular site being closest to it, and with a hierarchy of uses in descending order of rent-paying ability as distance increases from the nucleus. Thus, the business, retail, administrative, and entertainment functions, which depend upon convergence of transportation and thus accessibility to a maximum number of people, tend to cluster about the points of maximum accessibility, whereas industrial areas tend to develop along navigable waterways and railway lines. Residential areas develop between the industrial strips, which in turn are localized by railways. The highest densities of population develop along passenger transit routes.

All three of these descriptive theories of urban development depict the growth of cities as a gradual expansion of functional areas from the center toward the periphery. This expansion, and hence the size and form of a city at any given time, may be regarded as the resultant of two opposing forces: a centrifugal force derived from the attractions of the periphery and the repulsion from the central area; and a centripetal force derived from the convenience of proximity to the central area [34]. Although the centrifugal forces are increasing in strength in most American cities, owing to the increase of congestion in the urban core, and to the flexibility of transportation by automobile, the centripetal forces are generally stronger. The tendency of large cities to retain, and even increase, their primacy is evidence of the strength of the centralizing forces.

URBAN EXPANSION AND THE RURAL-URBAN FRINGE

The rapid expansion of cities in the present period, and the invasion of the rural areas by urban phenomena was brought about by radical changes in the technology of transportation. As long as the power for the movement of people and goods overland was furnished by horses, city sizes and the areas they served outside the city limits were relatively small. Urban areas were compact. But with the advent of suburban service on steam railroads in the latter half of the 19th century, suburban clusters began to appear around the railroad stations. The larger cities took a stellate or lobate shape, with protuberances along the axes of the radiating railroad lines, but with nuclei of denser development centering on the railroad stations, like beads on a string. With the growth of suburban and inter-urban electric railways, between 1890 and 1920, the areas between the earlier suburban nuclei filled in, for the electric car could stop at nearly any point. Railroad stations no longer monopolized accessibility. At the same time, electric street-car systems made it possible for people who worked in the central districts of a city to live farther away from the center. Multi-family apartment buildings became common along the main street-car routes in many cities. Still this expansion, both inside the city and outside of it, was tied to the rail lines, and the degree of accessibility dropped sharply away from such lines.

The automobile, bus, motor truck, and modern paved highways enabled cities to throw off the shackles of the steel rails. The period following World War I was marked by an accelerated rate of urban expansion, particularly on the fringes of cities and in the previously undeveloped areas between the radiating prongs of earlier growth. Both dormitory suburbs and industrial satellite communities within metropolitan areas shared in the expansion. As central cities grew, they absorbed many of the older nuclei in the outlying areas, which now remain as clusters of higher density, surrounded by areas of lower density.

This process of development and expansion has been examined in a large number of studies, some by geographers, some by workers in other disciplines [35]. The process was not orderly, for much of it was already well advanced before the inhabitants of a city or the planning officials were aware of what was happening. On the fringes of many cities much so-called dead land resulted from premature subdivision [36]. These sub-divided areas, often with improved streets and with public utilities in-stalled, could not be absorbed in productive urban uses, and at the same time their potentialities for rural-type uses were destroyed. In most in-stances, such areas are chronically tax-delinquent, and commonly the original owners have abandoned them, creating a pattern of obsolete platted small lots, with obscure titles, making reassembly and replatting extremely difficult. Building either did not begin or ceased before the areas were fully developed. Generally they became urban-fringe slums, particularly in smaller cities. Subsequent new development by-passed them for other land farther out from the urban centers, where land as-sembly, tax delinquency, and obsolete platting were not serious problems. This resulted in a series of "holes" interrupting the continuity of urban development and created serious financial problems for municipalities in the form of carrying charges on costs of the original improvements, maintenance costs of excessive mileages of streets and utility lines through areas producing little or no revenue, costs of fire and police protection in extremely low-density areas, and loss of tax revenue, particularly where dead land forced new development to spill over into suburbs or unincorp-orated areas beyond the municipal boundaries. Recent legislation at state and federal levels may present new opportunities for reclaiming dead land, replatting it, and putting it back into productive development.

Much attention has been given to the problem of excessive urban ex-pansion, particularly in relation to the problems of municipal planning and finance arising out of the lower densities and the extensive urban "sprawl" facilitated by the motor vehicle and modern highways. The discordance between the existence of a multiplicity of small municipalities and special-function administrative districts on the one hand, and the necessity for considering and planning the future development of entire metropolitan areas on the other, has given rise to a number of noteworthy metropolitan regional-planning studies [37]. In a few instances, the problems have been attacked with more or less success by official and unofficial metropolitan planning agencies. An important field for future investigation, although not necessarily in urban geography, is that of the legal and political mech-anisms by which metropolitan planning may be implemented. A few geo-graphers are directing their attention to governmental problems of fringe areas, for example, to the problems of annexation [38]. There is also op-portunity in problems of fringe areas and of metropolitan planning for joint studies by teams of workers in different disciplines [39].

LAND USES

Concepts and research procedures have also been developed in the study of urban land uses. The greater part of the area of most cities is used for residences, but each of the functions that a city performs occupies area and there are certain characteristic positions within the city framework commonly occupied by each functional area.

Residential land use has been investigated in considerable detail by a number of American geographers [40]. The growth of large-scale public housing projects since the 1930's, the rapid expansion of zoning, the real property inventories, the emergency war and postwar housing programs, and the recent activity in large-scale public housing and urban redevelopment under the National Housing Act of 1949 have spurred research into the nature and development of residential areas. The contributions of geographers have been especially important in the delimitation of blighted areas by field surveys [41]. These studies have, in far greater detail than the earlier ones of Burgess, described the blighted areas characteristically located on the fringes of the central business districts and adjacent to industrial areas. The resulting definitions and delimitations have taken on great practical significance, for federal and state legislation limits the use of public funds for redevelopment and housing projects to areas which are defined as blighted.

During the 1940's urban redevelopment caused many changes in city patterns. By urban redevelopment is meant the assembly of land in blighted areas, the clearing of titles, and the sale or lease of the assembled tracts to a private or public redeveloper at a written-down cost competitive with vacant land. Rebuilding is carried out according to an approved plan. Because of the adverse effects of blight in adjacent areas, in order to be eligible for loans or grants from government agencies, the redeveloped area must be large enough to form a complete neighborhood equipped with neighborhood services.

Neighborhoods, as visualized by the planners, are distinct units. They usually occupy about a quarter square mile, but their size may vary with the density of population. They should be bounded by arteries carrying traffic around rather than through them. The streets are designed to discourage through traffic. Each neighborhood should have an elementary school surrounded by appropriate open space, and located within walking distance from all parts of the area. It should be equipped with local shopping centers and other community services [42]. Several neighborhoods can be grouped into large communities which are surrounded by natural or artificial breaks in the continuity of residential development and that contain high schools, major outlying shopping centers, and other facilities [43].

Industrial districts, not the residential areas, form the economic framework of most cities. Such districts occupy several characteristic positions: they may be developed as strips along waterways and railroad lines; they may appear as strips or scattered individual plants along the circumferential belt lines that intersect the radiating trunk lines near or beyond the urban periphery; they may form clusters of industries in satellite towns beyond the continuously built-up urban area; or they may be scattered and intermixed in areas predominantly devoted to non-industrial uses.

The railroad pattern of a city is closely related to the pattern of industrial land use [44]. Typically the spider-web pattern of radial trunk lines and intersecting belt lines, and the bands of land occupied by factories for which the railroads constitute the axes, produce a series of cells within which residential communities develop. In hilly terrain the rail routes and associated industries are commonly crowded in the valleys.

Planned industrial districts are bécoming increasingly common [45]. It may prove possible to make use of blighted areas occupied either by obsolete factories or by deteriorated residential slums for the development of planned industrial districts. Geographers have participated as staff members of city planning agencies in the formulation of plans following comprehensive studies of the problems of industrial location [46].

Geographers have also studied business districts, both in central locations [47] and in outlying areas [48]. These applications of urban geography are important not only for planning purposes, but also in marketing research for the purpose of selecting store locations (chapter nine).

The decentralization of industry and business and the movement toward and beyond the urban periphery is an important trend in nearly all cities. This trend indicates the need for analysis of the centrifugal forces which cause the outward movement [49]. A fundamental conflict exists between the urge to seek more space and thus to spread the urban area more widely and to decrease the density of the center, and the urge to move into the central area because of the convenience of this kind of location.

SIZE OF CITIES

Much thought has been given to the optimum size of cities [50]. Mathematical generalizations have been developed to describe the relationships between city size and population [13], a concept that planners have long studied in terms of optimum population density for good living conditions. Some persons decry the increasing size of cities and advocate the "new towns" movement for planned urban decentralization. Some favor highly concentrated residential areas, with tall buildings covering only a small percentage of the land area, thereby giving light, air, and open space, but still preserving the advantages of compactness. The atomic bomb and other defense considerations may affect future planning relative to urban centralization versus decentralization [51].

The time-distance-cost relations of commuting between place of residence and place of work, as well as the effects of traffic congestion because of high urban densities are of major importance. The construction of highways, expressways, public mass transit systems, and parking and terminal facilities are expensive; nevertheless, if they are built in proper relation to each other and to the general pattern of land uses urban living can be made vastly more comfortable.

THE PROSPECT

Urban geography is a field to which a considerable number of professional geographers are now devoting themselves. The distinctive problems and procedures of urban study set this field off as a topical specialty, and one which is closely allied with urban sociology, land economics, and city planning. The field has both theoretical and applied aspects. Theory and application, however, should not be considered as distinct approaches, for theory separated from application in a field of this sort may be difficult to evaluate, and practical studies without the guidance of sound concepts can lead to thoroughly unreliable conclusions. In considering the prospect for urban geography nine aspects of city study in which further research is needed are here suggested as examples indicating the possible trends of the years ahead.

COMPARATIVE STUDIES OF CITIES

Urban geographers have been more concerned with individual cities than with the comparative treatment of two or more cities. Since different categories of phenomena have been applied at different degrees of generalization, depending on the individuals making the studies and on the conditions in the urban areas studied, comparisons have been difficult. Yet the development of the broader concepts of city patterns and functions demands comparative analysis. The need is for the formulation of basic categories and procedures to be applied to many cities. Two comparative studies suggest the possibilities of this approach: one by Raymond E. Murphy discussing York and Johnstown [52]; and one by Harland Bartholomew, a planner, applying a statistical analysis to the land-use areas and patterns of sixteen cities [53].

Urban geography, moreover, should not be restricted to occidental cities. There are many valuable contrasts in the forms, functions, and patterns of cities in different culture areas to examine. A number of American geographers have studied cities in Europe, Africa, the Far East, and in Latin America: among them should be mentioned George B. Cressey, Robert E. Dickinson, Norton S. Ginsburg, Robert B. Hall, Preston E. James, John B. Leighly, Dan Stanislawski, Glenn T. Trewartha, and Derwent Whittlesey [54]. The significance of cultural tradition in developing cities of similar character in diverse environments can only be weighed by such comparative analyses.

HISTORICAL EVOLUTION OF CITY FORMS AND FUNCTIONS

Once the streets of a city are laid out and lined with buildings, to make any changes is a major operation. In most American cities, a pressing problem is to find ways of using streets for automobile traffic that were originally designed for horses and wagons. Although the needs of a city and the functions it performs may change considerably, the frame of buildings and streets in which city living must go on remains unchanged. To understand the present conditions requires the historical perspective. The forms and patterns of cities, like those of rural areas, can be examined from the point of view of sequent occupance. Examples of historical urban geography are James' study of Vicksburg, Mississippi [55], and Stanislawski's paper on the origin and spread of grid-pattern towns [56].

SUBURBAN SETTLEMENTS

The expansion of residences, commercial establishments, and industrial plants from the crowded central parts of cities out toward the urban fringe is a notable process currently changing the character of cities. Outside the city there is the space needed for the efficient layout of factory buildings, or for the enjoyment of living in residential areas. And the urban spread is supported by the new freedom of movement given by the automobile and the truck. Although there are a few good examples of studies by geographers of the conditions in these "rurban" areas, much more needs to be done. The over-all view of urban development, illuminated by the vision of what a city might be as a place in which to live, work, and play, is needed to guide the planning and control of this process of expansion to the suburbs. In all this the attitudes of the city dwellers themselves cannot be overlooked. The geographer should be alert to apply to his own purposes the findings of sociologists and social psychologists regarding the intangible forces which underlie current urban change.

URBAN RIBBONS

This same process of expansion goes beyond the suburban settlements. Long ribbons of what is essentially urban development, both as regards the forms of buildings and the functions performed in them, extend far out into the rural areas along the main highways. This is a new phenomenon, and one which reflects another of the fundamental changes in the way of living of the people of the United States, who, more than any other people in the world, have adopted the automobile as an essential part of their culture. Urban ribbons are neither completely urban nor completely rural: they are completely highway oriented and have no effect on settlement patterns or land uses more than a few hundred feet away from the road. They are made up of roadside eating places, drive-in movies, and non-farm residences of workers who find their only employment in cities.

Urban ribbons are difficult to analyze by the usual geographic techniques. Because they are so narrow, and yet are extended so far, often all the way from one city to the next, it is difficult to find a scale of map suitable for plotting them. There is need for experimentation with new techniques of field study in order to bring these urban ribbons under the lens of geographic analysis.

Optimum City Size, and New Towns

City planners, architects, and sociologists have long been intrigued by the thought of building entirely new towns, properly planned for the efficient performance of urban functions and for the satisfaction of the needs of urban dwellers. The Garden Cities of Ebenezer Howard, the Broadacre City of Frank Lloyd Wright, and the skyscraper cities of Le Corbusier are examples of this movement. Yet the question remains unanswered regarding the extent to which the planner or architect can create urban growth in whatever spot he selects, or to what extent cities reflect much broader patterns of settlement, patterns of productivity, and patterns of population. The geographer could contribute more to the analysis of these problems than he has yet done.

Internal Relations Within Cities

The internal movement of people and goods within cities offers some important problems of observation and interpretation [57]. Movements of these kinds are probably the best index of functional interconnection that could be found. The measurement of transportation facilities and of traffic flows does not present serious difficulties, and when these measurements are plotted on maps they bring to light important relationships not clearly apparent from the examination of maps of purely static phenomena, such as the outlines of the functional areas or the patterns of streets.

Since World War II, origin and destination studies have been conducted by traffic engineers in a number of American cities. Some of these are so detailed that one can ascertain the residence of every person in each of the large down-town office buildings. The information gathered for each person is tabulated by zones and sub-zones within the city, and includes the person's starting point and destination, the type of conveyance he used, the time required for the trip, the purpose of the trip, and other facts [58]. Such data could be used to good advantage by geographers.

An analysis of the flow of telephone messages would also establish the internal and external interconnections of a city. The Bureau of the Census, for example, already makes limited use of statistics relating to such messages in determining the outer edge of the metropolitan area when other measures give an inconclusive answer. In such cases, where there are on the average less than four calls a month per phone from the outlying center to the metropolitan core, the outlying center is not considered to be a part of the metropolitan area. Undoubtedly other functional gradations could be determined within and without the city.

Considerable work has also been done by real estate and marketing specialists on pedestrian traffic as a basis for selecting store locations, but very little in utilizing such pedestrian traffic data in the preparation of flow maps or the geographic analysis of flows.

THE THIRD DIMENSION IN CITIES

Cities, and particularly American cities, cannot adequately be represented solely as two-dimensional phenomena. Many urban land uses are multi-functional, as, for example, where loft buildings house a variety of business establishments, or apartment buildings also contain business establishments on the ground floor, or various other combinations are found on the same piece of land and often in the same building. In commercial and high-density residential districts the multi-storied building is typical. Much work needs to be done in the development and application of field and mapping techniques for the study of the third dimension, involving building height, and differences in functions carried on in various parts of buildings. This is particularly important in the study of central business districts, where the skyscraper is a characteristic urban form. A few studies have been made of the vertical stratification of functions and of building forms to house the functions: for example, the paper by A. E. Parkins on the central business district of Nashville, Tennessee [47], and the paper by Eugene Van Cleef dealing with several cities in Ohio [59].

FUNCTIONAL AND LOCATIONAL INTER-RELATIONS

A neglected aspect of the structure of cities which presents interesting problems is that of the mutual connections between functional areas. Certain kinds of urban activities and certain kinds of land use owe their locations to the presence of other land uses or activities nearby. For example, printing establishments are commonly placed on the fringe of the central business district where the city's chief demand for printing is concentrated; or small shopping centers are commonly placed just outside a factory gate, or near a university, where customers come together in large numbers. In this same group of problems are those of residential location with reference to work locations. A city is made up of a great number of such locational relationships, the study of which would be greatly advanced by the use of geographic methods.

APPLICATIONS

These various aspects of urban geography can be examined as challenging problems for their own sake; or they can be attacked by geographers who are employed by businesses or planning agencies. Urban geography has many practical applications in the field of marketing research (chapter nine), and also in the field of military geography (chapter twenty-three).

Geographers can make a unique contribution to urban planning [60]. The kinds of studies that geographers make, their distinctive point of view regarding the areal associations of things, and their techniques of cartographic analysis and representation are all of fundamental importance to any planning projects. Many planners recognize that training in geography is one of the prerequisites to holding a responsible planning position. Among the specific geographic contributions to planning in recent years are: 1) applications of the concept of basic as distinguished from non-basic employment; 2) development of the techniques of classification and description of urban land uses; 3) delimitation of geographic areas, notably the urbanized area, for the use of the census takers as enumeration districts; 4) development of techniques of applying air photography and photo interpretation to city study (chapter twenty-five); and 5) development of methods for estimating future land requirements and their probable position within and around a city [61].

REFERENCES

1. GEDDES, P. The Civic Survey of Edinburgh, Edinburgh, 1911; idem. Cities in Evolution, London, 1915.
2. MUMFORD, L. The Culture of Cities, New York, 1938; idem. City Development: Studies in Disintegration and Renewal, New York, 1945.
3. TAPPAN, H. P. The Growth of Cities: A Discourse, New York, 1855.
4. EMERSON, F. V. "A Geographic Interpretation of New York City," Bulletin of the American Geographical Society, 40 (1908): 587-612, 726-738; 41 (1909): 3-20.
5. SALISBURY, R. D., and ALDEN, W. C. The Geography of Chicago and its Environs, Chicago, 1920.
6. JEFFERSON, M. "The Anthropogeography of Some Great Cities," Bulletin of the American Geographical Society, 41 (1909): 537-566; idem. "Great Cities of the United States, 1920," Geographical Review, 11 (1921): 437-441; idem. "Distribution of the World's City Folks," Geographical Review, 21 (1931): 446-465; idem. "The Great Cities of the United States, 1940," Geographical Review, 31 (1941): 478-487; idem. "The Law of the Primate City," Geographical Review, 29 (1939): 226-232.
7. LIEPMANN, K. K. Journey to Work: Its Significance for Industrial and Community Life, London, 1944.
8. BREESE, G. W. The Daytime Population of the Central Business District of Chicago with Particular Reference to the Factor of Transportation, Chicago, 1949.

9. DICKINSON, R. E. *City, Region and Regionalism*, London, 1947.
10. CHRISTALLER, W. *Die zentralen Örte in Süddeutschland*, Jena, 1935.
11. ULLMAN, E. L. "A Theory of Location for Cities," *American Journal of Sociology*, 46 (1940-1941): 853-864.
12. ZIPF, G. K. "The Hypothesis of the Minimum Equation as a Unifying Social Principle," *American Sociological Reveiw*, 22 (1947): 627-650.
13. STEWART, J. Q. "Empirical Mathematical Rules Concerning the Distribution and Equilibrium of Population," *Geographical Review*, 37 (1947): 461-485.
14. HOYT, H. "Is City Growth Controlled by Mathematics or Physical Laws?" *Land Economics*, 27 (1951): 259-262.
15. ROTERUS, V. "Effects of Population Growth and Non-Growth on the Well-Being of Cities," *American Sociological Review*, 11 (1946): 90-97.
16. KLOVE, R. C. "The Definition of Standard Metropolitan Areas," *Economic Geography*, 28 (1952): 95-104.
17. WRIGLEY, R. L. JR. "Urbanized Areas and the 1950 Decennial Census," *Journal of the American Institute of Planners*, 16 (1950): 66-70; United States Bureau of the Census. *Census Areas of 1950*, Series GEO, no. 1, Washington, 1951.
18. HARRIS, C. D., and ULLMAN, E. L. "The Nature of Cities," *Annals of the American Academy of Political and Social Science*, 242 (1945): 7-17.
19. HARRIS, C. D. "A Functional Classification of Cities in the United States," *Geographical Review*, 33 (1943): 86-99.
20. BOGUE, D. J. *The Structure of the Metropolitan Community: A Study of Dominance and Subdominance*, Ann Arbor (Michigan), 1949.
21. THOMAS, L. F. *Localization of Business Activities in Metropolitan St. Louis*, St. Louis, 1927.
22. HARTSHORNE, R. "The Twin City District: A Unique Form of Urban Landscape," *Geographical Review*, 22 (1932): 431-442.
23. HOYT, H. *The Economic Status of the New York Metropolitan Region in 1944*, New York, 1944.
24. Cincinnati City Planning Commission, *The Economy of the Cincinnati Metropolitan Area*, Cincinnati, 1946.
25. Philadelphia City Planning Commission. *Economic Base Study of the Philadelphia Area*, Philadelphia, 1949.
26. HOYT, H. *Economic and Housing Survey of the Orlando Metropolitan Region*, Orlando (Florida), 1946; idem. *A Report on the Economic Base of the Brockton, Massachusetts, Area*, Brockton, 1949.
27. ALEXANDER, J. W. *The Economic Life of Oshkosh*, Madison (Wisconsin), 1952; NELSON, H. J. *The Livelihood Structure of Des Moines, Iowa*, Chicago, 1949.
28. ROTERUS, V. "Stability of Annual Employment—Cincinnati and Selected Cities," *Economic Geography*, 23 (1947): 130-135.
29. BURGESS, E. W. "The Growth of the City," in PARK, R. E., BURGESS, E. W., and MCKENZIE, R. D. *The City*, Chicago, 1925: 47-62.
30. BLUMENFELD, H. "On the Concentric-Circle Theory of Urban Growth," *Land Economics*, 25 (1949): 209-212; DAVIE, M. R. "The Pattern of Urban Growth," in MURDOCK, G. P., ed., *Studies in the Science of Society*, New Haven (Connecticut), 1937: 133-161; HAUSER, F. L. "The Ecological Pattern of Four European Cities and Two Theories of Urban Expansion," *Journal of the American Institute of Planners*, 17 (1951): 111-129.
31. HURD, R. M. *Principles of City Land Values*, New York, 1924.
32. HOYT, H. *The Structure and Growth of Residential Neighborhoods in American Cities*, Federal Housing Administration, Washington, 1939.
33. RATCLIFF, R. U. *Urban Land Economics*, New York, 1949.
34. COLBY, C. C. "Centrifugal and Centripetal Forces in Urban Geography," *Annals of the Association of American Geographers*, 23 (1933): 1-20.

35. ANDREWS, R. B. "Elements in the Urban Fringe Pattern," *Journal of Land and Public Utility Economics*, 18 (1942): 169-183; BALK, H. H. "Rurbanization of Worcester's Environs," *Economic Geography*, 21 (1945): 104-116; DIEHL, L. F. "Major Aspects of Urbanization in the Philadelphia Metropolitan Area," *Journal of Land and Public Utility Economics*, 19 (1943): 316-328; KLOVE, R. C. *The Park Ridge—Barrington Area*, Chicago, 1942; MONCHOW, H. C. *Seventy Years of Real Estate Subdividing in the Region of Chicago*, Chicago, 1939; THROOP, V. M. *The Suburban Zone of Metropolitan Portland*, Chicago, 1948; WEHRWEIN, G. S. "The Rural-Urban Fringe," *Economic Geography*, 18 (1942): 217-228.

36. ASCHMAN, F. T. "Dead Land," *Land Economics*, 25 (1949): 240-245; CORNICK, P. H. *Premature Subdivision and Its Consequences*, New York, 1938.

37. *Regional Plan of New York and Its Environs*, 2 vols. New York, 1928-1931; New York Regional Plan Association, *Regional Survey of New York and Its Environs*, 9 vols., New York, 1927-1932; Regional Plan of New York and its Environs, *Regional Plan of the Philadelphia Tri-State District*, Philadelphia, 1932; Kansas City Planning Commission, *Metropolitan Area*, Kansas City (Missouri), 1947; Cincinnati City Planning Commission, *Metropolitan Master Plan*, Cincinnati, 1948; National Capital Park and Planning Commission, *Regional Aspects of the Comprehensive Plan*, Washington, 1950.

38. NELSON, H. J. "The Vernon Area, California—A Study of the Political Factor in Urban Geography," *Annals of the Association of American Geographers*, 42 (1952): 177-191.

39. ROTERUS, V., and HUGHES, H. "Some Effects of Population Changes on Municipal Services," *Public Management*, 29 (1947): 158-160.

40. APPLEBAUM, W. "A Technique for Constructing a Population and Urban Land Use Map," *Economic Geography*, 28 (1952): 240-243; JONES, W. D. "Field Mapping of Residential Areas in Metropolitan Chicago," *Annals of the Association of American Geographers*, 21 (1931): 207-214; MAYER, H. M. "Applications of Residential Data from the Chicago Land Use Survey," *Journal of Land and Public Utility Economics*, 19 (1943): 85-87; Chicago Plan Commission, *Residential Chicago*, Chicago, 1942; idem., *Forty Four Cities in the City of Chicago*, Chicago, 1942.

41. KLOVE, R. C. "A Technique for Delimiting Chicago's Blighted Areas," *Journal of Land and Public Utility Economics*, 17 (1941): 483-484; Chicago Plan Commission, *Master Plan of Residential Land Use of Chicago*, Chicago, 1943.

42. DAHIR, J. *The Neighborhood Unit Plan: Its Spread and Acceptance*, New York, 1947.

43. Cincinnati City Planning Commission, *Communities*, Cincinnati, 1947; Detroit City Plan Commission, *Detroit Master Plan*, Detroit, 1951; Kansas City Planning Commission, *Master Plan for Kansas City*, Kansas City (Missouri), 1947; Chicago Plan Commission, *Preliminary Comprehensive City Plan of Chicago*, Chicago, 1946.

44. MAYER, H. M. "Railroads and City Planning," *Journal of the American Institute of Planners*, 12 (1946): 2-18; ROTERUS, V., and Others. "Future Industrial Land Requirements in the Cincinnati Area," *Annals of the Association of American Geographers*, 36 (1946): 111-121.

45. WRIGLEY, R. L. JR. "Organized Industrial Districts with Special Reference to the Chicago Area," *Journal of Land and Public Utility Economics*, 23 (1947): 180-198.

46. Cincinnati City Planning Commission, *Industrial Land Use, Present and Future*, Cincinnati, 1946; Philadelphia City Planning Commission, *Industrial Land Use Plan for Philadelphia*, Philadelphia, 1950.

47. HARTMAN, G. W. "The Central Business District—A Study in Urban Geography," *Economic Geography*, 26 (1950): 237-244; PARKINS, A. E. "Profiles of the Retail Business Section of Nashville, Tennessee, and Their Interpretation," *Annals of the Association of American Geographers*, 20 (1930): 164-175.

48. MAYER, H. M. "Patterns and Recent Trends of Chicago's Outlying Business Centers," *Journal of Land and Public Utility Economics*, 18 (1942): 4-16; PROUDFOOT, M. J. "City Retail Structure," *Economic Geography*, 13 (1937): 425-428; *idem*. "The Selection of a Business Site," *Journal of Land and Public Utility Economics*, 14 (1938): 373-378.

49. DEWEY, R. "The Peripheral Expansion in Milwaukee County," *American Journal of Sociology*, 54 (1948): 118-125; DOUGLASS, H. P. *The Suburban Trend*, New York, 1925; FIREY, W. *Social Aspects to Land-Use Planning in the Country-City Fringe: The Case of Flint, Michigan*, East Lansing (Michigan), 1946; WHITNEY, V. H. "Rural-Urban People," *American Journal of Sociology*, 54 (1948): 48-54.

50. DUNCAN, O. D. "Optimum Size of Cities," in HATT, P. K., and REISS, A. J. *Reader in Urban Sociology*, Glencoe (Illinois), 1951: 632-645; KLABER, E. H. "Why the City of Medium Size Will Be the City of the Future," *The American City*, 48 (1933): 66-67; STEIN, C. S. *Toward New Towns for America*, Liverpool, 1951.

51. AUGER, T. B. "The Dispersal of Cities as a Defense Measure," *Journal of the American Institute of Planners*, 14 (1948): 29-35; IKLE, F. C. "The Effect of War Destruction upon the Ecology of Cities," *Social Forces*, 29 (1951): 383-391; National Security Resources Board, *Is Your Plant a Target?*, Washington, 1951.

52. MURPHY, R. E. "Johnstown and York: A Comparative Study of Two Industrial Cities," *Annals of the Association of American Geographers*, 25 (1935): 175-196.

53. BARTHOLOMEW, H. *Urban Land Uses: Amounts of Land Used and Needed for Various Purposes by Typical American Cities*, Cambridge, 1932.

54. CRESSEY, G. B. "Tungchow and Shatin, Two Chinese Communities," *Annals of the Association of American Geographers*, 27 (1937): 102; DICKINSON, R. E. *The West European City*, London, 1951; GINSBURG, N. S. "Ch'ing-Tao: Development in Land Utilization," *Economic Geography*, 24 (1948): 181-200; HALL, R. B. "The Cities of Japan: Notes on Distribution and Inherited Forms," *Annals of the Association of American Geographers*, 24 (1934): 175-200; JAMES, P. E. "Rio de Janeiro and São Paulo," *Geographical Review*, 23 (1933): 271-298; *idem*. "Belo Horizonte and Ouro Preto," *Papers of the Michigan Academy of Science, Arts and Letters*, 18 (1932): 239-258; LEIGHLY, J. B. "The Towns of Mälardalen in Sweden: A Study in Urban Morphology," *University of California Publications in Geography*, 3 (1928): 1-134; STANISLAWSKI, D. *The Anatomy of Eleven Towns in Michoacán*, Austin (Texas), 1950; TREWARTHA, G. T. "Japanese Cities: Distribution and Morphology," *Geographical Review*, 24 (1934): 404-417; WHITTLESEY, D. "Kano: A Sudanese Metropolis," *Geographical Review*, 27 (1937): 177-199.

55. JAMES, P. E. "Vicksburg: A Study in Urban Geography," *Geographical Review*, 21 (1931): 234-243.

56. STANISLAWSKI, D. "The Origin and Spread of the Grid-Pattern Town," *Geographical Review*, 36 (1946): 105-120.

57. MAYER, H. M. "Moving People and Goods in Tomorrow's Cities," *Annals of the American Academy of Political and Social Science*, 242 (1945): 116-128.

58. BARNETT, J. "Express Highway Planning in Metropolitan Areas," *Transactions of the American Society of Civil Engineers*, 112 (1947): 636-700; LYNCH, J. T.

"Traffic Planning Studies in American Cities," *Public Roads*, 24 (1945): 161-178; TRUEBLOOD, D. L. "Metropolitan Traffic Surveys Improved by Home-Interview Method," *Engineering News-Record*, 136 (1946): 146-150.
59. VAN CLEEF, E. "The Urban Profile," *Annals of the Association of American Geographers*, 22 (1932): 237-241; *idem. Trade Centers and Trade Routes*, New York, 1937.
60. ROTERUS, V. "Geographic Bases of Urban Planning," *American Institute of Architects Convention Seminar*, mimeographed, 1948; MAYER, H. M. "Geography and Urbanism," *Scientific Monthly*, 73 (1951): 40-42.
61. Cincinnati City Planning Commission, *Residential Areas: An Analysis of Land Requirements for Residential Development, 1945-1970*, Cincinnati, 1946; Philadelphia City Planning Commission, *Land Use in Philadelphia Metropolitan District*, Philadelphia, 1949.

POLITICAL GEOGRAPHY [A]

A. Original draft by Richard Hartshorne, based on an outline established by the Committee on Political Geography and on contributions from its members, including J. O. M. Broek, George Kish, Robert S. Platt, Malcolm J. Proudfoot, Stephen B. Jones, and Samuel Van Valkenburg.

ANALYSIS OF THE INDEPENDENT STATE
INTEGRATING FACTORS
INTERNAL ANALYSIS OF THE STATE AS A REGION
ANALYSIS OF EXTERNAL RELATIONS
Territorial Relations
Political Relations
Economic Relations
Strategic Relations
Summary

ANALYSIS OF OTHER POLITICAL AREAS
DEPENDENT COUNTRIES
AUTONOMOUS DIVISIONS
SUBDIVISIONS
POLITICAL ORGANIZATIONS AT HIGHER RANKS

TOPICAL STUDIES
BOUNDARIES
CAPITALS
TOPICS NEEDING STUDY
FORMULATION OF CONCEPTS

HISTORICAL POLITICAL GEOGRAPHY

APPLIED POLITICAL GEOGRAPHY

SUMMARY

CHAPTER SEVEN

POLITICAL GEOGRAPHY

AMONG THE CRITERIA used to distinguish homogeneous areas on the earth, the one most widely used is that of political authority. To most persons who went to school before World War I, the first picture of the world presented by maps showed the land areas divided solely into political units, each country in a solid color sharply contrasting with its neighbors. For many people, the political map is the only kind of map that is familiar, even if few take it quite so literally as Huck Finn on his balloon trip with Tom Sawyer: "We're right over Illinois yet. Illinois is green, Indiana is pink. . . . What's a map for? Ain't it to learn you facts?"

In disregarding the contrasts in the landforms, the cover of vegetation, the density of population, the types of economy, or other cultural features, political maps give an incomplete picture of reality. Nevertheless, for the study of the earth as the home of man, there is justification for placing strong emphasis on political areas. Political authority, organized in a system of independent, sovereign states, each with unique characteristics, is a dominating force in the way people live and the way they make use of the earth's resources. Whether one lives in a humid industrial area in Massachusetts or in a semi-arid ranching area in New Mexico may well be of less significance than the fact that in either case one is within the United States rather than in France, or in the Soviet Union [1: 238-239].

WHAT IS POLITICAL GEOGRAPHY?

Political geography, therefore, has long been recognized as an essential part of geography. It remains nevertheless one of the less-developed parts. Perhaps this is a result of the desire of geographers of the 18th and early 19th centuries to escape from the complete domination of political areas as the sole framework of what was called regional geography. Furthermore, to escape from subordination to history or to statecraft they emphasized the physical or natural factors that make for significant differences between areas, and, becoming increasingly imbued with the spirit of the rapidly developing natural sciences, they tended to ignore political boundaries almost completely [1: chap. 2].

European geographers found a much more congenial spirit of international scientific cooperation when they concentrated on aspects of geo-

graphy in which differences of opinion did not arouse nationalist senti-
ments. Geographers in the United States, on the other hand, living and
studying within the huge territory of their single nation, rightly regarded
its division into forty-eight more or less arbitrary units as of minor importance, but failed to recognize that this was far from true of the differences among the areas of independent states.

A more realistic viewpoint was forced upon geographers by the events
of World War I. Several geographers of the United States were called
upon by the government to make studies of European territorial problems,
both in preliminary work carried on at the American Geographical Society in New York and as members of President Wilson's staff at the
Peace Conference. None of them had had specific training in political
geography and none had published in this field. Essentially a new subject
to them, they applied their geographic training to it, but without being
able to draw upon any system of methods, terminology, or objectives.

Some of the geographers who participated in that work were stimulated
thereby to publish studies in political geography, but none of them turned
to this field as an area of specialization. In the two decades that followed,
only a few younger geographers concentrated attention on political geography. At one time, H. H. Sprout suggested that political geography
should be developed as a part of the field of political science [2]. Many
more geographers, however, though not specializing in this field, made
studies of problems in political geography as aspects of the broad treatment of specific areas of the world in which they were specializing. In
texts dealing with major areas of the world, such as the well-known volumes by George B. Cressey, Preston E. James, and Robert S. Platt, attention was paid to the political geography of the individual countries
studied.

World War II led to the introduction of many more courses in political
geography in graduate schools, liberal arts colleges, and teacher's colleges.
Interest in political geography at the secondary level is reflected in the appearance of a school text on geography and world affairs [3], written under the direction of a scholar of standing in political geography. The
number of scholars engaged primarily in research in this field, however,
remains small. In perhaps no other branch of geography has the attempt
to teach others gone so far ahead of the pursuit of learning by the teachers.

Since much of what has been published in political geography has resulted less from concern to develop a field of scholarship, and more from
a desire to contribute to an understanding of international problems, it is
not surprising that there has been little agreement on the nature and scope
of the field. At the time Isaiah Bowman wrote the first important work on
this subject in America there was no "body of principles or body of doc-

trine with respect to political geography," nor did he attempt to construct one.[B] Rather his book, *The New World* [4], represented a study of international relations as viewed by a geographer. In that field its methods of examining specific territorial problems of the world, country by country, opened a new path which contrasted with the methods of both diplomatic history and studies of the principles of international law and international relations. In the discussions of the specific problems, however, students found here no method or system demonstrating a distinctive character for political geography, with the important exception, generally recognized as the most outstanding contribution of the work, that an extraordinarily large amount of material was presented on maps.

In his review of the field early in the 1930's, Richard Hartshorne found in the publications of American or English geographers no clear basis for determining the purpose and scope of political geography. He sought to establish such a basis by introducing the system developed by contemporary German geographers on the foundations laid down a generation earlier by Friedrich Ratzel [5]. In the meantime, Derwent Whittlesey, who first among the geographers of the United States developed political geography as an academic subject of teaching and research, had released himself from the domination of the historical environmentalism of Ellen C. Semple and Harlan H. Barrows, a domination which was reflected in his earlier study of the historical relations of the United States and Cuba [6]. Whittlesey presented political geography as a study of areal differentiation based on political phenomena, a viewpoint that was in accord with the approach to geography as a whole then being adopted in the United States. Illustrated in various substantive writings, this view is outlined in a short article in 1935 [7], and a few years later in the introductory pages of his major work, *The Earth and the State* [8]. At the same time Samuel Van Valkenburg's *Elements of Political Geography* [9] presented a quite different view of the field.

GEOPOLITICS

With the outbreak of World War II, the interest in political geography was again focused on the immediate problems of international relations, particularly that of the geography of power. Many American geographers, as well as the general public and many military leaders, were impressed by exaggerated reports of the influence purportedly exerted on German strategic planning by the school of geopolitics which Karl Haushofer had developed in Germany. Consequently it became difficult to distinguish

B. From a letter written by Isaiah Bowman in 1937, in which he added that "there were political geographies, a bundle of nationalistic philosophies outside the scope of science."

between political geography and the supposedly new field of geopolitics.[c]

The difficulty in determining the distinction between the two fields was intensified by lack of any clear statement by the followers of geopolitics, whether in Germany or later in this country, as to the purpose and scope of that field. Haushofer's statements of his own views tended to add confusion rather than to clarify, though his basic purposes ultimately became quite clear.[D]

In his early attempt to analyze the German work in geopolitics, in 1935, Hartshorne discussed these difficulties, but concluded that geopolitics, as viewed by the more conservative members of the group, represented simply the application of the knowledge and techniques of political geography to the problems of international relations [5: 960-965]. But as it was soon evident that these problems require many other kinds of knowledge, geopolitics, having gained popular success as a catchword, became both broader in scope and narrower in purpose: namely, to apply all kinds of knowledge about foreign areas to the problems of foreign policy of the German state.

Whittlesey demonstrated a considerable correspondence between the conclusions presented in scattered writings of Haushofer and others of the group on the one hand and the actual strategy subsequently followed by Hitler on the other [10]. The German geographer Carl Troll, however, in his post-mortem on the school of geopolitics [11] makes it at least doubtful that this correspondence represented a direct cause and effect relationship. Further, if such a relationship did result, from early personal contact between Hitler and Haushofer, it is not clear which of the two had the greater influence on the other. There can be no question, however, that Haushofer and his group, through publication and by teaching of young journalists, made very important contributions to internal propaganda, supplying a pseudo-scientific rationalization for the Nazi policy of expansion.

Nevertheless, many students in this country, both geographers and political scientists, urged the development of a field of geopolitics purified of its nationalistic German ingredients. They sought to look back to H. J. Mackinder and A. T. Mahan rather than to Haushofer, but they overlooked the fact that neither of those individuals had attempted, any more

---

c. Whittlesey [8: 59] and others following him, used the adjective form "geopolitical" not in the sense used by the German students of geopolitics, but simply as a useful abbreviation of the cumbersome phrase "political-geographic."

D. Haushofer, whom the writer of this section found quite able to speak simple and clear German, wrote in a style extremely difficult for German as well as foreign readers, perhaps because of inability to write clear prose or possibly with the intention of obfuscating his reasoning while impressing his conclusions.

than had Haushofer, to provide a conceptual framework for a field of study, but had merely used what knowledge and techniques were available to them to draw conclusions for action on certain problems of power. N. J. Spykman, who had previously presented to his colleagues in political science detailed discussions of the importance of geography to foreign policy [12], adopted the term, and applied many of the concepts of geopolitics to his war-time studies of the situation and needs of the United States [13]. His conclusions as well as those of less influential students were strongly opposed by various critics. Hans Weigert concluded his analysis of the German school with an emphatic warning lest the American prophets of geopolitics should indoctrinate both military leaders and the public in this country with doctrines no less dangerous than those that had been propagated in Germany [14].

The disagreements that arose were not of the kind commonly involved in the attempt to determine facts and relationships. Rather, they stemmed from differences in fundamental assumptions of purpose and aspiration, assumptions seldom stated clearly either by the writer who held them or the opposing critics. Such failure to recognize basic assumptions is perhaps to be expected in a field whose nature and scope remains nebulous. One proponent, E. A. Walsh, while urging the development of "a future geopolitics, born of sanity and law and equity" [15: 34], provided no more definite concept of the field than "geopolitics, by which is meant a combined study of human geography and applied political sciences" [15: 2].

The last statement reflects the intellectual difficulty involved in defining any applied field. The economic geographer who attempts to apply his knowledge to any practical problem finds that his problem demands the consideration of matters not commonly regarded as included in geography. In the application of knowledge to concrete problems, the problems determine what shall be studied; definitions of fields of study as divisions of the whole field of knowledge become "academic" in the sense of not useful to the immediate purpose [1: 399-400].

Thus, the erection of a single bridge may involve problems not only of the physics and chemistry of metals and the aesthetics of architectural design, but also problems of hydrography, geology, economics of transportation, and labor. Bridge construction and statecraft both require the integration of information, principles, and techniques from many disciplines; it does not follow that either can be made into a discipline itself.

Finally, in the application of knowledge to concrete problems, the question of purpose is always involved. If the purpose can be assumed as accepted by all students concerned in the development of an applied field—as is commonly true in such cases as the application of medical knowledge to the prevention of disease, or the application of engineering knowledge to the construction of bridges—the answer to the question is universal and no problem arises. Just the opposite is the case in the application of knowledge to the concrete problems of international relations: every state concerned has a different purpose and these are most often in conflict. As an applied field concerned with specific problems, geopolitics would appear inevitably to be divided into as many different schools as there are independent states.

CONVERGENCE OF POLITICAL GEOGRAPHY AND POLITICAL SCIENCE

Nevertheless the convergence of students from geography and from other social sciences on the problems of war and peace must be welcomed and encouraged, not only for the aid it may give to statesmen in reaching tolerable conclusions on international problems, but also for the sake of the increasing depth of understanding in all fields that should result from such cross-fertilization.

The geographer, concerned with regions as defined in chapter two, has found in the political unit an example of an area that is homogeneous and cohesive in political terms, yet heterogeneous and perhaps not even cohesive in other terms. He studies the structure and functions of that area as a region homogeneous in political organization, heterogeneous in other respects. The political scientist, concerned with political processes, has found he must do more than develop generalizations independent of differences in different areas; the processes, which can only be partially described and analyzed in generalizations, must be studied as they operate in particular areas; he has been led increasingly to the study of what may be called regional politics.

Likewise in the field of international relations, grim reality has compelled students from all disciplines to focus on what may be called power analysis, the analysis of political units of power and the relations among them. Since these political units are defined by area and the relations among them are conditioned by space relations, geographers have long shown an interest in such problems. H. J. Mackinder presented nearly half a century ago a thesis of world power analysis and prognosis which for better or worse has become the most famous contribution of modern geography to man's view of his political world [16]. Mackinder's interest and purpose, it may be noted, were primarily political and practical and it is not surprising therefore that his hypothesis is much less firmly grounded than, for example, his more academic and geographic analysis of the foundation of Britain's seapower in the relation of Great Britain to "the British seas" [17].

Though many geographers have quoted and used Mackinder's thesis of the "heartland," Weigert is one of the few who have examined it critically [18]. It would appear to have greater interest for students of world politics, like Spykman, who not only examined it critically but also attempted a major revision [13]. The influence of Mackinder's thinking is likewise notable in the studies of the Atlantic Ocean and of the problems of national power by H. H. Sprout and M. Sprout [19]. In these examinations of regional politics a substantial amount of geography has been included. Owen Lattimore's war-time and postwar studies of regional politics in the Far East are focused on the search for solutions to current problems and are based on a specialized knowledge of the area [20]. Of a somewhat different kind, dealing with strategic problems of a single area of strategic unity, is the study of the Mediterranean by W. Reitzel [21].

An examination of these works reveals the almost unlimited variety of things that may have to be assessed if one is to reach a thorough evaluation of national power, a list that may quite literally include "shoes and ships and sealing wax, cabbages and kings."

Geographers can and have contributed information and techniques for the evaluation of many of these factors. In greatest volume, perhaps, have been studies of the physical and economic conditions within particular countries, studies prepared for governmental use and, therefore, unfortunately not published. One important aspect of production for national power, the occurrence and development of mineral resources essential for modern industry and war, has been most intensively studied and presented in well-known works of economic geologists such as C. K. Leith (chapters eight and eleven).

Geographers have also studied the strategic situation of major world areas, demonstrating the geographic technique of viewing power distribution in terms of space relationships. A number of such studies are included in the two symposia edited by Weigert, V. Stefansson, and R. E. Harrison [22]; and Stephen B. Jones has published several in the series of memoranda of the Yale Institute of International Studies.

Many other factors of national power, however, may be unfamiliar to the geographer. For the evaluation, say, of the strength or weakness inherent in the one-party dictatorship of the Soviet Union versus the two-party democracy supporting a strong executive in Great Britain, or the multi-party legislature dominating the executive in France, the knowledge and techniques of the student of government are essential. Therefore it is of the utmost practical importance that problems of international relations should be studied by the convergence of students from all of the disciplines that can contribute to their solution. Workers from various fields will and should seek to integrate the findings from the several disciplines in relation to specific problems, whether they seek this integration jointly in research teams or individually. The analysis of national

‣power represents a distinct area of convergence, not only of geography and political science but also of economics, anthropology, and psychology. It would seem well to identify this area of joint interest with a clear and simple name, such as "power analysis" rather than to obscure it by the all-embracing term, geopolitics, the origin of which is steeped in error, exaggeration, and intellectual poison.

Experience from the history of science indicates that each discipline participating in such a joint undertaking will be able to render its maximum contribution if it develops to the utmost its own body of principles and techniques. The formulation of concepts cannot be done in isolation, to be sure; but such formulations must be done independently. In the remainder of this chapter, therefore, attention is concentrated on political geography as an integral part of the whole field of geography.

POLITICAL GEOGRAPHY AS A BRANCH OF GEOGRAPHY

Even when the opinion of geographers concerning the nature and scope of political geography is examined there is still a wide divergence. This divergence is illustrated by the great variety of methods followed by the numerous geographers who contributed to the symposia edited by Weigert, Stefansson, and Harrison [22]. This divergence is even more striking in the book on *World Political Geography* by G. E. Pearcy, R. H. Fifield and associates [23]. In the introductory chapters by George T. Renner and Etzel Pearcy, political geography is identified with geopolitics purified of its irrelevant Nazi aspects. Renner attempts to summarize the different viewpoints of American political geographers in terms of three schools, but one may doubt whether the scholars he names would accept his interpretation of their writings. However, there is little indication that the writers of the subsequent chapters were guided by the views expressed in the introductory chapter.

The earlier attempts of Whittlesey and Hartshorne to determine the nature of the field have been mentioned previously. Somewhat different views have been offered by Stephen B. Jones [24] and J. K. Wright [25]. More recently, and partly as a result of preliminary discussions of the committee listed in footnote A of this chapter, Hartshorne presented in his presidential address to the Association of American Geographers in 1949, an appeal for "a more geographical political geography" in a paper [26] which emphasized the "functional approach" to the subject, thereby marking a major change as compared with the views he had published in 1935.

Discussions in the committee demonstrated not merely the existence of a variety of views regarding the scope of this field, which might well be a healthy situation. These discussions also demonstrated that geographers have tended to nibble at the more obvious aspects, such as boundaries, capitals, or international waterways, or, on the other hand, when endeavoring to cover the political geography of a whole state or larger area they have had in mind no systematic concept of what topics should be included, of what questions should be posed for which answers are to be sought. The committee therefore felt that, in reviewing the work that has been done and in seeking to offer guidance for future work, it was desirable not merely to determine the major topics that have been studied in political geography, but also to arrange these in an organized system that would make clear the relation of each topic to the whole and reveal gaps that need development. With such a framework any geographer interested in examining particular topics in political geography could see the relationship of his topic to the field as a whole and to the overall study of areal likenesses and differences—that is, to geography.

It is appropriate for geographers to begin, not with a verbal definition, but with a map. The primary facts in political geography are presented in the common political map. It was noted at the start of this chapter that such a map presents an important degree of reality in showing each sovereign state with a distinctive symbol covering the whole of its area; but it was also noted that such a map conceals important differences. We know that, in any particular case, some parts of a state may consist of barren mountains, others of fertile lowlands; some parts may be highly industrialized with primarily urban populations, others are largely rural and agricultural; there may be different languages spoken in different parts of the same state and diverse religions followed; portions of the state area may have recently belonged to another state and their populations may still feel themselves attached to the former state rather than to the one in which, as a result of war, they have been assigned; or indeed areas long recognized as parts of the state may be largely wilderness inhabited by scattered tribes who hardly know even the name of the state in which they are included on the political map.

While these cases refer to sovereign states, similar conditions may be found in political divisions at lower ranks. Thus the State of Tennessee is presented on the common political map as a homogeneous unit, and, on maps purporting to show political attitudes, it has appeared in the past as a part of "the solid South," blanketed as a terrain of one political party. Closer inspection, however, shows that over a third of the counties may have majorities of the opposing party, and the regions revealed by mapping these data show marked correspondence to the economic and racial differences in the several portions of the State: the Mississippi lowlands, the Nashville Basin, the Tennessee Valley, and the mountain region. These contrasts are significant not merely for those interested in interpreting or predicting election returns; the political differences, and the cultural, economic, and physical differences which underly them, present serious problems to the operation of the State of Tennessee as a coherent unit.[E]

In each of the cases cited in the previous paragraphs, the geographer analyzes the area concerned by the regional method. On the basis of a series of criteria relevant to political problems, he determines the regional differences and similarities within the area and by comparative study of the different regional systems revealed by each criterion, seeks for evidence of accordant areal relationships. By studies in historical geography he seeks the cause-and-effect relationships that are suggested by the application of the regional method.

SUBJECTS AND METHODS OF STUDY

Political geography, then, may be defined as the study of areal differences and similarities in political character as an interrelated part of the total complex of areal differences and similarities. The interpretation of areal differences in political features involves the study of their interrelations with all other relevant areal variations, whether physical, biotic, or cultural in origin.

Areal Differences in Political Character

One group of problems in political geography is concerned with the differences in attitudes of people in different areas regarding various political questions. The geography of political attitudes deals with non-material cultural phenomena to which relatively little attention has been paid by geographers in the United States.

There are, however, some exceptions to this widespread neglect. An outstanding exception is J. K. Wright's study of the voting habits of the American people as revealed by a compilation of election returns on a county basis for a long period of time [28]. Although Wright did little more than depict the facts, his work invites detailed studies of particular areas to interpret the significance of the facts shown. Thus, the breakdown of the "solid South" into a complex pattern of regional differences calls for comparative analysis with regional maps of such features as racial composition, agricultural systems, soils, and landforms. R. M. Crisler has pursued this subject in studies of political regionalism in Missouri [29].

A recent doctoral dissertation at the University of Wisconsin by R. H. Smuckler, a student of political science who was also trained in geo-

---

E. This particular example is chosen out of many that might have been used because the source from which it is taken, "Physiographic Influences in the Development of Tennessee," 1915, demonstrates an early recognition of the significance of political geography [27].

graphy, attempts the more complex task of mapping and analyzing the regions of isolationist attitude in the United States. Although this study confirms a general impression of concentration of isolationist attitudes in the Midwest, the marked variations within that area and the presence of equally pronounced isolationist districts in other parts of the country give strong reason for believing that the interior location of the area is not the basic factor. More important, it would appear, are differences among predominantly urban, small-town, and rural districts and differences resulting from different cultural and national origins. Detailed analyses to determine the significance of each of these, however, have not been carried out.

Until recently the data available for measuring political attitudes have been limited largely to election returns or to voting records of elected representatives. Today these could be greatly increased by use of the records of public opinion polls and "content analyses" of newspapers and publications. To use such data for geographic purposes, however, it is necessary that the results be tabulated in much smaller enumeration areas than is now the case. Few of these are broken down even by States within the United States, whereas Smuckler's study demonstrates that the totals for States obscure the significant relationships revealed by the smaller areas of Congressional Districts, and his further study of voting by counties within an electoral district makes clear that such a district is likewise too large.

AREAS ORGANIZED AS POLITICAL UNITS

The distinctive characteristic of political geography, however, arises from the fact that in practically all parts of the world today political organization represents not the organization of a particular kind or class of people but rather the organization of a particular area including, along with the other contents of the area, the people permanently resident within it. The legal establishment of any political unit, whether a sovereign state, a province, department, county, or township is essentially a declaration, in terms of the criterion of common government, that the area involved shall constitute a region as defined in chapter two. This is one situation in which the regional division of any area appears as a primary fact, rather than as a result of study [1: 402-404]. Whether this appearance, as presented by law, corresponds to reality is a question which will be discussed later.

It may be that a region established by political action corresponds to a region determined by other criteria, whether a uniform region, as, for example, in terms of the linguistic character of its people, or a nodal region, in terms of economic development and transport. In such a case, a cause-and-effect relationship is obviously probable, but only historical study can determine whether the political region was established to conform with the linguistic or economic region, or whether the latter was the result of a long-established political unity. In many other cases, however, there may be little correspondence between the regional division as defined politically by law, and other regional divisions of the same area as determined by other relevant characteristics.

The full reality of homogeneity of a political region is not expressed merely by the fact that the area has been legally established as a political unit of government. In some cases, indeed, the situation shown on the ordinary political map as sanctioned by law or international treaty may have little or no reality. Thus, Robert S. Platt told in 1935 of an Indian tribe in the upper reaches of the Amazon who live "in blissful ignorance of the fact that they inhabit a zone of international tension," an area at that time bitterly claimed as part of the political region of Peru and simultaneously of that of Ecuador; the inhabitants had no knowledge of the existence of those units or of what the difference between them might be [30: 273].

The degree of homogeneity of a political region is to be measured both in terms of the degree to which the various functions for which the region is organized politically operate uniformly over the area, and the degree to which political forces tie the area together into a coherent region. The first of these conditions is significant at all ranks of political division, the second is primarily important in the case of sovereign states, and is of decreasing importance at successive lower ranks of subdivision.

At all ranks of division, political units of area differ in the degree of regional homogeneity, as measured by the two conditions indicated. In no small part these differences are due to varying degrees of heterogeneity of non-political features within each unit. A large part of the study in political geography, therefore, is concerned with the determination of the degree of political homogeneity and its relation to the regional pattern in other factors.

Political geography, then, includes in addition to the study of the political character of areas, the study of the organization of areas into political units, examined individually and in relation to each other, and placed against a background of other relevant likenesses and differences of areas. It is in these terms that the remainder of this chapter will consider work in political geography. It is also in this respect that political geography is most closely interrelated with other branches of geography, a theme that Whittlesey has discussed in some detail [31]. The political organization of area into regional units, notably at the rank of the sovereign state, has profound effects on all aspects of geography in which man is a factor, including aspects commonly thought of as parts of physical geography, as in the geography of soils.

AREA STUDIES AND TOPICAL STUDIES

In the geographical study of political units of area as a special form of regions, the two methods common to all branches of geography are employed. In some studies attention is focused on the political organization found in a particular part of the world; in others political phenomena are studied generically as they occur in different parts of the world. Boundaries and capitals are the most obvious features in the political geography of any area, and no doubt for this reason they have been favored subjects of topical studies by geographers. Further analysis of the phenomena involved in the development of homogeneous political units of area may well reveal other topics of more fundamental importance.

The four sections of this chapter that immediately follow will consider problems that arise in studies of the division of the world into political units and in studies of individual units at various ranks of division. These considerations will suggest a number of topics requiring generic study, will point to the significance of historical studies, and will reveal possibilities for the application of political geography to current political problems, each of which matters will be summarized briefly in three subsequent sections.

THE WORLD DIVIDED INTO POLITICAL UNITS

Geographers recognize that the only complete whole with which they can deal is the whole surface of the earth. This One World has no corresponding political organization as a unit. It is organized, however, in terms of commonly accepted international law, into a single system of three kinds of areas. Some land areas are organized as independent, sovereign states, each of which is a unit region subject to no outside authority; others are organized as dependent areas, each under the control of one of the independent states. In legal theory these two forms of organization cover practically all the inhabited parts of the world; only Antarctica and a few uninhabited islands remain unassigned in this system. Finally the open seas are recognized as unorganized areas into which the authority of each of the organized states extends wherever ships or planes under such authority proceed. To solve the problems that arise when ships or planes of diverse state authorities come in contact in the seas, the independent states within the system have developed an elaborate structure of international law. To enforce this law, however, there is no single organization.

This state system is a product of recent times. It is new not only in covering the world, but in many parts of the world it has brought entirely new forms of political organization. In reading the narratives of early explorers, men familiar only with one kind of political organization, we are often misled by their use of such words as *king, state,* or *territory.* In many areas political organization was defined in terms of families and tribes, with only vague references to specific tracts of land. The relations of the Chinese empire to the peoples and governments of such areas as Korea, Manchuria, or Indo-China had no direct counterpart in the European scene.

The modern world system has resulted from the spread of European power or European ideas over all other parts of the world. For a long time the world system consisted largely of the state system of Europe together with areas elsewhere in the world organized by individual members of that system. Whittlesey's brief historical treatment of "the exploitable world" suggests the opportunity for valuable studies of the changing pattern of this system during recent centuries, a pattern that has changed markedly since he wrote in 1939 [8: 78-85]. But the countries that have changed from dependence under particular European states to independence, beginning over a century ago in the Americas and more recently in southern and southeastern Asia, have organized themselves after European models and have become members of a world system of states which represents an expansion of the former European state system. The same has also been true of countries, like Japan and China, the organization of which was never controlled by Europe.

Though the modern state system originated in Europe, it is notably different in character, as well as extent, from the systems that existed in Europe in earlier times, such as the feudal system of the Middle Ages. In some respects, indeed, the modern system is more akin to that of classical Greece; but on the other hand there is no modern counterpart of the Roman Empire.

The political organization of the world is constantly changing both in character and extent. Comparative studies of the geography of the political systems of the past might help us to recognize the trend of present changes. This subject is as yet almost untouched by geographers.

Mapping the World

A first step in the analysis of the present system of world political organization is to present it on a map. It is commonly assumed that this is done for us by the common political map. But such maps are by no means either so accurate or so realistic as is generally supposed.

To construct a map that clearly depicts even legal sovereignty involves problems that many map publishers avoid by purely arbitrary decisions. Current disputes over territories, as for example over the Falkland Islands, are often resolved by the cartographer with no appeal to authority. It is difficult, without complicated symbols, to show the status of a boundary like the present one between Poland and Germany, which is regarded as final by one party and only provisional by the other. The ill-defined boundaries of Tibet appear on most maps as no less precisely determined than those of Switzerland. Whether Antarctica is divided like a pie or remains politically unidentified depends on which foreign office the cartographer consults. S. W. Boggs found that at least nine symbols are required to show sovereignty over the territorial seas and continental shelves, and that a welter of new boundary disputes will arise if land boundaries are extended seaward without agreement as to the method of delimitation [32].

If it is difficult to map legal sovereignty with complete accuracy, obviously it is more difficult to make a world map of "political realities." The makers of such maps are driven to a variety of subjective decisions, the basis for which is not always clearly stated. M. Langhans, in 1926, published a world map showing degrees of autonomy in external and internal affairs [33]. In the main, this map was based on legal documents, but some highly subjective decisions obviously were made: thus the Ukraine and Sarawak appear in the same category.

In 1941, Hartshorne made a more complete break with political form and a shift toward reality as determined by effective political functioning [34]. Many of the details of his map are subject to question and few students today will find his categories acceptable. Nevertheless, this map may represent an important step in the direction of a more realistic political map of the world. A series of such maps covering a century or more of history would be very illuminating.

Maps of the world's legal systems have been published by J. H. Wigmore [35] and Whittlesey [8: 556-565]. Neither of these more than scratches the surface of the geography of the law, which remains a field in which much useful digging can be done. Whittlesey's black-and-white maps show only a single legal system in a given area, concealing the existence of compound systems such as the concurrence of Roman, Moslem, and local customary law in Indonesia. Wigmore's colors make it possible to show the coexistence of systems in a given area, but he follows political boundaries almost everywhere and so does not show enclaves of tribal law in remote parts of some organized countries.

HYPOTHESES OF THE POLITICAL DIVISION OF THE WORLD

Geographers have not, of course, been content to map the world political pattern but have sought to explain it in terms of other patterns of world differences. In particular they have attempted to find reasons for the political pre-eminence of certain parts of the world over others. This phase of political geography, however, remains richer in striking hypotheses than in well-demonstrated conclusions. Through much of the work runs the strain of environmentalism, inherited from the 19th century.

Thus Griffith Taylor presents the problem in terms that beg the answer: "Let us pose the urgent problems in a series of simple maps and diagrams. . . . Let us explain how the patterns depend on the environment" [36: 607]. After comparing the political map with world patterns of climates, landforms, natural resources, and races, he finds the foci of the political world in the humid, coal-rich lands of the northern middle latitudes.

Ellsworth Huntington also concluded that Europe and North America are likely to continue to dominate world affairs, but he based his conclusion primarily on the intellectual and physical energy of the population resulting from climatic conditions, reinforced by diet and perhaps by hereditary characteristics resulting from natural selection [37].

Mackinder, viewing the problem over a longer range of time, both past and future, saw in the pre-eminence of western Europe the special significance, during a particular epoch, of the maritime margins of the continental masses [16]. With the improved development of land communication, he foresaw the possibility of greater power development in the great interior lands, of which one, the "pivot area" or Heartland of Eurasia, was so placed as to offer the opportunity to dominate the world. Though Mackinder's hypothesis undoubtedly had a pronounced effect on Haushofer in Germany, and has been more widely discussed in popular writings than perhaps any other contribution of modern geography to common thought, critical discussion has focused chiefly on its forecast for the future. Unfortunately, few have attempted to examine critically or demonstrate adequately the detailed facts and relationships on which his premises are based.

Long before the writers discussed in the previous paragraphs, Ratzel had endeavored to lay the groundwork of principles in political geography that would explain the contemporary division of the areas of the world in terms of the past growth of individual states [38]. Viewing the state as the organization of a section of land and a section of people, originally quite small, he viewed its growth through processes of expansion over other small units and in competition with similar growing units. Ratzel, who had been trained as a biologist, thought of states as organisms, which therefore passed through a life cycle from birth through youth to maturity and finally to the decadence of old age. As a geographer he sought to find this cycle not merely in internal development, but more particularly in terms of area growth.

This theme was inherited by the German geographers who, following Kjellen in Sweden and Haushofer in Germany, developed the school of geopolitics. Semple, who brought so much of Ratzel's thinking to American geographers, had not introduced his political geography, primarily, we are told, because she found this thesis unacceptable. It has been introduced since, however, by Van Valkenburg; whether by transfer from Ratzel's writings or by analogy from the Davis physiographic cycle is not clear [9]. The thesis has been widely criticized not only because of a lack of demonstration that the life processes of any state have led inevitably to the characteristics that can be called old age and ultimately to dissolution, but, even more fundamentally, on the grounds that it is false to reason from a superficial analogy between a biological organism and a social organization operated by men, since men collectively through successive generations are at no time older than their predecessors.

APPROACHES NEEDED

For the purpose of attaining an understanding of the political divisions of the world, there appears to be a plethora of hypotheses. These hypotheses are all too commonly stated as though they were demonstrable theories. In contrast, the analyses are inadequate, both of the actual political pattern which the map of official political areas tends to obscure and of the patterns of the world distributions of many phenomena known to be relevant, or which appear likely to be relevant, to the political pattern. Studies are needed in the direction initiated by Langhans and Hartshorne, to provide a much more thorough analysis of the categories of political organization and their world distribution. The pattern or patterns thus developed need then to be compared with a large number of other world patterns. While many of these, such as patterns of natural conditions, population, production, and trade, have long been studied by geographers, other less familiar phenomena may be no less significant.

The overall data on population density for large political units are misleading, for they combine people living off the land, and therefore dependent on the amount of land available for agriculture, with people supported by urban occupations for whom such figures as acreage per capita of crop land have no significance. A more significant ratio is that of productive agricultural land to the population working on the land and supported directly by its production (chapters four and ten). Although such ratios are known to vary greatly in different areas of equal capacity for production, they nevertheless offer a method for providing comparable values that can be made into a world pattern. Hartshorne has taken certain preliminary steps in this direction in studies of the United States [39]. He has demonstrated the existence of a marked correspondence between the ratio of agricultural land to agricultural population on the one hand and certain well-known political differences on the other. The United States Department of Agriculture has published a similar study of agricultural relations in Europe [40: 22].

As pointed out in chapter four, studies of differential rates of growth of world population are needed. Unfortunately the available data are often given in estimates for the entire official areas of states, whereas a realistic pattern can be mapped only if it is possible to distinguish important areal differences within states, as in the case of the contrasted parts of such countries as Brazil or Indonesia.

The same is true with regard to one of the most important of world patterns for political geography, the pattern of differences in the levels of living. One of the most explosive issues in the world today is the growing realization among peoples of densely populated agrarian lands that, while they live on a bare margin of existence, peoples in other lands are enjoying a far higher and continuously rising standard of living. But only recently, through the work of Colin Clarke and others, are reliable data becoming available for the construction of a reliable world map.

STUDIES OF MAJOR WORLD SEGMENTS

In an age of global political relations and global wars, an age which began in the 18th century if not earlier, the political geographer is required to keep constantly in mind the world as a whole. But practical difficulties make it well-nigh impossible for him to study the whole pattern of world relationships at one time. Since these relationships must be seen as they are on the sphere, even the map, the geographer's distinctive tool, fails him in this purpose. Nor does the globe replace the flat map, since it is impossible to see more than half the globe at any one moment. To turn the globe around to see the other half is again like looking at the world mapped in but two hemispheres. As Boggs points out in his very illuminating study of "This Hemisphere" [41], one must consider an unlimited number of possible hemispheres, seen from all angles, as though one were to look at a globe both turning and shifting its axis, and at the same time to try to keep in mind all the different hemispheres seen.

It is often advantageous, therefore, to examine major world segments for the purpose of defining political regions of greater size but of lesser political homogeneity than that found in individual states. These major world segments are comparable to the realms defined in chapter two as the highest ranks of compages. The highest ranks of political organization may be identified on the basis of uniformity of particular political characteristics or of other cultural characteristics that are politically significant. Thus throughout nearly all of Latin America there is not only the obvious similarity of language among politically dominant groups but also similarity in certain inherited institutions of even greater importance for political organization, such as the role of the family, church, social classes, and army officers in the government, or the use of Roman law. In each of these respects almost all of Latin America stands in marked contrast with Anglo America. Against this background of similarity throughout Latin America, its division into many independent political units is to be compared with the patterns of population density and movement, racial composition, transport systems, and patterns of administrative organization inherited from colonial days, as James among others has suggested [42]. Similarily, J. O. M. Broek has analyzed the elements of unity and diversity in Southeast Asia [43], Hartshorne has examined East Central Europe [44], and J. S. Roucek has studied the Balkans [45].

A major world segment, or realm, may also be recognized on the basis of coherent unity if its various parts, however diverse in character, have closer political relations with each other, whether those of cooperation or of conflict, than they have with areas outside its limits. Thus the use of the Danube for navigation has for more than a century created a certain degree of regional unity among the otherwise very diverse countries through which it flows. George Kish has discussed the challenge to international planning which this situation presents [46]. A classic example, of still larger scale, is that of the Mediterranean realm, an area whose ancient homogeneity in cultural and political character was forever lost with the breakup of the Roman Empire and the later division between Christian and Moslem civilizations, but which is nonetheless bound together by sea connections into a single strategic theater [8: Chap. 9].

In the political-geographic study of such a realm, the student aims to determine the elements of regional homogeneity on the one hand and those of heterogeneity on the other, and to compare the regional pattern of each of these elements with the pattern of political organization. The existence of separate political units in an otherwise homogeneous region leads, in the modern world, to difficult problems in the functioning relations among the several states. This situation the political geographer must analyze. The method of studying these problems will be discussed in considering the analysis of individual political regions of greatest intensity, the independent states.

ANALYSIS OF THE INDIVIDUAL POLITICAL REGION[F]

Any specific area of the earth that has been established by law or international treaty as a political unit and in which a governmental system has

F. This and the following section are adapted, with modifications, from the more lengthy treatment in Hartshorne's "The Functional Approach in Political Geography" [26].

been organized and operated constitutes, in theory if not in full reality, a politically organized region. Such regions have certain aspects or features in terms of which they can be described. Furthermore, these features are created for the purpose of establishing political homogeneity. The successful operation of a political unit is directly related to the degree of political uniformity and functional coherence that is achieved. These characteristics can be studied geographically.

FEATURES OF A POLITICAL UNIT

There are certain aspects or features of any political unit with which a geographer is concerned. These features can be listed under four headings as follows:

1) *Morphology*. The politically organized region has a definite size and shape. It has an internal nodal structure, consisting of a center or focus, lines of communication from the center to all parts of the region, and features established to mark and control its boundaries;

2) *Dynamics*. On the basis of its internal structure, the political organization maintains a constant flow of authority between the centers of government and the different parts of the region; and also, though often far less directly, between the people of different parts of the region and the political organization;

3) *Location*. The politically organized region occupies a particular position in a pattern of similar regions of the same rank;

4) *External Relations*. On the basis of its particular position in the pattern of similar regions, the political unit has relations, as a unit of operation, with other regions both near and far.

HOMOGENEITY AS THE PURPOSE OF A POLITICAL REGION

The politically organized region is the product of a definite plan, formulated and put into effect with the conscious purpose of creating an area of political homogeneity in an area otherwise heterogeneous. Insofar as this kind of region is a product of planned human action it is to a certain degree analogous to the individual farm, plantation, industrial establishment, sales territory, or other cultural feature established for a specific purpose. It is to further the specific political purpose of developing uniformity and coherence that the kind of morphology and dynamics observed in political regions are built and maintained.

In most cases the size, shape, and location of a political region are inherited conditions that must, for better or for worse, be accepted. In this respect the politically-organized region differs from other kinds of planned regions. There are at least two cases in history that are exceptional in that a group of people first organized as a social unit, and subsequently sought an area of proper size, shape, and location to organize politically. These were the people of Israel under the leadership of Moses and their modern counterparts, the Latter-Day Saints who under Brigham Young founded the state of Deseret. George F. Brightman has presented an analysis of this interesting case of the organization of a political region in what was, at the time, almost a political vacuum [47].

The purpose of establishing regional uniformity and coherence is not an ultimate purpose but rather the means toward particular ends. These latter, the ultimate purposes of political organization, differ both at different ranks of organization and, in greater or less degree, among different units at the same rank. In particular, as will be discussed later, they differ greatly among independent states; in fact these differences constitute the justification for the independence of such units. But the inhabitants of any individual political unit can hope to attain their ultimate objectives only if they are able to create an adequate degree of regional homogeneity. Without such homogeneity the political unit that has been organized may not survive. Consequently this universal means for diverse ends is a generic purpose of all political regions, the importance of which at times even overshadows the ultimate ends. It is appropriate, therefore, to examine more closely the two elements of homogeneity: uniformity and coherence.

Uniformity

Political uniformity implies that with respect to the operation of all government functions all parts of the region are alike. In the methods and procedures of determining who shall govern, in the enactment of laws and in their enforcement, and in the administration of justice, the fundamental provisions of the constitution apply equally throughout the political region. If the region be subdivided into parts of lower rank, each of these participates on an equal footing in the overall government, and each is under the same form and manner of control. To the extent that particular areas within the whole politically organized region are treated differently by the overall organization, there is lack of uniformity and therefore lack of regional unity.

It follows from the above that the maximum degree of uniformity in detail is expected at the lowest rank of subdivision, decreasing as one moves to higher ranks of larger units; but even at the rank of the independent state, the common tendency is to seek as high a degree of uniformity as possible. In the absence of any higher form of political organization there is commonly no attempt to seek political uniformity among states within the same major world segment. In such associations as the British Commonwealth, to be sure, there are no doubt unorganized efforts to maintain as much similarity in basic constitutional structure and rights of individual citizens as the independent member states are willing to accept.

Coherence

The second element of homogeneity in a political region is its coherence. Political coherence implies that the internal functions of government are carried on successfully in all parts of the region. At the lowest rank of subdivision the units may be so small and the functions so limited that little difficulty is involved. At successively higher ranks, increasing size and increasing complexity of political functions make the problem increasingly important. At each of the ranks of subdivision it is also necessary that each individual region should have sufficient coherent unity to function as a unit in relation to the outside, but since these relations are chiefly those with the government of the larger region of which it is a part, they commonly present no particular problem. At the rank of the independent state, however, the political region must be able to function as a coherent unit in a variety of relationships with any outside forces with which it comes in contact.

In sum, every existing politically organized region represents the conscious attempt of a social group, beginning at some time in the past and continuing to the present, to create a particular kind of regional homogeneity which without this attempt would not exist. In some areas, to be sure, factors independent of political organizations may have produced a high degree of uniformity and coherence in social and economic conditions which greatly facilitates the attainment of political homogeneity. Nevertheless, in no area larger than a locality are such non-political factors sufficiently strong of themselves to produce the degree of political homogeneity necessary for the creation of a political region. Even in areas where the differences from place to place are slight, the difference between places in close proximity to each other and those more remote from each other is alone sufficient to create disunity unless overcome by the forces of political organization.

In the case of most political regions, however, and particularly at the higher ranks, political homogeneity exists in marked contrast to the underlying heterogeneity of non-political conditions. It is not to be assumed, however, that such contrast necessarily represents conflict. In the case of some factors it may have little or no significance. The difference between the level plain of eastern England and the more rugged terrain of the Weald, south of the Thames, would seem to have little relevance to the problem of political homogeneity, at least in modern times. In other cases, local differences in character of the land may lead to close interchange among districts of different economic production, as in the variegated lowland of northern France, which may actually further political unity. In many other cases, however, differences in character from place to place within the area, and a pattern of circulation that is discordant with the political region, may impose serious difficulties in the establishment of political homogeneity.

Geographic Analysis of the Political Region

The ideas summarized in the preceding paragraphs indicate the first major group of problems that the political region presents to the geographer. It is necessary to find a way to measure the degree of political uniformity and coherence that has been developed, and to note the differences in the degree of uniformity or coherence in different parts of the region. To interpret these conclusions of fact the geographer will study the regional structure and flow of political forces that make for political homogeneity in the region, and, by application of the regional method, he will determine the accordant and discordant relations of the various factors that aid or handicap the attainment of political uniformity and coherent unity.

In such an analysis, the external form of the region, its size and shape, will be found significant in relation both to the political structure developed and to the associated patterns of non-political elements. Until such a comparison has been made, however, it would be difficult to determine what aspects of the external form are significant in the political geography of the area. Hence, a procedure which starts with the size and shape of political areas because they are the facts first presented on the map, would appear in reality to be an indirect approach to the major problems. Indeed, in some cases it may prove quite a misleading approach. Thus, to start a study of Brazil with an analysis of the size and shape of the total national territory would tend to focus attention on the vast area of the Amazon Basin. Subsequent analysis would demonstrate that much of the Amazon Basin should not be included within the area effectively organized as the state of Brazil, but rather represents a dependent area under remote control from that state.

It is also true that the analysis of differences in degree of political homogeneity within a political unit may reveal important variations due to varying lengths of time in which different parts of the area have been together under one political authority. Full interpretation of such variations will require historical research concerning the evolution of the area of the region. In other cases, however, such differences inherited from the past may no longer have any significant relation to the present homogeneity of the region. It cannot be assumed that a full study of the evolution of the political unit is always necessary.

On the other hand, the existence of any political region in its present form presents a question of interest in its own right: how did this particular area come to be a political unit? In the case of subdivisions created by the larger units of which they are parts the problem is relatively simple; but in the case of independent states the processes involved have been highly complex and require careful study in historical political geography. Thus Whittlesey's historical studies of two anomalous districts in the Pyrenees, Andorra and the less-known Val d'Aran [48], lead to significant generalizations on the changing role of mountains in the political geography of Europe, generalizations which can be shown to apply also to the Alps.

It cannot, however, be assumed that the processes that have resulted in present morphology are necessarily interrelated with contemporary processes affecting the development of political homogeneity. Thus the difficulty of extending Spanish homogeneity across the crest of the Pyrenees into the Val d'Aran would be the same no matter what were the original reasons for the inclusion of this valley north of the crest in the political region to the south. In general, states have tended to accept any opportunity to secure additional territory regardless of whether such additions furthered or retarded the development of political uniformity and unity.

ANALYSIS OF THE INDEPENDENT STATE

The analysis of the independent state for the purposes of political geography involves three chief procedures. The strength of the integrating factors must be assessed, and the variations of strength from place to place plotted on maps. The internal structure of the state must be analyzed and its effect on the functioning of government interpreted. And the state must be studied from the point of view of its external relations. Each of these will be considered in turn.

INTEGRATING FACTORS

An independent state must be able to operate as an integrated individual in its external relations, for, if unable to operate in this way, its effectiveness is restricted and under severe stress it may disintegrate and vanish from the map.

Effective integration of a political region as an independent unit cannot be produced by government alone, no matter how well-designed and skillfully managed. It requires even more the conviction of integration in the minds of all groups in all areas of the region, a feeling, that is, of identification of themselves with the region as a whole and with its organization as a political state. This identification further must be accepted as stronger than any other forms of identification that might lead to conflict:

such as identification with a lesser part of the state, a section or a locality; or identification in religious communities overlapping many states; or identification with people of the same language overlapping into another state.

The geographer who wishes to measure the degree of coherent unity of a state as a political region must, therefore, determine what it is in the minds of the people that produces this conviction of unity. What are the integrating factors that make people identify themselves with a particular state, excluding or dominating feelings of identification with any other state? Under primitive forms of political organization, individuals identified themselves with the tribe, as a larger form of the family, as a group of people living in close juxtaposition and in cooperation with each other. What elements lead people to identify themselves with much larger entities, embracing people and areas they have never seen?

Throughout history, outstanding individuals in positions of leadership have been able to exert a strong personal appeal leading thousands and even millions of individuals to identify themselves as parts of the total organization of which the leader was the head. The principle of dynastic succession represented the conversion of this relationship to a particular mortal into a relationship to a concept that was immortal, the sovereign, continuously embodied by an individual unique only in birth. By this means a temporary regional political organization might be made a permanent one.

A much stronger basis for identification of the individual with his political region has been found in what is called "nationalism." At its most elementary level, nationalism appears as a conversion and enlargement of the concept of kinship: the individual identifies himself with, and therefore gives willing support to, the political organization and region which he feels belongs to him, and to which he belongs, because the people of all that region and those operating the organization are kin to him. Not that they are kin in the literal sense of common biological origin, but in the sense more meaningful to him that they are like him.

In some cases common religion may be accepted as the mark of kinship, but this is true only where religious affiliation determines not merely religious beliefs and activities but also a wide range of cultural and social customs and affiliations. The significance of this factor in the political division of India is well known; the complex regional pattern it presented has been analyzed in detail by J. E. Brush [49].

In African and Asian areas formerly under European colonial control, the sharp line drawn between individuals of European culture occupying positions of political and economic control, and all native peoples, a distinction commonly recognized by color of skin, tended to produce a feeling of kinship among all native, colored groups, regardless of major cultural differences among themselves. Once independence of outside control has been secured, the question whether this feeling of kinship, based primarily on a negative factor, will be adequate to produce a positive and permanent conviction of national unity is one of the larger questions facing these new states.

Most commonly the immediate evidence of likeness, accepted as kinship, is that of use of the same language. Since relatively few people anywhere in the world feel fully at home in more than one language, they feel kinship with those who speak that language, while those who speak a strange language not only seem strange on first contact, but remain permanently strangers. Hence identity of language has come to be the most common though not a universal basis of identification of the individual with the nation.

Simultaneously, the individual identifying himself with a group larger than the people he knows personally in his immediate locality, extends his feeling of sharing in the *land* in which he lives from the immediate vicinity with which he is familiar to a much larger area. Within the area of common language and under one political organization, his own, he can travel as one at home, and what he sees becomes part of his "homeland." If he does not actually travel there, he will hear of it from his kin or read of it in his own language. Hence the region occupied by his nation and organized into his political region, or state, becomes as a whole his fatherland.

The individual receives other cultural possessions from the politically organized region in which he lives and these lead him to identify himself with it. They include the memories and traditions of heroic leaders and heroic events. They include the richness of literary heritage produced in his national state. Since the world's literary culture is divided by language into many compartments only one of which is open to the average person, this factor tends to intensify the importance of language as the mark of the national unit.

If the factors that have been named tend to be the dominant ones at the elementary level of nationalism, and continue to function as important factors at all later levels, certain others may be of still greater importance in regions of more mature political development. If an area has had a long history as a highly unified political region, its people may find that the greatest values they hold in common are the political concepts, ideals, and institutions represented in their political organization and furthered by it. Since they constitute the one domain in which the political organization has supreme control, these values, as differing from those found in other states, are inseparably bound up in the particular state; to maintain them for the individual, his state must be maintained. It is in terms of such values that the particular state can make its strongest appeal to the ultimate loyalty of the people of all parts of its region.

The conclusion, therefore, is that while all states have in common the general purpose of establishing political uniformity and the maximum degree of coherent unity, each within its region, each state must seek to present to its people a specific purpose, or purposes, distinct from the purposes formulated in other states, in terms of which all classes of people in all the diverse areas of the region will identify themselves with the state that includes them within its organized area. This concept of a complex of specific purposes of each state has been called the "state idea" by various writers following Ratzel, or by others the *raison d'être*, or justification of the state [50].

It is not to be assumed that every area organized as an independent state actually embodies a distinct and well developed set of concepts or purposes representing its justification for existence. It is essential to the operation of the modern state system, essential indeed to the maintenance of stable human relations, that every inhabited area of the earth be included under some form of political organization. The people of a conquered area may object to the nature of the regime imposed on them and solely on the basis of that negative attitude, without any common positive purposes, seek to establish their area as an independent unit. Or the idea of the state may come from the outside: major powers may find it undesirable, from the point of view of power relations, to have the area organized as a dependency of any outside state and for that reason promote its organization as an independent unit. To accomplish this, to be sure, they must find some groups or individuals who accept that idea and are able to organize the area as a state.

Whatever the original reasons for the organization of a state, effective government will lead in time to the development of distinctive concepts penetrating into the diverse regions and peoples of the area. Ineffective government, on the other hand, will fail to provide the distinctive concepts that create, in the minds of the people, the conviction of identity with the region and with the state organization. In such a case, as in Spain, the state remains lacking in effective unity. Circumstances of geographical position or of historical chance may permit its continuance, or severe stress under outside attack may bring its disintegration, as in the case of Austria-Hungary [50]. Consequently every state, if only to attain its maximum of power in comparison with others, strives to secure universal acceptance of its distinctive state idea.

Universal acceptance of the specific state idea in all parts of the region is greatly facilitated if there is not only uniformity of political functioning throughout the region but also uniformity in political attitudes. Consequently there is a tendency in all states to produce, whether by force or persuasion, uniformity in the basic conditions and social institutions that influence political attitudes, including language, family, school, and church. Some states, on the other hand, have met the problem of local variations in cultural characteristics by permitting, under a federal system of organization, a considerable degree of variety among semi-autonomous subdivisions, in matters that do not adversely affect the unity of the state as a whole. Such variations within sub-regions of the state-region must not, of course, conflict with the *raison d'être* of the state. If this is a problem requiring constant adjustment in a federal state, it is by no means an insoluble one; indeed, the assurance of local control of local matters may form a part of the state idea, as is the case in the United States, where however that fact is not always recognized, whereas in Switzerland it is universally recognized as one of the essential elements in the *raison d'être* of that state.

To determine the particular character of the state idea of a specific state is no simple task nor is it one for which general training in geography prepares the student. Theoretically perhaps he might expect to find the answer in studies by students in other disciplines. Generalizations concerning the concept of state and nation, however, which have been the concern of most political scientists, will not help in this problem. What is needed is the distinctive concept of the particular country under study.

If the geographer cannot find the answer for the country he is investigating in the studies of historians, political scientists, or perhaps cultural anthropologists, he must seek it himself. Since the idea of the state in any particular country today is not the product of the moment, but is the constantly evolving product of a long past, this search will carry the student back into the history of the particular state, searching for those expressions of purpose that appear to have lasting influence.

INTERNAL ANALYSIS OF THE STATE AS A REGION

A state is a politically homogeneous region if in all its diverse areas there is homogeneity of attitudes and uniformity of relations between the political organization and the people. A state may be regionally homogeneous even though there are marked differences, even conflicts, in political attitudes among different social groups, provided these conflicts are fairly evenly distributed throughout the state. It may be said that such a state is heterogeneous, or lacking in uniformity, in the vertical sense, but homogeneous in the horizontal sense.

Thus France, which is regionally one of the most homogeneous of the larger states, has demonstrated since 1940 elements of national weakness based on social and economic cleavages that are relatively uniform throughout its area. Similar cleavages in Italy and Spain, on the other hand, are concentrated in certain areas and constitute, therefore, serious problems of regional disunity.

In some cases, as Hartshorne has suggested [34], regional disunity may be so marked that the territory legally included within the limits of the state consists in reality of one portion which constitutes the actual region of state organization, within which there may be even a high degree of unity, while other portions are essentially dependent areas, outside the state region.

Extreme cases of this kind are found in most of the countries that extend into the subarctic or tropical wildernesses, such as Canada or Brazil. Large portions of their territories are sparsely inhabited by primitive people who are regarded as alien to the population in the region of state organization, have no consciousness of identity with that state, and do not participate in its organization. In almost every respect, excepting those of contiguity of land connection (which may or may not be used for connections) and of legal definition, the relationship between the two portions of territory is similar to that between an independent state and an overseas colony or dependency. James uses the terms total national territory and effective national territory to express this distinction [51: 8-9].

Sometimes this represents a temporary stage in development. Thus, the territory of the United States during the 19th century included an originally vast but gradually shrinking area of Indian population which did not form part of the organized political region. After 1912 only the Indian reservations remained as vestiges of this formerly dependent territory and since the final granting of full rights of citizenship to the Indians in 1953, the distinction no longer applies even there. No doubt the states that now include similar areas in their official territories hope for a similar development, but that outcome cannot be assumed. Economic development, if it takes place, may result in increase in the indigenous population and continuance of the relationships of dependency on an essentially foreign state.

Such a situation, but still more complicated geographically, is found in certain Latin American countries, especially Peru and Bolivia, where densely populated districts of Amerindian culture persist, in which the inhabitants accept external controls imposed by the state that Spanish-Americans have organized but which are not truly integral parts of that state [51: 127-196]. A realistic map of the political character of such a country would show a diversified pattern.

When the organization of a state is based clearly on a national group that can be identified, a first step in geographic analysis is the comparison of the area occupied by people of that nation, the national region, with the whole area included in the territory of the state. If the national region extends beyond the region of the state, then the state is incomplete in its stated purpose of providing unity for the nation. If, on the other hand, the state includes portions of the national region of another state, especially of a neighboring one, its internal unity is disrupted.

While these conclusions are theoretically clear, it may be very difficult to secure reliable data, as Hartshorne has shown in studies of such areas in Europe [52]. The nationality of any individual cannot be determined by his origin, language, or any other objective fact about him, but depends rather on how he identifies himself. This may be extremely difficult to ascertain in the case of an individual living in a disputed territory not only because fear of reprisals discourages free expression, but also, in many cases, because the individual may change his attitude with changing conditions.

A fairly common situation is that in which the concept of over-all unity in a single region of political organization is accepted by all parts of the state, but the state idea has a relatively weak hold on certain parts. For example, to the French Catholic inhabitants of the Province of Quebec the values guarded by the regional unity and integrity of that province may be held to be more important than those represented by the larger regional unity of Canada. Much the same appears to be the case in many of the provinces of Spain. Western Australia has shown similar tendencies, not because of lack of political uniformity but rather because of lack of coherence resulting from physical remoteness and separatist economic interests.

In many cases, divergent viewpoints of a particular section of a state, although an element of weakness in its national unity, are restrained from disrupting the state because of economic, strategic, or other conditions that make separation an obviously impractical solution. No doubt it is for such reasons that Quebec is content to remain indefinitely within the Canadian union. A dissident group may feel itself to be nationally distinct from its neighbors but may recognize that it is too small to operate as an independent state and therefore may accept inclusion in a larger state. Nevertheless, the insistence of Eire on complete independence from the British Commonwealth warns us not to over-estimate the force of such considerations. Yet it seems that a dissident group in a central position within a state can scarcely contemplate secession, whereas one on the periphery of a state may be in a position to do so: compare, for example, the situation of São Paulo state of Brazil with that of Catalonia in Spain [53].

These examples indicate the significant relationships that may be brought out by comparing the pattern of political homogeneity of a state with the patterns of relevant differences in the character of the population or of regional disunity resulting from physical separation or diverging outside connections.

Even in states in which there is a high degree of regional homogeneity, based on full and even intensive support of the specific state purposes, there may be significant differences in interpretation of those purposes in different areas of the state. An obvious example is the United States, in which the fundamental principle of equality before the law, as expressed in the basic documents recording the purposes of this state, are differently interpreted in different sections of the country, differences that can be shown to correspond, though somewhat less directly than is commonly supposed, to areal differences in racial composition of the population [54].

Hence the political geographer is interested in the manner in which the internal organization of political authority is divided in sub-regions in order to permit different adaptations of government to different regional interests and attitudes. The maximum degree of divergence, within effectively unified states, is provided by the federal system as followed in Switzerland, the United States, Canada, and Australia. Many other states which have followed this in form, as have some Latin American states, and the Soviet Union, in practice have developed institutions and techniques of government that effectively destroy the federal principle. On the other hand, many states organized basically on a centralized, single-government system permit significant degrees of autonomy through delegation to local, indigenous, governmental units, as, for example, the English counties. The government of the United Kingdom, in addition, includes a variety of special cases of local autonomy at higher levels, as in Scotland, North Ireland, the Isle of Man, and the Channel Islands.

ANALYSIS OF EXTERNAL RELATIONS

All the associations of one state with other organized areas represent relations either of cooperation or of conflict, a theme which J. K. Wright has developed in a thought-provoking essay [25]. Relationships of both kinds may be grouped as territorial, political, economic, and strategic (i.e., military).

Territorial Relations

One of the most important topics in the political geography of a state is the extent to which it is in agreement or dispute with other states over the territory included within its borders. Attention to immediate concrete problems has caused most of the work in this field to concentrate on cases of dispute. It might be even more fruitful to study cases in which the people of adjacent states are in complete agreement on the extent of their mutual regions. Why, for example, is there agreement in Norway and Sweden that the existing division of the populated lowlands southeast of the Scandinavian Alps represents the correct and permanent limits of the regions of Norway and Sweden? Even cases which appear at first sight obvious, such as the limit between France and the United Kingdom, would merit more careful historical study, for though the present territorial limit is accepted as though ordained by nature, in earlier times the two states fought a long series of wars over claims of the kings of England to territories now in France.

Territorial disputes are often called "boundary problems," simply because the settlement of such a dispute requires the drawing of a precise boundary. This term would more properly be used to refer to the problems involved in the functioning of a boundary as a part of the structure of a state as a political region, namely in maintaining control of ingress and egress across the boundary, so that that line shall in fact "bound" the region. Boggs has considered certain of these problems in topical studies which will be considered later.

Another form of territorial problem arises from the use of the territory of one state for the activities of another. The most common problem of this sort is caused by the use of coastal waters by trading ships, or fishermen, of another state. Many states are concerned with the use of railroad facilities of another state, either in order to reach a third state, or, more importantly, to reach the sea. The development of air transport has raised a similar, but more complicated, problem, since there is no effective peaceful way of apprehending airplanes passing over the territory of an intermediate state. Finally, a special territorial problem arises when a state uses for military bases small pieces of another, and usually smaller, state.

Political Relations

The most obvious form of political relation of a state to an outside territory is that of direct political control, as over a colony, possession, dependency, or protectorate. Only some twelve of the ninety-odd independent states of the world have such colonial, or imperial, dependencies. One might, however, be justified in including in the same list the much larger number of states whose official territories include internal colonial areas of the types previously discussed.

The relation of sovereign states to their dependent areas includes a great variety of political forms. In analyzing these the political geographer will be concerned with the reality of the functioning relationships, whether or not those correspond to the legal forms proclaimed.

Less readily determinable than the areas recognized as dependencies by international treaty are those legally independent countries that are in fact under some degree of political control by an outside state. Such control may be limited to small districts of the country, used for military bases, or, though extending over the whole country, may be limited in function. In the latter case, particularly, the student may search in vain for any official documents that record the realities of the situation. The dependency of the satellite communist states of Eastern Europe on the Soviet Union cannot be read from their mutual-defense treaties.

Although the concept of independent sovereignty carries normally the assumption of equal political relations with all other independent states, many states have established special political relations with certain other states. The political geography of Canada or Australia can be understood only in terms of its membership in the British Commonwealth. For many years the United States had special political interest in Cuba and in certain of the other Caribbean countries. The Schuman Plan, though primarily economic in character, will result in each of its member countries having special political associations with the group in contrast with its political relations with other states. Likewise, while the North Atlantic Pact was primarily strategic in character, it contains political clauses which, if implemented, would tend to create a special political association of the United States and Canada with the states of Western Europe.

Economic Relations

International trade, services, and investments, whether operated by the government or by private individuals and corporations, constitute the economic relations of a state with other similar units. While the data here are the same as those used in the study of various aspects of economic geography, determination of their significance to the state area as a political unit requires a different type of analysis.

The first problem to assess is the extent to which the economy of one state depends on that of others. This must be broken down in terms of dependence on specific countries, classified according to their political association with the state under study and their location with respect to that state in the total world pattern.

Since economic dependence is necessarily mutual, the second problem is to determine for which of two states maintaining economic relations these relations are the more critical. Under normal peace-time conditions the answer depends on: 1) the relative importance of the foreign economic relations to the total economy of each country; and 2) the relative possibilities, in each case, of alternative areas as sources for needed supplies or services or for markets for surpluses. Geographers, who have been traditionally more concerned with production than with marketing, have commonly focused attention on the sources of supplies, but in normal peace-time economy the availability of markets is often the more critical matter. The classic case of this kind in modern times, the economic problem of an independent Austria, has recently been re-examined by George W. Hoffman [55].

In view of the increasing tendency of modern states each to control its own internal economy and, for that purpose, its outside economic relations, states are increasingly functioning as economic units and the economic relations with outside areas become increasingly important in the political-geographic analysis of a state.

Strategic Relations

In a world organized into mutually independent states among which the larger and stronger can, by use of force, destroy the political organization of the smaller and weaker, the existence of every state depends on its strategic situation in relation to others. Protection by military force against such possible attacks has from the beginning of states been one of their major external functions. Even if there are states that no outside power would wish to destroy, the fact that every populated area of the world can contribute something of strategic value to others, whether of productive power or only as a base for power, inevitably involves every state in strategic associations with outside areas. One needs only to think of Iceland today.

The strategic associations of any state are of three very different kinds: 1) those with states with which it is, or is likely to be, in open conflict; 2) those with states with which it cooperates in military alliance; and 3) those with the remaining areas of the world which may be presumed neutral. The great uncertainty in the attempt to analyze the total strategic situation of a state is the fact that its relations with any other state may change at any time from one of these three categories to either of the others.

The strategic associations of any one state with the other states and areas of the world form patterns of external connections, real or potential, and these, in turn, form integral parts of the total pattern of power in the international scene. In the analysis of these geographic patterns, the basic factor is the location of each state in relation to the total complex of the international power pattern formed by all the states and the strategic associations among them.

The location of a state, to be sure, remains constant. But the significance of that location varies with increase or decrease in the power capacity of the state itself, with similar changes in power of other organized regions, and with technological changes in transportation. It is an error, however, to assume, as many do, that the factor of distance has steadily decreased in strategic significance and is about to reach the zero point; in other words, that power available at one point can be just as effectively used against an enemy five thousand miles away as against one five miles away. One need only consider the differences in the recent war against forces supported by the Soviet Union if the scene of battle had not been in Korea, but say in Yucatán.

In general, it is still true that strategic relations are most important with neighboring states, provided any of these can command great power. On the other hand, it is also true, and has been so for at least two centuries, that power located anywhere in the world may be transported to operate, though in lesser strength, in any other area. Consequently each state must consider its full strategic situation in terms of possible strategic relations with any and all other parts of the world. This topic is discussed, from a different viewpoint, in chapter twenty-three.

Summary

The four different aspects of the external relations of a state with other areas of the world, are, of course, mutually interrelated. The territorial relations of a state with its immediate neighbors are commonly of first importance in determining the character of its strategic situation. An irreconcilable conflict over a border region, Alsace-Lorraine for example, forces recognition of the neighboring state as a possible enemy. Where no such conflict exists, as between the United States and Canada, the fact that both have approximately the same location in the pattern of strategic power, leads to close strategic associations of support, with or without specific treaty engagements.

Direct political association, notably in the case of dependencies, almost always involves close economic and strategic relations. The reverse, however, is not necessarily the case. Thus during the century following 1814, the United States had its most important economic connection with Great Britain but had no political connection with that state, other than the normal relations maintained with all other states, nor did it form any strategic association with Great Britain during that period. Eire is almost completely dependent on Great Britain both economically and strategically, but refuses to accept any special political or strategic association. For over a century the one area with which the United States did have a special political association was Latin America, where its economic interests were less than they were with Europe; on the other hand that relationship, as based on the Monroe Doctrine, was primarily an indirect result of the common strategic concern of the United States and the independent Latin American states to maintain independence from Europe.

ANALYSIS OF OTHER POLITICAL AREAS

There are various kinds of political homogeneity that may be used in defining political regions. In the two preceding sections the different attributes of political regions at different ranks have been presented, and the concepts and methods of analysis have been applied to the study of the independent state. But there are a number of other kinds of political regions in which the political geographer is interested. There is the problem of treating dependent areas which in many respects differ from the independent states. In many of the large nation states, moreover, there are major divisions that are autonomous in local affairs, such as the States of the United States; and these autonomous areas are further subdivided into local administrative units, such as the counties of the United States. Each of these kinds of politically-organized areas should be analyzed in a somewhat different manner. Finally, there is the problem of analyzing political organizations of a rank higher than that of the independent, sovereign state.

DEPENDENT COUNTRIES

The study of a dependent country differs in a number of significant respects from that of a sovereign state. The dependent country's evolution to its present areal form may have been due largely to competitive struggles between outside states and determined by power conflicts the issue of which was decided by events taking place thousands of miles away. So far as the dependent area itself is concerned, all that is significant is the end result of such rivalries. This, as Whittlesey has shown for specific cases in Africa, may produce political regions quite discordant with the existing cultural or tribal regions [56; 8: 127-159].

In any dependent area there is almost always a lack of unity of purpose in its political organization. This implies more than a mere conflict of purpose between the outside controlling state, the imperial power, and whatever indigenous political organization is permitted to exist. Because most areas of this sort were not politically organized regions prior to imperial control, there is commonly little uniformity or unity among the indigenous social and ethnic groups whose distribution forms a basic cultural pattern of political diversity within the area. These different groups may be striving for different political objectives, some through at least temporary alliance with the imperial authority, others in various forms of opposition, thus creating a pattern of regional variations in effectiveness of control.

Further, the representatives of the imperial power, official and unofficial, may disagree in terms not merely of current policy but of the ultimate objectives of the political organization of the area. Thus traders and other business men operating in the colony may have a different view of its ultimate future from that of the officials operating the government. Through the large corporations of the imperial country, which they serve, they may bring pressure on the home government against the government in the colony. The latter, on the other hand, though presumably functioning under direction of the colonial office of the home state, may in fact operate with a considerable degree of independence, even in conflict with principles determined by the home government. This is possible both because of remoteness from the home seat of government and because colonial officials tend to form an interested block committed to certain objectives regardless of changes in the government at home.

Finally, in those cases in which a considerable number of citizens of the imperial power have settled permanently in the dependent country, they form a powerful, special-interest group enjoying a position of social, economic, and political superiority to the great native majority. Their objectives may be quite different from either those of the home government or those of the local colonial officials, and are almost certainly in opposition to the objectives of the indigenous political groups.

Each of the several and often conflicting purposes pursued by the several authorities or special-interest groups will be found to vary in degree of intensity in different parts of the country. The geographer searches for accordant areal relations between these variations in intensity, differing for each specific purpose, and such other regional patterns as the capacity to produce exports, the spread of lands suitable for white settlement, or the characteristics of the native peoples.

Viewed solely from the side of the organizing powers, the purpose of the political organization of a colonial territory may approximate one of three contrasting types: 1) to maintain colonial control indefinitely; 2) to prepare colonial areas for ultimate independence; or 3) to incorporate the colonial areas as integral parts of the imperial state.

In the first case, only a minimum of political homogeneity is required, sufficient to allow maintenance of control. Political uniformity, though in some respects advantageous, may stimulate political unity and produce a dangerous challenge to imperial control. Under the dictum of "divide and rule," heterogeneity of diverse ethnic areas may therefore be encouraged. Experience, however, seems to indicate that this is a losing battle. The conditions prerequisite for effective government inevitably produce an increasing degree of political uniformity and regional coherence and the very effectiveness of imperial government produces common ties of opposition among otherwise opposing native groups and districts.

At the opposite pole is the purpose officially declared by Great Britain as its aim in colonial rule, namely, to develop colonial territories toward ultimate status as independent states, which it is hoped will choose to remain in association in the Commonwealth. The analysis of a colonial region in terms of this purpose would be much the same as the analysis of the internal aspects of an independent state. On the other hand, it is necessary to determine whether the indigenous political group or groups that are to take over the organization of the region as an independent state are in agreement on that purpose, or whether, as in India, their objectives might be better served by the division of the territory into two or more independent political regions.

The ultimate objective of the imperial state may be to incorporate the colonial territory into the political region of the home state on the same basis as that of present portions of that state. Presumably this is the purpose of the United States in respect to Alaska. In the former colonial areas of imperial Russia, the western parts of Siberia were gradually incorporated as integral parts of the state, and it is clearly the purpose of the Soviet Union to establish the same in respect to all its vast colonial domain. In the absence of reliable information on the actual operations of the Soviet government in such areas it is not possible to determine the degree to which this has been accomplished. Lattimore has considered the question in some detail in respect to the Yakut region [57].

France has recently incorporated certain small overseas colonial territories, its West Indian and Guiana possessions and Réunion, as departments of the home state, functioning through the parliamentary representation, like the other subdivisions of the state. This is the first instance in which a state has officially defined its political region as including portions separated from the main body by thousands of miles of open ocean. Even earlier, France had originated a similar policy in regard to its North African territories but while departments of France have been officially established in Northern Algeria, the actual structure of government there demonstrates still a high degree of colonial control.

It is significant that in seeking to incorporate its North African colonies France has not furthered the political unity of the existing regions, as inherited from indigenous governments of the past. Just as the first republic, following the Revolution, sought political homogeneity throughout France by destroying the historic provinces, replacing them by smaller departments to which little autonomous power was given in order to prevent the development of local regions of unity, so northern Algeria is not kept as a unit, but is divided into departments, which ultimately, one presumes, will have no overall regional organization other than that of all France.

Which of the ultimate objectives the imperial power pursues determines the present method of organization of the colonial territory. A major question is the degree to which the area is organized to function as a unit region separate from, though under the control of, the imperial power, or the extent to which its internal functions are determined in detail by the government of the controlling power, whether through a single colonial office, or, it may be, through separate agencies of each of the several departments of that government. Whittlesey has discussed the significant differences in this respect in parts of colonial Africa [8: 367-394], and Karl J. Pelzer has considered a problem that is similar in Micronesia [58].

The external relations of dependent countries are, generally speaking, controlled by the imperial power, though in many cases an indigenous government may share in determining these relations.

Internal organization and governmental operation are seldom carried on exclusively by the representatives of the imperial power. More commonly there is a combination of functions directed by the officials of the imperial power. In some cases these functions are handled directly by indigenous political organizations, some created by the imperial power, some adopted by it from traditional tribal systems; in other cases they are handled by agencies that include both native and imperial representatives. Such combinations vary not only in different empires but also in different parts of the same empire. In some colonial areas two separate forms of political organization function simultaneously in the same department of the government. Thus, in Tunisia, courts organized as part of the judicial system of France function in cases involving French citizens, while native courts operating under Moslem law handle cases involving only natives who have not become French citizens.

To determine the degree of dependence of the organization and functioning of the dependent country on that of the controlling state may be very difficult. A great variety of legal forms are recognized in imperial relationships; in the British Empire alone, the constitutional relationship is different for almost every colony. Further, the actual functioning relationship may differ greatly from the legal form. Nevertheless, the geographer who wishes to analyze the dependent country as an organization of area cannot avoid the necessity of determining what functions the colonial organization is intended to perform. To ignore this question would be as though one attempted to study the economic geography of a manufacturing plant established as a branch factory without considering whether it was planned to operate as an integrated unit or merely to carry out specific limited functions closely interrelated to the functions of the home establishment.

AUTONOMOUS DIVISIONS

The political geography of organized subdivisions of states represents a potential field to which very little attention has yet been given. In view of the great interest of American political science in the operation of the governments of the individual States of the United States, it seems somewhat surprising that American geographers have so largely ignored the significance of these areas as semi-autonomous units.

As long as American political geographers were concerned primarily with problems of boundaries and disputed territories, this field offered a great number of specific historical problems primarily of local interest only. Many of these are presented in a series of maps in the *Atlas of the Historical Geography of the United States* [59], and individual cases have been studied in more detail by S. W. Cushing [60], E. L. Ullman [61], G. F. Brightman [47], and B. E. Thomas [62]. Ullman, in addition has made a detailed examination of the effect of an interstate boundary on local development in the area through which it crosses [63].

Territorial disputes between the States of the United States in recent times have arisen chiefly where the boundary lines run through water bodies. In two cases in which such disputes were of considerable importance, and were finally settled by decision of the United States Supreme Court, geographers contributed studies of practical value: Bowman's, of the Red River boundary, which cuts through an oil field ultimately divided between Texas and Oklahoma [64]; and Lawrence Martin's, of the Michigan-Wisconsin boundary between two cities on the Menominee River and in Green Bay [65]. It does not seem likely, however, that many additional opportunities will occur for work of this kind.

There seems little gain, other than intellectual exercise, in considering imaginary changes in the territorial divisions of the United States among its forty-eight units. Whereas nearly every war results in territorial changes, the limits of the States of the United States may well be assumed to be immutable, short of a revolution the character or results of which could not be predicted.

On the other hand, the problems in each of the States which result from the particular geographic structure of its area are very real and significant. In the American federal system, these units are much more than subdivisions, like the departments of France; in a variety of respects each functions independently of the federal government. Furthermore, the existence of these forty-eight units operating under a common framework provides a valuable laboratory for political geography. For whereas every national state differs in significant degree in major functions or purposes, the major functions of these State units are largely similar. But each must carry out these similar functions over an area unique in geographic character.

American geographers have long been aware of certain outstanding problems presented by certain particular States. A familiar theme, more important in an earlier period but still significant, was the problem found in each of the South Atlantic States from Virginia to Georgia of establishing regional unity in an area that in each case included two contrasting sections, often in conflict, a portion of the Coastal Plain and a portion of the Piedmont. B. E. Thomas has considered the problem of maintaining connections between the physically separated parts of Idaho [66], and many students have commented on the problems of maintaining common interest over the two radically different regions of Washington, sharply separated by the Cascades. But these are only the more obvious cases; every State has its peculiar problems arising from its particular structure and location. Years ago a journalist pointed to the special problem of New Jersey, as a political region whose two main economic foci were in two outside States. Pennsylvania is obviously split between two industrial regions clearly separated from each other and each focused on its own metropolis. On the other hand, although North Dakota is economically homogeneous, the emphasis on one major surplus product, wheat, presents grave fiscal problems in times of depression.

It is a paradoxical result of our school system that while we have had numerous books on the economic geography of particular States, no one of which functions as an economic unit, no thoroughgoing study has been made of the political geography of even one of the States, although every State operates as a political unit. Here is an opportunity for research of practical value requiring no foreign travel.

Similar opportunities are to be found in the relatively few other nations that function under a federal system, such as Canada and Australia.

SUBDIVISIONS

The first rank of subdivision of autonomous divisions, the rank at which the units are determined and controlled by the larger organization of which they are but a part, is represented in the United States and England by counties, and in France by departments. The political geographer is interested, first, in the method and pattern of such subdivisions. While changes in the boundaries of subdivisions are legally possible in all three of these countries, the pattern developed in the past appears to be firmly implanted. A few changes are noted from time to time in the counties of the United States; discussion of a proposed reorganization of counties in Oregon led Stephen B. Jones to prepare a study of the geographic structure of that State in terms of nodal regions recognized in practice by various governmental and business agencies, a study that led to conclusions significantly different from those that had been reached by less careful consideration of the problem [67]. Similar studies were made by G. H. Hanson in Utah [68], and are reviewed by Ullman for the Pacific Northwest [69].

A general analysis of the units of government in the several States of the United States, with conclusions regarding the optimum size of counties, was published in 1934 by W. Anderson [70]. In most parts of the United States the county has little legislative power, but serves, rather as a subdivision of the administrative and judicial branches of the State government. The persons performing these functions in the counties are selected by local ballot, not by the State governments; for this reason, no doubt, the counties and larger cities are the basic area units of American political parties. In most of the counties, it has been said, local government tends to remain under the control of the same party over long periods of time. This provides a faithful core of party workers who can normally be counted upon in every election, regardless of wider State or national issues. This situation suggests a number of opportunities for geographic studies, whether in terms of individual counties selected as case studies, or of differences among the counties of a State or group of States or of the whole country.

The next lower subdivision, the township, is of such minor functional importance in most parts of the United States as to appear to offer little material for geographic analysis. In New England, however, these units, there called "towns," represent regions of political organization more important than the counties. In view of the notable changes in economy, in the pattern of population, and in the interchange of goods with other parts of the country that have taken place since the New England towns were first established, a fine opportunity is offered for the comparative study of such changes and their effects in areas small enough to permit full field coverage. A student might well expect to find in miniature many problems which would be difficult to analyze in larger areas.

The political geography of cities is also a field but little touched by American geographers, including those specializing in urban geography (chapter six). The fact that the political city seldom coincides with what has been called the geographic city may have discouraged such studies. But this very situation presents city governments with one of their most difficult problems, to which Harold Mayer has given considerable attention. Characteristically the large American city is an economic and social entity whose political organization controls but a part of the area involved in its economic and social organization. At the same time, city governments are expected to supply an increasing number of economic services to all the population entering the city in day-time hours, including the large number residing in areas organized in completely separate political units. H. J. Nelson reports on an extreme case of political independence of a small portion of an urban center, leading to almost exclusive use for industry [71].

In addition to the standard area units of government, there are in the United States a number of territorial organizations for special purposes which may either include particular groups of standard units or even cut across them. As examples, each quite different from the other, we may mention the Port of New York Authority, the Tennessee Valley Authority, or, at a lower level, a school district that includes several townships, or parts thereof. A current dissertation in geography by R. M. Beveridge, at the University of Illinois, analyzes problems arising from local regional differences in the development of a consolidated school district, problems involving annexation and secession of pieces of townships.

Finally, there is an important field for geographic study represented by areas which, while not officially organized as political regions, nevertheless are found to function politically as such, with greater or less degree of unofficial organization. In the more primitive areas of the world such unofficial political regions may in fact represent the only effective political organization present. Thus Robert S. Platt has shown that in the upper reaches of the Amazon such local organizations may even straddle and ignore the legal international boundaries of sovereign states [30], and many Indian communities in the Andes show similar independence of official political organization.

Even within highly organized states, unformalized political organizations of regions may play a significant role; a notable case is the political block that re-forms its ranks in the American Congress whenever the special regional interests of the South appear to be at stake. On the other hand, it should not be assumed that this case represents an example of a general situation, as though the United States consisted of a group of such functioning regions; Walter Kollmorgen has warned against easy acceptance of popular ideas on this subject [72].

POLITICAL ORGANIZATION AT HIGHER RANKS

Geographers have long recognized the need for studies of organized political areas at ranks above that of the individual sovereign state. The most common is that of an "empire," composed of one independent state and its dependent areas. As generally understood, such units have the peculiar structure of widely separated parts, connected only across the international seas; but, as previously noted, the compact territory officially credited to Brazil, Canada, or the Soviet Union, represents in each case such a composite of state and dependent territories.

The Commonwealth of Nations (formerly called the British Commonwealth) represents a different type, until recently the unique case of a loosely organized, but nonetheless effective, association of independent states. Since various of its independent members also hold dependent territories, the total organization is extremely complex. An adequate analysis of the political geography of such an organization of area must involve more than a series of studies of each of its parts: it must describe and interpret the interrelations among those parts that account for the total system.

In a different category are a variety of "leagues" or associations of independent states, each organized for specific and limited purposes. (One does not include here, however, the ordinary military alliance which commonly represents no more than an agreement on a particular aspect of foreign policy.) A clear example is the Arab League, a regional organization of states having some common internal interests as well as external purposes, which has attempted to function as an inter-state organization. The North Atlantic Treaty Organization, NATO, has perhaps gone farthest in integration of military forces and has at least charter authority to develop as an economic and political association of states. Whereas such a transoceanic association, involving countries in three continents, seems to the layman as non-regional in character, geographers should be among the first to demonstrate the intrinsic regional unity of the countries bordering the North Atlantic.

When the student of political geography attempts to look beyond the present to the foreseeable future, as surely he must, he finds a host of problems in interstate organization that call for study. Perhaps the most obvious and pressing is the question of the unification of Western Europe, or at least the development of a greater degree of integration than now exists. Solution of this problem, for which the need from the strategic as well as the economic viewpoint is well-nigh imperative, will require creative imagination of the highest order, since it will involve laying the groundwork for a system of regional organization essentially new in history. Techniques of political-geographic analysis could well be applied to this area as a partial basis for considering proposed methods of synthesizing its political diversity into a greater degree of regional unity.

In the international field certain regional associations of states function without official organization. Thus the Scandinavian states, bound together by proximity, and by linguistic and political similarity and cultural heritage, have at various times cooperated as a loose unit in international affairs, and this cooperation continues even though they are now split within and without the North Atlantic Treaty Organization. The officially recognized Organization of American States is in many ways less of an international force than is the unofficial Latin American block. Yet the latter tends to be pulled apart by divergent economic and strategic interests, a subject on which R. S. Platt has commented [73]. In each such case the geographer is presented with a set of contrasting patterns of centripetal and centrifugal forces resulting from elements of regional homogeneity and heterogeneity, of which the total resultant varies, in relation to specific outside problems, from strong regional unity to complete regional disunity.

Finally, we are living in the age of the great experiment in political organization of the world, the experiment which began in 1919 with the League of Nations and has been resumed since 1945 with the United Nations. In comparison with all other official organizations of areas into regions, this attempt to organize the world into a single region is still extremely weak in structure and function. Like the Commonwealth of Nations, but unlike federal states, its structure is that of an association of governments of political regions, with no direct authority reaching inside the regions themselves; and unlike the Commonwealth it has no historic tradition and no similarity of political institutions to bind it together.

Founded primarily for the negative purpose of preventing catastrophe, the world organization will have difficulty in finding universal appeal until and unless it can convince the peoples of all parts of the world that it has prevented such a catastrophe. In the meantime, however, it has developed numerous less striking functions, contributing to the general welfare in matters of universal interest, such as health and living standards.

That these contributions do not lead to world-wide conviction that every part of the world has an important stake in the world political organization is probably not so much due to criticism of the weakness of structure of that organization, of which so much is heard, but rather reflects the fact that, in spite of all that is said and can be demonstrated regarding the unity of the world in terms of economic life, health, freedom, and security of the individual, relatively few people as yet in any part of the world are conscious of the reality of this world community for which a political organization is needed.

This sketch of the current situation of the United Nations is sufficient to suggest numerous specific topics for geographic analysis. The primary function for which the organization was established, that of suppressing international conflict, presents difficult and complex problems in the world strategy of power used to prevent or crush war. The United Nations appears to be lifted above the surface of the earth, as though it operated without reference to differences in specific location, reaching conclusions on the basis of multiple translation of speeches at meetings held in New York, Paris and elsewhere. A realistic view of capacity to maintain international security requires an analysis of power units that can be placed at its service by its individual members, power units distributed in specific locations and must be evaluated in terms of space relationships to each other and to the location of possible or actual theaters of action.

TOPICAL STUDIES

Throughout the discussion of the study of regions of political organization in the previous sections there has been recurrent mention of particular elements, whether of structure or function, whether material or non-material, that were found to be significant in the analysis of the homogeneity of individual regions or in the relations among them. To further the evaluation and interpretation of these elements as factors both in relation to the individual regions and in relation to the regional political organization of the world, there is need for comparative studies focused on specific elements or topics as they are found in many regions, or over the entire world.

Such comparative studies should lead to generic concepts and categories, facilitating precise description and evaluation. The study of the processes in which a particular element is involved in different situations may make possible the construction of hypotheses, may even permit the demonstration of principles of cause and effect, which in the individual case of a single country can only be more or less intelligently surmised.

As in all other branches of geography, therefore, topical studies are no less necessary than area studies. The latter were considered first because the tentative analysis of areas should indicate which of the great number of possible topics are likely to be of most importance.

BOUNDARIES

This has not been the procedure by which the geographers have selected topics for study in political geography. The most notable development of topical study has been devoted to the problem of boundaries, whether because they are an obvious feature or because of predominant interest in the practical problems of territorial disputes.

Whatever the reasons for the attention given to this topic, the results appear to be of more lasting value than has been achieved in any other aspect of political geography. Building on the earlier work of various European geographers, as well as on their own experience in the field or in government service, Hartshorne [52], Boggs [74], and Stephen B. Jones [75; 76] have constructed generic concepts and categories of international boundaries that are far more precise and useful than any that were previously available; in particular, dispensing, one may hope, with the common but often misleading distinction between "natural" and "artificial" boundaries [77]. The latter concept, which has been influential in French political thought, as N. J. G. Pounds has pointed out [78], has also been examined critically by Broek [79] and by Eric Fischer [80].

The most thorough and detailed examination of a particular type of political boundary is provided by Boggs' series of studies on the problems associated with boundary lines drawn through water bodies [32; 81]. Of a different category is Hartshorne's attempt to provide a systematic approach to the study of territorial disputes in areas long occupied by peoples of different nationalities, an approach which analyzes the regional overlap of factors pertinent to the determination of homogeneous political regions [82]. The ideas and procedures developed in these studies are summarized and amplified by Stephen B. Jones in his masterly book on boundary making [83]. The studies of boundaries have provided not only intellectual stimulus, but have contributed in important ways to the practical needs of boundary makers.

CAPITALS

A second topic that has been well developed in European studies in political geography has to do with the location of capital cities. British geographers, such as James Fairgrieve and Vaughan Cornish, have given much attention to this subject. The American geographers, on the other hand, have largely neglected capitals. Perhaps the latter have felt that the location of capitals in the United States, whether federal or State, was geographically less interesting than in countries where capitals appear to have become established because of intrinsic advantages of location rather than by legislative fiat; as though any seat of government could be located other than by conscious decision of the government. Or, perhaps, they have had less interest in a topic which appears to be unrelated to current practical problems and the study of which requires intensive historical research. Whittlesey, however, has demonstrated that in areas where the political region of a nation differs notably from the territory officially included in its state, as in many countries of East Central Europe and of Latin America, the study of capitals as the foci of national interest and organization provides a useful approach to the analysis of political regions [8: 195-234, 461-475].

TOPICS NEEDING STUDY

A survey of the problems involved in the political geography of individual areas reveals a large number of topics less obvious than boundaries and capitals, but which may be no less important to an understanding of the political differentiation of areas. Few of these, however, have been examined systematically. Thus, various students have designated an area of particular location and political function by one or more of the following terms: hearth, cradle, heart, core, nucleus, kernel, or focus, but there is no agreed understanding of the concept involved. Even such older terms as "buffer state" or "pioneer zone" remain ambiguous in meaning, because of lack of systematic study to determine categories.

One topic emerges from the discussion of the geography of contemporary states as most in need of comparative, systematic study: namely, the character and ingredients of a nation. Not the significance of nationalism in general to the development of states, which has been well developed by political scientists, but the analysis of what differences in nations are significant to the organization of political regions as units independent of each other. For the nation is not merely a social group; it is also a phenomenon of area, commonly monopolizing a core area of complete regional homogeneity in this respect, though extending in many cases into marginal areas of hetereogenity. Certain characteristics may be common to all nations and national areas; certain others are found in some but not all nations; and each nation has its unique qualities or purposes that differentiate it from all others.

FORMULATION OF CONCEPTS

In political geography, as in other branches of geography, generic concepts and principles may be borrowed from the related systematic sciences. As political scientists develop increasing concern for the study of individual states, rather than of the state as an abstraction, geographers may expect more aid from their work. Geographers may also look to the work of cultural anthropologists and social historians for light on the determination of distinctive differences among national groups.

One great difficulty in topical studies in political geography when focused on aspects of independent states is the small number of cases. There are considerably less than one hundred existing specimens of this species and each of them is so different from all the others in both its political and its geographic aspects that it must be placed in a class by itself. The number of cases can, to be sure, be considerably increased by borrowing from the past, but the farther one digs back the greater are the differences in the frame of reference: the purposes of states and consequently the relations among them were basically different in past periods.

It has been suggested, therefore, that more progress might be made by focusing attention on studies of subdivisions of particular states. Thus the United States offers forty-eight units each having essentially the same political functions and purposes, but each occupying a unique geographic area. Much greater numbers can be utilized if one works at a lower rank of subdivision. It seems likely that such systematic studies might lead to sound generalizations of principles and it is conceivable that at least some of these would then prove useful in the study of independent states.

HISTORICAL POLITICAL GEOGRAPHY

In any of the branches of geography, as pointed out in chapter three, it should be a truism that every study will involve, repeatedly, the use of the historical approach. Human beings, farms, factories, cities, nations, and states do not appear in any area instantaneously; nor is their character at any given time the product of present circumstances only. Much that exists now can only be understood in terms of past conditions that may no longer exist. In all of the previous sections of this chapter it was assumed that interpretation of the present political geography requires repeated dips into the history behind it.

In a different category, however, are those studies in which the focus of interest is in the past. If political geography is of value toward an understanding of the present character of the political world, it is of no less value in understanding that of any time in the past. If such studies lack the practical importance of contributing directly to the solution of vital problems, they are of no less intellectual interest and scientific value; indeed, the absence of an overpowering concern for immediate decisions provides a clearer atmosphere for objective analysis and, in many cases, makes available to the scholar a large amount of data which in the case of present problems may still be locked in "classified" documents and unpublished diaries. Finally, study of the political geography of past periods increases enormously the materials available for comparative studies in topical political geography.

Because political geography in America has been stimulated largely by current world problems, the vast storehouse of materials from past periods has been little exploited. The most notable exceptions are studies by Whittlesey, such as the chapters in *The Earth and the State* which concentrate on the political geography of Western European countries during the medieval period [8: Chaps. 5, 7, 10, 13, 16], and certain of the works by Lattimore [84].

Other American geographers have occasionally presented studies focused on the political geography of a past period, such as Hartshorne's studies of the Franco-German boundary of 1871 [85], or his paper, published only in abstract, on the regional heterogeneity of the Austro-Hungarian empire [50]. The interest of the Berkeley school of historical geography (chapter three) is represented in the field of political geography by Dan Stanislawski's two studies of certain areas in Mexico, one in pre-Columbian time, another in the colonial period [86].

Relatively little has been done in the political geography of earlier periods of United States history, though the chapter on the United States in Whittlesey's volume is largely concerned with past periods [8: Chap. 16]. Of the many problems of political organization of regions in our past history, two would appear to be of special interest at a time when there is great interest in the integration of independent units into larger regional wholes. The historian Merrill Jensen has demonstrated that before the adoption of the Constitution of 1789, the United States was not a mere association of independent states unable to function as a unit, as many earlier writers have claimed, but in many respects operated as a single state [87]. His study might well be tested by geographic analysis to assess the degree of regional uniformity and unity that existed under the Articles of Confederacy. Likewise it would be fruitful to subject to similar examination the Confederacy of the South, during its brief period of existence, to peer under the cloud of political hysteria and war in the hope of determining in what degree the establishment of that temporary political unit represented regional homogeneity within itself and regional distinctness from the northern part of the country.

A somewhat different method of inquiry is the study of any particular area through a series of periods, as S. B. Jones and H. Mehnert have considered Hawaii's position in the Pacific [88]. Thus one might analyze the political geography of the region centering on the Mid-Danube in the early period of formation of Austria and Hungary, then in the period of maximum Turkish advance, then in the period of the dual monarchy, followed by the secession states after World War I, and finally, for the present, in the period of Soviet domination. A somewhat less embracing study, which might well provide insight into the difficult problem of determining the distinctive features of a particular nation, would be a geographic examination of, say, French nationalism during the centuries of its development. Such a study would seek from historical sources, to determine the elements of that complex both as they developed in character and as they spread in regional extent. It should throw light on such questions as to what degree the French nation was developed by the state, to what degree the national region spread beyond the contemporary area of the state, to be followed later by extension of the state.

APPLIED POLITICAL GEOGRAPHY

Political geography in the United States has been predominantly concerned with the application of geographic knowledge to specific and contemporary political problems, primarily problems of international relations. Many of the studies of this kind have been made for agencies of the United States government. Most of them have not been published; others have been published for academic purposes or to inform the general public concerning current problems.

One of the earliest of these applications is the report of a United States Commission appointed by President Cleveland to investigate the dispute between Venezuela and Great Britain over the boundary between Venezuela and British Guiana. The report was published in Washington in 1896-1897 [89].

Political geography received its first major impetus, however, as a result of the public interest in territorial settlements following the upheaval of World War I. Leon Dominian's now classic study of *Frontiers of Language and Nationality in Europe* [90], published in 1917, was written primarily to provide a basis for anticipated revisions in the regional organization of Europe. Other American geographers, as noted earlier, contributed to the background materials used by President Wilson at the Peace Conference. One of them, Bowman, was stimulated by his work there to incorporate brief summaries of the territorial disputes of all parts of the world in *The New World* [4], a volume which, however, went far beyond that type of problem.

During the ensuing two decades of official American withdrawal from European affairs, such service within the government was rendered almost exclusively by the Geographer of the State Department, and was concerned primarily with territorial problems in Latin America and detailed boundary problems between the United States and its immediate neighbors, Canada and Mexico. As a result of his work on a large number of such problems, Boggs published in 1940 a volume of lectures on International Boundaries [74], and it was largely through his influence that Stephen B. Jones prepared his handbook on *Boundary Making* as a guide to treaty-makers and boundary commissioners [83].

As the United States became involved in World War II, a large number of geographers were drawn into government service. Although most of these worked on military problems, a large number of studies of territorial problems were made by the greatly augmented staff of geographers in the State Department, by Bowman as special adviser to the President, and toward the end of the war, by geographers in the Office of Strategic Services, some of whom were subsequently transferred to the State Department and to other government agencies.

In the meantime a few geographers in academic circles had published analyses of individual cases of territorial disputes, either before or after official settlement. Hartshorne examined the consequences of the partition of Upper Silesia, in an attempt to develop sound techniques for examination of such a problem in the field [52]. R. S. Platt's discussion of "Conflicting Territorial Claims in the Upper Amazon" illuminated the disruptive effects of an international dispute resulting from ill-defined boundary lines in tropical-forest areas remote from developed regions [30]. The problems resulting from the addition of the South Tyrol district of German population to Italy following World War I have been re-examined in terms of the situation following World War II by both Guido Weigend [91] and George Hoffman [92]; the latter has also published a detailed analysis of the demands of the Netherlands for changes along its boundary with Germany [93]. Under the direction of Samuel Van Valkenburg, students at Clark University have completed a number of doctoral dissertations on specific problem areas, at least one of which, C. C. Held's study of the Saarland, has been published in part [94]. The boundary problem of greatest intensity in the Far East, that in Korea, has been examined in two studies by Shannon McCune [95].

Of studies of a less technical character, designed for general and adult education, mention may be made of Van Valkenburg's surveys of problems in several larger realms, published in the Headline Series of the Foreign Policy Association [96].

Since the field of political geography is far wider than the study of territorial disputes, it offers greater opportunities for the application of scholarly research to practical problems than has hitherto been appreciated. Wherever it is politically feasible to change the areas of political units, as of provinces, or counties, there is opportunity and need to apply whatever knowledge has been acquired in political geography to practical problems. We have already noted Stephen B. Jones' contribution to the discussion of the division of Oregon into new counties, at a time when that appeared to be a current issue [67].

A greater number of opportunities may be expected in the field of government planning. Wherever such plans overlap existing political units, the problem of determining what extent of territory to include in any particular plan and the problem of organizing that territory into a functioning unit are matters on which the student of political geography should be able to throw light.

Perhaps the greatest opportunity may come in the field of urban government. It now appears likely that the problems resulting from the expansion of cities into suburban areas and over satellite cities may result less often in simple enlargement of the central city by annexation than in the establishment of some form of overall government, with limited functions, for metropolitan districts, within which the previous political units may retain considerable autonomy. Since every such case presents unique problems not to be solved by universal rules, the geographer trained in the study of local differences and interrelations, should be able to play a helpful role in analyzing the specific situation and in recommending solutions.

SUMMARY

A great part of the work in political geography has been produced by geographers who are not specialists in this field, but who have either been intrigued by its problems or have been called on to make studies in it, either because they possessed special knowledge of problem areas or because they held certain positions in government service. Consequently the literature in the field shows more substance than organization or method. It has, therefore, seemed desirable not to divide the material in this chapter in terms of inventory and prospect, but rather to present an outline of topics that need to be analyzed and interpreted. Most of the topics pertain to areas organized politically as regions, regions both in the sense of uniformity and in the sense of coherent unity.

The citations noted in the text give some indication of the development of studies in the various parts of political geography, and the sum total of these references may stand as an inventory. It is clear that the greater part of the field as a whole remains little developed. The discussion of the various undeveloped topics suggests opportunities for future work in terms of a systematic framework that should enable successive students to build upon the work of their predecessors. Cultivation of a field of knowledge may be initiated by students who work wherever need arises or where interest and opportunity beckon, but the continued production of studies not related to each other or to any organization of knowledge will not produce, by sheer accumulation even over indefinite time, a useful and reliable body of knowledge. In geography, as in all branches of science, knowledge should increase at a geometric rate made possible by the establishment of fundamental concepts, principles, and methods within an organized framework, on the basis of which subsequent students can progress far more rapidly and surely than did their predecessors.

In such a development of an organized body of knowledge in political geography, important contributions will be made not only by those who will specialize in this field. Every geographer concerned with phenomena of area that result from differences in the character and work of man in different places will find that the interpretation of his data involves, among other factors, the political organization of the areas he studies. More especially, those who dig deeply into the geography of particular areas, as surely every geographer must do at one time or another, will have both the opportunity and the need to analyze the pattern of politically organized regions and their interrelations. Such an analysis is not merely an end chapter, not to say postscript, in the study of the geography of an area; in view of the direct and indirect effects of political organization on all patterns of human development, the political geography of an area is an essential and intrinsic part of its total geography. Every area specialist therefore has need of political geography and has opportunity to contribute to its development.

REFERENCES

1. HARTSHORNE, R. "The Nature of Geography . . . ," *loc. cit.*, ref. 3, chapter one.
2. SPROUT, H. H. "Political Geography as a Political Science Field," *American Political Science Review*, 25 (1931): 439-442.
3. JONES, S. B., and MURPHY, M. F. *Geography and World Affairs*, Chicago, 1950.
4. BOWMAN, I. *The New World: Problems in Political Geography*, New York, 1921; 4th edition (revised), 1928.
5. HARTSHORNE, R. "Recent Developments in Political Geography," *American Political Science Review*, 29 (1935): 784-804, 943-966.
6. WHITTLESEY, D. "Geographic Factors in the Relation of the United States and Cuba," *Geographical Review*, 12 (1922): 241-256.
7. ———. "Political Geography: A Complex Aspect of Geography," *Education*, 50 (1935): 293-298.
8. ———. *The Earth and the State*, New York, 1939.
9. VAN VALKENBURG, S. *Elements of Political Geography*, New York, 1939.
10. WHITTLESEY, D. *German Strategy of World Conquest*, New York, 1942.
11. TROLL, C. "Geographic Science in Germany during the Period 1933 to 1945," *Annals of the Association of American Geographers*, 39 (1949): 128-135.
12. SPYKMAN, N. J. "Geography and Foreign Policy," *American Political Science Review*, 32 (1938): 28-50, 213-236; SPYKMAN, N. J., and ROLLINS, A. A. "Geographic Objectives in Foreign Policy," *ibid.*, 33 (1939): 391-410, 591-614.
13. ———. *America's Strategy in World Politics*, New York, 1942; *idem. The Geography of Peace*, New York, 1944.
14. WEIGERT, H. W. *Generals and Geographers: The Twilight of Geopolitics*, New York, 1942.
15. WALSH, E. A. "Geopolitics and International Morals," in WEIGERT, H. W., and STEFANSSON, V., eds. *Compass of the World* . . . , New York, 1944: 12-39.
16. MACKINDER, H. J. "The Geographical Pivot of History," *Geographical Journal*, 23 (1904): 421-437; *idem. Democratic Ideals and Reality*, New York, 1919, 1942.
17. ———. *Britain and the British Seas*, Oxford, 1906, 1930.
18. WEIGERT, H. W. "Mackinder's Heartland," in WEIGERT, H. W., STEFANSSON, V., and HARRISON, R. E., eds. *New Compass of the World* . . . , New York,
19. SPROUT, H. H., and SPROUT, M. "Command of the Atlantic Ocean," in *Encyclopedia Britannica*, 1946, vol. 2, p. 637; *idem. Foundations of National Power*, New York, 1945, 1951.
20. LATTIMORE, O. *The Situation in Asia*, Boston, 1949.
21. REITZEL, W. *The Mediterranean: Its Role in America's Foreign Policy*, New York, 1948.
22. WEIGERT, H. W., and STEFANSSON, V., eds. *Compass of the World: A Symposium on Political Geography*, New York, 1944; WEIGERT, H. W., STEFANSSON, V., and HARRISON, R. E., eds. *New Compass of the World: A Symposium in Political Geography*, New York, 1949.
23. PEARCY, G. E., FIFIELD, R. H., and associates. *World Political Geography*. New York, 1948.
24. JONES, S. B. "Field Geography and Postwar Political Problems," *Geographical Review*, 33 (1943): 446-456.
25. WRIGHT, J. K. "Training for Research in Political Geography," *Annals of the Association of American Geographers*, 34 (1944): 190-201.
26. HARTSHORNE, R. "The Functional Approach in Political Geography," *Annals of the Association of American Geographers*, 40 (1950): 95-130.
27. GLENN, L. C. "Physiographic Influences in the Development of Tennessee," *The Resources of Tennessee*, April, 1915: 44-63.
28. WRIGHT, J. K. "Voting Habits of the United States," *Geographical Review*, 22 (1932): 666-672.

29. CRISLER, R. M. "Republican Areas in Missouri," *Missouri Historical Review*, 42 (1948): 299-309; *idem.* "The Little Dixie Region of Missouri," *Summaries of Doctoral Dissertations*, Northwestern University, 17 (1949): 352-356.
30. PLATT, R. S. "Conflicting Territorial Claims in the Upper Amazon," in COLBY, C. C., ed. *Geographic Aspects of International Relations*, Chicago, 1938: 243-276.
31. WHITTLESEY, D. "The Impress of Effective Central Authority upon the Landscape," *Annals of the Association of American Geographers*, 25 (1935): 85-97.
32. BOGGS, S. W. "Delimitation of Seaward Areas under National Jurisdiction," *American Journal of International Law*, 45 (1951): 240-266; *idem.* "National Claims in Adjacent Seas," *Geographical Review*, 41 (1951): 185-209.
33. LANGHANS, M. "Karte der Selbstbestimmungsrechtes der Völker," *Petermann's Mitteilungen*, 72 (1926): 1-9.
34. HARTSHORNE, R. "The Politico-Geographic Pattern of the World," *Annals of the American Academy of Political and Social Science*, 218 (1941): 45-57.
35. WIGMORE, J. H. "A Map of the World's Law," *Geographical Review*, 19 (1929): 114-120.
36. TAYLOR, G., ed. *Geography in the Twentieth Century*, New York, 1951.
37. HUNTINGTON, E. *Mainsprings of Civilization*, New York, 1945.
38. RATZEL, F. *Politische Geographie*, München, 1897.
39. HARTSHORNE, R. "Agricultural Land in Proportion to the Agricultural Population in the United States," *Geographical Review*, 29 (1939): 488-492.
40. United States Department of Agriculture, Office of Foreign Agricultural Relations, *Agricultural Geography of Europe and the Near East*, Washington, 1948.
41. BOGGS, S. W. "This Hemisphere," *United States Department of State Bulletin*, 12 (1945): 845-850.
42. JAMES, P. E. "Geopolitical Structures in Latin America," *Papers of the Michigan Academy of Science, Arts and Letters*, 27 (1941): 369-376.
43. BROEK, J. O. M. "Diversity and Unity in Southeast Asia," *Geographical Review*, 34 (1944): 175-195.
44. HARTSHORNE, R. "The United States and the 'Shatter Zone' of Europe," in WEIGERT, H. W., and STEFANSSON, V., eds. *Compass of the World . . .*, New York, 1944: 74-88.
45. ROUCEK, J. S. "The Geopolitics of the Balkans," *American Journal of Economic and Social Affairs*, October, 1946: 365-377.
46. KISH, G. "TVA on the Danube," *Geographical Review*, 37 (1947): 274-303.
47. BRIGHTMAN, G. F. "The Boundaries of Utah," *Economic Geography*, 16 (1940): 87-95.
48. WHITTLESEY, D. "Trans-Pyrenean Spain: The Val d'Aran," *Scottish Geographical Magazine*, 49 (1933): 217-228; *idem.* "Andorra's Autonomy," *Journal of Modern History*, 6 (1934): 147-155.
49. BRUSH, J. E. "The Distribution of Religious Communities in India," *Annals of the Association of American Geographers*, 39 (1949): 81-98.
50. HARTSHORNE, R. "The Tragedy of Austria-Hungary: A Post-Mortem in Political Geography," *Annals of the Association of American Geographers*, 28 (1938): 49; *idem.* "The Concepts of 'Raison d'Être' and 'Maturity' of States," *ibid.*, 30 (1940): 59.
51. JAMES, P. E. *Latin America*, Revised Edition, New York, 1950.
52. HARTSHORNE, R. "Geographic and Political Boundaries in Upper Silesia," *Annals of the Association of American Geographers*, 23 (1933): 194-228; *idem.* "The Polish Corridor," *Journal of Geography*, 36 (1937): 161-176.
53. JAMES, P. E. "Forces for Unity and Disunity in Brazil," *Journal of Geography*, 38 (1939): 260-266.
54. HARTSHORNE, R. "Racial Maps of the United States," *Geographical Review*, 28 (1938): 276-288.

55. HOFFMAN, G. W. "The Survival of an Independent Austria," *Geographical Review*, 41 (1951): 606-621.
56. WHITTLESEY, D. "Reshaping the Map of Africa," in COLBY, C. C., ed. *Geographic Aspects of International Relations*, Chicago, 1938: 127-159.
57. LATTIMORE, O. "Yakutia and the Future of the North," in WEIGERT, H. W., STEFANSSON, V., and HARRISON, R. E., eds. *New Compass of the World . . .*, New York, 1949: 135-149.
58. PELZER, K. J. "Micronesia: A Changing Frontier," *World Politics*, 2 (1950): 251-266.
59. PAULLIN, C. O. *Atlas of the Historical Geography of the United States*, New York, 1932.
60. CUSHING, S. W. "The Boundaries of the New England States," *Annals of the Association of American Geographers*, 10 (1920): 17-40.
61. ULLMAN, E. L. "The Historical Geography of the Eastern Boundary of Rhode Island," *Research Studies of the State College of Washington*, 4 (1936): 67-87.
62. THOMAS, B. E. "Demarcation of the Boundaries of Idaho," *Pacific Northwest Quarterly*, 40 (1949): 24-34; idem. "The California-Nevada Boundary," *Annals of the Association of American Geographers*, 42 (1952): 51-68.
63. ULLMAN, E. L. "The Eastern Rhode Island-Massachusetts Boundary Zone," *Geographical Review*, 29 (1939): 291-302.
64. BOWMAN, I. "An American Boundary Dispute: Decision of the Supreme Court of the United States with Respect to the Texas-Oklahoma Boundary," *Geographical Review*, 13 (1923): 161-189.
65. MARTIN, L. "The Michigan-Wisconsin Boundary Case in the Supreme Court of the United States," *Annals of the Association of American Geographers*, 20 (1930): 105-163; idem. "The Second Wisconsin-Michigan Boundary Case in the Supreme Court of the United States," *ibid.*, 28 (1938): 77-126.
66. THOMAS, B. E. "Boundaries and Internal Problems of Idaho," *Geographical Review*, 39 (1949): 99-109.
67. JONES, S. B. "Intra-State Boundaries in Oregon," *The Commonwealth Review*, 16 (1934): 105-126.
68. HANSON, G. H. "The Geographic Factor and its Influence in Utah Administrative Units," *Yearbook of the Association of Pacific Coast Geographers*, 1937: 3-8.
69. ULLMAN, E. L. "Political Geography in the Pacific Northwest," *Scottish Geographical Magazine*, 54 (1938): 236-239.
70. ANDERSON, W. *The Units of Government in the United States*, Public Administration Service, Publication No. 42, Chicago, 1934.
71. NELSON, H. J. "The Vernon Area, California: A Study of the Political Factor in Urban Geography," *Annals of the Association of American Geographers*, 42 (1952): 177-191.
72. KOLLMORGEN, W. "Political Regionalism in the United States: Fact or Myth," *Social Forces*, 15 (1936): 111-122.
73. PLATT, R. S. "Latin America in World Affairs," *Journal of Geography*, 40 (1941): 321-330.
74. BOGGS, S. W. *International Boundaries: A Study of Boundary Functions and Problems*, New York, 1940.
75. JONES, S. B. "The Forty-Ninth Parallel in the Great Plains: The Historical Geography of a Boundary," *Journal of Geography*, 31 (1932): 357-368.
76. ———. "The Cordilleran Section of the Canada-United States Borderland," *Geographical Journal*, 89 (1937): 439-450.
77. HARTSHORNE, R. "Suggestions on the Terminology of Political Boundaries," *Mitteilungen des Vereins der Geographen an der Universität Leipzig*, 14/15 (1936): 180-192.

78. POUNDS, N. J. G. "The Origin of the Idea of Natural Frontiers in France," *Annals of the Association of American Geographers*, 41 (1951): 146-157.
79. BROEK, J. O. M. "The Problem of Natural Frontiers," *Frontiers of the Future*, University of California, Berkeley, 1941: 3-20.
80. FISCHER, E. "On Boundaries," *World Politics*, 1 (1949): 196-222.
81. BOGGS, S. W. "Problems of Water-Boundary Definition," *Geographical Review*, 27 (1937): 445-456.
82. HARTSHORNE, R. "A Survey of the Boundary Problems of Europe," in COLBY, C. C., ed. *Geographic Aspects of International Relations*, Chicago, 1938: 163-213.
83. JONES, S. B. *Boundary Making: A Handbook for Statesmen, Treaty Editors, and Boundary Commissioners*, Washington, 1945.
84. LATTIMORE, O. *Inner Asian Frontiers of China*, American Geographical Society Research Series, No. 21, New York, 1940.
85. HARTSHORNE, R. "The Franco-German Boundary of 1871," *World Politics*, 2 (1950): 209-250.
86. STANISLAWSKI, D. "Tarascan Political Geography," *American Anthropologist*, 49 (1947): 46-55; *idem.* "The Political Rivalry of Patzcuaro and Morelia: An Item in the Sixteenth Century Geography of Mexico," *Annals of the Association of American Geographers*, 37 (1947): 135-144.
87. JENSEN, M. *The New Nation: A History of the United States during the Confederation, 1781-1789*, New York, 1950.
88. JONES, S. B., and MEHNERT, H. "Hawaii and the Pacific," *Geographical Review*, 30 (1940): 358-375.
89. *Report and Accompanying Papers of the Commission Appointed by the President to Investigate and Report Upon the True Divisional Line Between the Republic of Venezuela and British Guiana*, Washington, 1896-1897.
90. DOMINIAN, L. *The Frontiers of Language and Nationality in Europe*, New York, 1917.
91. WEIGEND, G. C. "Effects of Boundary Changes in the South Tyrol," *Geographical Review*, 40 (1950): 364-375.
92. HOFFMAN, G. W. "South Tyrol: Borderland Rights and World Politics," *Journal of Central European Affairs*, 7 (1947): 285-308.
93. ———. "The Netherlands Demands on Germany: A Post-War Problem in Political Geography," *Annals of the Association of American Geographers*, 42 (1952): 129-152.
94. HELD, C. C. "The New Saarland," *Geographical Review*, 41 (1951): 590-605.
95. MCCUNE, S. "Physical Bases for Korean Boundaries," *Far Eastern Quarterly*, (1946): 272-288; *idem.* "The Thirty-Eighth Parallel in Korea," *World Politics*, 1 (1949): 223-232.
96. VAN VALKENBURG, S. *European Jigsaw*, Headline Series, Foreign Policy Association, New York, 1945; *idem. Pacific Asia, ibid.*, 1947; *idem. Whose Promised Land? ibid.*, 1948.

THE GEOGRAPHY OF RESOURCES[A]

A. Original draft prepared by J. Russell Whitaker under the direction of Raymond E. Murphy, chairman of the Committee on Economic Geography. Among those particularly helpful in manuscript revisions were the chairman, William Applebaum, Joseph A. Russell, and Alfred J. Wright.

CHAPTER EIGHT

THE GEOGRAPHY OF RESOURCES

THE UNITED STATES has provided geographers with a fruitful field for the study of resources. It has furnished ample opportunity for examining the commercial development of natural resources in a land which was essentially unused at the time of the first settlement by Europeans; and it has also furnished a laboratory for recording the processes of resource destruction. During the present century Americans have been active in the movement to counteract the destruction of natural resources and to maintain them at levels commensurate with current and anticipated needs. The United States has also furnished data for the analysis of human as well as natural resources.

WHAT ARE RESOURCES?

Resources have been defined by Erich Zimmermann as "those aspects of man's endowment and environment upon which people are dependent for aid and support" [1:1]. Unfortunately, one is all too likely to limit such a definition to those aspects that lie within one's own particular interest. Viewed broadly, it is possible to recognize at least three chief categories of resources: 1) those derived from the physical and biotic conditions of the land, called natural resources; 2) those derived from the population in the sense of man-power, or human resources; and 3) those derived from the attitudes, objectives, and technologies of the people, called cultural resources.

Natural resources are the materials provided originally by non-human agencies. The word "resources" however, implies something which has proved useful to man. As J. O. M. Broek has stated, "an element of the natural environment becomes a resource only when man recognizes it as useful" [2:322]. Zimmermann has developed more fully than anyone else the concept of resources as "purely functional, inseparable from human wants and human capabilities" [3:10]. Sauer remarked that "the term 'resources' implies the determination that the thing is useful and therefore a cultural achievement" [4:18; 5:3]. It is helpful in this analysis to follow Zimmermann [3:13] in speaking of natural conditions as essentially neutral material into which man drives a wedge of culture. In so doing, he converts some aspects of the neutral material into resources, but at the same time he meets new resistances. This functional approach to earth conditions is not new to economic geographers; indeed, it is the very essence of their approach. Significant in this connection is the heavy reliance of Zimmermann's book, especially the first edition, on the work of economic geographers.

If taken literally, the definition of resources as framed by Zimmermann must also include man himself. Certainly men provide aid and support to other men. Within certain limits, moreover, human resources can be substituted for those derived from the earth. To distinguish earth-derived resources from human is so useful in both theory and action that it seems desirable to continue to recognize these two categories.

The third category of resources results primarily from what man himself does, and in large degree from his learned behavior. Accordingly these are called cultural resources, fitting the terminology into the customary usage of the word "culture" by social scientists. Cultural resources may be divided into two groups: the elements of material culture, such as houses, roads, tools, and machines; and those of non-material culture, such as economic, social, and political institutions, religious ideas, and group attitudes.

As in all classifications, this one is useful if not followed too rigidly. One can think of resources that fall rather neatly into each category; on the other hand, there are easily identifiable resources that are clearly gradational. Virgin soil is a natural resource for farmers; roads are a cultural resource; a canalized river, or a soil modified by use, lies somewhere between natural resources and cultural resources. Similarly there are gradations between other classes. A highly-skilled technician is a human resource, but he also reflects a large measure of technical (non-material) culture; and he can do his work only with the tools and machines (material culture) at hand.

In actual practice it is customary to focus attention on one of these resource classes and to have the others less sharply in mind, or to include only their margins. The term " natural resources" thus commonly includes not only the original earth conditions that have become significant to man, but also earth conditions which have undergone significant modifications. Soil is considered a natural resource, even though man may have greatly changed it; and similarly other resources that have been greatly modified are included in the term. In brief, the label "natural resources" is interpreted broadly enough to include what Zimmermann has called natural-cultural resources [3:83].

In rounding out the definition of resources, two other points should be considered briefly. Fundamental is the fact that resources must be considered in combination. Not only do these major groups of resources function together, but each group also is normally a composite of several elements. To illustrate the first point, the development of New England is completely misjudged unless both human and cultural resources are given due weight. It is a truism among geographers, too, that iron ore is of relatively little use unless it can be assembled with coal (coke) and limestone at relatively low cost. Thus the distributions of resource combinations are of critical significance.

A second consideration applies particularly to the work of geographers and further restricts the scope of the term "resources" in this analysis. In general, those American geographers who have been interested in resources have focused attention on the natural (and natural-cultural ones), and the other categories, man-power and material and non-material culture, have been marginal. More precisely, we have considered man as the agent, conditioned by his non-material culture, utilizing the natural resources to produce his material culture. An assessment of past performance indicates that geographers have focused their attention chiefly on the earth conditions, both original and man-modified, as valued and used by particular peoples. In the discussion that follows the term "resources" will be used in this restricted meaning.

RESOURCES IN AMERICAN GEOGRAPHY

Geographers recognize that all of the phenomena with which they deal are also the concern of persons in other disciplines. This is certainly true of resources. Not only that, but resources are becoming the center of an organization of fact, principle, and procedure that cuts across established disciplines; or, to put it another way, an area of endeavor into which various specialists have moved, there to work together under a different frame of reference from that of the disciplines of their original preparation. This new field is one of policy and action as well as of theory, and it is unified by purposes and history. It should be recognized, accordingly, that many American geographers who study resources have been working not only in their own discipline and among their own brethren but that they commonly cooperate in an organization of scholars and practitioners that transects their field; an organization which is nevertheless built on geographic lines because of its attention to the association of phenomena in particular places.

Bases of Concern and Competence

Of primary importance in the study and management of resources is the concern of the geographer with individual elements of the earth's surface (more precisely, of the life layer of the earth) as these elements are distributed over the earth and as they are associated in particular places. Geographers are concerned in physical geography with earth conditions, whether resources or not; and the utilization, destruction, and betterment of these earth conditions are the concern of the economic and political geographer.

Fundamental to the geographer's approach is his traditional interest in man's dependence on nature. It falls to the geographer, among students of human behavior, to analyze these relationships as they exist in space and in time. Yet the geographer well knows that the identification of a dependence of man on nature is but the beginning of his problem. He has still to analyze man's wants and capacities in order to understand how the given dependence came about.

Geographers have been among the foremost workers in earth science in the study of man's effect on nature. Moreover, geographers are concerned with all changes in nature, whether those changes affect the current usefulness of earth conditions or not. The physical geographer would certainly disavow any restriction of his study of the earth to matters clearly of functional significance. On the other hand, the economic geographer quite logically is primarily interested in the changes that increase or decrease the resource value of an area.

Since the days of George P. Marsh [6], who became interested at least one hundred years ago in the modification of earth conditions by man and who relied heavily on European geographers, many Americans have given special attention to resource destruction and betterment. It is perhaps unfortunate that geographers look to Gifford Pinchot for the phrase "conservation of natural resources," [7:319-326] and fail to look still farther back to distinguished scholars, Europeans as well as Americans, who worked on the problem but did not use this label. The geographer will recall not only the work of Marsh and Nathaniel S. Shaler [8], but also, at a later period, that of Carl O. Sauer [9; 10] and others [15; 19; 24].

A major contribution of geographers in dealing with resources emerges from their attention to the interrelations of associated phenomena in particular places. Geographers have been able to bring broad perspective to problems of resource-use planning. The geographic approach facilitates the investigation of resource conflicts and the planning of cooperative combinations of resources. Geographers can do much to clarify conflicts and to suggest remedies: for example, in the conflict between strip mining and farming; in the struggle between those who favor water conservation and those who want more cheap grazing land; or in the differences between promoters of unrestricted hunting and fishing privileges and the proponents of game and fish preservation. One of the more active participants in this work in the last twenty years told a member of this committee that as late as 1930 he found little awareness among non-geographers of the various interrelations involved in water-resource planning. There was, for example, virtually no recognition among non-geographers of the unity of individual watersheds, whereas such unities are of the very essence of the geographer's thinking.

ESTABLISHED PLACE OF RESOURCES IN AMERICAN GEOGRAPHY

In American geography resource study appears in at least three connections: 1) as a part of the study of man-earth relations; 2) as a part of the practical work of land classification and use inventory; and 3) in the study of the patterns of economic production and trade.

The history and the theory of man-resource relations have made up a significant sector of geographic thought. They constituted a part of the analysis of geographic influences (E. C. Semple), geographic adjustments (H. H. Barrows), and areal relationships. On these themes virtually all students of the philosophy of geography have had their say, from the earliest American geographers to the contributors to this volume.

The inventory and appraisal of natural resources has been a major concern of American geographers. From the initial settlement of our country, and from the very beginning of geographic interest in America, one of the tasks constantly besetting Americans has been the making of inventories and appraisals for the areas of settlement. The somewhat crude and undifferentiated analyses of early years gave way in the late 1800's to the work of specialists on individual resources, and the overall evaluation suffered somewhat. In the last thirty years, however, geographers have been utilizing the findings of specialists and applying these findings to the study of resource complexes in specific localities.

To understand the functional significance of resources, study of the time sequence is essential. Historical geography, in which the changing significance of resources in specific areas can be analyzed, has thrown new light on man's relation to the land and has developed new concepts to guide geographic research. Regional geography, in its revival in the 1920's, sharply directed attention to the resource base of specific areas, giving consideration not only to regional advantages but also to regional liabilities. Here, also, the time perspective brought new illumination to regional analysis.

Somewhat eclipsed during the 1920's by regional geography, political geography, cultivated with renewed and increasing vigor in the 1930's and 1940's, took as one of its principal themes the national endowment of the countries considered. This emphasis on national endowment, the resource inventories of politically organized areas, is in accordance with the mid-20th century preoccupation with the goals of national autarchy; it is still a phase of the study of the resource base of specific areas or regions, only here the region is defined in political terms (chapter seven).

Quite at the other end of the geographical scale is detailed land classification, a theme developed in chapters ten, sixteen, and twenty-four. Workers in land classification have accepted a conclusion that many geographers have long held, namely, that land should be classified in terms of specific uses; in other words, that land classification should conform to the concept of a resource as something useful for a specific purpose under specific conditions. Noteworthy work in this country in land classification was carried out for the Michigan Land Economic Survey in the 1920's, for the Tennessee Valley Authority in the 1930's, and, more recently, for the Puerto Rico Rural Land Classification Project, which is discussed in later chapters.

Some geographers have been watching the rising standards of living over the world in relation to the various limits that appear to be set by natural resources. Moreover, geographers are reminded by some members of the profession that the limits to resources are cultural, as well as natural. As some opportunities disappear with the filling up of hitherto unoccupied lands, others are created by technical advances. Even though one accepts the "closed-space" thesis of Frederick Jackson Turner and others, James C. Malin warns that a closing out of opportunities for expanding resources should not be assumed [11].

The geographer's contribution to the inventory and appraisal of resources is thus many-sided and wide-ranging. It runs the scale from R. J. Russell's study of desert rainfall [12], to the recent study, under E. A. Ackerman's direction, of Japan's natural resources, a thorough inventory and evaluation of that nation's resources in terms of its needs and capacities [13].

Much of the work of geographers with resources is a normal phase of their study of production and trade. It occasionally happens, moreover, that an investigator who is primarily concerned with economic production may devote a considerable part of his time to preliminary work on resources. In addition, there are certain phases of the productive use of resources that give special attention to the resources themselves, notably the so-called extractive industries, in which the product is a part of the resource actually removed from its place in nature. Any study of the forest industry or the fishing industry of necessity gives a great deal of attention to the resources being exploited.

RESOURCES AS A SPECIAL FIELD

The study of natural or earth-derived resources, in original and in altered form, is the aspect of resource study most commonly cultivated by geographers, presumably because they normally are trained not only in the economic side of their field but also in the physical side. Research in this field makes use of geographic methods that are oriented toward special objectives.

METHODS

In an investigation of natural-resource problems, the usual methods of the geographer are all utilized: field observations, interviews, statistical and cartographic analysis, and the study of historical documents (chapter twenty-four). It is perfectly clear to any geographer who analyzes such a paper as E. W. Miller's on strip mining [14] that all of these methods are essential. As in other phases of geography, the relative balance will depend on the specific problem, as well as on the bent of the investigator.

In view of the scope of resource analysis and its importance in action programs, it is desirable to stress the increasing degree to which geographers can and do depend on the work of specialists in related disciplines. As previously noted, geographers have moved from the somewhat general, undifferentiated study of earth conditions of Marsh's day through a period of fragmentation and specialization to a second period of comprehensive views, based at present not only on the investigator's own observations but also on those of others. Moreover, as research is carried out by groups of workers, whether under the direction of governmental agencies or not, it becomes increasingly possible for the geographer who operates as a resource generalist actually to direct lines of investigation by co-workers in related disciplines. Here is an opportunity for team research consistent with research procedures developed during the period of World War II.

ACHIEVEMENTS AND OPPORTUNITIES

The foregoing analysis of the place of resources in American geography as a whole identifies or implies some of the principal achievements and inviting prospects in this field. These will be stated more precisely now in terms of: 1) long-time developments; 2) strong developments since World War I; 3) good beginnings needing vigorous support; and 4) major opportunities along little-developed lines.

LONG-TIME DEVELOPMENTS

At least three long-time developments in the field of natural resources were identified in the above analysis. These are:

1. From the beginning of American geography there has been sustained inquiry into the nature and geographic distribution of natural resources;

2. Throughout the 19th century and to the present, American geographers have been concerned with resource appraisal for specific areas;

3. For more than a century, study of the depletion and conservation of natural resources has continued, but with uneven emphasis. It now appears that the main line of descent from Marsh, the principal of the pioneers, crossed to Europe and passed through Alexander Woeikof and Ernst Friedrich to Jean Brunhes, and from Brunhes back to this country to appear in the works of many of our contemporaries [15: 73-95]. The scope of Marsh's study, world-wide and richly elaborated, has been matched in no subsequent work, although occasional essays, as by Sauer [4; 9; 10], are in the Marsh tradition. The main American stem out of the Marsh root, however, has been concerned with the natural-resource base of the United States.

Strong Developments Since World War I.

During World War I the efforts of American geographers were channeled to a large degree into wartime activities (chapter twenty-three). Moreover, the interest of non-geographers in resources was likewise subordinated to the exigencies of war. The return of geographers to academic posts and the shift of the attention of some American political leaders back to internal problems set the stage for a strong development along many lines.

1. The systematic analysis of the natural resources of the United States and their conservation expanded rapidly and became an important part of school and college education in this country, with geographers taking a leading place in writing and teaching about this subject [16; 17; 18; 19].

2. During the 1930's there developed a keen appreciation of the interrelation of various items in resource complexes, an appreciation to which regional geographers and participants in regional planning have made major contributions.

3. The region as an operational unit in resource management has received much attention. The regional approach to American problems was greatly accelerated during the 1920's and 1930's, springing up almost independently in various disciplines. It is not too much to say, however, that geographers have made notable contributions to this aspect of resource study, in terms not only of the equipment of regions but also in setting up geographic frames for resource management.

4. Another strong development in the period immediately following World War I was the growth among geographers of an awareness of resource management as an action field, and an appreciation of the opportunities and responsibilities of geographers, particularly as resource generalists. This is a revival of the spirit of John Wesley Powell appearing in the work of H. H. Barrows, K. C. McMurry, C. C. Colby, and others.

Some are content to stop with making clear the pertinence of geographic findings to particular policies, but others are concerned primarily with the actual guidance of regional development. In referring to the rise of like purposes in France, Jean Gottman writes, "Demangeon asks of a geographical study that it draw conclusions as to the possible improvement of existing conditions" [20:128]. And this is becoming more and more a general tendency in the geographer's endeavor to achieve his part in what Isaiah Bowman called "the creative experiment."

5. Among the major developments in the period between the two world wars was the contribution to the field of water resources, into which geographers have moved with exceptional vigor in the last twenty years. Perhaps one might take as a landmark of early achievements the report of the Mississippi Valley Committee, on which Barrows had a place [21]. Indications of more recent achievements in this field were the presence of Gilbert F. White on the President's Water Resources Policy Commission; the direction of Volume II of the Commission report, a voluminous work, by Ackerman; and the cooperation of a number of other geographers in the work of the Commission [22]. No discipline has grown up to cover all of the aspects that come to the fore in any functional analysis of water problems. It is into this gap that many geographers have moved with considerable vigor and notable results.

Good Beginnings Needing Vigorous Scholarship

Turning to promising lines of study and application in which as yet relatively little has been accomplished, one moves gradually from an evaluation of achievements to an identification of open gates and inviting vistas, with corresponding room for greater difference of opinion. Four suggestions in identifying good beginnings needing vigorous scholarship follow.

1. There is work to be done in the topical study of each specific kind of resource. The definition and the mapping of vegetation types are especially needed (chapter nineteen). Although the native vegetation has undergone wholesale removal over large parts of the world, there are still vast areas for which such information makes a highly significant addition to our knowledge of resources. Similarly geographers and others are working on water resources with inadequately formulated concepts for guidance and with essentially no body of literature that describes water resources in a systematic way, whether for the United States or the world.

2. Americans need to appraise resources in the outlying parts and dependencies of the United States. The long-time, vigorously supported analysis of the resources of the continental United States has not been

matched by corresponding achievements in other parts of our national domain. Geographers are moving rapidly to close this gap for Puerto Rico, and significant beginnings have been made in Alaska; but much remains to be done.

3. Case studies are needed at every hand, studies of the functioning of particular resource combinations, of resource depletion, and of resource management, whether successful or unsuccessful by current standards. In discussing river floods, Gilbert White pleads for intensive studies of adjustment to floods in this and other countries, in order to identify the conspicuously successful and unsuccessful adjustment for each important type of floodplain and floodplain occupance. An example is the comprehensive study of drainage in the lower Mississippi by R. W. Harrison and W. M. Kollmorgen [23].

Especially needed for shaping policies and action programs are case studies, by independent investigators, of controversial aspects of resource development. Too commonly the assessment of resource needs and resource treatment has been made by persons who have not been free, because of the avowed interests of the organization or persons for whom they are working, either to carry out disinterested inquiry or to publish the results of their findings. Geographers not only have the competence and interest for making such case studies, but a large proportion of them are essentially impartial.

4. Geographers might well expand their study of resource problems to include great geographic realms, such as the rainy tropics. Karl J. Pelzer and Earl P. Hanson have worked in this direction, as have also Charles E. Kellogg and Robert L. Pendleton. As yet, however, no one has provided us with a comprehensive and comparative study of the impact of shifting cultivation on soil resources for the rainy tropics as a whole.

Some Major Opportunities Where Little Has Been Done

American geographers who have entered the profession within the last ten years appear to have given relatively little attention to resources as a special field, however much they have been concerned with resources as relevant to an understanding of productive occupance or even of urban development. A check of doctoral dissertations and of articles in *The Geographical Review, The Annals of the Association of American Geographers,* and *Economic Geography* substantiates this conclusion. It would appear, therefore, that there is ample room for many more American geographers to direct their attention to some of the problems that resources present. A list of major opportunities along lines where relatively little has been done can not be much more than an identification of the more obvious openings.

1. Investigations of the recreational value and use of resources and of the scenic aspects of resources is a new field that is described more fully in chapter nine.

2. As the study of mineral geography moves forward, there is an opportunity to analyze the impact of mining on other resources and the development of policies which will take social as well as market values into account (chapter eleven). There has been a tendency to leave the field of non-renewable resources to the engineer and to the geologist. While in no sense wishing to belittle the contributions of workers in other disciplines, one must emphasize also the need for the geographic study of mineral resources. Over forty years ago, C. R. Van Hise wrote in the preface of his book on conservation that he hoped that it would not be long until there would be separate handbooks dealing with each major group of resources in terms of their destruction and conservation [24:v]. Gradually through the years these gaps have been filled, save one: there has been no comprehensive attempt to deal with the conservation of mineral resources in book-length detail.

3. Historical physical geography, including natural resources, has been little cultivated. What, precisely, has happened to soils or waters over a considerable length of time? Indeed, we have increasing reason to wonder regarding the nature and extent of modification by the American Indian of the so-called natural conditions of this country [5: 1-18]. All too commonly geographers have rather naively assumed that the white man found a virgin continent. The English schoolmaster, Thomas Arnold, clearly phrased the basic question when he wrote, "How much we want a physical geography of countries, tracing the changes they have undergone either by such violent revolutions as volcanic phenomena, or by the slower but not less complete changes produced by ordinary causes: such as alteration occasioned by enclosure and drainage, alteration in the course of rivers, and in the level of their beds, alteration in the animal and vegetable products of the soil, and in the supply of metals and minerals, noting also the advance or retreat of the sea, and the origin and successive increase in the number and variation in the line of roads, together with the changes in the extent and character of the woodland" [25:66]. Such a stimulating demand upon the potentialities of geographic study has still not been adequately met.

4. There is need to enlarge the scope of resource-management studies to include the comparative investigation of small areas and the study of areas larger than localities but smaller than countries. As yet, relatively little is known regarding the reasons why national policy and resource use are not equally effective in all parts of the United States. There is a clearly urgent need for materials on a regional and a State level. In the

United States a large measure of resource management is the function of the individual States. In a number of them educational programs to match that responsibility have been developing rapidly. It is still true, however, that only a comparatively few States have attempted to provide suitable teaching materials for the resources of the State as a whole, a notable exception being Florida [26]; and even scarcer are teaching materials which cover the resources of State subdivisions. It would seem, in this connection, that the examination of the contrasts resulting from varying State policies would be a rewarding line of research combining political geography and the resource field.

5. The identification and description of types of resource associations and of types of geographic patterns of resource depletion and conservation, and the mapping of those types, have scarcely begun. Although there are world-wide climatic classifications, classifications of land use, and many others, the classification and mapping focused on these additional phases of resource analysis have lagged far behind.

6. Through American geography runs a reasonably adequate emphasis on natural factors in man's relation to the earth; but relatively slight attention has been given to cultural factors. This same weakness appears in the special field of resource management. Indeed, this presents an even greater need because many of the more vigorous and productive students of resources are in the natural sciences, where their bent, training, and habits of formulating problems may leave cultural factors out of consideration. The geographer should give much attention to these factors in man-resource relations, and should most certainly be able to evaluate culture as an aid or hindrance to desired changes in resource use. There is no good reason why economic geographers should not move vigorously to fill this gap.

7. Most of this section, beginning with the concern and competence of geographers with resources, has dealt with earth resources, both unaltered and as modified by man. The preoccupation of geographers with natural resources harmonizes with the development of American geography as a whole, but it does not appear to be justified by the logic of the discipline. There is now no good reason why geographers should fail to give direct sustained attention to the areal differentiation of human resources and of resources of material and non-material culture. Much of our failure properly to judge the vitality of the economy of regions stems from lack of attention to human and cultural resources. A word of warning is appropriate, however. Where man is directly involved, as in the study of human and cultural resources, objectivity is vastly more difficult to attain than in the study of earth resources. But this very difficulty presents a challenge. Work along this line would surely reward geographers who are competent to deal with man's handiwork and with man himself.

REFERENCES

1. ZIMMERMANN, E. W. "What We Mean by Resources," Chapter 1 in DRUMMOND, L., ed. *Texas Looks Ahead*, Austin (Texas), 1944.
2. BROEK, J. O. M. "The Relations Between History and Geography," *Pacific Historical Review*, 10 (1941):321-325.
3. ZIMMERMANN, E. W. *World Resources and Industries*, 2nd ed., New York, 1951.
4. SAUER, C. O. "Early Relations of Man to Plants," *Geographical Review*, 37 (1947): 1-25.
5. ———. *Agricultural Origins and Dispersals*, American Geographical Society, New York, 1952.
6. MARSH, G. P. *Man and Nature; or Physical Geography as Modified by Human Action*, New York, 1864.
7. PINCHOT, G. *Breaking New Ground*, New York, 1947.
8. SHALER, N. S. *Man and the Earth*, New York, 1905.
9. SAUER, C. O. "Theme of Plant and Animal Destruction in Economic History," *Journal of Farm Economics*, 20 (1938):765-775.
10. ———. "Destructive Exploitation in Modern Colonial Expansion," *Proceedings of the International Geographical Congress*, Amsterdam, 1938, Vol. 2, Section III C: 494-499.
11. MALIN, J. C. "Space and History," *Agricultural History*, 18 (1944):65-72, 107-126.
12. RUSSELL, R. J. "The Desert-Rainfall Factor in Denudation," *Report of the 16th International Geological Congress*, 1933, Vol. 2, Washington, 1936:753-763.
13. General Headquarters, Tokyo, *Japanese Natural Resources*, Tokyo, 1949.
14. MILLER, E. W. "Penn Township: An example of Local Governmental Control of Strip Mining in Pennsylvania," *Economic Geography*, 28 (1952):256-260.
15. WHITAKER, J. R. *The Life and Death of the Land*, Nashville, 1946.
16. PARKINS, A. E., and WHITAKER, J. R., eds. *Our Natural Resources and Their Conservation*, New York, 1936.
17. RENNER, G. T. *Conservation of National Resources*, New York, 1942.
18. SMITH, G-H., ed. *Conservation of Natural Resources*, New York, 1950.
19. WHITAKER, J. R., and ACKERMANN, E. A. *American Resources, Their Management and Conservation*, New York, 1951.
20. GOTTMANN, J. "Vauban and Modern Geography," *Geographical Review*, 34 (1944):120-128.
21. Public Works Administration, *Report of the Mississippi Valley Committee*, Washington, 1934.
22. The President's Water Resources Policy Commission, *Report*, (3 vols.) Washington, 1950.
23. HARRISON, R. W., and KOLLMORGEN, W. M. "Drainage Reclamation in the Coastal Marshlands of the Mississippi River Delta," *Louisiana Historical Quarterly*, 30 (1947): 1-57.
24. VAN HISE, C. R. *The Conservation of Natural Resources in the United States*, New York, 1910.
25. FITCH, J. G. *Thomas and Matthew Arnold and Their Influence on English Education*, New York, 1897.
26. BECKER, H., and others. *Florida, Wealth or Waste*, Florida State Department of Education, Tallahassee (Florida), 1946.

THE FIELDS OF ECONOMIC GEOGRAPHY

DEVELOPMENT OF ECONOMIC GEOGRAPHY IN AMERICA[A]
THE CENTRAL THEME
BASIC PROCEDURES
RELATION TO ECONOMICS
APPLICATIONS OF ECONOMIC GEOGRAPHY
SUBDIVISIONS OF ECONOMIC GEOGRAPHY

MARKETING GEOGRAPHY[B]
SCOPE
    Presenting Market and Marketing Data
    Evaluating Markets
    Delineating Trading and Selling Areas
    Selecting Channels of Distribution and Business Locations
TECHNIQUES

RECREATIONAL GEOGRAPHY[C]
THE PHENOMENA OF RECREATION
GEOGRAPHIC LITERATURE ON RECREATION
ECONOMIC RESEARCH IN RECREATIONAL GEOGRAPHY
THE PROSPECT

OTHER FIELDS OF ECONOMIC GEOGRAPHY[A]

[A]. Original draft by Raymond E. Murphy.
[B]. Original draft by William Applebaum.
[C]. Original draft by K. C. McMurry and Charles M. Davis.

CHAPTER NINE

THE FIELDS OF ECONOMIC GEOGRAPHY

ECONOMIC GEOGRAPHY can hardly be treated as a research field in the same sense as the fields discussed in other chapters of this book. It is, rather, a group of fields. All economic geography has a common base in that it deals with man's means of gaining a livelihood; but there are so many different aspects of this basic interest and so many different procedures in the analysis of the different aspects that no one person could become equally competent in all of them. To a certain extent, economic geography is like physical geography in that continued growth tends more and more to break it into component topical specialties.

This process of subdivision has not gone so far in economic geography as it has in physical geography. Some professional geographers still maintain an interest in economic geography as a whole. Furthermore, the subdivisions that have emerged as separate topical specialties are not yet numerous, and there are many aspects of the subject that have not been cultivated by professional geographers but in which the application of geographic method could yield results of great significance.

DEVELOPMENT OF ECONOMIC GEOGRAPHY
IN AMERICA

The most rapid development of economic geography and the most rapid subdivision into component specialties took place during and after World War I. Among the earliest geographic writings in the United States were papers in economic geography, and in general geographic works it was customary to include sections on the geography of economic processes. The maps and reports on land quality based upon the great surveys of the Western States could be considered as contributions to this general field. Furthermore, the beginnings of the intensive study of small areas were made between 1912 and 1915, as described in chapters ten and twenty-four.

During the first two decades of the 20th century the standard works on economic geography dealt with trade in commodities. The economic geographers were then concerned with tons of coal, bushels of wheat, bags of coffee, and a country with a small foreign trade received scant attention. Most of the publications were based on statistical sources.

Among the professional geographers whose names are associated with this period of commodity studies are G. B. Roorbach, J. Russell Smith, and R. H. Whitbeck. The maps of world agricultural production by V. C. Finch and O. E. Baker (discussed in chapter ten) represented a culmination of this work.

World War I proved a great stimulus to economic geography. The geographers responded to a demand for knowledge about sources of food and raw materials in the world as a whole, and for an understanding of the economic problems of different countries. The new magazine *Economic Geography*, published by Clark University, was started in 1925, and in it there is a record of the changing character of research in this field. The economic geographers in considerable numbers turned their attention to agricultural problems and to the classification of land in terms of potential use. Land-classification studies were undertaken in greater and greater detail on maps of larger and larger scales. From experience in the out-of-door study of agricultural areas emerged many of the ideas incorporated in chapter two of this book on the regional concept, and many of the field techniques discussed in chapter twenty-four. The detailed field studies led quite naturally toward attempts to put geographic methods to practical use, a move that received great impetus from the need for economic planning growing out of the depression. In addition to the studies of agricultural geography, discussed in chapter ten, and those treated in other chapters dealing with the different aspects of economic geography, there were a number of works on relatively small areas in which the economic picture as a whole was drawn [1; 2; 3; 4].

As the years went by, economic geography grew in breadth and depth. It became apparent that here was a whole group of fields rather than just one, and a distinct literature for each of the individual branches began to appear. These specialties became more and more definite and were focused upon more specific aims. The needs of military and economic planning during World War II stimulated a trend toward applied studies in economic geography.

General economic geography ceased to exist as a research specialty, if, indeed, it could ever have been so regarded. As Finch put it "No one is so versatile that he may claim a complete mastery of this field" [5: 207]. More and more the scholar who claims penetration in his research thinks of himself as a specialist in land utilization, in resources, in manufacturing geography, in the geography of transportation, or in some other special aspect, and he makes no pretense to competence in research in economic geography as a whole. With this recognition has come the realization that each requires a particular background on the part of the investigator.

THE CENTRAL THEME

There is a central theme, however, common to all the topical specialties within the general limits of economic geography. Economic geography has to do with similarities and differences from place to place in the ways people make a living. The economic geographer is concerned with economic processes especially as manifested in particular places modified by the phenomena with which they are associated.

In earlier days, the patterns of economic activity were considered chiefly as influenced by the conditions of the natural environment. This older, deterministic approach, as in other branches of geography, is being replaced by a more balanced treatment of all the factors that are deemed to be relevant to such a problem, whether they are natural or cultural in origin.

BASIC PROCEDURES

The research procedures of all aspects of economic geography are dependent to a considerable degree on statistical data. The most important sources of information are the publications of the Bureau of the Census and of other statistical agencies. The geographer, however, is becoming less and less content with statistics compiled for large enumeration areas (chapters four and six), and is turning more and more to direct field observation for the precise location of the enumerated items. It is now common practice to.gather and organize data in map form, not merely to supplement or illustrate statistical material, but also for the purpose of analyzing areal relations.

The economic geographer should be experienced in the use of both cartographic and statistical methods of analysis. The cartographic, or regional, method involves the search for accordant areal relations through the matching of mapped patterns, as explained in chapter two. By this procedure associations of phenomena relevant to economic problems are identified, and the degree of correspondence between areally-associated phenomena can be measured in geometric terms. The statistical method, which is basic to economics, also can be applied to the examination of areal relations, although for this purpose statistical methods are often inconclusive because of the large size of the enumeration areas for which data are usually gathered. A more important use of statistical analysis, however, is for the identification of patterns and relationships that recur in the course of time. The geographer who fails to gain competence in the use of statistical procedures is thus severely handicapped in the search for the causal connections among the areal relationships that he reveals cartographically.

RELATION TO ECONOMICS

Economics is the systematic social science that most closely touches upon the field of economic geography. The same kind of close relationship should exist here between these economic aspects of geography and the related systematic discipline as exists between geomorphology and geology, or between climatology and meteorology. Many economists are concerned with the formulation and evaluation of laws or principles through the examination of isolated processes; but many are also concerned with the application of general understandings to the problems of specific places or specific businesses. To map or describe the distribution patterns of economic activities is not, of itself, geography. But if similarities and differences from place to place are to be described and their significance weighed in terms of causes and consequences, geographic methods can properly be applied. Unfortunately, however, cooperation between economists and economic geographers is not common. Each has tended to follow his own procedures with little regard for the help each could give to the other. In this the geographer has been as much at fault as the economist.

Most economic geographers lack adequate training in economics. Unlike economic geology or economic botany, where the adjective denotes merely an interest in the practical application of the discipline, economic geography, as already pointed out, deals in large part with man's means of gaining a livelihood, and this is also of prime concern to the economist. The conclusion is inescapable that, although the economic geographer should be first of all a geographer, a comprehensive grounding in economics is essential. "Absence of economics in the economic geographer's training is only comparable to absence of geology in the training of the geomorphologist." [6: 40].

APPLICATIONS OF ECONOMIC GEOGRAPHY

To many the term "economic geography" suggests utility and practicality. While there are theoretical aspects of these fields as elsewhere in geography, it is nevertheless true that economic geography has many opportunities for the practical application of its research procedures to a variety of administrative and business problems. What government policy could be formulated to stimulate a bettter use of this particular piece of land? Where can factories or other investments in productive capacity be placed to greatest advantage? Where can sales outlets be located in order most effectively to serve a given market? What developments can be suggested in order to increase the employment possibilities in an area? These and many other practical questions of public and private policy can be approached by geographic procedures. Geographic analysis can provide additional perspective for the clarification of issues involved in decisions of policy.

SUBDIVISIONS OF ECONOMIC GEOGRAPHY

Some of the distinct topical fields into which economic geography has been subdivided in America now include so many extraneous matters that they can no longer be considered as lying wholly within the scope of economic geography: for example, urban geography (chapter six), and the geography of resources (chapter eight). Of the several fields definitely within that scope, some have been well cultivated, such as agricultural geography (chapter ten), and the geography of manufacturing (chapter twelve); and some, though relatively new, are now being vigorously developed, such as the geography of mineral production (chapter eleven), and transportation geography (chapter thirteen). Each of these has its own concepts, its own distinctive procedures, and a literature of its own, and is cultivated as a specialty by a number of professional geographers.

But there are many other fields of economic geography. Some of these exist only in the imagination as logical subdivisions yet to be explored. Two, now beginning to emerge, deserve special discussion here because of their potential importance: marketing geography and recreational geography.

MARKETING GEOGRAPHY

The well-developed topical fields of economic geography are, with the exception of transportation, concerned with the production of material goods. In a commercial economy, as distinct from a subsistence economy, however, production represents only one side of economic geography. Goods must not only be transported from the areas of production to those of consumption, but they also must be transferred from the hands of producers, by collection and subsequent distribution, into the hands of consumers. This is the function of marketing. It is a vital part of the modern capitalist system; it is a major factor in the differential growth of cities and in the changing economic aspect of rural areas. Nevertheless, marketing has until recently been considered by geographers as hardly more than incidental to the various topics concerned with material production. If, however, problems of marketing form a necessary part of the study of production they must be considered as far more than incidental, in view of the wide range in value, represented by price, between the product as it leaves the producer and the same material as purchased by the consumer.

Furthermore, if one considers the large section of the working population that is engaged exclusively in marketing functions, the large part of the urban landscape that is devoted to the structures of wholesale and retail trade, and the complex channels of distribution that lead from producing to consuming areas, it becomes even clearer that there is place and need in economic geography for special attention to marketing geography. Such a development would not only greatly increase understanding in the other fields of economic geography, but would also help the geographer working in the marketing field to contribute significantly to solutions of problems in the actual business of marketing [7].

SCOPE

Marketing geography is concerned with the delimitation and measurement of markets and with the channels of distribution through which goods move from producer to consumer.

A market consists of the existing or potential consumers of goods and services within an area. Consumers are individuals, business establishments, and institutions. All business activity is concerned with markets because it is geared to meet present and potential demands of consumers. In studying markets, the geographer is primarily concerned with where these markets are. He is interested in the distribution of individual consumers and in the magnitude of actual or potential sales within specific areas. Such areas may be rural communities, branch sales territories, metropolitan districts, salesmen's territories, trading areas; or they may correspond to political areas, such as countries and their various administrative subdivisions. The market for some goods and services is highly specialized with a relatively small number of users. For other goods and services, the demand is widespread. Some products are consumed essentially where produced, others at great distances from the source of production. Differences in demand can be traced to economic, social, technological, and other societal factors, and they are affected also by the factor of location and by the physical and biotic conditions of the total environment.

Channels of distribution are the various economic organizations that perform functions in the transfer of goods and services from producer to consumer. As defined by the Census of Business (1948) these include wholesale, retail, and service establishments. Both markets and channels of distribution are material entities located within specific mappable areas on the earth's surface. But they are not static. Change is continuous in number, size, requirements, function, organization, and location.

In the study of channels of distribution, the marketing geographer is primarily concerned, again, with the location of these channels. Having ascertained this from available sources, the geographer can go further. He can determine the trading areas served by the various channels. He can examine the transportation network, the facilities used, and the advantages that might be gained by changes. Also, he can evaluate the arrangement and character of several competing channels of distribution.

The best place to develop the field of marketing geography is in business. The marketing geographer should either work for business full time or should seek to conduct research for business on a part-time consultant basis. Many problems in marketing geography cannot be studied without the private information and facilities possessed by business. Similarly the application and evaluation of ideas and methods needs the laboratory of actual business operations.

The marketing problems to which the geographer can best apply the concepts and procedures of his discipline may be grouped, for convenience, under four headings: 1) presenting market and marketing data; 2) evaluating markets; 3) delineating trading and selling areas; and 4) selecting channels of distribution and locations for wholesale, retail, or service establishments.

Presenting Market and Marketing Data

The mapping of relevant data regarding markets and the marketing process is a contribution in itself [8]. Although such cartographic presentation is only a first and relatively simple step in geographic study, it is so often omitted or done crudely in business establishments that the expert work of a geographer is, in contrast, impressive.

The nature of the information considered relevant depends on the kind of marketing problem. In a majority of market studies the number, composition, purchasing power, and distribution of the population are basic. But in certain cases other kinds of data are needed. For example, a distributor of block salt is interested in the distribution of the cattle population, and a manufacturer of milking machines in the distribution of dairy cows. In the marketing of products and services used only by industrial plants or business organizations the distribution of the particular users of each product or service is essential information.

Marketing data include channels of distribution: wholesale, retail, and service establishments. These also embrace warehouses, storage facilities of various types, headquarters and branch offices of large companies, salesmen's territories, delivery routes, and the like. The distribution of these by type, size, and ownership, and the trading or service areas of each, are all necessary marketing data that the geographer can and should map.

Many of the phenomena that the geographer maps in other kinds of studies, such as terrain, soil, vegetation, climate, land use, mineral resources, industries, settlements, types of farming, arteries of traffic and transportation, or communication systems, bear upon marketing in various ways. At present, however, neither the geographer nor the marketing specialist fully understands this significance. For example, it is generally recognized that climate and weather affect consumption and buying habits, but little has been done to measure these factors quantitatively. Obviously, the marketing geographer should consider for presentation all factors known to be significant.

Evaluating Markets

The fundamental purpose in evaluating markets is to forecast sales potentials for a product or a service. The sales experience of the past and many other indices are employed as measurements. For new products there may be no past sales experience to serve as a guide, and conclusions have to be inferred entirely from other types of data.

In evaluating markets a dynamic approach is necessary. The expansion of industrial and agricultural production, the growth and shift of population, the discovery or exhaustion of natural resources, the spread or eradication of pests and disease, the occurrence of floods, hurricanes, and other disasters all contribute to change the market potentials of one area or another. Some of these changes may be slow and gradual, as the overall economic decline of a large section of the country. Others may be rapid, as when the tapping of a new source of irrigation water permits a wave of new settlement in a dry area. Some changes represent long-time trends, others are only temporary fluctuations; some are related to the normal passage of the seasons, others to abnormal or cyclical variations in weather conditions, still others to technological and other cultural changes. Small initial impulses, however, may set off chain reactions in market potentials. The marketing geographer must concern himself not only with the arrangement of relevant factors at the present time, but also with the arrangement of these factors in the past and with the forecast of future conditions, both short-term and long-term.

The regional concept and the regional method are directly applicable to the practical problem of evaluating markets. Markets can never be studied as isolated phenomena; here business men are accustomed to weighing the complex of factors associated in an area and arriving at forecasts on which they stake the success or failure of their ventures. The geographer's experience in dealing with complex areal differentiation by defining categories, matching associated patterns, noting trends and variations in the operation of the relevant processes, can be used to good advantage in the study of business problems. The regional systems that the geographer defines and plots on maps include areas homogeneous in terms of specific criteria. They are based on the cartographic analysis of continuities and discontinuities, and on the identification of relevant criteria and the disregard of internal differences found to be irrelevant. Through the definition of market-area regions it is possible to distinguish between those elements of consumption or demand for a product or service that are ubiquitous and those that are peculiar to certain areas.

Delineating Trading and Selling Areas

The trading area of a business establishment is the territory from which it draws its customers or in which it sells its products. Since wholesale, retail, and service business establishments tend to be grouped together in clusters or shopping centers, each such cluster has its own trading area. Furthermore the trading areas can be considered as forming a hierarchy of ranks: the trading area of a neighborhood shopping center, of the commercial core of a town, of the whole commercial hinterland of a metropolis. In the terms suggested in chapter two these are nodal regions of different ranks. The generalized commercial hinterland of the large city can be subdivided into the specific trading areas of specific businesses or shopping centers.

The intensity of coverage within a trading area is not uniform. Knowledge of these intra-territory differences is important for effective focusing of marketing effort. The impact of competition from without is strongest on the buying behavior of customers living along the periphery. Since the configuration of a trading area does not remain permanently fixed, the delineation of its boundaries must be checked and revised periodically. Studies of various customer-buying habits are frequently conducted in delineating and evaluating trading areas, and for the purpose of discovering significant changes in buying patterns. Changes in channels of distribution and modes of travel affect customer-buying habits.

Selling areas, or sales territories, are areal units delineated and set up for business administrative purposes so that sales efforts can be directed effectively. A selling area may be a milkman's route, a salesman's territory, or the larger domain of a major division of a far-flung business.

The basic objective in dividing the marketing functions of a large business into sales territories is to reduce marketing costs and to achieve maximum sales at a profit. Adequate geographic coverage without overlapping, at a reasonable cost, and with fair compensation to salesmen or to other agencies of distribution, must be considered.

Trading and selling areas outlined on a map represent a generalization of earth-space of the kind commonly conceived by geographers with respect to a variety of other phenomena. Geographers are also concerned with the techniques of selecting sample areas for intensive study for the purpose of revealing the components of larger regional divisions. In business terms, sampling areas are used not only to reveal significant intra-territory variations, but also as experimental laboratories in which to determine specific consumer characteristics. They are used for the conduct of experiments in the introduction of new products and in the development of new methods of distribution and new devices for sales promotion. The results serve to guide decisions on marketing policy.

Selecting Channels of Distribution and Business Locations

To a considerable extent the success of a manufacturer or producer depends on the channels of distribution that he selects for selling his goods. Many products used by industry are sold directly. Most goods, however, follow a less direct course of distribution and go through several channels before reaching the consumer.

The manufacturer must make two fundamental decisions in selecting his channels of distribution. He must first decide through what type of channels his goods are to be sold, and then he must choose the specific establishments of each type that are to handle his products. The channels of distribution selected must give him satisfactory market coverage and distribution costs.

After the original selection of channels has been made, the manufacturer must continue to amplify and revise his choice. One dare not stand still in a dynamic environment.

Finally, there is the problem of selecting locations for wholesale, retail, and service establishments. This problem involves two considerations: 1) to select the region, city, or shopping center which offers opportunities; and 2) to choose the particular site or sites for setting up an office, warehouse, or store. Here it is not enough merely to look for areas of opportunity and for suitable sites, but one must also undertake to estimate in advance potential sales and operating costs for each proposed installation. A company's own experience in similar locations, or competitors' experience, are valuable guides. But this implies a comparative knowledge of markets, trading areas, and competition. Without such knowledge, comparative locational characteristics cannot be identified and evaluated quantitatively. And without quantitative yardsticks there can be no scientific approach for estimating in advance the potentials of a given location.

Techniques

The marketing geographer employs all research techniques that are applicable to his problem. Some of the techniques useful in marketing studies, such as cartographic presentation and the identification and analysis of areal relations, are the geographer's special forte. Others, such as the statistical method, cost analysis, the use of the questionnaire, and consumer-behavior research, he borrows from other disciplines.

Just as skill in geographic techniques is not acquired without a solid background in geography, so it is necessary for the marketing geographer to obtain at least a working knowledge of the principles and methods of economics, marketing research, statistics, and the theory and practice of accounting. Having done this, he should blend these techniques with those of geography. Better research and more valuable results can be derived from such cross-fertilization.

Cartography as a tool has practically unlimited possibilities in presenting market and marketing data. Good as his present skills are, the marketing geographer needs to improve his methods of analysis and to develop new cartographic techniques. The devices of color (or shading), dots, and isopleths are all excellent within their limitations, but experimentation, ingenuity, and imagination are needed in their application. Present methods in delineating market areas, trading areas, and sales territories are still crude.

There is a substantial, descriptive geographic literature on the location of industrial and commercial establishments. But little in this literature enables a geographer to select a location and evaluate its potentials quantitatively in advance of the installation of a business establishment. Specific techniques and quantitative standards of measurement are needed. A beginning has been made in developing such research tools, but only a beginning, and not all of this by geographers. Marketing geography will achieve its full stature only when a substantial number of properly qualified geographers apply their talents and energy to developing this field.

RECREATIONAL GEOGRAPHY

A second newly-emerging field of economic geography is that of recreational geography. Recreation, including tourism, achieves significant proportions only in those segments of the occidental world that are relatively prosperous. In the United States the occasional refreshment of mind and body that is obtained by a change in the character and location of daily work and play is considered by many people to be a necessity. In no other parts of the world are so many people financially able to seek this refreshment at more or less regular intervals. The mental and bodily states before and after recreation may be difficult or impossible to observe, measure, or plot on a map, but the recreational activities together with the phenomena, both natural and cultural, that facilitate recreation can be observed, measured, and mapped. Recreational phenomena serve to differentiate or characterize areas; some areas are predominantly recreational in economic character.

The phenomena associated with recreation are widespread and the study of them involves many of the branches of geography. Few geographers, however, have devoted themselves primarily to the geography of recreation as a topical specialty. This is a relatively new field, which existing literature covers in scanty fashion. Because many of the published studies deal only with the economic aspects of recreation, it is included among the economic fields; if and when it is more fully developed, however, it may well emerge as a distinct topical field.

The importance of recreational studies is suggested by the increasing attention being paid to recreation in American society. The trend toward shorter working periods, higher incomes, and more time for leisure, together with the tradition of tourism, indicates an increase of investment in recreational facilities and an enlargement of the opportunities for employment in recreational services. Already there are large sections of certain States in which the income from recreational services greatly exceeds all other sources of income. The demand for investigations of the conditions and possibilities of recreational development seems certain to grow.

THE PHENOMENA OF RECREATION

Recreational phenomena are of three kinds: 1) the peculiar associations of physical and biotic conditions that people think are conducive to recreation and which, therefore, constitute natural recreational resources; 2) the structures and other facilities that represent capital investment for recreational purposes; and 3) the recreational activities themselves.

Recreational facilities have form, function, and pattern of distribution. When they assume a certain degree of prominence in a landscape, or in the economy of an area, that area may be characterized as recreational. Such an area may be simply the recreation room of a house; or it may be the playground in a neighborhood, or the theater district of a city; or it may be an extensive area, such as the Florida coast, northern New England, the northern parts of the Lake States, or the National Parks. Geographers may interest themselves in recreational areas of any size, but so far have given attention principally to such units as States, counties, or cities.

GEOGRAPHIC LITERATURE ON RECREATION

Among the earliest of the writings on recreational geography in the United States was a paper by K. C. McMurry, published in 1930, discussing the significance of recreational land use in northern Michigan in connection with the work of the Michigan Land Economic Survey [9]. He pointed out the types of land useful to hunters and fishermen, indicated some of the tools of recreational research, and stressed contributions that geographers might make to this field. In 1935 Robert M. Brown attempted to survey the general field of recreational geography [10]. He described the phenomenon of tourism and pointed out several techniques for measuring its magnitude, patterns, and economic value. Some twelve years later E. C. Prophet pointed to the continuing need for adequate statistics and for the development of research procedures [11].

Following McMurry's paper there have been several studies of recreation in specific areas: by Prophet and G. F. Deasy in Michigan counties [12; 13]; by R. B. Greeley in New England [14]; by Clifford M. Zierer in western United States [15]; and by Edwin J. Foscue in the Great Smokies, the Rockies, and in Mexico [16]. As a group, these papers point out the general character of recreation, its forms, and its economic importance in the several areas studied.

One of the earliest attempts to make a systematic survey of recreational forms and values is contained in a bulletin by W. O. Hedrick based primarily on the returns from questionnaires sent to township assessors in Michigan [17]. From these he produced maps of various items of the recreational plant and described the environmental features that seemed significant in their distributions. He also isolated the taxes paid by recreational property owners to local government units.

Between 1947 and 1951 several surveys were made by State agencies with the intent of identifying recreational-travel patterns and expenditures in the States [18; 19; 20; 21; 22]. These are based upon questionnaires filled out by tourists, who stated, among other facts, the origins, routes, and destinations of their travels, the types of accomodations they used, and expenditures they made. The chief contributions of this group of studies have been to outline the patterns of tourist travel, to identify the facilities used, and to appraise expenditures.

Published works intended to depict the character and distribution of the phenomena connected with recreational land use were initiated in 1933 by Stephen B. Jones, who compared the features of a recreational town with a mining town in British Columbia and defined recreation regions in the Canadian Rockies [23]. A. W. Booth analyzed the lake-shore characteristics and the nature of the lakes themselves in terms of recreational values in the state of Washington [24]. This study is interesting for its application of techniques to the problem of recreational mapping. R. I. Wolfe studied the source areas of occupants of summer cottages in Ontario, using home addresses secured from local postmasters [25].

The most inclusive study of the recreational values and distributions within a State is published in three papers dealing with New Hampshire. The first of these, by A. S. Carlson, is general in nature and deals with the recreational features in the six recognized recreational districts of the

State [26]. The other two, by Snydor Hodges, were published as official reports by the State Planning and Development Commission [27; 28]. One of these is an intensive study of the tax rolls in the recreation districts and points out the financial contributions of recreational property to both the local and the State governments. The other deals with the locations and classifications of recreational facilities and the expenditures by recreationalists within them, and is based on field interviews.

ECONOMIC RESEARCH IN RECREATIONAL GEOGRAPHY

From the published literature it appears that the primary interest of geographers in recreation has been in its appraisal from a monetary point of view. This is to be expected from the fact that recreation in its broader aspects is supported by public funds expended for such purposes as road maintenance, facility development, and fish and game propagation and protection. State governments are interested not only in justifying such expenditures but also in appraising them as investments that produce monetary returns.

The difficulties of research into recreational values are those of definition and statistics. The question of who is a tourist has not been solved for statistical purposes, and reliable techniques have yet to be devised for computing the portion of the travelling public's expenditures attributable to recreational attractions. Primary recreational establishments such as motels can estimate the proportion of their total business that is derived from tourists, but proprietors of filling stations or grocery stores seldom can make accurate estimates.

Lack of detailed, comparable statistics presents one of the major problems facing the research worker in recreational geography. The United States Bureau of the Census has given serious consideration to the collection of recreational statistics, but so far has been unable to formulate satisfactory definitions for identifying recreationalists or their activities.

THE PROSPECT

Clearly a good start has been made in the study of the economic aspects of recreational geography. With improved data and with a more effective use of questionnaire and interview techniques it should prove possible to estimate the monetary value of recreation in the economy of an area. But much work remains to be done before general concepts of wide application can be established.

It may be that undue emphasis has been given to recreation in rural areas. Some geographers hold the opinion that the more important recreational foci are to be found in cities—more important, that is, in terms of monetary value. According to Prophet, the greatest tourist attraction of

Michigan is Detroit, rather than the unsettled lake area of the northern part of the State. It is probable that more tourists are attracted by the recreational facilities and features of Boston than by the facilities and features of much larger rural recreational areas. The dollars spent for recreation in a small area on Manhattan Island make this area one of the greatest recreational foci of the whole country. Notwithstanding the implications of these facts, little or no work has been done towards evaluating the recreational significance of cities that are not primarily recreational in function. Chauncy D. Harris points out in his classification of cities of the United States (chapter six) that there is no satisfactory statistical criterion for identifying cities that have recreation as their principal function [29]. Urban mappers have delineated recreational areas as one of the categories in functional mapping but nothing has been done toward identifying the patterns of recreational movement within cities. On the whole, recreational study offers many opportunities to urban geographers.

There is great need in the new field of recreational study for careful analyses of recreational areas for the purpose of depicting and explaining the patterns both of recreational structures and recreational activities. Such primary information and preliminary interpretation were contained in the papers by both Prophet and S. B. Jones. But these early papers have not been followed by additional studies. It seems obvious that only by detailed observation and mapping can the features that characterize recreational areas be discovered, the kinds of lands involved be determined, and the real values of recreation both in a material and a social sense be understood.

The small amount of work that has been done toward analytic morphology in recreational areas leads the careful student to suspect that many of the current ideas concerning the location of recreational areas in relation to site factors may not be valid. The fact is that recreation is so largely personal and subjective that it is to be sought wherever one finds it rather than in any specific combinations of environmental conditions.

OTHER FIELDS OF ECONOMIC GEOGRAPHY

Marketing geography and recreational geography do not by any means exhaust the list of potential specialties that may ultimately take shape in the general domain of economic geography. Although studies of international trade were popular early in the present century, not much has been done in this field recently. The analysis of factors relevant to such trade certainly involves geographic concepts and methods, and for the properly qualified scholar with competence in both geographic and economic techniques this could be a rewarding subject. E. A. Ackerman's monograph on the fishing industry of New England suggests another type of contribution a geographer can make to a single segment of the economic life [30]. Almost nothing has been done about the geographical aspects of forest utilization, and very little with regard to the geography of hydro-electric power.

In a still broader sense, there is need for the services of the geographer in connection with the formulation of national economic policy. Geographical concepts and methods are applicable to the evaluation of the economic capacity of nations, our own as well as others. The point of view of a scholar experienced in the geographic discipline would be of great value to the policymakers at the highest level.

REFERENCES

1. TREWARTHA, G. T. "A Geographic Study of Shizuoka Prefecture, Japan," *Annals of the Association of American Geographers*, 18 (1928): 127-259.
2. HALL, R. B. "The Yamato Basin, Japan," *Annals of the Association of American Geographers*, 22 (1932): 243-292.
3. STARKEY, O. P. *Economic Geography of Barbados*, New York, 1939.
4. MURPHY, R. E. "The Economic Geography of a Micronesian Atoll," *Annals of the Association of American Geographers*, 40 (1950): 58-83.
5. FINCH, V. C. "Training for Research in Economic Geography," *Annals of the Association of American Geographers*, 34 (1944): 207-215.
6. ROBERTSON, C. J. "Economic Geography and Geonomics," *Geography*, 35 (1950): 38-41.
7. APPLEBAUM, W. "The Geographer in Business and his Requisite Training," *The Professional Geographer*, April, 1947.
8. GOODMAN, M. E., and RISTOW, W. W. *Marketing Maps of the United States, An Annotated Bibliography*, (Second Revised Edition), Washington, 1952.
9. McMURRY, K. C. "The Use of Land for Recreation," *Annals of the Association of American Geographers*, 20 (1930): 7-20.
10. BROWN, R. M. "The Business of Recreation," *Geographical Review*, 25 (1935): 467-475.
11. PROPHET, E. C. "The Tourist and Resort Industry," *Annual Report of the Michigan Academy of Science, Arts and Letters*, 1947: 41-51.
12. ———. "Some Types of Recreational Land Use in Michigan," *Papers of the Michigan Academy of Science, Arts and Letters*, 22 (1936): 385-395.
13. DEASY, G. F. "The Tourist Industry of a 'North Woods' County," *Economic Geography*, 25 (1949): 240-249.
14. GREELEY, R. B. "Part-Time Farming and Recreational Land Use in New England," *Economic Geography*, 18 (1942): 145-152.
15. ZIERER, C. M. "Tourism and Recreation in the West," *Geographical Review*, 42 (1952): 462-481.
16. FOSCUE, E. J. *Gatlinburg: Gateway to the Great Smokies*, Dallas (Texas), 1946; idem. *Taxco: Mexico's Silver City*, Dallas (Texas), 1947; FOSCUE, E. J., and QUAM, L. O. *Estes Park: Resort in the Rockies*, Dallas (Texas), 1949.
17. HEDRICK, W. O. *Recreational Use of Northern Michigan Cut-Over Lands*, Agricultural Experiment Station, Michigan State College, Special Bulletin 247, East Lansing, 1934.
18. STEDMAN, G. H. "Business Aspects of Vacation Travel," *New York State Commerce Review*, 1 (1947): 7-13.
19. Minnesota Department of Business Research, *Steps to Better Vacations in Minnesota*, St. Paul, 1948.

20. LANNING, V. H. *The Wisconsin Tourist,* Wisconsin Commerce Studies, 1, Madison (Wisconsin), 1950.
21. BIGELOW, L. E. *Survey of Expenditures of 1949 Summer Vacationers in Massachusetts,* Massachusetts Development and Industrial Commission, Boston, 1950.
22. LANZILLOTTI, R. F. *The Washington Tourist Survey,* Recreational Resource Studies, 17, Pullman (Washington), 1951.
23. JONES, S. B. "Mining and Tourist Towns in the Canadian Rockies," *Economic Geography,* 9 (1933): 368-378; *idem.* "Recreational Regions of the Canadian Rocky Mountains," *Bulletin of the Geographical Society of Philadelphia,* 34 (1936): 50-72.
24. BOOTH, A. W. *The Lakes of the Northeastern Inland Empire: A Study of Recreational Sites,* Bureau of Economics and Research, 5, Pullman (Washington), 1948.
25. WOLFE, R. I. "Summer Cottagers in Ontario," *Economic Geography,* 27 (1951): 10-32.
26. CARLSON, A. S. "Recreational Industry of New Hampshire," *Economic Geography,* 14 (1938): 255-270.
27. HODGES, S. *Recreation Property in New Hampshire: 1945,* New Hampshire State Planning and Development Commission, Concord, 1948.
28. ———. *The Vacation Business in New Hampshire: 1946,* New Hampshire State Planning and Development Commission, Concord, 1949.
29. HARRIS, C. D. "A Functional Classification of Cities in the United States," *Geographical Review,* 33 (1943): 86-99.
30. ACKERMAN, E. A. *New England's Fishing Industry,* Chicago, 1941.

AGRICULTURAL GEOGRAPHY[A]

A. Original draft by Harold H. McCarty with the assistance of Raymond E. Murphy and the other members of the Committee on Economic Geography. Special mention should be made, also, of the contributions of Carleton P. Barnes, Loyal Durand, Jr., and Walter M. Kollmorgen.

CHAPTER TEN

AGRICULTURAL GEOGRAPHY

AGRICULTURAL GEOGRAPHY is one of the more actively developed branches of geography in the United States. As a field of topical specialization it is widely recognized, although even a hasty examination of its content shows many overlaps and many intimate connections with other branches of the field. Agricultural geography and soil geography were developed together and in some cases by the same people. Field workers interested in agriculture and soils can be given a large measure of credit for the development of the regional concept and the method of regional study as presented in chapter two, and many of the field techniques described in chapter twenty-four were derived from experience in the study of agricultural problems. Agricultural geography touches closely on studies of climate, water, resources, and rural settlement. In the general field of economic geography agricultural studies were the first to attract the attention of a considerable number of professional geographers.

An inventory of the work done in this branch of geography reveals two contrasts in approach. There is the contrast between topical and regional studies, and the contrast between the intensive study of small areas and the broader study of large areas. In the agricultural field a topical study focuses attention on a particular commodity, such as wheat, whereas a regional study focuses attention on the crop combinations and farm problems of a particular area. In the world-wide development of geographic thought, the commodity approach is generally associated with the British, who had great need for information concerning sources of food and raw materials for use in developing their Commonwealth economic system. The regional approach, on the other hand, reached its greatest European development on the continent where geographers were responding, in part at least, to a call for the better utilization of agricultural resources within their own national areas.

In the United States both approaches have been developed. The commodity approach was almost universal in American textbooks until the early 1920's, and the earlier studies in commercial geography dealt largely with the production and exchange of commodities. Studies of individual commodities are still numerous. Yet research interests in the United States have tended to follow the area approach. This seems quite fitting in a nation whose major economic problems and policies have more often been concerned with the development of areas than with the procurement of commodities. The major research products of American agricultural geography have appeared as statements explaining the manner in which certain sets of natural and human resources are used to produce certain kinds of agricultural products in specific parts of the world. The main emphasis has been on production, particularly on establishing the nature of conditions surrounding production in specific areas, or of the conditions favorable to instituting production in areas not currently devoted to that purpose. Generally speaking, if an American geographer has been concerned with measures to increase the supply of wheat, he has thought first of all in terms of producing wheat rather than of buying it. He has then studied natural and social conditions in areas devoted to wheat production and, with that evidence in hand, has set about discovering other areas in which these conditions prevail, or could be established, in order to determine where new supplies of wheat might be obtained. Analytical studies in agricultural geography, even when dealing with one commodity, have nearly always been concerned with particular areas.

Probably the most distinctive feature of American geography, compared with geography elsewhere, is its emphasis on the intensive field study of small areas. The ideas and procedures of detailed field study were formulated by workers in agricultural geography. In the examination of the areal associations of agricultural land use with the underlying qualities of the land, geographers came to grips with the problem of defining meaningful categories and of measuring the degree of correspondence between associated phenomena. From such close observation of the similarities and differences from place to place on the earth came the first appreciation of the need for defining a hierarchy of regions, each rank with a different degree of generalization, as described in the chapter on the regional concept and the regional method. One of the chief problems, however, is still the development of valid procedures for relating the intensive study of small areas with the more highly generalized treatment of large ones.

FIRST STEPS IN MODERN AGRICULTURAL GEOGRAPHY

First steps in the development of modern agricultural geography in the United States were taken between 1910 and 1920. Evidence of an interest in field observation can be found in the abstracts of papers delivered before the early meetings of the Association of American Geographers. Most of these papers dealt with foreign areas, and usually they included written, but not precisely mapped, observations concerning what Richard

E. Dodge, in 1910, called "the relations between slope, exposure, drainage, ground water, weather, climate, and plant and animal production" [1].

LARGE-SCALE STUDIES

The beginnings of the large-scale field mapping of phenomena relevant to agricultural problems were made by a group of geographers at the University of Chicago. In 1915, Wellington D. Jones and Carl O. Sauer published a discussion of the methods of intensive field study, and out-lined a system of categories of agricultural land use to be applied on large-scale maps [2]. At about this same time Jones offered the first course in field training for graduate students in geography to be given by an American university. During the decade from 1915 to 1925 interest in the field study of small areas spread rapidly, especially among the geographers of the Midwest. The field techniques developed in that period and the geographers who were involved in the field conferences and in the experimental mapping projects are discussed in chapter twenty-four. It should be emphasized here that this was the first time in the United States that geographers made the large-scale field map an essential part of the record of a research undertaking. On such maps, they plotted categories of agricultural land use and categories of slope and soil.

SMALL-SCALE STUDIES

Meanwhile the food and raw material problems of World War I had stimulated an interest in studies of agricultural distributions on a global scale. One of the most widely used documents to result from this wartime attention to world sources of food was the *Geography of the World's Agriculture* prepared by Vernor C. Finch and O. E. Baker and published by the United States Department of Agriculture in 1917 [3]. This atlas presented the most complete array of mapped information regarding agricultural commodities ever to appear in the United States.

Although this atlas stimulated a few studies of specific commodities, its chief use was to provide new data for the study of areal relations between crops and features of the physical environment. In 1921, O. E. Baker emphasized the importance of physical conditions underlying the differences in agricultural production throughout the United States [4]. Two years later, in 1923, Baker presented the agricultural part of a series of papers on the geography of the Great Plains [5], followed in 1925 by a study of the world's potential supply of wheat [6], and in the years from 1926 to 1933 by a series of papers on the agricultural regions of North America [7]. Baker's work established him as the leading agricultural geographer of the period.

The several articles on the agricultural regions of North America were a part of a program of similar studies covering all of the occupied continents. Over a period of years that started in 1925 the newly-established magazine *Economic Geography* published an agricultural-region series that eventually covered the whole world. In this series Olaf Jonassen studied Europe, Çlarence F. Jones studied South America, Griffith Taylor studied Australia, Samuel Van Valkenburg studied Asia, and Homer L. Shantz studied Africa [7; 8; 9; 10; 11; 12]. The regions developed in these studies were plotted on small-scale maps with a high degree of generalization.

In recent years several papers have appeared in which the ideas developed in this series are carried on either to the further refinement of the regional outlines, or to further subdivision in terms of specific commodities. Prominent among these are Loyal Durand's studies of the dairy regions [13], Earl B. Shaw's study of swine production in the Corn Belt [14], Merle Prunty's report on changes in the outline of the United States Cotton Belt [15], and John C. Weaver's study of barley in the United States [16].

THE SEARCH FOR UNIFORM CRITERIA

It soon became evident that the series of articles on agricultural regions lacked comparability from one continent to another. To a considerable extent the differences in the methods of defining regions resulted from differences in the availability of statistical data. But to a certain extent the lack of comparability was due to a lack of agreement among the authors regarding the relative importance of different criteria. The search began for a system of uniform criteria to be used in the definition of categories of agricultural regions throughout the world.

This search was in large part stimulated by geographers who had experience in the field study of small areas. On maps of a topographic scale, on which specific fields and farms could be plotted, categories of agricultural land use were defined by specific criteria [17]: why could not the more highly generalized regions be similarly defined?

The search for uniform criteria led not only to the examination of a wide variety of quantitative values and ratios, but also to the consideration of many criteria which could be mapped precisely but could not be treated quantitatively.

QUANTITATIVE CRITERIA

Wellington D. Jones in a paper published in 1930 [18] proposed methods for the more precise measurement of agricultural land occupance through the use of ratios and the plotting of significant isopleths. Jones suggested criteria such as the percentage of land in crops, livestock units per hundred acres in crops, the ratio of gallons of milk to acres in crops, and the percentage of total livestock represented by work animals. Ratios of this sort could be computed from statistical data commonly published for most countries where census data are available at all. Once plotted on maps, the patterns of variation could be precisely recorded through the use of isopleths. On the basis of the four criteria suggested above, Jones defined three types-of-farming categories: commercial grain farming, dairy farming, and commercial livestock farming.

After 1930, the use of isopleths in research studies of agricultural areas became increasingly common. At first there was an attempt to find specific critical isopleths to serve as regional boundaries, thus transforming the continuities (which the isopleths reveal) back into discontinuities (using the isopleths as boundaries). The errors involved in this procedure became apparent as experiment continued (see chapter one, pp. 10-11). In 1935, applying techniques advocated by Jones, Richard Hartshorne presented a study of the United States dairy areas in which he used measures such as the number of gallons of milk produced per 100 acres of crop land to determine the degree of specialization in dairy production [19]. Techniques employed by Hartshorne were later applied to peninsular Ontario by J. R. Whitaker and to the State of Mississippi by Clyde F. Kohn [20]. In 1935, Hartshorne collaborated with Samuel N. Dicken to produce maps of agricultural regions of North America and Europe that were based on uniform criteria [21]. Lack of data prevented the authors from extending the scheme beyond Western Europe.

In the period since Jones' stimulating paper on the use of ratios and isopleths, a variety of different kinds of measurement have been tried. Quantitative criteria for the definition of categories of agricultural land use have involved measurements of areas, of population, and of income.

Land-area units have been most commonly used to measure quantities of agricultural production. Absolute quantities may be expressed in terms of the number of acres (or other area units) devoted to specific purposes within a particular area; but since relative measures are generally desired, ratios such as the percentage of land in farms, percentage of farm land in crops, percentage of crop land in a certain crop or crop combination, number of animals per square mile, or quantity of production per acre are commonly employed. O. E. Baker used criteria such as the percentage of land in crops to define the western boundary of the Great Plains region (1926), bushels of corn produced per square mile for the Corn Belt (1927), the ratio of value of vegetables and fruits to value of all crops for his Middle Atlantic Trucking Region (1928), and the ratio of arable land acreage to acreage of hay and pasture for his Hay and Dairy Belt (1928) [7]. Baker used one criterion for boundary determination, but Wellington Jones (1930) advocated multiple criteria, and considered livestock units as well as crops in his computations. Hartshorne and Dicken (1935) likewise used both multiple criteria and ratios of livestock to cropland in their uniform classification of agricultural regions in Europe and North America. More recent studies have generally followed these latter procedures.

Relating quantities and types of agricultural phenomena to units of population rather than units of area has been employed rarely in agricultural geography. In 1939, Hartshorne produced a series of maps relating farm population to quantities of arable land, value of land, and value of farm products in the eastern United States [22]. He thus was able to introduce important social connotations concerning agricultural scales of living into the literature of agricultural geography. Occasional maps labeled per capita rather than per acre have appeared in later studies, but their volume has not been great, in spite of the frequently stated opinion that the social aspects of agricultural geography are worthy of additional emphasis.

Value of production has been used with considerable success as a measure of types of farming in the United States. Here the obvious advantage lies in the fact that the volume of all sorts of agricultural production may be stated in terms of price, and the ratios derived therefrom are good indicators of the sources of farmers' purchasing power. In highly commercialized agricultural economies, such as those found in the United States, data of this sort are of great social and economic significance. Such measures were favored by O. E. Baker, but it was not until after the Census of 1930 that sufficient information became available to prepare a map of the agricultural regions of the United States based on farm income data [23]. The resulting map of types of farming, revised from time to time by the United States Department of Agriculture, is still frequently used by students of agricultural geography [24].

Much has been written in criticism or justification of these various measures. In practice, the land-area unit has been justified most frequently because it is the only available unit in many parts of the world. Critics have pointed out that it makes no allowance for variations in productivity from one piece of land to another. Per capita measures overcome this difficulty but they obscure individual variations in productivity such as those attributable to differences in inherent ability, training, or the availability of machinery. Income measures, since they are expressed in monetary terms, provide the best opportunity for the comparison and summation of production from a variety of sources, such as crops, animals, and personal services. In year-to-year comparisons, however, income data must be adjusted for changes in price levels, and there is also the problem of selecting a suitable base period. More important is the fact that since these data are not available in sufficient detail for most parts of the world, few studies for areas outside the United States have appeared.

The selection of units of measurement obviously depends on the kind of problem being investigated. If, for example, the purpose is to provide a background for planning specific uses for various types of land, these units are likely to be land-unit areas. Such units are also useful in places where land qualities are relatively uniform and areal variation in productivity can be considered as of little consequence. If, on the other hand, the main interest is in how people make a living, or in per capita output, measurement may well be in terms of numbers of persons whose livelihood is derived from specific ways of making a living. Studies intended for use in connection with welfare programs are often based on these types of data, which are best calculated to measure the economic status of the inhabitants of an area, rather than the uses of land itself or the volume of production. Investigators who use income data see virtue in computing the gross product of an area in order to determine the importance of various elements in its economy. Since gross income includes production attributable to land, labor, and capital, it commonly is defended as the best measure of total production (or any phase of total production) in an area. In any case, measurement criteria must be chosen in the light of the objectives of the study, and there appears to be no single measure best suited to all possible objectives.

Criteria for the Definition of World Regions

Derwent Whittlesey, who worked closely with Wellington Jones in the search for uniform criteria for the definition of agricultural regions, published his proposed division of the earth into major agricultural regions in 1936 [25]. His 13 categories of regions are defined on the basis of: 1) crop and livestock association; 2) methods of production; 3) intensity of land use; 4) disposal of the products; and 5) the ensemble of structures used in the farming operations. Although Whittlesey presented his classification as a "target for criticism" its popularity has persisted. His map, in revised form, appears currently in *Goode's World Atlas* [26]. . .

The criteria used by Whittlesey and others who have worked along similar lines include a variety of non-commodity elements, which are nevertheless of great value in distinguishing between different types of farming. The most widely used of these is the distinction between a farm economy that is self-sufficient, known as a subsistence economy, and a farm economy in which the farmer sells what he produces and buys what he consumes, in other words a commercial economy. There are, of course, many combinations of these. Another set of criteria has to do with the distinction between intensive and extensive production, usually measured in terms of the amount of labor and capital applied to a given unit of land. In this case the kinds of tools and machines used are considered in defining categories. A third set of criteria has to do with types of management. It is often desirable to know whether the farm units are operated by the owners, by tenants, or by farm laborers who are paid in wages or compensated in other ways. Many students also find it useful to include the nature of the farm buildings: their forms, sizes, arrangement, and functions in the farm system (see chapter five). Whittlesey advocates including all these kinds of criteria, and envisions a classification of world regions based on a weighted average of a wide variety of component elements. With the inclusion of all aspects of the occupance of agricultural areas such regions tend to approach the compages described in chapter two.

By 1936, the profession had been provided with systems of criteria for the definition of categories of world agricultural regions. The ideas developed by that date have been and must continue to be tested, evaluated, and amplified. But before these ideas can be applied to the construction of a world system of regions, a vast amount of detailed work remains to be done, and a great volume of new statistical data must be collected. The way has been charted for the approach to agricultural geography through detailed study and also through broadly-conceived world regions, but the connections between the two have not been built.

THE DEGREE OF GENERALIZATION

This involves the problems of the degree of generalization and of regional hierarchies already discussed in chapters one and two. The hierarchy of regions suggested for compages involves the construction of homogeneous and cohesive areas at different degrees of generalization by a process of aggregation-subdivision (pp. 47-51). At the topographic scales, on which specific fields and farms can be seen, localities and districts can be studied with a minimum of generalization. At the chorographic scales, where specific fields must be generalized in broader associations, provinces can be examined. And at still smaller scales and greater degrees of generalization it is possible to treat realms. Many topographic studies have been made in the field of agricultural geography and a considerable effort has been devoted to the definition of categories for world regions with the rank of provinces or realms. The problem is to connect the two extremes.

Intensive Studies of Small Areas

The great bulk of the work in agricultural geography that has been done, and probably will be done in the near future, is concerned with the intensive study of small areas. Before any valid system of world regions can be established by process of aggregation the basic building blocks of such a hierarchy must be made available. These are the large-scale studies in which the specific forms and patterns fashioned by man in his agricultural use of the land can be precisely mapped and analyzed by the regional method.

The study of small areas was an important factor in undermining the now generally discarded concept of environmental determinism. As experience with this kind of work accumulated, it became increasingly clear that an adequate interpretation of the agricultural patterns could only be gained by broadening the base of the inquiry. It was necessary not only to seek relationships between agriculture and the physical earth, but also to give balanced consideration to many economic, social, and political conditions that geographers had formerly overlooked. The regional method of seeking accordant relations among associated phenomena can be applied just as effectively to the correlation of agriculture with markets or prices as to the correlation of agriculture with soils or water.

Among the earliest papers to give a more balanced treatment of the factors relevant to agricultural patterns were those of Charles C. Colby dealing with the Fresno Raisin District and with the Annapolis-Cornwallis Apple District [27; 28]. Preston E. James analyzed the factors involved in the localization of coconuts in Trinidad [29], and Glenn T. Trewartha applied detailed methods to the study of the cheese industry in a Wisconsin county [30]. Walter M. Kollmorgen has provided examples of agricultural communities where the societal background of the people sheds more light upon their farm practices than does their present physical environment [31; 32], and Otis P. Starkey has interpreted the distribution of sugar cane in the West Indies in terms of changing prices [33].

Attention was first directed toward the identification of associations of features in the field study of small areas. The actual mapping of conditions out-of-doors led to the recognition of recurring associations, such as forests on steep slopes, pastures on intermediate slopes, crops on gentler slopes. Not that all such associations had been overlooked before; certain more obvious relationships of this sort had been observed and gave support to the concept of environmental determinism. But when they were mapped on topographic scales the degrees of correspondence among them could be measured with a degree of precision comparable to that achieved by statistical procedures.

A considerable stimulus to the recognition of associations in the field came from the development of the so-called fractional code system of notations for mapping (described in chapter twenty-four). This system, in which the elements of the physical land are represented by digits in the denominator and the elements of the cover by digits in the numerator, was first conceived by Charles C. Colby, and was tested in an elaborate study of Montfort, published in 1933, by V. C. Finch [34]. As mapped in the field, areas are outlined that are homogeneous with respect to the symbols of the fractional code; with a change of any one of the elements of the symbol a boundary line is drawn and a new region established. Any one symbol represents an association of features.

The fractional-code system has been tested and elaborated in several land-classification and land-use surveys discussed later in this chapter. Its use permits the rapid recording in the field of a variety of associated patterns and the later separation of these patterns and analysis of their areal relations. As a result of this and other techniques the land-use map has become an essential part of any study in agricultural geography [35; 36]. Such maps are now sufficiently precise to provide a basis for planning for the better use of specific fields and farms.

A Hierarchy of Agricultural Regions

Very little attention has been paid as yet to the problem of establishing a hierarchy of agricultural regions comparable to that suggested in chapter two for compages. There is little or no correlation between the intensive studies of small areas and the more general studies. The fact is that there are not enough trained workers to extend the coverage of detailed studies fast enough to meet the need for them. It has been urged that more attention be paid to chorographic studies at intermediate scales which are not so highly generalized that they cannot be tied back to their component details. If the categories used at intermediate scales (such as 1/500,000 or 1/1,100,000) are defined in terms of associations of related features, the specific characteristics of each association can be revealed by a sampling technique. In the Corn Belt, for example, it might not be feasible to map individual fields of corn, wheat, oats, or pasture at such scales; but it would be possible to group in one association all the area involved in the characteristic combination of crop land and pasture. A few properly-placed sample studies done in detail would then reveal the components of the association.

There has been some experimentation with sampling in the field of agricultural geography. In 1934, J. Sullivan Gibson published a study of a county in Kentucky. He first divided the county into regions and then surveyed square-mile areas selected from each division [37]. In 1935,

Samuel N. Dicken published a paper on Central Florida in which he made use of a different technique for defining his sample areas [38]. Similar techniques were used in investigations of the Louisiana sugar-cane plantations by Edwin J. Foscue and Elizabeth Troth, and of an area in southeast Nebraska by Walter Hansen [39; 40]. The traverse as a sampling technique was tried by G. Donald Hudson in Scotland [41], and by Robert S. Platt in Latin America [42]. Malcolm J. Proudfoot outlined the basic procedures for sampling with traverse lines [43]. The question remains whether better results can be obtained by making use of geometric or statistical methods for selecting samples, or by depending on the trained judgment of the field investigator.

THE REGIONAL METHOD

At topographic and chorographic scales it is possible to apply the regional method to the analysis of areal relations. This method, as described previously, involves: 1) the selection of criteria relevant to a stated objective; 2) the use of these criteria in defining categories of either single or multiple features; 3) the plotting on maps of the areas which are homogeneous in terms of the criteria; and 4) the matching of maps drawn with the same degree of generalization for the purpose of identifying and analyzing areal relations.

A fundamental step in a study in agricultural geography is the construction of a map of agricultural land use at whatever scale is most appropriate. This map is not a direct factual picture of the agricultural land use, like a photograph, for it is already generalized on the basis of selected criteria and to this extent reflects the judgment of the investigator. But if it is dated, documented, and annotated with respect to the criteria used in its compilation, it is then subject to evaluation by others. Furthermore, it is likely to be more eloquent than a photograph, just as a landscape drawing may be more expressive than a photograph of the same scene because the artist has selected relevant features and eliminated irrelevant details. Obviously the construction of such maps calls for a considerable degree of training, experience, and intellectual honesty.

The Search for Accordant Relations

With the map of agricultural land use in hand, questions at once arise, if they have not arisen during its construction, regarding the meaning of the areal differentiation of agriculture that it portrays. Accordant areal relations with other kinds of phenomena are sought. Some are obvious or become apparent from the field observation of the agricultural patterns. If the fractional code system is used, certain associations of land use with underlying land qualities have already been mapped. The map of agriculture is matched with the map of soil, or with complexes of soil, slope, and drainage. Various degrees of correspondence among these elements are noted, and may be measured. In modern agricultural studies, as indicated previously, an attempt is made to reach a balanced analysis that normally considers not only the physical character of the land, but also economic, social, and political conditions. The discovery of accordant relations, however, does not prove a causal connection. If the analysis of areal relations shows that in a very high proportion of the total area *where a is present, x, y, and z are also found,* there is a strong indication that the four are causally connected. To find this connection may be the next step.

THE INVESTIGATION OF PROCESS

Those who search for causes and consequences soon find themselves involved in the study of processes. What are the processes operating through time that have resulted in the observable association of phenomena now visible on the maps? Of course, the geographer does not proceed to this point in his research study without forming some hypothesis regarding the areal association of factors affecting agriculture. He is familiar, or should be, with the processes described in the related systematic sciences: the botanist or agronomist for example has established in a controlled experiment the existence of a certain relationship between crop growth or yield and elements such as rainfall, humidity, temperature, or soil character. With such findings in mind, the geographer selects those relationships that seem most nearly applicable to his study of a particular place. But he also is conscious of the fact that the processes described by the systematic scientists as a result of controlled experiment are here operating in the presence of other factors that may or may not produce important modifications of the ideal sequence of cause and effect.

There is, in fact, much to be said for the investigation that deliberately confines itself to the exhaustive consideration of just one element, for which adequate data may be obtained. Such special aspects of agricultural land use as ownership, tenancy, and land values have been the basis for a number of valuable studies in the field of agricultural geography [44; 45; 46; 47; 48]. Investigations of the areal relations between one or two climatic features, such as rainfall and temperature, and the yields of particular crops, have added much to the understanding of observed associations [49; 50]. Research has also been directed to a particular agricultural boundary, or part of a boundary [51]. Even though such studies fall far short of a complete explanation of the agricultural occupance of the areas investigated, they constitute the beginnings of such explanations, and show what may be achieved after similar examinations of other factors have been added to the literature. In other words, they provide the materials from which more useful hypotheses may be constructed for future use. This is the toilsome process by which geography acquires the means of understanding the agricultural occupance of particular areas.

There remains the technical problem of determining the degree of correspondence between agriculture and any possibly or probably related factors. Practically, this is a process of correlation, in which the analyst observes similarity of pattern between the distribution of his agricultural data and that of each factor with which he compares it. In nearly all cases, these correlations have been made visually, by comparing maps or plotting both sets of data on the same map. Where similarities of patterns are observed they are noted. A few authors, however, aware of the possibility of differences of opinion concerning these visual correlations, have pointed to the desirability of substituting measurement for judgment. Statistical devices for measuring the degree of correlation may be used where both sets of data are completely quantified (as in rainfall and crop yields), but where one factor (such as soil quality) has not been quantified reliance must be placed on graphic analysis. It appears that geometric procedures for measuring these types of correlations may easily be devised.

LAND CLASSIFICATION AND USE SURVEYS

Mention has been made of the close connections between small-area research and certain large-area government-sponsored projects designed to gather data for the planning of the more rational use of lands for agricultural or other purposes. In three of these surveys American geographers have had ample opportunity to test the systems of classification that had been developed during the 1920's. These are the Michigan Land Economic Survey, the surveys of the Tennessee Valley Authority, and the Rural Land Classification Program of Puerto Rico. It is interesting, also, to compare these with the similar Land Utilisation Survey of Great Britain, which was undertaken during the 1930's.

Earliest of these government-sponsored projects was the Michigan Land Economic Survey, which was started in 1922. The project came into existence because of a need for adequate information on which to base a State-sponsored program to determine which of the lands of northern Michigan might best be devoted to agricultural, recreational, or other uses [52]. In essence, the Michigan survey gathered data concerning resources, such as soil quality, slope, vegetation, and similar natural features. It employed specialists in the systematic fields to classify these various features, but in addition used geographers in mapping and analyzing the information gathered. The problems encountered in this work stimulated thinking with regard to schemes for land classification that came into being after 1925 [53].

In a sense, the second major American survey, the land survey of the Tennessee Valley Authority, was a lineal descendant of the Michigan project. Like the Michigan survey, it was concerned essentially with an appraisal of resources. This appraisal, in turn, was for the use of a central planning agency, which had a variety of problems to solve: general objectives, such as how to alter the agricultural economy to provide a higher level of living for the inhabitants; regional problems, such as the effect on the regional economy of the abandonment of certain areas to be used as reservoirs; administrative problems, such as determining limits for purchase of areas to be used for reservoirs and other public purposes. The TVA program employed a large number of geographers, who had the advantage of a considerable amount of prior academic training on the problems that were involved. The fractional-code system for field mapping had come into existence by this time as well as a considerable body of speculation concerning other technical aspects of the land-appraisal problem. The TVA geographers, faced with the problem of surveying a large area in a relatively short time, succeeded in adapting a fractional-code technique to that work. Representative small areas were chosen for experimental mapping and on the basis of these tests it was determined that air photographs could be used as a base for recording field observations. Field experience also demonstrated the practicability of mapping on a relatively small scale, covering relatively large areas quickly [54; 55].

The Land Utilisation Survey of Great Britain had been started in the early 1930's. As compared with the American surveys, the emphasis was placed on the classification of land uses rather than on the ensemble of resources [56]. Field notations were made on large-scale maps of the Ordnance Survey of Great Britain. A code system was used which gave detailed types of land use but only broad categories of resources. It was some years later that British geographers became concerned over a more specific appraisal of the resource aspects of land-use problems [57].

More recently, a lineal descendant of both the land-resources and the land-utilization surveys appeared in the Rural Land Classification Program of Puerto Rico, in which the Department of Geography of Northwestern University played an important role. In this survey, field observers indicated both land-use characteristics and major resources. The survey also classified lands according to their agricultural usefulness [58]. Field tests again indicated the desirability of altering the notation scheme somewhat to meet local conditions; but the basic features of earlier plans were retained. The resulting data have proved a valuable source of research and planning material.

In summary, it appears that the government-sponsored land surveys, as well as a few surveys sponsored by private agencies, have demonstrated the practical value of geographical methods. Furthermore, these surveys have benefited the field of geography. In the first place, they have provided valuable facilities for testing the procedures that had been developed academically, but for which adequate testing facilities had not been available. In the second place, the land surveys have provided a considerable number of geographers with specific experience in the field administration of complex programs of geographical research. They have shown the need for a more adequate theory of agricultural occupance, a more detailed set of principles and laws that might be carried into the field to make the work of the agricultural geographer not only simpler but also more precise. Not the least of the benefits of these programs has come from the researches of students who encountered problems in the field and sought solutions for them in their individual research programs [59]. In these ways, as is so often true in scientific advancement, applied phases of the subject have given considerable stimulus to the general development of the field.

PROSPECT

For the future, it appears that the need for research in agricultural geography is very likely to become acute. So long as the need for food continues to be a problem, man is going to require information that will indicate not only how but also where food supplies are to be produced. Throughout most of human history the spectre of starvation has been a constant threat to mankind. Only in our recent history has technology advanced so rapidly that new increments in a growing population have seemed to be in no danger of starvation [60]. With the present relatively high rates of population increase and with continued extension of the life-span, as well as the growing recognition of dietary inadequacies in most parts of the world, it seems reasonable to conclude that in the not too distant future the food-population ratio may again become a matter of serious import.

More immediate, perhaps, is the likelihood that modern nations, bent on improving levels of living for their people, will demand more precise information as to where supplies of food and raw materials may be most effectively obtained. Whether such campaigns will take the form of governmental programs or programs initiated by private enterprise, the need for information on which to base them is likely to be insistent.

In general, these research needs point to an appalling deficiency of information for nearly all parts of the world. Of immediate interest to agricultural geography are the programs for the world land-use survey, sponsored for the Eastern Hemisphere by the International Geographic Union, and for the Western Hemisphere by the Union in cooperation with the Commission on Geography of the Pan American Institute of Geography and History. The maps of the world land-use survey are to be prepared with uniform categories and on scales not smaller than 1/1,000,000. They should have far greater practical and professional utility than the maps of the agricultural-region series for two reasons: 1) they should be strictly comparable, because of the use of uniform criteria and of a basic system of standard categories; and 2) they could be used for the analysis of areal relations because the scale of mapping is not too small for this purpose. The meaning of the land-use maps would be greatly enhanced by the simultaneous mapping, at the same degree of generalization, of soil, slope, drainage, climate, and of various economic, social, and political features pertinent to an understanding of agriculture. The practical utility of such surveys and of the geographic analyses which could be made from them will undoubtedly be more widely recognized outside the profession of geography when a larger number of completed maps and reports are made available.

Over and above these considerations of food and raw material supplies, the generalizations provided by the study of land-use data may well give the American people a basis for understanding the circumstances, problems, and ways of life of a majority of the world's people. The significance of such knowledge in the interest of world peace is obvious.

REFERENCES

1. DODGE, R. E. "Geography and Agriculture," (abstract), Annals of the Association of American Geographers, 1 (1911): 144.
2. JONES, W. D., and SAUER, C. O. "Outline for Field Work in Geography," Bulletin of the American Geographical Society, 47 (1915): 520-525.
3. FINCH, V. C., and BAKER, O. E. Geography of the World's Agriculture, Washington, 1917.
4. BAKER, O. E. "The Increasing Importance of the Physical Conditions in Determining the Utilization of Land for Agricultural and Forest Production in the United States," Annals of the Association of American Geographers, 11 (1921): 17-46.
5. ———. "The Agriculture of the Great Plains Region," Annals of the Association of American Geographers, 13 (1923): 109-167.
6. ———. "The Potential Supply of Wheat," Economic Geography, 1 (1925): 15-52.
7. ———. "Agricultural Regions of North America," Economic Geography, 2 (1926): 459-493; 3 (1927): 50-86, 309-339, 447-465; 4 (1928): 44-73, 399-433; 5 (1929): 36-69; 6 (1930): 166-190, 278-308; 7 (1931): 109-153, 325-364; 8 (1932): 325-377; 9 (1933): 167-197.
8. JONASSON, O. "Agricultural Regions of Europe," Economic Geography, 1 (1925): 277-314; 2 (1926): 19-48.
9. JONES, C. F. "Agricultural Regions of South America," Economic Geography, 4 (1928): 1-30, 159-186, 267-294; 5 (1929): 109-140, 277-307, 390-421; 6 (1930): 1-36.
10. TAYLOR, G. "Agricultural Regions of Australia," Economic Geography, 6 (1930): 109-134, 213-242.

11. Van Valkenburg, S. "Agricultural Regions of Asia," *Economic Geography*, 7 (1931): 217-237; 8 (1932): 109-133; 9 (1933): 1-18, 109-135; 10 (1934): 14-34; 11 (1935): 227-246, 325-337; 12 (1936): 27-44, 231-249.

12. Shantz, H. L. "Agricultural Regions of Africa," *Economic Geography*, 16 (1940): 1-47, 122-161, 341-389; 17 (1941): 217-249, 353-379; 18 (1942): 229-246; 19 (1943): 77-109, 217-269.

13. Durand, L. Jr. "The Migration of Cheese Manufacture in the United States," *Annals of the Association of American Geographers*, 42 (1952): 263-282; *idem.* "Dairy Region of Southeastern Wisconsin and Northeastern Illinois," *Economic Geography*, 16 (1940): 416-428; *idem.* "Cheese Region of Northwestern Illinois," *Economic Geography*, 22 (1946): 24-37; *idem.* "Recent Market Orientations of the American Dairy Region," *Economic Geography*, 23 (1947): 32-40; *idem.* "The Lower Peninsula of Michigan and the Western Michigan Dairy Region: A Segment of the American Dairy Region," *Economic Geography*, 27 (1951): 163-183.

14. Shaw, E. B. "Swine Production in the Corn Belt of the United States," *Economic Geography*, 12 (1936): 359-372.

15. Prunty, M. Jr. "Recent Quantitative Changes in the Cotton Regions of the Southeastern States," *Economic Geography*, 27 (1951): 189-208.

16. Weaver, J. C. *American Barley Production: A Study in Agricultural Geography*, Minneapolis, 1950.

17. Jones, W. D., and Finch, V. C. "Detailed Field Mapping in the Study of the Economic Geography of an Agricultural Area," *Annals of the Association of American Geographers*, 15 (1925): 148-157.

18. Jones, W. D. "Ratios and Isopleth Maps in Regional Investigation of Agricultural Land Occupance," *Annals of the Association of American Geographers*, 20 (1930): 177-195.

19. Hartshorne, R. "A New Map of Dairy Areas of the United States," *Economic Geography*, 11 (1935): 347-355.

20. Whitaker, J. R. "Agricultural Gradients in Southern Ontario," *Economic Geography*, 14 (1938): 109-120; *idem.* "Distribution of Dairy Farming in Peninsular Ontario," *ibid.*, 16 (1940): 69-78; Kohn, C. F. "Development of Dairy Farming in Mississippi," *ibid.*, 19 (1943): 188-195.

21. Hartshorne, R., and Dicken, S. N. "A Classification of the Agricultural Regions of Europe and North America on a Uniform Statistical Basis," *Annals of the Association of American Geographers*, 25 (1935): 99-120.

22. Hartshorne, R. "Agricultural Land in Proportion to Agricultural Population in the United States," *Geographical Review*, 29 (1939): 488-492.

23. Elliott, F. F. *15th Census of the United States: Census of Agriculture, Types of Farming in the United States*, Washington, 1933.

24. Bureau of Agricultural Economics, U. S. Department of Agriculture, *Generalized Types of Farming in the United States*, Washington, 1950.

25. Whittlesey, D. S. "Major Agricultural Regions of the Earth," *Annals of the Association of American Geographers*, 26 (1936): 199-240.

26. Goode, J. P., and Espenshade, E. B. Jr., eds. *Goode's World Atlas*, Chicago, 1953.

27. Colby, C. C. "The California Raisin Industry—A Study in Geographic Interpretation," *Annals of the Association of American Geographers*, 14 (1924): 49-108.

28. ————. "An Analysis of the Apple Industry of the Annapolis-Cornwallis Valley," *Economic Geography*, 1 (1925): 173-197, 337-355.

29. James, P. E. "Geographic Factors in the Trinidad Coconut Industry," *Economic Geography*, 2 (1926): 108-125.

30. Trewartha, G. T. "The Green County, Wisconsin, Foreign Cheese Industry," *Economic Geography*, 2 (1926): 292-308.

31. KOLLMORGEN, W. M. "A Reconnaissance of Some Cultural-Agricultural Islands in the South," *Economic Geography*, 17 (1941): 409-430; 19 (1943): 109-117.

32. ———. *The German-Swiss in Franklin County, Tennessee; A Study of the Significance of Cultural Considerations in Farming Enterprises*, Washington, 1940.

33. STARKEY, O. P. "Declining Sugar Prices and Land Utilization in the British Lesser Antilles," *Economic Geography*, 18 (1942): 209-214.

34. FINCH, V. C. *Montfort—A Study in Landscape Types in Southwestern Wisconsin*, Geographic Society of Chicago, Bulletin No. 9, Chicago, 1933.

35. MURAKOSHI, N., and TREWARTHA, G. T. "Land Utilization Maps of Manchuria," *Geographical Review*, 20 (1930): 480-493.

36. KOSTANICK, L., and PRUNTY, M. JR. "Soils and Farm Economy about Mount Warner, Massachusetts," *Economic Geography*, 18 (1942): 173-187.

37. GIBSON, J. S. "Land Economy of Warren County, Kentucky," *Economic Geography*, 10 (1934): 74-98, 200-216, 268-287.

38. DICKEN, S. N. "Central Florida Farm Landscape," *Economic Geography*, 11 (1935): 173-182.

39. FOSCUE, E. J., and TROTH, E. "Sugar Plantations of the Irish Bend District, Louisiana," *Economic Geography*, 12 (1936): 373-380.

40. HANSEN, W. "Dissected Drift Plain of Southeastern Nebraska," *Economic Geography*, 12 (1936): 381-391.

41. HUDSON, G. D. "Agricultural Pattern of East Lothian, Scotland," *Economic Geography*, 14 (1938): 16-22.

42. PLATT, R. S. *Latin America, Countrysides and United Regions*, New York, 1942.

43. PROUDFOOT, M. J. "Sampling with Traverse Lines," *Journal of the American Statistical Association*, 37 (1942): 265-270.

44. CUTSHALL, A. "Problems of Land Ownership in the Philippine Islands," *Economic Geography*, 28 (1952): 31-36.

45. DAVIS, C. M. "Land Ownership in Middle Park, Colorado," *Economic Geography*, 17 (1941): 169-179.

46. MURPHY, R. E. "Landownership on a Micronesian Atoll," *Geographical Review*, 38 (1948): 598-614.

47. ———. "Land Values in the Blue Grass and Nashville Basins," *Economic Geography*, 6 (1930): 191-203.

48. TREWARTHA, G. T. "Land Reform and Land Reclamation in Japan," *Geographical Review*, 40 (1950): 376-396.

49. ROSE, J. K. "Corn Yields and Climate in the Corn Belt," *Geographical Review*, 26 (1936): 88-102.

50. ACKERMAN, E. A. "Influences of Climate on the Cultivation of Citrus Fruits," *Geographical Review*, 28 (1938): 289-302.

51. GIBSON, L. E. "Characteristics of a Regional Margin of the Corn and Dairy Belts," *Annals of the Association of American Geographers*, 38 (1948): 244-270.

52. BARNES, C. P. "Land Resource Inventory in Michigan," *Economic Geography*, 5 (1929): 22-35.

53. HUDSON, G. D. "Methods Employed by Geographers in Regional Surveys," *Economic Geography*, 12 (1936): 98-104.

54. HUDSON, G. D., MILLER, H. V., and others. "Studies of River Development in the Knoxville-Chattanooga Area," *Economic Geography*, 15 (1939): 233-270.

55. HUDSON, G. D. "The Unit Area Method of Land Classification," *Annals of the Asociation of American Geographers*, 26 (1936): 99-112.

56. STAMP, L. D. "The Land Utilisation Survey of Britain," *Geographical Journal*, 78 (1931): 40-53.

57. ———. "Fertility, Productivity, and Classification of Land in Britain," *Geographical Journal*, 96 (1940): 389-412.

58. Northwestern University, Department of Geography, *The Rural Land Classification Program of Puerto Rico*, Northwestern University Studies in Geography, No. 1, Evanston, 1952.
59. JAMES, P. E. "Trends in Brazilian Agricultural Development," *Geographical Review*, 43 (1953): 301-328.
60. ZIMMERMANN, E. W. *World Resources and Industries*, 2nd ed., New York, 1951.

THE GEOGRAPHY OF MINERAL PRODUCTION [A]

A DISTINCT FIELD
CENTRAL THEME OF MINERAL GEOGRAPHY
DESCRIPTION AND INTERPRETATION

SOURCES THAT SUPPLEMENT FIELD WORK

WORK THAT HAS BEEN DONE

FRONTIERS OF INVESTIGATION
STUDIES OF INDIVIDUAL MINERAL INDUSTRIES
STUDIES OF MINING REGIONS
CHANGES THROUGH TIME
MINING SETTLEMENTS
SOCIAL PROBLEMS OF MINING REGIONS
DELIMITING MINERAL-PRODUCING REGIONS
MINERALS IN WORLD AFFAIRS

[A] Original draft by Raymond E. Murphy. Criticisms and suggestions were received from S. R. Abrahamson, W. Applebaum, P. E. James, H. H. McCarty, E. W. Miller, R. S. Platt, J. A. Russell, W. H. Voskuil, J. R. Whitaker, and A. J. Wright.

CHAPTER ELEVEN

THE GEOGRAPHY OF MINERAL PRODUCTION

MINERALS ARE of ever-increasing importance in modern civilization. Strikingly uneven in their distribution over the earth's surface, they are at the same time unlike other resources in being exhaustible and nonrenewable. Mining[B] commonly creates scenes of intense activity and produces great wealth from a limited area for a short time. No other land use is so ephemeral. Cut a forest, and it will spring up anew. Mine a coal seam and carry its fragments to market, and the coal is gone forever. A sense of importance in studying mineral production comes from an awareness of the finality of mining as much as from the knowledge that the mined product has great value.

A DISTINCT FIELD

Mineral production is of interest to workers in several disciplines. In this field the economic geographer meets the geologist, the mining engineer, and the mineral economist. Each makes his own distinctive contribution. The geographer concentrating on the minerals field is not, like the economic geologist, primarily interested in the origin of mineral deposits. Neither is his chief interest in methods of production, as is the mining engineer's, nor in the collection and analysis of mineral statistics, which is the business of the mineral economist. Instead, he focuses on the spatial patterns and associations of mineral production. He is responsible, more than are these others, for the examination of mining as a part of the total geographic complex of particular regions.

Other branches of economic geography deal to some degree with minerals, but in none of the others are mineral-production patterns the center of interest. Mining is a form of land utilization, but the areas it occupies are so limited that it generally is accorded little attention in land utilization surveys. The student of resources is concerned with minerals, but he is more likely to think of them in a broad resource program than to study individual mining regions or the patterns of mineral production. Moreover, the geography of mineral production grades into the geography of manufacturing, since most of the world's minerals constitute raw materials

B. The term "mining" is used throughout this chapter to include recovery of such mineral resources as petroleum and natural gas as well as solid mineral products.

for factories and the location of mineral production is an important factor in industrial location. Thus the geography of mineral production has a number of near relatives. Yet it also has its own distinctive core.

Central Theme of Mineral Geography

Although the economic geographer working in the field of mineral production deals primarily with a mineral product or with the economic activity of mining, he always does so in terms of an areal setting. Through the study of mining patterns in relation to the patterns of other relevant phenomena, he attempts to gain a deeper perspective with regard to the problems of mineral production than would be possible through the study of mining in isolation from its setting. He is concerned not only with the phenomena of mining themselves but also with the geologic pattern, the flow of mineral products to the factory or consumer, the concentrations of people who depend directly or indirectly upon mining for a living, the political patterns that affect the mineral-production patterns, and other spatial distributions that help to explain the phenomena upon which he focuses.

The student of mineral geography does not confine his interest to the spatial distributions of the present. Mineral-production patterns have changed with time and so, too, have the areal distributions relevant to mining. The patterns of mining today are understandable only when the study is grounded upon a knowledge of past changes. A study of mineral production in the past may also furnish a key to an understanding of areas where mining is no longer significant. For example, the development of mining has often been responsible for an influx of people. Though in some cases these folk left when the mining ceased, in many instances they remained and developed other means of gaining a livelihood. Thus mining, though itself ephemeral, has operated to establish permanent communities.

Description and Interpretation

Like all geographic work, the study of spatial distributions associated with mineral production involves description and interpretation. A fundamental part of the description may be expressed in the form of maps of mineral production. Here there is a marked analogy to manufacturing: in both mining and manufacturing a great deal of wealth comes from very small areas. This gives rise to special cartographic and mapping problems. It is difficult, for example, adequately to represent either mineral production or manufacturing on the maps of a general land use survey, since these activities are likely to have an economic importance out of all proportion to the surface area they occupy. Obviously, symbols ordinarily must be on different scales than used to depict the facts of agriculture.

Scale also functions in interpretation. It is relatively easy to gain an understanding of the details of local patterns and production, since they reflect such simple factors as the relative accessibility of minerals and the methods used for their recovery. The arrangement of iron mines on the Mesabi Range of Minnesota exhibits a linear pattern, because the mines are localized by a belt of iron-bearing rock, and the oil wells in northern Pennsylvania show a checkerboard pattern of local distribution that is attributable directly to the "five-spot" system of secondary oil recovery.

Similar factors also help to explain the broad mineral-production patterns that may be seen on maps covering larger areas on smaller scales. The belt of coal mining in the Appalachian Plateau corresponds to the trend of a broad geosyncline with its rich coal seams, and the distribution of coal production throughout the United States and throughout the world conforms to a considerable degree with the occurrence of rocks of certain favorable ages.

But unfortunately such simple and obvious explanations are inadequate. Of course the mineral values have to be present or there can be no mineral production, but that still leaves questions of time and place. Why were some mineral deposits that are now of tremendous importance neglected until a few years ago? Why does the mineral-production map fail to disclose the actual distribution of mineral resources for large sections of the earth's surface?

No simple answer will suffice. The whole problem is tied up with the rapid increase in industrialization. The use of minerals has been pyramiding in modern times. H. L. Keenleyside reported that "the quantity of mineral products consumed between 1900 and 1949 far exceeds that of the whole preceding period of man's existence on earth" [1: 3]. With demands have come technological developments that have constantly changed the patterns of mineral production. Deposits formerly considered valueless have become workable through the use of new techniques of ore treatment. Minerals unused a decade ago are suddenly so much in demand in industrial processes that they are listed as "critical." New prospecting devices and techniques, such as the airborne magnetometer and the Geiger counter, are bringing increased efficiency into the search for minerals. The student of mineral geography, if he is to understand the broad aspects of the mineral-production pattern, should be aware of the rapidity with which the minerals picture is changing.

With the great increase in production and use of minerals has come a change in financing. Capital requirements for mineral extraction have become so great that small deposits are of interest only in the case of particularly rare commodities. Moreover, minerals remain unused in countries that lack adequate capital for exploration and development; and yet some countries put up barriers against the inflow of such capital, fearing the loss of control which foreign investment might entail. C. A. Fisher has pointed out that "an atlas of maps showing the distribution of capital, by country of origin, and by area of investment, for decennial intervals during the past hundred years" would be of tremendous value in making contemporary economic geography intelligible [2:79]. Such maps would be most useful in the interpretation of the world's mineral-production patterns.

To understand area contrasts over the earth's surface the geographer studying mineral production needs to go much further. He must take account of the great contrasts in the world's economies that aid in explaining why the mineral deposits of some areas are undeveloped and of others largely unknown. He should be aware of the great differences between production under communism and in a capitalist society. Under the former, the needs of the state and the supply of labor determine whether an operation is to be carried on or not. Under capitalism, production costs normally must be sufficiently beneath market prices so that a profit results. If, under capitalism, an increase in demand for a mineral commodity or a decline in supply results in a rise of prices, it becomes profitable to work lower-grade deposits and more feasible to obtain capital for investment in exploratory activities, which ultimately lead to increasing available reserves. The geographer should take cognizance, too, of the growing tide of nationalism that is making it more difficult to carry on the world-wide operations that have characterized capitalistic economy. Still another complicating factor is the effect of war and war preparations, which focus demand upon strategic and critical minerals.

SOURCES THAT SUPPLEMENT FIELD WORK

The mineral geographer supplements his field work with whatever published and unpublished materials may be available. Published reports, many of them in technical magazines, are vital to his work, and statistical data play a prominent role in his investigations.

On a world scale, production data are reasonably adequate. They are available in the *Minerals Yearbook* of the United States Bureau of Mines, the chief official collector of mineral-production data in this country, and in various world statistical volumes. As with other statistics, mineral data are lacking for some countries because of inadequate machinery for collection and for others because of deliberate withholding of such information.

Statistics of mineral production within the United States are published in the *Minerals Yearbook*. For a number of mineral products, only State totals are released. Certain types of mineral production are more widespread, however, and for these better localizing data are given. Petroleum figures are published by leading fields within States; iron-ore totals are given for the larger mines; and, for the most part, the data for copper, lead, and zinc are localized by districts within States. For coal production, chiefly because of its wide areal spread, figures are given by counties.

For the geographer interested in local mineral production other sources of data are available. For example, several States have bureaus of mines which publish figures on tonnages and employment, sometimes in greater detail than do the federal agencies. Such information is useful in supplementing federal data, thus serving better to localize the mining activity.

Valuable data are also published by such trade associations as the American Petroleum Institute, the American Gas Association, the Iron and Steel Institute, and the Bituminous Coal Association.

For the geographer, data on the movement of minerals are particularly important. Annual figures for anthracite are available. For crude and refined petroleum products such data are reasonably complete. The United States Bureau of Mines formerly published from time to time detailed statistics on the movements of bituminous coal within the United States, but because of lack of authorization and curtailment of funds such data have not been collected since 1946. Some information of this sort for the United States can still be gleaned from the Interstate Commerce Commission reports; and the Economic Commission for Europe does a reasonably good job with statistics of movements from coal fields to consuming nations for Europe. It would be extremely useful to geographers, however, to have comprehensive international and intranational data for all of the mineral products.

WORK THAT HAS BEEN DONE

The geography of mineral production has attracted only a few American geographers. Not more than ten articles in this field have appeared in the *Geographical Review* during the magazine's existence, and fewer than this in the *Annals of the Association of American Geographers*. The number has been somewhat greater in *Economic Geography*, as might be expected from the specialization of the magazine, but the attention paid to mining in the geographic literature in general has been vastly less than that accorded to agriculture.

Why has this subject attracted so few investigators? There is the very obvious answer that agriculture is much more widespread and that even where mining is important it has little surface expression in proportion to the value of production. Doubtless, too, many have felt that they lacked the technical background that studies of mineral-production geography seemed to involve. The inattention to this phase of geography may also have been due in part to a disposition on the part of American geographers to leave the problems of mineral production to the geologist.

The last tendency may explain why geographers were overshadowed for years in their contributions in this area of research by economic geologists with leanings toward mineral economics. The books of C. K. Leith, for example, have been outstanding in this field [3; 4]. Geology and geography were long associated at the University of Wisconsin, as they have been at so many universities, and Leith, who was chairman of the combined departments, undoubtedly was influenced by his geographic colleagues. Nevertheless, he was a geologist. So, too, was H. Foster Bain, the author of a particularly challenging article that appeared in the 1920's [5]. These contributions and several others of the same type [6; 7; 8:423-567] may lack something when measured by the geographic yardstick, but they show a breadth that geographers in this field are still short of attaining.

Two mineral atlases have appeared over the years, both presenting the occurrence, production, and use of the world's minerals. The first of these was the work of geologists, mining engineers, and mineral economists [9]. In the preparation of its successor the geographer and the mineral economist have both taken part [10]. *World Geography of Petroleum*, like the new atlas of mineral production, is a joint product of geologists and geographers [11].

Studies in the geography of mineral production which can be credited exclusively to geographers have been sporadic. A few illustrations will serve to indicate the directions this work has taken.

The commodity, or commodity-in-area, approach appears to have been the most common. In this group are studies of the aluminum industry [12], coal mining [13], copper [14], natural gas [15], and various other commodities. In these studies the area covered ranges from the mining district or the oil camp to the continent or even the entire world. In some cases the resource rather than the industry has received the emphasis [16]; in a few instances transportation of the product has been the major focus [17]; and E. W. Miller has used the commodity approach in a statistical study comparing the mineral production of nations [18]. For the most part these studies are informational and do not represent anything new in method or any real specialization; but there are exceptions. One of these is J. W. Frey's work on petroleum [11:354-374; 19; 20:378-401], a subject on which he has become a recognized authority. Walter H. Voskuil's work is another exception, since its author has dealt consistently with mineral-production problems for a number of years [21; 22; 23].[c]

c. See also his various bulletins issued by the Illinois Geological Survey.

Studies by geographers of mineral-producing regions have been almost as common as commodity articles. Here, again, the emphasis has varied greatly from one writer to the next, just as the areal unit discussed has varied. For the most part the products are regional studies with emphasis upon the mineral factor [24; 25; 26; 27; 28; 29; 30]. Some writers have stressed the contemporary patterns of occupance [31; 32; 33]. Sequent occupance has been emphasized in at least one case [34], and J. Russell Whitaker used the methods of historical geography in his detailed study of a Michigan iron mining community [24]. Oddly enough there seems to be no case in which mining as an occupation forms the central theme.

Not infrequently problems of mineral conservation have been brought into the commodity and regional studies, and the several books in the general field of conservation have sections dealing with mineral problems. But, aside from some work by E. W. Miller on the problems of conservation associated with strip mining [35; 36], no geographer appears to have concentrated on this aspect of the minerals field.

Both within individual regions and for larger areas the patterns of mineral production are constantly changing. Recognition and measurement of shifts are well within the province of the geographer. R. E. Murphy and H. E. Spittal have made a start on the measurement of such changes through an adaptation of the centrographic method to shifts in coal production [37; 38], and S. R. Abrahamson, an economic geographer employed by the United States Bureau of Mines, has used the same technique in studying the effect of movements of the world center of petroleum production on pricing systems [39].

It was pointed out in an earlier chapter that a knowledge of economics should be part of the equipment of every economic geographer. But for the economic geographer who specializes in mineral production more is needed. For serious work in this specialty a background in geology and mineralogy is essential. It is the economic geographer with a background in economics, geology, and mineralogy who is likely to make substantial contributions in this field, particularly if he knows something of mineral-production methods or is willing to persist in his topical specialty long enough to build up such knowledge.

FRONTIERS OF INVESTIGATION

Much work remains to be done in the geography of mineral production. Some potential lines of investigation in this field include studies of: 1) individual mineral industries; 2) mining regions; 3) changes through time; 4) mining settlements; 5) social problems of mining regions; 6) delimiting mineral producing regions; and 7) minerals in world affairs.

STUDIES OF INDIVIDUAL MINERAL INDUSTRIES

If mineral geography is to play a larger role outside the classroom, students in this field must specialize, and one possible line of specialization is on individual mineral industries. For example, the coal industry of the United States should be re-evaluated. There has been a great deal of technological research on coal in recent years. Is the geographic pattern of the industry being fundamentally changed thereby? The constantly-changing picture of coal production within the United States and the movement of coal to markets should furnish material for a number of worthwhile studies. It is to be hoped that in the future some geographers will have so far distinguished themselves in this subject that their services will be considered indispensable when national planning involving coal is contemplated.

The development and shifts of a number of other mineral industries, also, might well be studied. Iron ore mining, petroleum production, the copper industry, the zinc industry, and various others should be considered from the standpoint of their evolution, how they have moved and are moving, and how they are likely to move in the future. A single industry, or even certain aspects of a single industry, could well form the subject for a lifetime of research. Obviously, there is no necessity for confining such studies to the United States, but just as obviously there is much work of this sort that could be done within the country's borders.

STUDIES OF MINING REGIONS

More studies should be made of individual mining regions of all sizes. The investigation of mining regions reveals the interplay of the mining process with other elements of the local setting, the people who work in the mines, the houses they live in, the transportation pattern, the other industries that are present, and the many other items that go to make up the unique character of the region. The historical perspective is essential to giving depth to such work. There are many regions that might well be studied: the Tri-State Zinc District, the Iron Ranges of Minnesota, a copper mining center of the Southwest, or any one of hundreds of other mining districts and communities. Although several such studies have been made, they constitute only a little more than a good start.

Ideally, studies of individual mining regions should go hand in hand with studies of the mineral industries. The two approaches are appropriately correlated, since anyone examining the zinc industry, for example, could gain much from research on individual zinc-mining regions.

CHANGES THROUGH TIME

Mining regions change through time just as do other types of areas. New areas of mineral production have characteristic features and problems; so do mature and old regions. Studies of the characteristics of mining regions of different ages are well worth while. Age studies have been carried on from the geological point of view [6:62-90], and, as earlier pointed out, geographers have studied the historical geography of mining regions and sequent occupance in mining areas where mining has been replaced by other activities, but more work of this sort is needed. Aside from knowledge for its own sake, they might yield much that would be useful in planning. Possibly some of the problems that plague mining communities in their last stages of existence might be avoided if more were known about the whole aging process.

Mining Settlements

In some coal-mining areas the typical settlement is the company-built village with a company store or commissary. In others, a cluster of company houses with urban spacing but no store is characteristic. And in still others the towns are politically separate, almost entirely dependent upon mining for their existence. Or, the independent mining town may have a section that is completely company-built and company-owned. In some of the company-owned towns the company now is selling its houses to the miners; in others paternalistic control is still maintained. Such settlement phenomena are worth studying by geographic methods as well as by the methods of the sociologist.

Social Problems of Mining Regions

Various social problems that accompany mining also need to be studied geographically. Why, for example, is an area of unusual mineral wealth so often a marginal one in terms of economic well-being? The Welsh coal fields and our own Appalachian coal fields were real problem areas during the depression years of the early 1930's, and, even in more prosperous times, the natural wealth of such areas may not find adequate reflection in the lives of their citizens.

How can the importance of mining in the economy of an area be studied? Employment in mining obviously is not enough to measure the significance of the industry, since transportation workers, merchants, and various others who are not directly employed in mining are nevertheless largely supported by the mining activity. What has been the significance of the copper industry to Arizona or of the oil industry to Oklahoma? Perhaps some investigator will be able to set up criteria that will answer this question.

Related to this problem is the attempt to measure the value of an exported mineral product to the nation in which it originates. Take the example of copper mined in Chile and exported to the United States. It would indeed be interesting if we could develop some means of comparing the returns to the originating country in terms of taxes, wages, and other forms of income, with the returns to the consuming country.

Another intriguing social problem is presented by the relations of certain population groups to mining. The importance of Welsh people in the anthracite industry of Pennsylvania and the considerable number of Cornish men in the copper country of Keweenaw Point suggest interesting lines of research. To what extent did immigrants bring their special interest in mining with them to the United States?

Delimiting Mineral Producing Regions

Depicting the world's mineral production presents its problems. No one has recognized types of mining regions for individual countries, or for the world, that are at all comparable with those evolved for agriculture. The development of such a system presents serious difficulties. Mining is limited to points on a map, though the importance of these points may be out of all proportion to areal extent. Other difficulties are the seeming heterogeneity of the picture: one mining area often produces a variety of minerals; and there are great variations in the richness of deposits within short distances. Nevertheless, a better job of depicting mineral-producing regions than has been done thus far should be possible.

Minerals in World Affairs

Minerals are also increasingly important in world affairs. No modern nation can hope to be economically strong without controlling a supply of the more essential minerals. But here again the picture is by no means static. The discovery of new mineral sources and the decline of old ones require periodic re-evaluation of the mineral position of nations; and so, too, do new developments in technology and the rise of "new" minerals. Possible effects on mineral production of the rising tide of nationalism; tariffs in relation to minerals; taxation of mineral deposits; the place of mineral production in "Point 4" programs; the desirability of stockpiling critical minerals: on such problems the point of view of the geographer should be of value.

REFERENCES

1. KEENLYSIDE, H. L. "Critical Mineral Shortages," *Proceedings of the United Nations Scientific Conference on the Conservation and Utilization of Resources,* Vol. 1, Plenary Meetings, Lake Success (New York), 1950: 38-46.

2. FISHER, C. A. "Economic Geography in a Changing World," *Institute of British Geographers Publication,* 14 (1948): 69-85.

3. LEITH, C. K. *World Minerals and World Politics,* New York, 1931.

4. LEITH, C. K., FURNESS, J. W., and LEWIS, C. *World Minerals and World Peace,* Brookings Institution, Washington, 1943.

5. BAIN, H. F. "The Third Kingdom: Some Reflections on our Mineral Heritage," *Geographical Review,* 18 (1928): 177-195.

6. TRYON, F. G., and ECKEL, E. D., eds. *Mineral Economics: Lectures under the Auspices of the Brookings Institution,* A. I. M. E. Series. New York, 1932.

7. LOVERING, T. S. *Minerals in World Affairs,* New York, 1943.

8. ZIMMERMANN, E. W. *World Resources and Industries,* 2nd. ed. New York, 1951.

9. United States Geological Survey, *World Atlas of Commercial Geology, Part I: Distribution of Mineral Production,* Washington, 1921.

10. VAN ROYEN, W., and BOWLES, O., eds. *The Mineral Resources of the World,* New York, 1952.

11. PRATT, W. E., and GOOD, D., eds. *World Geography of Petroleum,* American Geographical Society Special Publication No. 31, New York, 1950.

12. COLLIER, J. E. "The Aluminum Industry of the Western Hemisphere," *Economic Geography,* 20 (1944): 229-257.

13. CARMIN, R. L. "The Coal Mining Industry of Guernsey County, Ohio," *Economic Geography* 19 (1943): 292-300.

14. BIRCHARD, R. E. "Copper in the Katanga Region of the Belgian Congo," *Economic Geography,* 16 (1940): 429-436.

15. PARSONS, J. J. "The Geography of Natural Gas in the United States," *Economic Geography,* 26 (1950): 162-178.

16. COLLIER, J. E. "Aluminum Resources of the United States," *Economic Geography,* 24 (1948): 74-77.

17. VOSKUIL, W. H. "Bituminous Coal Movements in the United States," *Geographical Review,* 32 (1942): 117-127.

18. MILLER, E. W. "Some Aspects of the Mineral Position of Eight Principal Industrial Nations," *Economic Geography,* 26 (1950): 133-143.

19. FREY, J. W. "The World's Petroleum," *Geographical Review,* 30 (1940): 451-462.

20. National Resources Committee, *Energy Resources and National Policy,* Washington, 1939.

21. VOSKUIL, W. H. *Minerals in Modern Industry,* New York, 1930.

22. ———. "Coal and Political Power in Europe," *Economic Geography,* 18 (1942): 247-258.

23. ———. *Postwar Issues in the Petroleum Industry,* Business Studies No. 3, Urbana (University of Illinois), 1946.

24. WHITAKER, J. R. "Negaunee, Michigan: An Urban Center Dominated by Iron Mining," *Bulletin of the Geographical Society of Philadelphia,* 29 (1931): 137-174, 215-240, 306-339.

25. BRIGHTMAN, G. F. "Cuyuna Iron Range," *Economic Geography,* 18 (1942): 275-286.

26. MILLER, E. W. "Economic Geography of the Bradford Oil Region," *Economic Geography,* 19 (1943): 177-187.

27. MURPHY, R. E. "Southern West Virginia Mining Community," *Economic Geography,* 9 (1933): 51-59.

28. MURPHY, R. E., and MURPHY, M. "Anthracite Region of Pennsylvania," *Economic Geography,* 14 (1938): 338-348.

29. ZIERER, C. M. "An Ephemeral Type of Land Occupance," *Annals of the Association of American Geographers,* 26 (1936): 125-156.

30. ————. "Broken Hill: Australia's Greatest Mining Camp," *Annals of the Association of American Geographers*, 30 (1940): 83-108.

31. PLATT, R. S. "South Range, Keweenaw Copper Country: A Mining Pattern of Land Occupancy," *Economic Geography*, 8 (1932): 386-399.

32. ————. "Pattern of Occupance in the Marcaibo Basin," *Annals of the Association of American Geographers*, 24 (1934): 157-173.

33. ————. "Mining Patterns of Occupance in Five South American Districts," *Economic Geography*, 12 (1936): 340-350.

34. HOFFMEISTER, H. A. "Central City Mining Area," *Economic Geography*, 16 (1940): 96-104.

35. MILLER, E. W. "Strip Mining and Land Utilization in Western Pennsylvania," *Scientific Monthly*, 69 (1949): 94-103.

36. ————. "Penn Township: An Example of Local Governmental Control of Strip Mining in Pennsylvania," *Economic Geography*, 28 (1952): 256-260.

37. MURPHY, R. E., and SPITTAL, H. E. "Movements of the Center of Coal Mining in the Appalachian Plateaus," *Geographical Review*, 35 (1945): 624-633.

38. MURPHY, R. E. "Wartime Changes in the Patterns of United States Coal Production," *Annals of the Association of American Geographers*, 37 (1947): 185-196.

39. ABRAHAMSON, S. R. "The Shifting Geographic Center of Petroleum Production and its Effect on Pricing Systems," *Economic Geography*, 28 (1952): 295-301.

THE GEOGRAPHY OF MANUFACTURING[A]

DISTRIBUTION OF MANUFACTURING
THE AREAL DISTRIBUTION OF MANUFACTURING
CHANGES IN THE DISTRIBUTION OF MANUFACTURING

STATISTICAL MATERIALS
CRITERIA USED TO MEASURE DISTRIBUTION
STATISTICAL AND CARTOGRAPHIC DATA ON MANUFACTURING
GENERAL STATISTICAL MEASURES OF LOCATION

LOCATION THEORY AND LOCATION TYPES
LOCATION THEORY
LOCATION TYPES

STUDIES OF SPECIFIC INDUSTRIES

STUDIES OF INDUSTRIAL AREAS

EXPLANATION AND PREDICTION OF INDUSTRIALIZATION

PROBLEMS NEEDING STUDY
SIGNIFICANCE OF INDUSTRIES TO REGIONAL DEVELOPMENT
DISTRIBUTION OF MANUFACTURING AND OF MANUFACTURING AREAS
IMPACT OF NEW TECHNOLOGY

A. Original draft by Chauncy D. Harris, with suggestions and criticisms from John W. Alexander, William Applebaum, John E. Brush, Richard Hartshorne, Robert A. Heil, Preston E. James, J. Granville Jensen, Clarence F. Jones, H. H. McCarty, E. Willard Miller, Harold V. Miller, Raymond E. Murphy, James J. Parsons, Victor Roterus, Bernard H. Schockel, Thomas R. Smith, Edward L. Ullman, and Alfred J. Wright.

CHAPTER TWELVE

THE GEOGRAPHY OF MANUFACTURING

MANUFACTURING OCCUPIES little space on the earth's surface but is vitally important to man's livelihood, security, and progress. It exhibits sharp localizations and close areal associations and functional interrelations with other phenomena. Manufacturing commonly is associated with a high density of population, high levels of living resulting from the multiplication of human productivity by the application of inanimate power, good transportation facilities, great urban markets, large volumes of foreign trade, a high degree of interdependence, and large political and military power.

Interest in the geography of manufacturing in the United States dates back to the foundation of the republic. Alexander Hamilton proposed to Congress the establishment of a federal manufacturing city because of the importance of manufacturing in the promotion of the general welfare; the city was to be Paterson, New Jersey. In 1853 the first volume of the new *Bulletin of the American Geographical and Statistical Society* contained an informative account by a New York merchant of cotton manufacturing in the leading countries of the world [1].

Not until the quarter-century 1903-1927, however, did professionally trained geographers begin to develop an interest in manufacturing and to consider how manufacturing should be treated geographically. Studies of this period were concerned mainly with factors in the location of manufacturing in specific areas.

In 1927 three major papers marked the beginning of present methods in manufacturing geography. Sten de Geer described quantitatively and cartographically the areal extent of the American Manufacturing Belt [2]; Richard Hartshorne attempted a general quantitative location theory for manufacturing [3]; and Robert S. Platt formulated a geographic classification of industries based on location types [4].

The quarter-century since 1927 has been characterized by a regular flow of manufacturing studies by geographers. There has been little consideration, however, of the nature and potentialities of the field of manufacturing geography as such. As yet no American book has appeared devoted solely to the geography of manufacturing, but one is in preparation. In recent years courses in the geography of manufacturing have been offered in several universities.

293

The following account is concerned primarily with the contributions of American geographers to the geography of manufacturing. Contributions by geographers of other countries or by economists, planners, engineers, and business men are mentioned here only insofar as they are closely related to work by Americans. Geographers are interested in where manufacturing is located and in methods of outlining manufacturing areas and of measuring the intensity of concentration; they are interested in the complex inter-relationships between individual manufacturing establishments or between concentrations of such establishments and the communities of which they are a part; they are interested in the areal relations between manufacturing of different kinds and such items as raw materials, power, labor, markets, and transportation facilities; and they are interested in seeking an understanding of the factors that have led to industrial concentrations in particular places in the past and at present, and in the forecast of future trends in industrialization. As in the other fields of economic geography discussed in the preceding chapters, workers in the field of manufacturing geography choose between two approaches: 1) the study of the areal relations of an individual establishment or of a specific kind of industry; or 2) the study of areas in which manufacturing is a predominant feature of the economy. In either case geographers are concerned with areal relations, associations, and the processes operating in particular places to produce industrialization.

The participation by geographers of the United States in the study of the geography of manufacturing will be summarized under seven major headings: 1) distribution of manufacturing; 2) statistical materials; 3) location theory and location types; 4) studies of specific industries; 5) studies of industrial areas; 6) explanation and prediction of industrialization; and 7) problems needing study.

DISTRIBUTION OF MANUFACTURING

An important concern of students of manufacturing geography is to portray the distribution of manufacturing on maps. This requires generalization, for only on very large-scale maps can the individual plants be plotted to scale, and the economic importance of a plant bears no necessary relationship to its physical size or the area it occupies. The geographer seeks criteria whereby he may define meaningful categories of manufacturing regions (in the sense defined in chapters one and two). In the United States, manufacturing enterprise is concentrated in the Manufacturing Belt, one of the greatest concentrations of its kind in the world. This belt extends from the east coast, from north of Boston to south of Baltimore, westward across New York and Pennsylvania to the Lake States. Exactly how is this belt to be defined? How are manufacturing concentrations outside of the belt to be identified and plotted on maps? And since geographers are concerned with more than the static picture of contemporary areal differentiation, they are also interested in changes in manufacturing distribution and in methods of measuring and mapping these changes.

THE AREAL DISTRIBUTION OF MANUFACTURING

The pioneer work on the quantitative delimitation of the American Manufacturing Belt was carried out by the Swedish geographer Sten de Geer and reported in a paper published in 1927 [2]. His method of defining the area was based on numbers of wage earners employed in manufacturing. He used figures for towns of 10,000 or more population only, and he defined such a town or city as industrial if it contained at least 1,000 wage earners employed in manufacturing. Having defined the manufacturing area, he examined the possible factors that had led to its development and its position.

Two aspects of Sten de Geer's study have been challenged by American students: 1) the use of data for cities of over 10,000 population only, for it is clear that, although manufacturing is primarily an urban function, it extends also to the suburbs, to towns of less than 10,000, and even to certain otherwise rural areas; and 2) the use of data on employment only, for other kinds of criteria can be used to give additional perspective to the definition of manufacturing areas. It has been pointed out that de Geer's map does no more than reflect the general distribution of urban population, since any large city would have enough workers in industry to be classed as an industrial city.

Several American geographers have attacked the problem of defining manufacturing areas. In 1937 Helen M. Strong published a map of the distribution of manufacturing in the United States on the basis of power used in county units [5]. In 1938 Alfred J. Wright evaluated the various criteria that had been proposed and concluded that an effective summation of the total effort put into manufacturing activity (labor, capital, and power) could be derived from data on the value added in the manufacturing process[6]. Like de Geer he used towns of 10,000 or more population, but by including the major manufacturing districts listed in the Census of Manufactures he also covered suburban areas within these districts. Also in 1938 Clarence F. Jones noted that the restriction of criteria to towns of 10,000 or over resulted in the elimination of a significant segment of manufacturing enterprise: more than half of such enterprise in South Carolina; two-fifths of it in Texas [7]. Like Strong he recommended the use of county units. He plotted on three separate maps the number of persons employed in manufacturing, the amount of power used, and the value added by manufacture, from which he prepared a generalized map of the areal distribution of manufacturing.

Meanwhile Richard Hartshorne had attempted to determine where manufacturing had produced the distinctive characteristics generally attributed to the Manufacturing Belt [8]. Much manufacturing is ubiquitous in that it is found in every urban community, and is local in function in that its services are limited to the needs of the communities of which it is a part. Such manufacturing does not contribute to areal differentiation except when considered on very large-scale maps. In order to estimate the employment in non-local industries, Hartshorne, assuming that about 10 per cent of the total population was engaged in local industrial activities, subtracted this amount from the total number of wage earners in each city. He then defined an industrial city as one in which at least 500 workers were employed in non-local manufacturing. Chauncy D. Harris continued to work along the same lines as Hartshorne, attempting to ascertain areas in which manufacturing is the key activity [9]. He classified industrial cities on the basis of the ratio of employment in manufacturing to employment in retail and wholesale trade. H. H. McCarty defined the limits of the Manufacturing Belt on the basis of counties having more factory workers than farmers, counties with per capita or per county manufacturing greater than the national average, and counties in which the ratio of manufacturing to wholesale trade exceeded the national average [10].

In addition to these attempts to define manufacturing areas in the United States, it is interesting to note that, just as the first quantitative delimitation of the American Manufacturing Belt was made by a European geographer, the first similar approach to the delimitation of the European manufacturing areas was made by the American geographers, Chauncy D. Harris and Burton W. Adkinson. A similar study of Japanese industrial areas was done by Thomas R. Smith.

CHANGES IN DISTRIBUTION OF MANUFACTURING

The patterns of distribution of manufacturing activities are dynamic. The movements of industry may be from the central parts of cities to suburbs or to nearby points within the same industrial area, or from one industrial area to another, or from industrial areas to non-industrial areas, as from the American Manufacturing Belt to the South or the West. McCarty investigated migrations for the period 1914-1937 and discovered that the industries showing the greatest relative shifts between states tended to be either raw-material-oriented or market-oriented [11]. His statistical analyses failed to confirm any general flight of industry from the larger cities to rural areas. Plants built during World War II were concentrated in metropolitan districts, which alone could provide quickly the large reserves of labor needed for huge industrial enterprises [12]. Alfred J. Wright measured and described the shifts in the relative importance of the major sections of the United States in manufacturing 1919-1939 according to several criteria, and during the war period 1939-1945 as measured by increased facilities [13]. John W. Alexander also analyzed the remarkable industrial expansion of 1939-1947 [14].

STATISTICAL MATERIALS

Studies of manufacturing geography, like studies in other economic fields, are dependent on statistical data. They also make use of statistical procedures in finding correlations and recurrent patterns in these data. What are the quantitative criteria that may be used to measure the distribution, relative importance, and rate of growth or decline of manufacturing in an area? What are the sources of statistical and cartographic information? And what are the statistical indices of localization, dispersion, or areal association?

CRITERIA USED TO MEASURE DISTRIBUTION

All statistical data for the measurement of manufacturing distribution suffer from limitations, whether the data be of labor, value added, total value, power, physical quantity of production, or productive capacity.

Some geographers favor the use of data on labor. For example, Hartshorne makes use of such data because they give a quantitative view of how people make a living in an area. In the United States there are two kinds of data on labor, those based on employment and those based on occupations. The former are reported in the Census of Manufactures, the latter in the Census of Population. Employment figures are reported by each industrial establishment; but a rule prohibits the Bureau of the Census from revealing figures that would make it possible to calculate data concerning any one plant. This rule as applied to employment seems scarcely necessary, since employment figures are common knowledge in the locality and are often reported to other federal or State agencies. Employment data are also reported in other countries of the world, and were used for the construction of a map of the industrial regions of India by Glenn T. Trewartha and J. L. Verber [15].

At least three geographers, McCarty, A. J. Wright, and Victor Roterus, have indicated a preference for data on the value added by manufacture as the best criterion for defining manufacturing areas [16]. Data of this kind are available for the various areas of the United States, but are usually not available in other parts of the world. In some cases the total value of manufactured products can be substituted for value added, but it is often misleading as an indicator of manufacturing activity in a given locality because it includes the values added by earlier processing, sometimes in different places. All criteria based on value are difficult to apply where two or more countries are being studied or where different periods are being compared.

There are still other kinds of data that can be used where statistics of employment and of value are lacking. The physical output provides an excellent base for comparisons between countries or between different historical periods when such data are used for industries with standardized products, such as the iron and steel industry. Industrial capacity is also useful in connection with industries where the product is standardized and where the data are available by individual plants. The amount of floor-space available has been used as an indicator of relative capacity in the aircraft industry, where the size of units and prices differ widely from plant to plant. Generally, data on power consumed are not good indices of manufacturing.

STATISTICAL AND CARTOGRAPHIC DATA ON MANUFACTURING

In the United States the Bureau of the Census is the basic source of statistical materials, although in a few States there are valuable directories of industrial concerns that can be used to supplement the federal census data. From time to time the Bureau of the Census has published summary statements and maps of the localization of specific industries. Both the Bureau of Foreign and Domestic Commerce and the Bureau of the Census have published tables on manufacturing industries on a county basis. Such tables are valuable but are difficult to summarize by inspection. Meredith F. Burrill produced a handy atlas of manufacturing industries, showing the distribution (by States) of each of the major industries of the United States (1929), based on total value of product for each industry [17]. A series of maps on a county basis is contained in *The Structure of the American Economy* [18].

GENERAL STATISTICAL MEASURES OF LOCATION

Attempts have been made to measure quantitatively the extent of localization, of dispersion, and of areal association of manufacturing both as a whole and for individual industries in the United States [19]. Suggested measures of dispersion and localization, respectively, are the coefficients of scatter and of localization. The coefficient of geographic association is designed to measure the degree of correspondence in area of any two activities or phenomena, such as manufacturing and total population, or flour milling and wheat production, or publishing and total population. Actually, the coefficient of localization is merely a special case of the coefficient of geographic association, in which the distribution of a given manufacturing industry is compared with the general distribution of all manufacturing. Often a more valid comparison is that of the distribution of a particular industry with the distribution of population, as in market-oriented industries, or with the distribution of the sources of specific raw materials, as in raw-material-oriented industries. If the object of study is not a national industry but an area, the location quotient is useful in measuring the relative intensity of location of various industries.

LOCATION THEORY AND LOCATION TYPES

Geographers and economic historians have pointed out the changing significance of location factors. Many industries owe their original location to factors that are no longer important; the persistence of such obsolete locations has been described as reflecting geographic inertia. The successive roles of water power, coal, and electricity as power sources and localizing factors have been described often. Studies of present localization patterns may be grouped under location theory and location types.

Location Theory

Virtually all geographic studies of manufacturing note sources of raw materials, sources of power and fuels, labor, markets, transportation connections with raw materials and markets, capital, management, and individual enterprise. These factors are here significant insofar as they play a role in the localization of manufacturing in specific areas, as in the role of available female labor in the attraction of textile industries to areas of heavy industry.

Several geographers have contributed to location theory. In 1909, R. H. Whitbeck presented the first important contribution by an American geographer to an appraisal of location factors in manufacturing industries [20]. He demonstrated the falsity of certain commonly accepted ideas concerning the causes of localization, for example, that the pottery industry of Trenton, New Jersey, was tied to the presence of local china-clay deposits. Walter S. Tower in 1911 analyzed the general factors in industrial localization [21]. In 1927 Richard Hartshorne proposed a quantitative method of appraising the relative importance of raw materials, fuel, markets, and labor, and emphasized the key role of relative location or locus [3]. His ideas were similar to those of Weber [22; 23]. Recently George T. Renner noted the effect of agglomerative economies, which he calls conjunctive or disjunctive symbioses, depending on whether or not there is a functional relationship among the areally associated industries [24]. Charles C. Colby recognized centrifugal and centripetal forces affecting industrial localization within cities [25].

Economists have contributed much to the formulation of concepts or principles of industrial localization. They generally have approached the problem in four steps: 1) analyzing cost factors in individual industries; 2) attempting to measure the variations in these costs from place to place; 3) erecting hypotheses to explain where the industries should be located; and 4) testing these hypotheses by applying them to particular areas. Geographers, on the other hand, have generally studied the actual distributions first and then attempted to explain them. Among the noteworthy economic contributions to the theory of industrial localization within cities are those by R. M. Haig and R. C. McCrea [26]. General location theory was introduced to America through the writings of Alfred Weber [23], and has been considerably refined by Edgar M. Hoover [27]. Much work remains to be done, however, to bridge the gap between economic theory and the actual observed distributions. This involves, among other things, the analysis of location types or of specific industries.

LOCATION TYPES

In 1927 Robert S. Platt essayed a classification of the industries of Puerto Rico based on location types [4]. Three years later McCarty classified the manufacturing industries of Iowa on the basis of location factors [28]. He distinguished between industries localized near raw materials, those localized near markets, those oriented toward sources of power, and those attached to supplies of labor. He also distinguished a fifth type in which two or more of the localizing factors were important. The United States National Resources Committee, analyzing a series of plants involved in successive steps in the manufacture of a single product, recognized that the plants in which the first steps were taken were localized near the raw materials, whereas those which completed the final processing were localized near the consumers [18:41-47].

STUDIES OF SPECIFIC INDUSTRIES

Except in textbooks, most of the studies that American geographers have made of specific industries deal with such industries within specified countries or parts of countries. World-wide views of certain major kinds of industry have been prepared by J. Russell Smith [29] and Erich W. Zimmermann [30]. Treatments of specific industries have been confined largely to those which involve the simplest relation to the localizing forces.

There have been numerous studies, for example, of the location factors in the iron and steel industry. Since the raw materials used in iron and steel plants are bulky and few in number, the exact cost of transportation for all raw materials can usually be determined for the actual or potential centers of production, and since the finished products, in part at least, are bulky, standardized, and competitive, transportation costs to market can also be calculated in most instances. The most detailed study of an American iron and steel district is that by John B. Appleton on the Calumet District of Chicago [31]. Otis W. Freeman in 1923 listed the factors in the location of iron and steel mills [32]. Hartshorne's critical summary of location factors in this industry analyzes variations in the quality of the raw materials and in the relative location of raw materials and markets, defines location types and illustrates them by specific examples; and C. Langdon White has analyzed the location factors in many individual iron and steel districts of the United States [33].

There have been some studies of other specific industries. Four published doctoral dissertations examine the location factors in the cotton textile industry. Rollin S. Atwood analyzed the factors of location in the textile mills of Lancashire, England [34]; J. Herbert Burgy undertook a similar analysis of those in New England [35]; Thomas R. Smith studied those of Fall River, Massachusetts [36]; and Ben F. Lemert turned his attention to those of the Southern Appalachian Piedmont [37]. W. Glenn Cunningham made a detailed case study of the aircraft industry, which has undergone rapid location changes in recent times [38].

STUDIES OF INDUSTRIAL AREAS

The earliest American papers on the geography of manufacturing were concerned with manufacturing in specific places. After defining the kinds of manufacturing in an area and establishing the patterns of distribution, these papers analyzed the areal and functional relationships between the industries and a variety of environmental conditions. Relationships were sought not only with the so-called natural environment, but also with such elements as water supply, power, availability of labor, taxes, zoning and planning regulations, and other similar matters. The connections between the area under investigation and other parts of the country or of the world were studied; and the flow of materials was followed from farms and mines to factory, from factory to factory, and from factory to market.

Notable among the early papers on manufacturing geography were those by R. H. Whitbeck, Malcolm Keir, and George B. Roorbach. Whitbeck arrived at the concept that as a manufacturing area develops, its industries become more and more diversified and its dependence on local raw materials becomes less and less; thus Wisconsin industries in 1912, in contrast with those of New Jersey, were closely tied to local raw materials [20; 39; 40]. Keir recognized that in New England the industries that arose in the period of water power were localized at inland sites, whereas those that arose later in the period of steam power (from waterborne coal) were on the coast [41]. Roorbach noted that both local resources (water power) and accessibility to coal, raw materials, and markets were important in the rise of industries in the Mohawk Valley [42].

The more recent trend in geographic studies of specific areas is revealed by four doctoral dissertations and articles based thereon, by Alfred J. Wright, Alden D. Cutshall, Herman F. Otte, and John W. Alexander. Wright studied the industrialization of the Middle Miami Valley (Ohio) at various stages in the development of the valley (canal era, railroad era, highway era) and attempted to recognize the forces in the changing locations [13]. Cutshall's investigation of the Lower Wabash Valley (Indiana and Illinois) noted the decline of smaller industrial centers with the decline in rail service, and the decreasing importance of local raw materials [43]. Otte appraised the possibilities of the development of manufacturing in the Tennessee Valley of Northern Alabama, a Southern area, not part of the Manufacturing Belt [44]. Alexander emphasized the human factors responsible for the early indigenous development of manufacturing in the Rock River Valley on the western edge of the Manufacturing Belt, where slightly higher transportation costs are offset by slightly lower labor costs [45; 46; 47].

Less extensive investigations have been made of areas on the edge of the Manufacturing Belt or outside it altogether. J. Russell Whitaker probed the northern edge of the Manufacturing Belt in Canada [48]. Freeman and H. F. Raup attempted to appraise the trends in the Pacific Northwest [49]. More recently James J. Parsons surveyed the industrialization of the two industrial areas of California with many market-oriented industries, and that of the Gulf Coast of Texas and Louisiana with fuel-oriented industries [50; 51].

Papers by American geographers on foreign industrial areas are few. Preston E. James mapped and described the distribution of industries in transition from local handicraft to urban factory production in São Paulo, Brazil [52]. George W. Hoffman appraised the raw material position of Austria and the potentialities of industrial development and stability [53]. Hartshorne investigated Upper Silesia [54], and Harris and Norman J. G. Pounds, the Ruhr [55; 56]. The outstanding analysis of a foreign industrial area, however, is that of Japan by John E. Orchard [57].

Individual industrial cities have been examined by many. Margaret T. Parker's book on Lowell, Massachusetts, summarized the evolution of industries in that city [58]. Bernard H. Schockel's monograph on Evansville, Indiana, ascertained, mapped, and described the trade patterns of the materials used in Evansville factories and of the products of these factories [59]. Harold M. Mayer organized research projects on the industrial areas of Chicago and Philadelphia [60]. Albert G. Ballert has continued the Chicago series for the Chicago Plan Commission. Robert L. Wrigley described organized industrial districts as illustrated by examples in Chicago [61]. Roterus discussed the future industrial land requirements in the Cincinnati area [62; 63].

EXPLANATION AND PREDICTION OF INDUSTRIALIZATION

It is one thing to recognize and define the American Manufacturing Belt, and another thing to explain it. Although Sten de Geer made a laudable attempt to explain its location and limits, to this day a thoroughly critical, comprehensive, and balanced evaluation of the factors in the localization and development of manufacturing in this belt is lacking.

Textbooks frequently contain elementary remarks on the differences of industrialization in various parts of the world, but little research has been done on the factors underlying such differences. James listed the conditions handicapping industrialization in Latin America [64]. Orchard sought to determine the factors inhibiting the industrialization of China as compared with that of Japan and to appraise the potentialities of Japan for industrial development [65; 66]. At the onset of rapid Soviet industrialization under the five-year plans, S. S. Visher attempted to evaluate the potentialities of the Soviet Union in manufacturing [67]. Nevertheless there is not yet available a really satisfactory geographic analysis of the processes that lead to differences in industrialization over the globe, nor of the possibilities for manufacturing development in nonindustrialized areas. The current interest in the underdeveloped areas of the world may well yield facts and concepts of great value; but the subject is difficult and complex and will achieve important results only when much research has been devoted to it.

PROBLEMS NEEDING STUDY

Some of the problems that need to be investigated, whether for the purpose of formulating general understandings or of answering specific practical questions, are discussed under three heads: 1) the significance of industries to regional development; 2) the distribution of manufacturing and of manufacturing areas; and 3) the impact of new technology.

SIGNIFICANCE OF INDUSTRIES TO REGIONAL DEVELOPMENT

1. What is the contribution of manufacturing to community, regional, or national development? What is the relationship of various types of manufacturing to the stability of employment, as studied, for example, by Victor Roterus [68]? What is the number of persons supported in local commercial and service establishments for each person in basic industry?

Harold V. Miller has used a ratio of two workers in commerce and service to one in basic industry. If each worker supports two dependents, it follows that for each increase of 100 workers in basic industry in a community there will be a total increase of 900 [69]. Miller's paper is a step in the right direction, but further research is needed to measure this "multiplier effect" of basic industry more precisely. What, then, is the relationship of manufacturing developments to the growth of population and to the increasing size of cities [70]?

2. What are the effects of industry on the agriculture of an area? What is its role in draining surplus labor off the farm and thus maintaining a higher rural standard of living and a higher degree of mechanization, both of which are favorable to the development of large rural markets for industrial goods? What is the role of industrial markets in the intensification and specialization of agriculture?

3. What are the factors in the rise of industrialization in various areas? Why have some areas become industrialized thoroughly or rapidly and others only slightly or slowly? What is the relative importance of markets, natural resources, transportation, individual initiative, labor, cultural tradition, or of various combinations of these or other factors varying in time and place?

4. What are the possibilities for industrialization in new areas? The desire for industrialization is widespread. The urge for industrialization affects non-industrial areas in the United States and also in certain underdeveloped areas the world around. What are the industrial possibilities of the Missouri Valley, the Mountain States, the Pacific Coast, or Texas? What types of industries can be developed in these areas?

5. What are the inter-regional relations of industries in terms of actual flow of raw materials and finished goods?

DISTRIBUTION OF MANUFACTURING AND OF MANUFACTURING AREAS

1. What are likely shifts in areas of industrialization? Which types of manufacturing are migrating, which dispersing, and which concentrating?

2. What are the relative rates and extents of dispersion of industry in the United States under a capitalist system with uncoordinated governmental influence and in the Soviet Union under highly centralized governmental ownership, planning, and operation?

3. What role should public policy play in industrial localization? Should government policy encourage the bringing of new industries to existing centers of labor, or should labor be brought to new sources of power, as in the Gulf Coast, or should power be moved to existing centers, as by new natural-gas pipelines? Should governmental policies aim at well-rounded regional development or at regional specialization? The

exploratory work of Edgar M. Hoover in this field needs to be pushed further [27].

4. What are the likely or desirable effects of military and security considerations on the location of industries in large cities, in established industrial areas, and in coastal or frontier areas? Comparative studies of countries of continental dimensions such as the Soviet Union and the United States and of the small countries of Western Europe might be instructive.

5. How can localized and ubiquitous industries best be separated in statistics, in mapping, and in analysis?

THE IMPACT OF NEW TECHNOLOGY

1. What effect on the geography of manufacturing can be anticipated from certain pending changes or developments in the political field? For example, what will happen as colonial areas secure their independence? What differences in the industrial structure of Europe are likely to come as a result of the Schuman Plan or the establishment of a European Union? In the United States what would be the effect of a zonal freight-rate pattern, as in parcel post?

2. And there are many impending changes in technology and many proposed engineering works which cannot fail to have an impact on the industrial picture. What developments can be foreseen in Canada and the United States as a result of the construction of the St. Lawrence Seaway? What would be the effect of the construction of a belt conveyor for coal and iron ore between the Great Lakes and the Ohio River, or the establishment of a nation-wide grid of electric power lines, or the extension of natural-gas pipe lines? What changes can be forecast as a result of the depletion of high-grade iron ore in the Mesabi Range, or the development of new sources of ore in Labrador or Venezuela, or the opening of new oilfields? What might be the impact of new technologies in the use of low-grade iron ore, or in the more efficient use of fuels, or in cokeless steel production? What will the new synthetics, such as orlon, rubber, and plastics, do to the older established industries?

3. The biggest question of all has to do with the impact of atomic power [71]. The nature and patterns of manufacturing industry were revolutionized as a result of the invention of methods for the use of controlled inanimate power, starting with James Watt's steam engine in 1769. The developments that followed came so fast that they could scarcely be described, let alone anticipated. Now the use of atomic energy could lead to a similar radical change in the processes and patterns of industrialization. What might these changes be? Can geographers and others who are interested in the geography of manufacturing find ways to foresee these

impending developments and perhaps even to plan for them? Can scholars give some sort of guidance to the business men and public administrators who must be responsible for private or public policy?

REFERENCES

Chauncy D. Harris has prepared an extensive bibliography of manufacturing studies, available in mimeograph from the Department of Geography, University of Chicago.

1. DUDLEY, J. G. "A Paper on the Growth, Trade, and Manufacture of Cotton," *Bulletin of the American Geographical and Statistical Society*, 1 (1853): 105-194.
2. DE GEER, S. "The American Manufacturing Belt," *Geografiska Annaler*, 9 (1927): 233-359.
3. HARTSHORNE, R. "Location as a Factor in Geography," *Annals of the Association of American Geographers*, 17 (1927): 92-99.
4. PLATT, R. S. "A Classification of Manufactures, Exemplified by Porto Rican Industries," *Annals of the Association of American Geographers*, 17 (1927): 79-91.
5. STRONG, H. M. "Regions of Manufacturing Intensity in the United States," *Annals of the Association of American Geographers*, 27 (1937): 23-43.
6. WRIGHT, A. J. "Manufacturing Districts of the United States," *Economic Geography*, 14 (1938): 195-200.
7. JONES, C. F. "Areal Distribution of Manufacturing in the United States," *Economic Geography*, 14 (1938): 217-222.
8. HARTSHORNE, R. "A New Map of the Manufacturing Belt of North America," *Economic Geography*, 12 (1936): 45-53.
9. HARRIS, C. D. "A Functional Classification of Cities in the United States," *Geographical Review*, 33 (1943): 86-99.
10. MCCARTY, H. H. *The Geographic Basis of American Economic Life*, New York, 1940.
11. ———. "Industrial Migration in the United States, 1914-1927," *Iowa Studies in Business*, 7 (1930): 1-79.
12. MCLAUGHLIN, G. E. "Industrial Expansion and Location," *Annals of the American Academy of Political and Social Science*, 242 (1945): 25-29.
13. WRIGHT, A. J. "Recent Changes in the Concentration of Manufacturing," *Annals of the Association of American Geographers*, 35 (1945): 144-166; idem. "The Industrial Geography of the Middle Miami Valley, Ohio," *Papers of the Michigan Academy of Science, Arts and Letters*, 21 (1935): 401-427.
14. ALEXANDER, J. W. "Industrial Expansion in the United States, 1939-1947," *Economic Geography*, 28 (1952): 128-142.
15. TREWARTHA, G. T., and VERBER, J. L. "Regionalism in Factory Industry in India-Pakistan," *Economic Geography*, 27 (1951): 283-286.
16. ROTERUS, V. "Value Added by Manufacture and its Significance," *Virginia Economic Review*, 1 (1948): 1-3.
17. BURRILL, M. F. *An Atlas of Manufacturing Industry*, Norman (Oklahoma), 1932.
18. United States National Resources Committee, *The Structure of the American Economy*, Washington, 1940.
19. United States National Resources Planning Board, *Industrial Location and National Resources*, Washington, 1943.
20. WHITBECK, R. H. "Specialization in Industry by Certain Cities, With Particular Reference to Trenton, New Jersey," *Journal of Geography*, 8 (1909): 32-38.
21. TOWER, W. S. "Some Factors Influencing the Location and Migration of Industries," *Bulletin of the Geographical Society of Philadelphia*, 9 (1911): 64-81.
22. HARTSHORNE, R. "The Economic Geography of Plant Location," *Annals of Real Estate Practice*, 6 (1926): 40-76.

23. WEBER, A. *Alfred Weber's Theory of the Location of Industries*, (English edition with introduction and notes by Carl J. Friedrich), Chicago, 1929.
24. RENNER, G. T. "Geography of Industrial Localization," *Economic Geography*, 23 (1947): 167-189.
25. COLBY, C. C. "Centrifugal and Centripetal Forces in Urban Geography," *Annals of the Association of American Geographers*, 23 (1933): 1-20.
26. HAIG, R. M., and McCREA, R. C. *Major Economic Factors in Metropolitan Growth and Arrangement: A Study of Trends and Tendencies in the Economic Activities within the Region of New York and Its Environs*, Regional Plan of New York and Its Environs, Vol. 1, New York, 1927.
27. HOOVER, E. M. *The Location of Economic Activity*, New York, 1948.
28. McCARTY, H. H. "Manufacturing Trends in Iowa," *Iowa Studies in Business*, 8 (1930): 1-79.
29. SMITH, J. R. *Industrial and Commercial Geography*, New York, 1913; third edition with PHILLIPS, M. O., 1946.
30. ZIMMERMANN, E. W. *World Resources and Industries*, New York, 1951.
31. APPLETON, J. B. *The Iron and Steel Industry of the Calumet District: A Study in Economic Geography*, University of Illinois Studies in the Social Sciences, 13, no. 2, 1927.
32. FREEMAN, O. W. "Two Dozen Causes for the Location of Blast Furnaces and Steel Mills in the United States," *Journal of Geography*, 22 (1923): 144-148.
33. HARTSHORNE, R. "Location Factors in the Iron and Steel Industry," *Economic Geography*, 4 (1928): 241-252; WHITE, C. L. "Geography's Part in the Plant Cost of Iron and Steel Production at Pittsburgh, Chicago, and Birmingham," *ibid.*, 5 (1929): 327-334; WHITE, C. L., and PRIMMER, G. "The Iron and Steel Industry of Duluth: A Study in Locational Maladjustment," *Geographical Review*, 27 (1937): 82-91.
34. ATWOOD, R. S. *The Localization of the Cotton Industry in Lancashire, England*, Geography Series, University of Florida Publications, 1, no. 1, 1930.
35. BURGY, J. H. *The New England Cotton Textile Industry; A Study in Industrial Geography*, Baltimore, 1932.
36. SMITH, T. R. *The Cotton Textile Industry of Fall River, Massachusetts; A Study of Industrial Localization*, New York, 1944.
37. LEMERT, B. F. *The Cotton Textile Industry of the Southern Appalachian Piedmont*, Chapel Hill (North Carolina), 1933.
38. CUNNINGHAM, W. G. *The Aircraft Industry: A Study in Industrial Location*, Los Angeles, 1951.
39. WHITBECK, R. H. "Industries of Wisconsin and their Geographic Basis," *Annals of the Association of American Geographers*, 2 (1912): 55-64.
40. ———. "Manufacturing Industries of Wisconsin," *Journal of Geography*, 12 (1914): 257-262.
41. KEIR, M. "Some Responses to Environment in Massachusetts," *Bulletin of the Geographical Society of Philadelphia*, 15 (1917): 121-138, 167-185.
42. ROORBACH, G. B. "Geographic Influences in the Development of the Manufacturing Industry of the Mohawk Valley," *Journal of Geography*, 10 (1911): 80-86.
43. CUTSHALL, A. D. "Industrial Geography of Lower Wabash Valley," *Economic Geography*, 17 (1941): 297-307.
44. OTTE, H. F. *Industrial Opportunity in the Tennessee Valley of Northwestern Alabama*, New York, 1940.
45. ALEXANDER, J. W. *Geography of Manufacturing in the Rock River Valley*, Wisconsin Commerce Papers, 1, no.2, 1949.
46. ———. "Manufacturing in the Rock River Valley—Location Factors," *Annals of the Association of American Geographers*, 40 (1950): 237-253.
47. ———. "Rockford, Illinois: A Medium-Sized Manufacturing City," *Annals of the Association of American Geographers*, 42 (1952): 1-23.

48. WHITAKER, J. R. "The Great Lakes–St. Lawrence Lowland: Industrial Heart of Canada," *Journal of Geography*, 33 (1934): 329-339.
49. FREEMAN, O. W., and RAUP, H. F. "Industrial Trends in the Pacific Northwest," *Journal of Geography*, 43 (1944): 175-184.
50. PARSONS, J. J. "California Manufacturing," *Geographical Review*, 39 (1949): 229-241.
51. ——. "Recent Industrial Development in the Gulf South," *Geographical Review*, 40 (1950): 67-83.
52. JAMES, P. E. "Industrial Development in São Paulo State, Brazil," *Economic Geography*, 11 (1935): 258-266.
53. HOFFMAN, G. W. "Austria: Her Raw Materials and Industrial Potentialities," *Economic Geography*, 24 (1948): 45-52.
54. HARTSHORNE, R. "The Upper Silesian Industrial District," *Geographical Review*, 24 (1934): 423-438.
55. HARRIS, C. D. "The Ruhr Coal-Mining District," *Geographical Review*, 36 (1946): 194-221.
56. POUNDS, N. J. G. *The Ruhr: A Study in Historical and Economic Geography*, Bloomington (Indiana), 1952.
57. ORCHARD, J. E. *Japan's Economic Position: The Progress of Industrialization*, New York, 1930.
58. PARKER, M. T. *Lowell: A Study of Industrial Development*, New York, 1940.
59. SCHOCKEL, B. H. *Manufactural Evansville, 1820-1933*, University of Chicago (Department of Geography), 1947.
60. Philadelphia City Planning Commission, *Industrial Land Use for Philadelphia in Relation to Metropolitan Area Development*, Philadelphia, 1950.
61. WRIGLEY, R. L. JR. "Organized Industrial Districts with Special Reference to the Chicago Area," *Journal of Land and Public Utility Economics*, 23 (1947): 180-198.
62. ROTERUS, V. *Industrial Land Use: Present and Future*, Cincinnati City Planning Commission, 1946.
63. ROTERUS, V., KEYES, S., and VAN SCHAAK, R. "Future Industrial Land Requirements in the Cincinnati Area," *Annals of the Association of American Geographers*, 36 (1946): 111-121.
64. JAMES, P. E. "The Significance of Industrialization in Latin America," *Economic Geography*, 26 (1950): 159-161.
65. ORCHARD, J. E. "Can Japan Develop Industrially?" *Geographical Review*, 19 (1929): 177-200.
66. ——. "Contrasts in the Progress of Industrialization in China and Japan," *Political Science Quarterly*, 52 (1937): 18-50.
67. VISHER, S. S. "Russian Industrialization," *Journal of Geography*, 31 (1932): 68-77.
68. ROTERUS, V. "Suitability of Economic Activities in Relation to the Local Economy," *Journal of the American Institute of Planners*, 13 (1947): 29-31.
69. MILLER, H. V. "The Growth and Development of the Morristown, Tennessee, Community," *The Professional Geographer*, 8 (1948): 28-40.
70. HARRIS, C. D. "The Cities of the Soviet Union," *Geographical Review*, 35 (1945): 107-121.
71. JONES, S. B. "The Economic Geography of Atomic Energy, A Review Article," *Economic Geography*, 27 (1951): 268-274.

[A] Original draft by Edward L. Ullman with the collaboration of Harold M. Mayer.

CHAPTER THIRTEEN

TRANSPORTATION GEOGRAPHY

TRANSPORTATION is a measure of the relations between areas and is therefore an essential aspect of geography. The economic relations and connections between areas are reflected in the character of transportation facilities and in the flow of traffic. Transportation is a major part of the "geography of circulation," as the latter term is used by the French geographers. In fact, this term, although not widely used in English, seems to be the most desirable and all-inclusive label for the subject. Geography is concerned with all spatial connections and interactions, including communication and transportation. The study of transportation or circulation thus cuts across all fields of human geography. It sheds light on such diverse problems as the diffusion of cultural traits or the resistance offered by political boundaries. The field of transportation geography in the United States is undergoing a new development, although it is still in the stage of seeking the most effective procedures for making measurements, for carrying out geographic analyses, and for arriving at useful concepts of wide application.

A number of geographers, both in Europe and America, are coming to recognize that the study of the connections between areas and of spatial interchange can provide a new and deeper insight into the meaning of areal differentiation. They see transportation as providing a key for measuring the significance of likenesses and differences among places on the earth. For those geographers who view the core of geography as primarily the analysis of spatial interaction, the study of transportation and, in the broader sense, of circulation as a whole is of crucial importance [1]. P. R. Crowe insists that geographers are more concerned with currents of men and things moving than with static transport forms, as such [2]. Transportation facilities are examined primarily as indicators of the degree of connection and as patterns of spatial interchange.

NATURE OF TRANSPORTATION GEOGRAPHY

Students of transportation geography have been primarily concerned with its economic aspects. Transportation stands as one of the four traditional components of economic geography: primary production, manufacturing, marketing, and transportation. The ease or difficulty of moving things over the earth, the volume of traffic, and the distances involved have profound effects on the nature and arrangement of economic activities of all kinds.

Changes in the technology of transportation during the past two centuries have resulted in a great reduction of the cost per unit of moving goods from one place to another. This, in turn, has made possible the economic specialization of areas, the substitution of interdependence for self-sufficiency, and has produced the notable areal differentiation in forms of production and land use that characterizes the modern world. Transportation overland used to be slow and costly. Bulk movements of raw materials at all comparable to the shipments which come daily to a modern steel plant were unknown. Each area of concentrated settlement, therefore, had to produce most of its own fuel, food, and other necessities, and trade was restricted largely to luxury items that could stand the high cost of shipment. To be sure, the relative cheapness of transportation by water permitted a certain amount of crop specialization even in sailing-ship days: as witness the dependence of Athens on the wheat lands of what is now the Ukraine, and the dependence of Rome on grain shipped by sea from Egypt and other parts of North Africa. The partial dependence of Great Britain on specialized producing areas overseas began before the development of modern forms of transportation. But the railroad for the first time in history provided cheap bulk transportation overland and set the stage for the appearance of specialized agricultural regions in areas remote from navigable water. For the first time on a world-wide scale it became possible to take advantage of locally favorable conditions of land, labor, and capital for the specialized low-cost production of goods destined for distant markets. Improvement of transportation thus radically altered the scale of geography by substituting larger and more distant regions of production for smaller ones nearby. This contrasts with von Thünen's famous model of concentric producing regions around a city in the *Isolierte Staat* of 1827.

Transportation consumes an important part of the world's energy. In a modern industrial country like the United States, an estimated 20 percent of the labor force is directly or indirectly employed in the operation, servicing, manufacturing, and selling of transportation and communication. In a primitive society, equipped with little or no machinery, the daily output of energy is great, as each person laboriously moves things from place to place within a small area; but the volume and distance of movement are small; the scale and range of spatial relations are likewise small. Among peoples of similar cultural status, the amount of energy given to circulation varies from place to place with the natural endowment and with differences in economy. Iowa, for example, ships a smaller tonnage of agricultural products than that shipped by either of its neighbors, Nebraska or Kansas, mainly because Iowa farmers feed their corn to cattle and hogs on the farm. The result is a product of lighter weight but greater value, and a consequently smaller demand for transportation. In certain places, transportation is the major industrial employer: for example, in a railroad town such as Altoona, Pennsylvania. In a country like Norway, which provides shipping service for the whole world, the proportion of the total population employed in transportation is also quite high.

The actual physical facilities for transportation may be examined and mapped for geographical, as well as engineering, purposes. In many cases, where data on the flow of traffic are inadequate, the facilities provide the best indicator of the volume and route of movement. The transportation geographer needs some familiarity with the technical or engineering aspects of such facilities if he is to determine the significance of, for example, the type and depth of railroad ballast, the weight of rails, or the types of railroad signals, or if he is to evaluate the capacity of railroad yards or of a port. Similarly the interpretation of the significance of a net of highways requires technical knowledge both of highway engineering and also of the geographic method.

The study of rates and costs is an aspect of transportation geography that touches on the field of economics. Just as the effective study of transportation facilities calls for some familiarity with engineering, so also the study of rates and costs calls for some understanding of the findings of economic and business research. The specialist in transportation geography, equipped with an adequate background in these related fields, is in a position to make a unique contribution to the interpretation of the factors in, and the nature of, spatial interchange.

RESEARCH IN TRANSPORTATION GEOGRAPHY

In view of the obvious importance of transportation to geography, it is surprising how little professional work concerning it has been done by geographers. Transportation studies in the United States have been made mostly by scholars in other professional fields, and from non-geographic viewpoints. Geographers should not count on using these studies secondhand to arrive at significant geographical objectives. Although specialists in transportation geography are few, the importance of their contribution has already been recognized by economists and others.

Specialists in this field must attack their problems with quite inadequate statistical data especially in the United States and Britain. Such data as are available are usually too general to lend themselves easily to the kind of analysis required for geographic purposes. This situation results in part from: 1) war-time restrictions on the publication of this kind of informa-

tion, 2) the competitive relations between private carriers, and, finally, 3)
from a general lack of interest in localized or geographic data in collection
agencies in the past, although this may well be remedied in part in the future, as will be shown later.

In the two sections which follow some of the previous studies in the
transportation field in Europe and America are noted, but without any
attempt at systematic coverage. Certain of the more important recent
works are listed in connection with the discussion of the objectives of
transportation geography.

EUROPEAN STUDIES

The Europeans have done more in the field of transportation geography
than have their colleagues in the United States. This may be a result, in
part, of the fact that statistical data of the kind geographers need are more
readily available. German writers have produced several works on *Ver-
kehrsgeographie*, many of which are now out of date and have been criticized as being merely encyclopedic and of little penetrating value. Notable original studies on land transportation, however, have been made by
Erwin Scheu, Helmut Haufe, and others [3; 4]. In France, a detailed and
lengthy work by H. Lartilleux describes the French railroads, not only
those of continental France but also of the outlying parts of the country
and of its colonies [5]. This four-volume work contains a vast amount of
information on track layout, signal equipment, grades, and other physical
characteristics. Information of this sort that can be specifically located on
a map is not usually available for publication in the United States. It was
originally intended to extend this compendium of information regarding
the French railroads to other forms of transportation and to other parts of
the world. The difficulties of obtaining the necessary data, however, make
the completion of the work doubtful. The stimulating study by Robert
Capot-Rey, published in 1946, is less encyclopedic but is restricted to certain aspects of land transportation as related to settlement and environment
and does not attempt to cover economic and traffic features [6]. There
are also several excellent French sudies of local transportation facilities
and of the flow of traffic, such as René Clozier's *La Gare du Nord* [7],
and works on other aspects of rail transport by Maurice Pardé. In England
important transportation studies have been made by S. H. Beaver, K.C.
Edwards, A. C. O'Dell, and others. Notable also are the works of O. G.
Jonasson in Sweden and Aage Aagesen in Denmark [8].

More attention has been devoted to ocean trade and to ports in European
studies than has been given to other aspects of transportation. A. J. Sargent's earlier books on seaways and ports, as well as work by L. Mecking
on Japan, particularly come to mind [9; 10]. In 1952, F. W. Morgan published a brief, general treatment of ports and harbors [11], and O. Hölcke

completed a monumental study on the port of Stockholm's freight traffic [12]. W. E. Boerman in the Netherlands has contributed marine studies some of which are noted, along with studies by other European geographers, in the report of the Commission on Industrial Ports of the International Geographical Union [13]. The Norwegian geographers, Tore Ouren and Axel Sømme, published an important analysis of trends in trade and shipping between the two world wars, and André Siegfried pictured the results of the opening of the Suez and Panama Canals on the patterns of world transportation [15].

STUDIES IN THE UNITED STATES

The changes in the patterns of world trade that followed the opening of the Panama Canal stimulated some of the earliest works in the field of transportation geography in the United States. A pioneer work in this field was the book on ocean commerce by J. Russell Smith, published in 1905 [16]. Although a number of general texts on ocean trade and transportation have appeared, geographers in the United States had not probed deeply into this subject until they did so for the various government agencies during World Wars I and II (chapter twenty-three).

On the other hand, there are a number of studies of individual ports by American geographers. The relatively greater attention paid to ports is in part a recognition that such places are significant focal points in the pattern of world trade, and are sufficiently compact to permit field investigation; but it is also in part a result of the greater availability of statistical data useful for geographic purposes concerning the movement of goods in and out of ports. Of the many papers published on ports in the United States, four may be cited as representative [17; 18; 19; 20]. A few geographers have participated officially as directors and staff members of organizations that produced plans for port development [21]. Geographical studies of ports are usually concerned either with the movement of goods between a hinterland and the other parts of the world reached by water routes, or with the port facilities as an aspect of urban land use. During World War II, new techniques of port description, new kinds of maps, and new ways of estimating port capacities were worked out, and applied to ports around the world that had never previously been examined in such detail. Unfortunately, most of these war-time research monographs are unavailable.

A considerable number of geographers have also worked on various aspects of railroad patterns and facilities. Mark Jefferson produced a striking portrayal of the railroad patterns of the several continents [22], in which he identified various degrees of railroad density by such terms as railweb, railnet, rail tentacle, or riverlink. Edward L. Ullman classified railroads of the United States according to their importance based on the number of tracks and the system of signalling [23]. Of direct practical importance

are the studies of railroad terminals in relation both to the ease of movement of goods in and out of cities and to the functional patterns of cities.
Harold Mayer's study of the railroad pattern of metropolitan Chicago
[24] and of the passenger terminal problem of central Chicago [25] are
among the geographical contributions to problems of this sort. Geographers, whose point of view leads them to see the areal associations of economic activities, land uses, and transportation facilities, are beginning to
apply their methods to the practical problems of planning urban and
metropolitan development and of providing for the effective service of
metropolitan areas by transportation systems.

A few studies by geographers in the United States also deal with various
aspects of air transportation. Professional geographers have been employed
by the airlines and by regulatory agencies in the planning of new routes
and in estimating the traffic potential of areas not previously served. The
establishment of long-distance, and especially trans-oceanic, airlines has
modified the accessibility pattern of major portions of the earth, and these
changes have stimulated some studies comparable to those that appeared
many decades earlier regarding the patterns of ocean transportation [26;
27]. Airports have been examined in terms of the patterns of connection
with other areas and also as a land-use category in the city [28; 29; 30].
The effect of airports on other urban uses of land has received little
attention.

In the literature of American geography there are almost no studies of
the interrelations among the various forms of transportation, or of the
effects of transportation on other aspects of human geography. Two exceptions, both relating to the problem of supporting cities, are a study of
Salt Lake City by Chauncy D. Harris and one of Mobile by Edward L.
Ullman [31; 20].

OBJECTIVES OF TRANSPORTATION GEOGRAPHY

Studies in various aspects of transportation geography are undertaken
for a number of different objectives. In order to establish the connections
between areas and the nature of the spatial interchange, it is necessary to
find some way of measuring and mapping the flow of traffic, including its
volume and speed of movement and its origin and destination. Closely
related to the study of traffic flow are geographical analyses of rates and
rate structures, studies of the effect of terrain and other environmental
conditions on transportation, and conversely studies of the effect of the
changing technology of transportation on the significance of environmental circumstances and on the process of economic development of areas
on the earth. Transportation studies also are related closely to the concepts
of geography as a whole, and such studies contribute to the formulation

and evaluation of geographic theory. Transportation studies also are practical and can be applied to specific current problems.

MEASUREMENT AND MAPPING

Transportation geography, due to the lack of proper data, is still in the stage of searching for better methods of measurement and mapping. Many old conclusions and concepts are unreliable because they are based on insufficient quantitative data, such as the erroneous idea that the New York Central, rather than the Pennsylvania, has the heaviest traffic between east and west. In the absence of direct data, such as traffic flow on railroads, other indicators can be used to construct meaningful maps. For example Ullman's map of the railroads of the United States classifies lines according to number of tracks and type of signalling, the best quantitative indicators of flow and of relative importance available from records [32; 23].

Statistics that give the volume of traffic moved along a specific route and that make possible the mapping of the origin and destination of flows are particularly difficult to obtain in the United States. To be sure, some commercial research organizations, such as H. H. Copeland and Son, produce traffic-flow maps of individual railroad systems, but the great expense of such maps and the fact that they are confidential has precluded their regular use in geographic studies, except those undertaken for specific railroad companies. It is worth noting, however, that Copeland and Son recently granted permission for the reproduction of a small-scale version of a traffic-flow map of the United States in a forthcoming publication by Ullman. Some individual railroads have prepared their own traffic-flow maps; information of this sort for two Canadian railroads has been published in fugitive government reports. The voluminous statistics published in the United States by the Interstate Commerce Commission, however, are not broken down into sufficient detail to be useful for many geographic purposes.

On the other hand, statistics of geographical value do appear from time to time. The Interstate Commerce Commission in 1947 started the issue in mimeographed form of an annual one-percent sample of rail waybills that makes it possible to chart rail commodity movements by States of origin and destination. Ullman has mapped and briefly interpreted some of these data in two papers soon to be published. A sample of the maps is offered in Figures 9 and 10, showing the shipments of animals and products to and from the State of Iowa, with the width of the lines proportional to the volume of traffic. Briefly, these two representative maps show for the first time graphically and quantitatively: 1) the generally greater volume of shipments to nearby points than to more distant points, a reflection of the friction of distance; 2) the greater volume of outbound over inbound

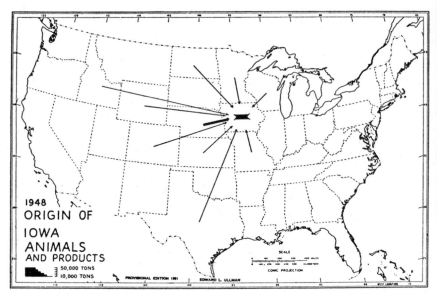

Fig. 9. Origin, by states, of animals and products shipped into Iowa by rail, 1948. Width of lines is proportionate to volume (of short tons of 2000 lbs.) on Figures 9 and 10.

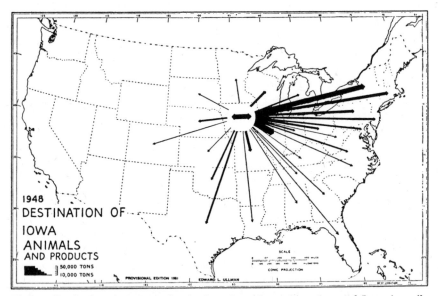

Fig. 10. Destination, by states, of animals and products shipped out of Iowa by rail, 1948. Source of data for Figures 9 and 10 is Interstate Commerce Commission's 1% sample of rail traffic reported in *Carload Waybill Analyses*, 1948, Washington, D. C. (statements: 4838, October, 1948; 492, January, 1949; 498, March, 1949; 4920, June, 1949).

traffic, reflecting Iowa's role as the leading animal producer, chiefly hogs and hog products, in the United States; 3) the heavy movement to the markets of the industrial northeast, which is sufficient to counteract the friction of distance (note especially the shipments to the deficit meat areas of New York and New England); 4) the small but distinctive flow to the rising California market, a new feature of American economic geography; and 5) the general west-to-east movement in the United States, a fundamental feature of American economic geography, reflecting heavy raw materials moving to eastern industrial areas, generally outweighing the backflow of lighter-weight industrial products.

For studies of port traffic, useful data are frequently available. Yet the student is often presented with data that are not exactly what are needed. For example, statistics on the value of shipments or the tonnage of vessels calling at a port are less significant for many purposes than figures of actual weight loaded and unloaded. In other cases only foreign exports and imports are shown, as though a shipment from New York to Cuba were of greater geographic significance than one from New York to California by water. Admittedly, no one total measure is perfect. For example, combining tons of liquid bulk (principally petroleum), dry bulk (coal, wheat, ore, etc.) and general cargo, combines unlike items. Liquid bulk can be pumped and requires little in the way of port facilities; dry bulk is generally handled by machinery in large quantities, and a given tonnage of dry bulk is usually of less significance to a port than the same tonnage of general cargo.

More difficult to obtain are data on the routes of movement. Examples of new mapping are presented in Figures 11, 12, and 13, which show in preliminary form the foreign trade of the United States by coastal areas for dry bulk, liquid bulk, and total tonnage for 1938; and in Figure 8, which shows the complete flow to and from Mobile, Alabama, for both foreign and domestic trade. Such maps provide at a glance a much clearer characterization of the actual oceanic connections than could be gained from pages of words or statistical tables. As yet direct data are too fragmentary to permit the preparation of similar maps for world traffic flow. However, other kinds of information could be used to provide something more or less equivalent. The British Chamber of Shipping, for example, in 1936 and 1937 published maps showing the position of all British ships at sea at a given time. Trade routes, and the volume of movement along them, stand out fairly well on these maps [33]. Some adaptation of this method might be used to provide world maps of the flow of ocean traffic.

At the time of writing, the Maritime Administration, the Corps of Engineers, and the Bureau of the Census are planning to publish tonnage statistics by routes on a coastal-area to coastal-area, or a port-to-port,

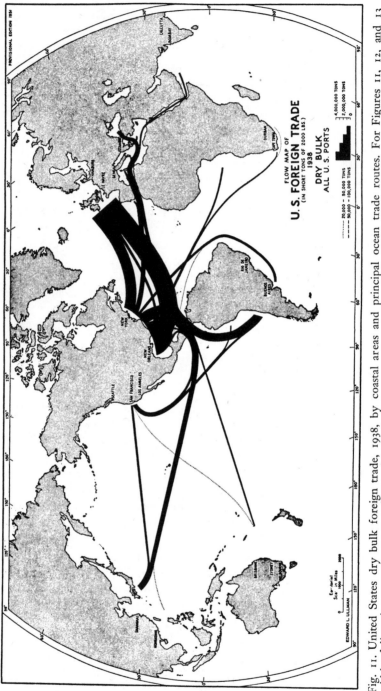

Fig. 11. United States dry bulk foreign trade, 1938, by coastal areas and principal ocean trade routes. For Figures 11, 12, and 13 width of lines is proportionate to volume of actual weight tons of cargo moved in both directions; data are compiled from U. S. Maritime Commission *Report No. 2610*, September 30, 1939, and Report No. 42-37, 1939. (Washington, D. C.).

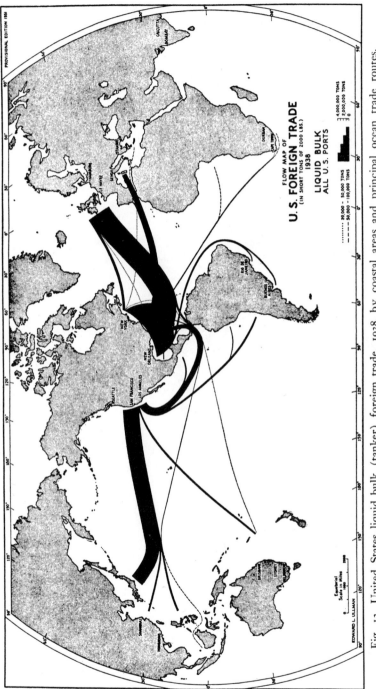

Fig. 12. United States liquid bulk (tanker) foreign trade, 1938, by coastal areas and principal ocean trade routes.

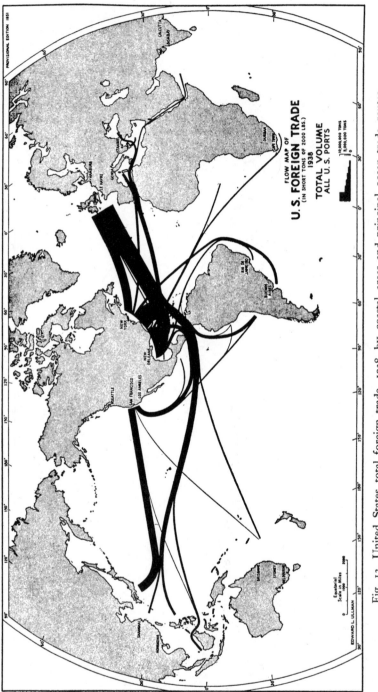

Fig. 13. United States total foreign trade, 1938, by coastal areas and principal ocean trade routes.

basis, arranged by major commodities for all the ocean trade of the United States. Geographical analyses of the pattern and meaning of such trade will then be enormously facilitated. Unfortunately, the disclosure rule will prevent some of these data from being made public, as, for example, on the Seattle-to-Alaska route where one steamship line is dominant. Already the Corps of Engineers, and to some extent the Maritime Administration, publish figures by commodities of all ocean trade from ports of the United States, but the routes, origins, and destinations of these flows are not shown. Facilities available in ports of the United States are described in detail, but not interpreted geographically, in the *Port Series*, published jointly every few years by the Corps of Engineers and the Maritime Administration.

Inadequacy of data also makes the mapping of traffic flow on inland waterways difficult. However, it is possible to make rather general flow maps from figures available in manuscript from the Corps of Engineers, as Donald Patton has done for the United States as a whole. Great Lakes flow maps for major commodities are contained in reports of the Corps of Engineers, and at least two geographers have mapped and analyzed aspects of the grain trade [34; 35]. Recent inland-waterway studies include, among other things, an analysis of the commerce of the Ohio River and the Illinois Waterway [36].

In spite of the availability of statistics on highway traffic in the United States, geographical studies of highways are few. Yet some 85 percent of the total volume of intercity passenger traffic moves by highway. There are masses of statistics on highway traffic flow, and, within the last few years, data have been gathered on origins and destinations for most metropolitan areas. Nearly every State and county and most municipal highway authorities produce tabulations and maps of highway and street traffic flow. Since World War II, under the leadership of the United States Bureau of Public Roads, scores of cities and metropolitan areas have conducted origin-destination surveys, which include all trips made to, from, and within the respective survey areas, classified by origin and destination in terms of zones of varying size, with maps and tabulations by time of day, purpose of trip, elapsed time, type of conveyance, and, in some instances, by age, sex, and occupation of the person involved. Surveys of this sort should prove very useful for geographers in the study of the functional and dynamic organization within urban areas, although inadequate for analyses of larger areas (chapter six). Most such surveys are carried on primarily for highway-planning purposes; consequently, the tabular and cartographic forms in which the data are presented are not always those of maximum utility for geographic research. Geographers can do much to secure the kind of information they would find most useful by participating in the planning of such surveys.

A number of other aspects of transportation and communication are also likely to engage the geographer's attention in the future. Statistical data, for example, are quite good for aviation: flow maps of the number of flights can readily be plotted from timetables. The Civil Aeronautics Administration has already published highly informative studies on American air traffic. For pipe-lines, which carry an increasing amount of American oil and gas and which already stretch from coast to coast, covering more route-miles than do railroads, statistics and maps on pipe diameter have been published by the Federal Power Commission; from these and other data it might be possible to construct a flow map, although help from individual private companies would also be required to get information regarding the percentage of capacity used, and for precise data on origins and destinations. In the communications field, long-distance telephone message data can often be obtained from regional telephone company files. Such information, plotted on a map, furnishes a very useful measure of the connections between areas, both urban and rural.

To conclude, the problem of securing adequate data remains to be solved, but the many specific new sources just noted indicate that prospects for improvement appear fairly bright. Also many fugitive data that have always been available could be discovered with persistent, hard work especially if one knew what to look for. The first proposed federal census of transportation should also provide more data if and when it is made. One flaw in the picture is the apparent tightening up of possible leaks in the disclosure of data when three or fewer operators are involved; federal agencies are becoming more efficient in enforcing the disclosure rule and holding back data at the same time that their efficiency enables them to obtain more. The poor geographic scholar thus scarcely knows where he stands. The rule particularly affects geography since geography is especially concerned with localized data. The rule, likewise, adversely affects many other legitimate interests, while not affording much, if any, real protection to individual companies. Furthermore, as the law now stands it generally protects areal monopolies—which certainly was not the intent of Congress in framing it.

On the world scene, the United Nations is steadily increasing the detail of its comparative transportation statistics, as are also some specialized agencies, such as the International Railway Union. The period is rapidly coming to an end when lack of data could be used as an excuse for neglecting this field; yet transportation geographers should take steps to see that their needs are understood in the hope that the collecting agencies can provide more of the kind of information, specifically localized, that are needed for geographical purposes.

TRANSPORTATION RATES

One of the determining factors in spatial interchange obviously is the cost of, and the rate charged for, moving things from one place to another. Only in general terms are freight rates directly proportional to distance. Rate structures are so complex that to generalize them into significant geographic patterns is extraordinarily difficult. Freight rates in the United States, and in most parts of the world, vary with each individual commodity shipped between every two points. So-called "class-rates" probably apply to less than 10 percent of the traffic moved.

A basic practice in making commodity rates is to offer a low rate when a sufficient volume between a producing point and a consuming area is available. Because of this practice, small shippers often claim that they have no opportunity to develop volume unless they are favored initially by rates low enough to generate business. An examination of American freight rates indicates that large-volume movements generally have lower rates than small-volume movements, whether it be rates on corn from the cash-grain area of central Illinois and cattle from the cattle-producing areas of western Illinois to major markets [37], or the low trainload rates on the movement of alumina from Baton Rouge and Mobile to aluminum-refining centers in the Pacific Northwest. Practices such as these do not, therefore, appear necessarily to work against specialization based on natural advantages, but rather seem to accentuate and tend to perpetuate such areal differences. Perhaps this is the most significant general statement that can be made about the geographical aspects of freight rates at the moment and is advanced as a new hypothesis worthy of further testing.

Another well-established practice is to charge lower than total-cost rates (sufficiently high to cover out-of-pocket or variable costs) on low-value commodities which cannot afford to pay high costs, and to recoup the difference on higher-value commodities able to stand the charges. Such a practice would tend to result in low-value bulk commodities moving longer distances than they might otherwise do and higher-value commodities moving shorter distances. This topic and other considerations in transport geography in relation to monopolistic and economic theory have been thoughtfully discussed by E. F. Penrose [38].

How much the monopolistic-rate practices cited by Penrose actually affect major flows in the United States is difficult to determine. The largest-volume haul of one commodity in the United States, for example, is coal from West Virginia and Virginia to the Midwest and the eastern seaboard. The three principal carriers originating this movement (Norfolk and Western, Chesapeake and Ohio, and Virginian) obtain the overwhelming bulk of their revenue from this one commodity, so it is unlikely that their rates do not cover total costs; in fact their coal rates in the past undoubtedly covered more than coal's allocated share of total costs and resulted in these three roads being the most profitable railroads in the United States.

Changes in technology likewise have probably differentially affected the relative costs of moving bulky, as opposed to high-value, commodities. How have the introduction of pipelines, coal and ore unloaders, and other labor-saving devices, especially in a period of rising labor costs, affected bulk movements in comparison to package freight, or general cargo, which is not susceptible of mechanization? Both rates and technological practice in rail and water shipping also appear to combine to promote volume moves and the concentration of flows; but the introduction of trucking relatively favors dispersion and higher-value goods.

In spite of all these difficulties, it would be worth while to attempt some geographic generalization of rate patterns. Obviously, for example, deep-water shipping in large volume is cheaper than overland movements, and thus cost distances over the oceans are not so great as over land. Greater speed, and insurance costs on expensive goods may, however, change this general picture. In any case, rate structures can be studied from a geographic point of view [37]. In the meantime, analysis of actual flows shows that rate structures do not produce as many freak or unnatural volume movements as some might suppose from the complex rate structures existing. The friction of distance is still very real, and the production advantages of certain areas great, although the actual distance may be distorted in various ways.

RELATION OF ROUTES TO ENVIRONMENTAL CONDITIONS

Many of the geographic contributions of two or three decades ago emphasized unduly the "influence" of environment upon the location of transportation routes. Perhaps the major influence of environment might be conceived as the effect of contrasting environments, as we have noted, in creating a potential demand for interchange. Thus in the United States the major flows are strikingly across the grain of the country, particularly across the north-central Appalachians. In larger terms, regional traffic potential appears a more important determinant of transportation than actual restrictions to movement encountered along a route.

Nevertheless, the physical environment plays an important role in man's choice of routes for his transportation services. Climate affects the choices; but probably more important as an engineering factor is the role of grades and consequently of terrain, as expressed in the relative costs of overcoming terrain friction. This terrain friction, along with distance, is a principal element in the cost of transportation and hence in the effectiveness with which communication and interchange within and among regions can take place. Isolating the effect of the environment, however, is difficult. Volume and type of traffic, or terminal facilities, or various organizational conditions, may each add more to the cost of operation than so-called natural conditions.

The various media of transportation differ widely in their ability to overcome terrain friction. This ability is measured in terms of both capital cost and of energy input. Of all the forms of land transportation, if canals be excluded, railroads are the most sensitive to grades; hence the choice among alternative routes is more restricted for a railroad line than it is for a highway. Areal and route studies of relative terrain friction offer an opportunity for the transportation geographer to contribute to a problem usually investigated by engineers and economists. What, for example, is the precise effect of terrain friction on the cost of transportation and the stage of economic development in two contrasted countries, such as the Netherlands and Switzerland? We do not know, but procedures might be developed to make it possible to measure environmental effects.

EFFECTS OF THE CHANGING TECHNOLOGY OF TRANSPORTATION

The industrial revolution might be called the transportation revolution. Thus water transport was preeminently the cheap way of moving goods prior to 100 years ago, and cities, industries and commercial activities tended to cluster near the oceans or navigable waterways. The application of the steam engine to transportation revolutionized land transport and produced fundamental changes in the accessibility of one place to another. At the same time the steamship made the oceans and the world almost one as far as transport costs alone were concerned. Most of the technological changes in transportation have tended to favor long hauls and larger volumes, and have thus created larger facilities and route ways. The automobile and truck in the 20th century, however, brought a great advance to local, dispersed movements of passengers and freight, just as the telephone has done in communication.

In the process of the transportation revolution, the actual flow of goods has increased incalculably. What fragmentary figures we have seem to indicate that movements along the great trade routes of the past were mere trickles compared to modern transport flows. The role of different features of the natural environment has also changed in the process. Thus a mountain range is a different phenomenon to horses, yaks, canal boats, steam locomotives, diesel engines, automobiles, trucks, jeeps, airplanes, pipelines, electric wires, or radio.

All of this provides a fertile field for study of the past and for projections of, at least, near-future changes, many of which are on us almost before we recognize them. Assessing the role of these new inventions and developments on regional differentiation and growth, or on centralizing or decentralizing areal tendencies, is a difficult but pressing task. Geographers should address themselves to it.

GEOGRAPHIC THEORY AND SOCIAL PHYSICS

Transportation geography is inextricably related to other ꞈspects of geography. The approach of the transportation geographer underlines the need for seeing areal differentiation on the earth not only as a geometric design of overlapping patterns but also as a design of things in motion. This design, in many cases, tends to be the reverse of that for static regions, inasmuch as movement may be greatest across the borders of contrasting regions whose very differences promote interchange. Thus transportation or circulation is obviously an essential element in the formation of a nodal region, as defined in chapter two (sometimes known as an organizational or functional region).

The concepts of transportation geography are, likewise, involved in the distinction, proposed by Richard Hartshorne, between "kinetic" barriers, such as a mountain range which promotes terrain friction, and "static" barriers, such as an empty or sparsely populated area [39; 40]. In this connection S. Daggett and J. P. Carter reasonably make the following observation: ". . . the transcontinental rate structure distorts geographical and topographical relationships in the Far West in ways which geographers seldom stop to appreciate" [41]. They mean by this that sharp increases in rates occur both in the western mountains ("kinetic" barriers) and also out on the level Great Plains ("static" barriers). Such practices, however, fit logically the concepts of the barriers noted above and should surprise no geographer, although the practices might be presumed to contradict the views sometimes imputed to geographers by others.

Some geographers and other social scientists have become interested in recent years in "social physics." As related to geography this involves the search for general laws expressable in the form of mathematical equations concerning areal relationships and spatial interchange. Insofar as areal relationships can be quantitatively expressed in statistics concerning flows of traffic and communications, flow phenomena taking place over circulation routes may prove to be excellent indicators of the nature of these relationships. The measurement and analysis of traffic flow, therefore, should constitute an important geographic contribution to "social physics".

The mathematical formula used to describe the volume of spatial interchange between two areas is analogous to the formula used to present the law of gravity, as follows: goods move between regions in direct proportion to the size (measured by population or probably better by some index of production and consumption volume) of the two regions and in inverse proportion to their distance apart (with distance probably having

an exponent not far from 1 or 2); the formula is often written $\dfrac{P_1P_2}{d}$ or

population of place 1 (P_1) times population of place 2 (P_2) divided by distance (d) (chapter six, p. 149).

Does this general law in fact illuminate the problems of spatial interchange with which geographers must deal? Consider, for example, an analysis of the traffic flow by railroad between thirteen States of the United States and the other States. In all cases traffic within a State, and between a State and its neighbors, is heavy, as would be expected from some sort of gravity formula. In addition, the industrial belt of Northeastern United States draws traffic from the whole country, as indicated on the maps of Iowa (Figs. 9 and 10), and certain other areas of specialized production or consumption show up as generating surprisingly long hauls. California, thus, is both a specialized producer and a market with ties across the country; Florida is a specialized off-season fruit and vegetable producer with ties to the Northeast and Midwest; Washington exhibits similar ties for its shipments of forest products.

All of these movements within the United States are only superficially described by the previously-mentioned formula. Localized, specialty commodities so dominate the trade of certain States that the total traffic flow of these States is distorted in relation to any gravity norm. For example, virtually no flow of forest products, the dominant commodity export of the Pacific Northwest, occurs between that region and the South, because, of course, both produce forest products. The formula might provide a measure of the breaking point in the Midwest between southern and Pacific-Northwest suppliers or more generally an index of variation; but the variations in many cases appear more important than the generalization, and the variations may well be more clearly presented on maps of origin and destination.

From this one example it should not be concluded that such mathematical formulae are useless to geography. For movement of passengers, long distance telephone calls, bank checks, rumors, and other phenomena, they have apparently been found valid between certain points by such scholars in other fields as G. K. Zipf, J. Q. Stewart and S. C. Dodd [42: 471-473; 43: 375-415], and in the 19th century E. G. Ravenstein applied the formula in producing his classic "law" of migration.

APPLIED TRANSPORTATION GEOGRAPHY

As in other branches of geography, the development of basic concepts leads, directly or indirectly, toward the application of these concepts to the study of practical problems of man on the earth. Transportation geographers, although few in number, have made important applications of their point of view and their procedures to practical questions. Indeed, it was the practical problem of forecasting potential traffic and changes in trade routes resulting from a proposed trans-Isthmian Canal that led to some of the earliest transportation studies in the United States. It is curious that more geographers have not followed along the lines suggested by this pioneer contribution of J. Russell Smith. Perhaps the study of traffic potential and route changes was given more emphasis in the period of railroad building and pioneer settlement in this country. One of the founders of the American Geographical Society, H. V. Poor, produced famous railroad studies, some parts of which were geographical in concept and execution [44: 25-27].

Transportation geography was applied to many problems of logistics and transport-planning both in World Wars I and II and continues to this day on a permanent basis. Numerous studies of this sort in the civilian and military agencies of the government have remained as "classified" and so are not available for professional scrutiny. It remains difficult, therefore, fully to assess the contribution these studies actually made to transportation geography as a field of theoretical and practical knowledge.

Perhaps more important in the long run are the contributions being made in increasing number by transportation geographers as staff members and consultants for transportation companies, and for federal, State, or local planning or regulatory agencies. Airline route-extension planning, port development, pipeline extension, highway and street planning, and numerous other aspects of transportation development are in need of the services of persons trained in the geographical concepts and procedures. But the nature and extent of these contributions will inevitably be related to the validity of the general concepts that are formulated and to the success achieved in securing the kind of basic statistical information that is required.

REFERENCES

1. ULLMAN, E. L. "Human Geography and Area Research," *Annals of the Association of American Geographers*, 43 (1953): 54-66.
2. CROWE, P. R. "On Progress in Geography," *Scottish Geographical Magazine*, 54 (1938): 1-19.
3. SCHEU, E. *Deutschland's wirtschaftsgeographische Harmonie*, Breslau, 1924.
4. HAUFE, H. *Die geographisches Struktur des deutschen Eisenbahnverkehrs*, Berlin-Leipzig, 1931.
5. LARTILLEUX, H. *Géographie des chemins de fers français*, Tome I of *Géographie universelle des transports*, Paris, 1947.
6. CAPOT-REY, R. *Géographie de la circulation sur les continents*, Paris, 1946; FROMONT, P. "Les chemins de fers et l'agriculture," *L'Année Ferroviaire*, (1948): 63-96, contains extensive bibliography of French studies.
7. CLOZIER, R. *La Gare du Nord*, Paris, 1940.

8. AAGESEN, A. *Geografiske Studier Over Jernbanerne: I Danmark* (Geographical Studies of the Railways of Denmark), Copenhagen, 1949.
9. SARGENT, A. J. *Seaways of the Empire*, 2nd ed. London, 1930; *idem. Seaports and Hinterlands*, London, 1938.
10. MECKING, L. *Japan's Häfen*, Hamburg, 1931.
11. MORGAN, F. W. *Ports and Harbours*, London, 1952.
12. HÖLCKE, O. *Varutrafiken over Stockholms Hamn* (Freight Traffic Through the Port of Stockholm), Stockholm Business Research Institute, Stockholm, 1952.
13. *Preliminary Report of the Commission on Industrial Ports*, International Geographical Union, New York, 1952.
14. OUREN, T., and SØMME, A. *Trends in Inter-War Trade and Shipping*, Publication no. 5, Geographical Series, Norwegian University School of Business, Bergen, 1949.
15. SIEGFRIED, A. *Suez and Panama*, New York, 1940.
16. SMITH, J. R. *The Organization of Ocean Commerce*, Philadelphia, 1905.
17. WALKER, F. "The Port of Bristol," *Economic Geography*, 15 (1939): 109-124.
18. BALLERT, A. G. "The Major Ports of Michigan," *Papers of the Michigan Academy of Science, Arts and Letters*, 35 (1949): 137-157.
19. TAVENER, L. E. "The Port of Southampton," *Economic Geography*, 26 (1950): 260-273.
20. ULLMAN, E. L. *Mobile: Industrial Seaport and Trade Center*, University of Chicago, Chicago, 1943.
21. MAYER, H. M. "Development Problems of the Port of the Delaware," *Annals of the Association of American Geographers*, 38 (1948): 99-100; *Where Two Great Waterways Meet*, First Biennial Report of the Chicago Regional Port District Board, Chicago, 1953.
22. JEFFERSON, M. "The Civilizing Rails," *Economic Geography*, 4 (1928): 217-231.
23. ULLMAN, E. L. "The Railroad Pattern of the United States," *Geographical Review*, 39 (1949): 242-256.
24. MAYER, H. M. "Localization of Railway Facilities in Metropolitan Centers as Typified by Chicago," *Journal of Land and Public Utility Economics*, (1944): 299-325.
25. ———. "The Railway Terminal Problem of Central Chicago," *Economic Geography*, 21 (1945): 62-76.
26. VAN ZANDT, J. P. *The Geography of World Air Transport*, Brookings Institution, Washington, 1944.
27. WHITTLESEY, D., WILLIAMS, R. L., and DAWSON, M. D. "Aviation in Africa," *Air Affairs*, 1 (1947): 439-451.
28. PLATT, R. S. "Problems of Our Time," *Annals of the Association of American Geographers*, 36 (1946): 1-43.
29. TAAFFE, E. J. *The Air Passenger Hinterland of Chicago*, University of Chicago, Chicago, 1952.
30. PEARCY, G. E., and ALEXANDER, L. M. "Pattern of Commercial Air Service Availability in the Western Hemisphere," *Economic Geography*, 27 (1951): 316-320; *idem*. "Pattern of Air Service Availability in the Eastern Hemisphere," *Economic Geography*, 29 (1953): 74-78.
31. HARRIS, C. D. *Salt Lake City: A Regional Capital*, University of Chicago, Chicago, 1940.
32. ULLMAN, E. L. *United States Railroads Classified According to Capacity and Relative Importance* (map), New York, 1951.
33. ———. "Mapping the World's Ocean Trade: A Research Proposal," *The Professional Geographer*, vol. 1, no. 2 (1949): 19-22.

34. JONES, C. F. "The Grain Trade of Montreal," *Economic Geography*, 1 (1925): 53-72.
35. HARTSHORNE, R. "The Significance of Lake Transportation to the Grain Traffic of Chicago," *Economic Geography*, 2 (1926): 274-291.
36. BECHT, J. E. *Commodity Origins, Traffic, and Markets Accessible to Chicago Via the Illinois Waterway*, University of Illinois, Urbana (Illinois), 1952.
37. ALEXANDER, J. W. "Freight Rates as a Geographic Factor in Illinois," *Economic Geography*, 20 (1944): 25-30.
38. PENROSE, E. F. "The Place of Transport in Economic and Political Geography," *Transport and Communications Review*, 5 (1952): 1-8.
39. HARTSHORNE, R. "Suggestions on the Terminology of Political Boundaries," *Mitteilungen des Vereins der Geographen an der Universität Leipzig*, (1936): 180-192.
40. ULLMAN, E. L. "Rivers as Regional Bonds: The Columbia-Snake Example," *Geographical Review*, 41 (1951): 210-225.
41. DAGGETT, S., and CARTER, J. P. *The Structure of Transcontinental Freight Rates*, Berkeley (California), 1947.
42. STEWART, J. Q. "Empirical Mathematical Rules Concerning the Distribution and Equilibrium of Population," *Geographical Review*, 37 (1947): 461-485.
43. ZIPF, G. K. *Human Behavior and the Principle of Least Effort*, Cambridge, (Mass.), 1949.
44. WRIGHT, J. K. *Geography in the Making: The American Geographical Society, 1851-1951*, American Geographical Society, New York, 1952.

CLIMATOLOGY[A]

THE GROWTH OF CLIMATOLOGY IN THE UNITED STATES
Lorin Blodget and the Smithsonian Institution
The Signal Service
State Weather Services
The Weather Bureau
Unofficial Climatology

APPLICATIONS OF CLIMATOLOGY
The Statistical Treatment of Climatic Data
Agricultural Climatology
Other Applications

PROSPECT

[A]. By John Leighly

CHAPTER FOURTEEN

CLIMATOLOGY

CLIMATOLOGY IN THE United States began in the first half of the 19th century, in the period when the term "climatology" itself emerged. A wide net of observing stations was then established, through the combined efforts of private persons and official agencies. The men who established this net assumed that they were providing, by the compilation of the results of standardized observations, the means of solving the problems of climate. Most of our climatologic literature is written from that point of view, on the assumption that the data of such observations speak for themselves. As soon as specific questions regarding the impact of climate on inanimate objects and organisms are asked, however, it becomes evident that these data are mute: that impact is too complicated to be defined by measurements standardized to facilitate comparisons on a continental scale. The literature that attempts to define quantitatively the apparently obvious but in fact elusive relations of climate to other phenomena at the surface of the earth is smaller, but growing.

Because climate impinges on so many of the things with which science and practical affairs are concerned, interest in and investigation of it have arisen at many points in our society. This diffuseness of interest makes for a corresponding heterogeneity in our climatologic literature, and for difficulty in organizing an account of it. The greater part of such an account must therefore be written in the form of historical narrative.

THE GROWTH OF CLIMATOLOGY IN THE UNITED STATES

The period before 1850 may appropriately be called the cradle-age of our climatology. Most of our climatologic incunabula are included in a series of publications issued by the Surgeon-General's office of the United States Army [1]. The concern of responsible medical officers with weather and climate arose from the prevalence among the widely-scattered troops of the Army of diseases apparently related to climate. Specifically, men stationed in the Gulf states suffered from malarial and dysenteric ailments more than did those in the northern states. The most important initial stimulus to climatology in our country was thus derived from the deficiencies of medical science in the early 19th century. The observations made at military posts formed the only general body of climatologic data accumulated before the Smithsonian Institution established its net of observing stations in 1849.

The results of these observations were made known in a series of publications which with time became ever bulkier [2]. Out of them came the first general work on the climate of the United States, by Samuel Forry (1811-1844) [3], who had earlier worked up the *Meteorological Register* and the *Statistical Report on the Sickness and Mortality in the Army of the United States*. In his *Climate of the United States*, the most important of our climatologic incunabula, Forry stood firmly on the foundation laid by Alexander von Humboldt. He used Humboldt's terminology and Humboldt's definition of climate: "the aggregate of all the external circumstances appertaining to each locality in its relation to organic nature." Forry's book contains as a frontispiece the first temperature map of the United States. It bears winter isotherms for 26° and 41° Fahrenheit, summer isotherms for 65° and 81°, and the annual mean isotherm of 51.75° (annual mean for Fort Vancouver on the Columbia River), all drawn schematically and without regard to relief. Forry emphasized, as he had earlier in the publications of the Medical Department, the effects on the distribution of temperature of what we should today call continentality, and of the Great Lakes.

Lorin Blodget and the Smithsonian Institution

There was no sharp break as the 19th century passed its meridian; even the persons involved in the study of climate remained in part the same. Forry died untimely; but the next name that is written large in the history of our climatology at the middle of the century, that of Lorin Blodget (1823-1901), is linked with the work of the Medical Department of the Army. With Joseph Henry's initiation of an organized net of observing stations in 1849, under the auspices of the Smithsonian Institution, the investigation of climate moved on to a larger stage. From the time he became secretary of the newly founded Institution, Henry (1799-1878) directed its funds into work in meteorology. Aside from current observations, which, in accordance with Henry's interests, were primarily undertaken for what would today be called synoptic purposes, the working up of earlier observations was kept in mind. The first larger result of this more strictly climatologic interest among the publications of the Institution was James H. Coffin's memoir on the winds of the Northern Hemisphere [4].

As the material from the observing stations accumulated, it became necessary to employ someone to take charge of working them up. Henry engaged Blodget for this task, and Blodget entered on his duties at the end of 1851. A year and a half later Blodget began to report publicly on his work. At the meeting of the American Association for the Advancement of Science held in June, 1853, he read four papers based on the observations on which he was working [5]. And on the same occasion the obverse of Blodget's activity, the displeasure of his superior, Henry, was exhibited in public. Henry reminded his and Blodget's hearers that the material of which Blodget was speaking was collected by certain public agencies, and that these agencies should be credited with it. Friction between Blodget and Henry ended with Blodget's dismissal at the end of 1854, when Henry employed Coffin in his place.

Two long papers by Blodget now appeared, which were later incorporated, in revised and amplified form, into his justly famous *Climatology of the United States* [6]. The first is a work on the agricultural climatology of the United States, published in 1854 [7]. From this came the chapters of his *Climatology* on the general character of the climate of the various sections of the country, the comparison of these climates with those of Europe, and the capacity of the American climate for agricultural staples. The two other essential chapters in his book, those on the distribution of temperature and precipitation with their accompanying charts, are revised from a contribution to the *Army Meteorological Register* for the years 1843-1854 [8].

Joseph Henry's disapproval of Blodget's work followed him as these treatises appeared. In the first of a series of papers on the general subject of meteorology in its connection with agriculture, published in the agricultural section of the *Report of the Commissioner of Patents*, 1855 to 1858 [9], Henry praised the tabular part of the *Army Meteorological Register* published in 1855, but, without mentioning Blodget's name, decried the temperature maps as "premature publications, constructed from insufficient data, and on a principle of projection by which it is not possible to represent correctly the relative temperature in mountainous regions."

One can share Henry's adverse opinion of the drawing of isotherms on Blodget's temperature maps. Blodget evidently had no notion of the character of isarithms as indices of the distribution of a continuous variable over a surface; he permitted his to terminate on the map, to bifurcate and anastomose. But what Henry was objecting to in the passage cited was Blodget's treatment of data from the mountainous West, where the scattered stations were all situated in valleys. Blodget used the same procedure that Forry had, fifteen years earlier: he drew his isotherms with reference to the valley stations, explaining that the distribution of temperature shown referred to the valleys, and made no pretense of representing the temperature of elevated tracts. With his second paper in the agricultural reports Henry included a temperature map of the United States of his own construction, and in his text called particular attention to his reduction of temperature to sea level at a rate of 3 °F per thousand feet elevation. It is clear from the preface to Blodget's Climatology that what would today be called synoptic and climatologic investigation were separated within the Smithsonian Institution while he was employed there. According to his account, his first assignment was the reduction of recently received observations synoptically, in the manner exemplified by the contemporary work of Elias Loomis. But he found this work unsatisfying, and the climatological reduction of the observational material much more rewarding. And in his papers in the reports of the Commissioner of Patents, Henry, too, made a distinction between the knowledge that can be derived from the statistical treatment of observations of the weather and the possibility of predicting the movement of storms; thus implying that both branches were being pursued at the Smithsonian Institution.

Blodget's Climatology remains, in spite of its faults, the great monument of American climatology in the 19th century. Its author brought together the material he had published elsewhere, revised his maps, and skillfully fashioned the whole into a rounded scientific and literary fabric. He gave his readers, in truth, God's plenty: tables of observations, maps, quotations from travelers, forceful verbal characterizations. Above all he exemplified the sanguine spirit of the mid-century citizen of the Republic, surveying the imperial expanse of its area, contemplating its immensities, and prophesying its magnificent future. The master to whom he looked for inspiration and example, as had Forry before him, was Humboldt. He confessed this debt in his preface, where he rather ostentatiously flaunted an appreciative letter from the master, in which he undoubtedly found abundant compensation for the slights he had suffered at Joseph Henry's hands. Humboldt's clear-eyed and comprehensive view of nature was highly congenial to the spirit of the Old Republic, which in 1857, like the aged Humboldt himself, was hastening toward dissolution. It was from Humboldt that Blodget caught the urge to compare the climates and potentialities of the new world with those of the old. In Humboldt's writings, too, he could find precedents for his tabular and graphic presentations of numerical material. Tables and maps might be improved in the future, and the observations subjected to a closer scrutiny; but no later decade could provide the intellectual atmosphere in which such a work as Blodget's could be written. It is no accident that the Climatology is contemporaneous with Walt Whitman's "Song of the Broad-Axe"; both are flowers of the waning years of our intellectual golden day.

Blodget's successor, James H. Coffin (1806-1873), compiled the data for the joint publication of the Smithsonian Institution and the Patent Office issued in two volumes in 1861 and 1864 [10]. He is better remembered for his two memoirs on the winds: that on the winds of the Northern Hemisphere [4] and his second, on the winds of the whole globe [11], unfinished at his death and completed by other hands. As early as 1858 Henry had employed Charles A. Schott, since 1855 chief of the computing division of the Coast Survey, on the computation of special collections of observations. It fell to Schott (1826-1901), a master of computing techniques and familiar with the pitfalls of observational data, to work up the accumulated material on precipitation and temperature [12]. Schott's maps represent an immense improvement over Blodget's, not only because he had more material to work with than Blodget had had, but also because he was more attentive to the gradation of quantities over the surface of the earth and more conscious of the character of isariths than Blodget ever learned to be.

As examples of the combination of observations, Schott's memoirs have never been surpassed in the history of American climatology. In reducing the data on precipitation he not only drew maps, but also investigated the annual march, and computed amounts per day with precipitation, the probable error of means, and secular fluctuations. In his memoir on temperature he showed the daily march, comparing it among groups of stations by means of Fourier analysis to three harmonic terms. He subjected the annual march at 46 stations to the same kind of analysis, and examined the records for possible recurrent irregularities in daily temperature and for secular change. In view of Joseph Henry's pronounced preference for the reduction of temperature to sea level on maps, it is of interest to note that Schott's isotherms represent unreduced surface temperature; and that in his text he argues strongly for this manner of presenting temperatures and against reducing them to sea level.

THE SIGNAL SERVICE

Meanwhile, in 1870, an official weather service had been established in the Signal Service of the Army. But there is continuity from the earlier organizations: Colonel (later Brigadier General) Albert J. Myer (1827-1880), to whom the organization of the new service was entrusted, had earlier served in the Medical Department. The new service took over entire the system of voluntary observers that Henry had created, and the Smithsonian standards of instrumentation and observation. All records in the possession of the Smithsonian Institution were handed over to the Signal Service in 1873. The Signal Service was not instructed by law to concern itself with climate, but its responsibility to "commerce and agriculture" linked it with the tradition that had expressed itself in Blodget's and Henry's writings for the Commissioner of Patents before the federal Department of Agriculture was established in 1862. Climatologic papers appear early among the publications of the Service. A set of maps of mean monthly temperature was issued in 1881, and two years later maps and tables of monthly and annual precipitation [13]. A new tendency appears in these: restriction of the summary to a fixed period and to the Service's net of stations, except for the use of records from Army posts in the work on precipitation. Observations were now sufficiently abundant to obviate the need for scraping together all available ones without regard to their homogeneity.

The Signal Service was assigned its largest climatologic task in its last few years as a weather service, when it was called upon for information concerning the dry western parts of the country in connection with proposed federal legislation on behalf of irrigation. In the years 1888 to 1890 it supplied a series of reports, mostly by W. A. Glassford, though on the title pages generally credited to A. W. Greely (1844-1935), the last Chief Signal Officer before the weather service was transferred to the Department of Agriculture [14]. The most important of these reports were those on the rainfall of the Pacific slope and on the climatology of the arid regions. The latter contains extensive tables and maps of temperature and precipitation by seasons and critical months separately for Arizona, New Mexico, California and Nevada, Colorado, and Utah. An interesting innovation is a map of California and Nevada bearing generalized isarithms of annual evaporation in inches as measured with the Piche evaporimeter. The measurements used were a part of those on which Thomas Russell had based his evaporation map of the entire United States [15], which E. N. Transeau later used in his attempt to express the moisture element in climate by the ratio of precipitation to evaporation [16].

These reports present an estimate of the potentialities of the dry West for agriculture greatly different from those expressed by Blodget and Henry a generation earlier. The older authors thought that only a small fraction of the dry area could ever be cultivated, since the supply of water for irrigation was obviously limited. Glassford, on the contrary, writing in 1890 with a background of experience in the West, estimated the area that could be cultivated far higher. In discussing Arizona he aimed to demonstrate "that the measured amounts [of rainfall] are sufficient to supply water for the irrigation of much more land than the acreage known to be available," a popular exaggeration against which John Wesley Powell argued in vain at an irrigation congress held a little later [17]. As for evaporation in its bearing on the use of water for irrigation, Glassford held it to be "relatively so small in comparison with the total fluid contents of the actual and projected storage basins that it may be economically disregarded as a vanishing quantity."

STATE WEATHER SERVICES

In 1881 the second chief of the Signal Service in its capacity as a weather service, W. B. Hazen (1830-1887), initiated an expansion of the net of observing stations by soliciting the aid of the governors of all the States in organizing State weather services. His appeal met with great success. Before the national service was transferred to the Department of Agriculture as a civilian bureau in 1891, almost all the States had established such services, and the last one was organized in 1892. The State services were as a rule articulated with the State agricultural experiment stations, through which their observational data were frequently published. They were gradually absorbed by the Weather Bureau.

The State service that was conceived most broadly was that of Maryland, founded in 1891. Its first director, William Bullock Clark (1860-1917), planned for it an exceedingly inclusive program, which was realized in a series of impressive monographs. Its first Report that embodied this program, published in 1899, contained four important contributions: "A General Report on the Physiography of Maryland" by Cleveland Abbe, Jr. (1872-1934), son of the scientific pillar of the Signal Service and the early Weather Bureau; "The Aims and Methods of Meteorological Work, Especially as Conducted by National and State Weather Services," by the elder Cleveland Abbe (1838-1916); "A Sketch of the Progress of Meteorology in Maryland and Delaware," by Oliver L. Fassig (1866-1936); and an "Outline of the Present Knowledge of the Meteorology and Climatology of Maryland," by F. J. Walz, who at the time was in charge of the Weather Bureau Office in Baltimore. Walz's monograph contains by far the handsomest series of climatic maps issued by any State weather service: no other State had within its boundaries so competent a map printing firm as A. Hoen and Company of Baltimore. The Maryland service issued two additional reports comparable in bulk and quality with the first. Volume 2, issued in 1907, is a monumental work on the climate and weather of Baltimore by O. L. Fassig, a model of its kind; Volume 3, 1911, reflecting Clark's broad conception of the function of the Service, is concerned with the plant life of Maryland.

Nothing remotely comparable was achieved by any other State weather service. Their meager publications consist almost wholly of observational data, and the task of publishing these passed quickly to the Weather Bureau. When the services disappeared as administrative entities, they left behind them a mild interest in climate among the State agricultural experiment stations. Most of these have issued publications on the climates of the respective States, few of which are more than bare tabular compilations of observations. The authors of a few State climatographies have attempted to extract meaning from the material they have had to work with. Two that deserve mention are an older report on Wisconsin [18] and a more recent one on Florida [19]. The two latest ones issued by State agricultural agencies, on Kansas [20] and Illinois [21], make a more serious attempt than most of their predecessors to present the information on weather and climate that is significant for agriculture. In other States climatographies have been published by State geological surveys [22] and State universities [23]. S. S. Visher's *Climate of Indiana* contains by far the most elaborate presentation of climatic information ever attempted for a State. On hundreds of tiny maps Visher has shown by isarithms many quantities computed and counted from the climatic record.

THE WEATHER BUREAU

In transferring the national weather service to the Department of Agriculture, the Congress recognized what the public expected of the service. Among the duties with which the new bureau was charged was "the taking of such meteorological observations as may be necessary to establish and record the climatic conditions of the United States." Accordingly, climatology occupies a prominent place in its publications. Of its numbered series of bulletins, issued between 1892 and 1913, the one that still commands the greatest interest is one by Cleveland Abbe, written in 1891 but not published until 1905, on the relations between climate and crops [24]. It is mainly a digest of literature, but it contains a good deal of Abbe's own critical thinking. He was not satisfied with what he found in the literature, and appealed for a "climatic laboratory," using an expression uttered by A. de Candolle in 1866. Both he and de Candolle would welcome the realization of their wish in our time [25].

In its quarto series of bulletins, designated by letters, the Bureau summarized and refined the constantly more numerous and lengthening records of climate. In Bulletin C, 1894, the whole mass of records of precipitation to the end of the regime of the Signal Service was summarized in the most thorough presentation since the second edition of Schott's memoir. Bulletin D, 1897, bearing the indifferent title "Rainfall of the United States," is a critical supplement to its predecessor. Its author, Alfred Judson Henry (1858-1931), attempted in it to sharpen the concepts used in discussing precipitation. He was concerned with such matters as the length of record required to attain a dependable normal; the types of distribution of precipitation through the year; the amount of precipitation that falls during the crop-growing season, a motive that runs through all our later climatology; the secular variation of precipitation in different parts of the country; the synoptic conditions associated with rainy and dry months; the relation of precipitation to the relief of the land surface; and the frequency of occurrence of intense precipitation.

The Bureau's most ambitious effort to summarize what was known of the climate of the country, except in purely tabular or cartographic form, Bulletin Q, appeared in 1906 under the title "Climatology of the United States," also by A. J. Henry. By this time official observations had been accumulating for a generation. The hope of having an observing station in each county, expressed by W. B. Hazen when urging the establishment of State weather services, had not been realized. But Bulletin Q has its tabular material arranged so as to refer any county to a representative station. This tabular material, which includes more pertinent facts than are usually found in comparable works, occupies by far the largest part of the thousand quarto pages of the book. But what impresses one today is Henry's introductory text. Here Henry expressed his dissatisfaction with the organization of the material he was constrained to use. "The monthly means of temperature, pressure, the total monthly precipitation, etc.," he wrote, "indicate in a general way merely the dominating influence of the month. In the final analysis, however, one should not fail to examine the daily weather charts, since similar monthly mean values do not necessarily represent one and the same type of weather conditions." And again: "It would also be a great help to the better understanding of seasonal variations in the weather if the usual statistics of temperature and rainfall distribution, amount of cloudiness, etc., were classified and arranged according to the several types of weather that prevail in the United States."

What Henry was asking for was thus a synoptic climatology. He could not give his presentation the form he desired, but he did what he could to give meaning to his climatic maps by describing in synoptic terms and by illustrative weather maps the synoptic situations that are responsible for spells of typical and atypical weather. The synoptic situations he envisaged as appropriate to the grouping of observations are those in which the weather is dominated respectively by a cyclone or an anticyclone, and possibly a transitional type. It is scarcely necessary to remark that Henry's appeal for a synoptic climatology had no perceptible effect. He would have had to invent a new routine for the reduction of observations, and this he did not do. The abyss between the investigation of storms and the handling of large numbers of observations so as to obtain a normal, which first opened in the Smithsonian Institution a half-century earlier, remained unbridged.

It is not practicable to enumerate here all the Weather Bureau's lettered Bulletins. They provided a sound numerical basis for a climatology of the country, especially a climatology for the weather forecaster of the time: readily accessible means, computed for different units of time, with which current weather could be compared.

Bulletin W, originally entitled "Summaries of Climatological Data by Sections," is the one of the Weather Bureau's lettered series that has been known by its serial designation better than any other. In one of the three editions in which it has appeared it has for forty years been the standard source of climatic data wanted for any purpose. It is purely an assemblage of numerical data. The original edition summarized the record to about 1910, the later ones, respectively, to 1920 and 1930. The last edition no longer bore the serial letter, but its users have continued to call it "Bulletin W." That no later revisions have been made is sorely regrettable, since for the more recent record one must go to the serial "Climatological Data," which in recent years, with the addition of new material and reproduction by offset instead of letter-press printing, has become so bulky as to be the despair of librarians.

By far the best set of climatic maps prepared in the Weather Bureau was issued by a different bureau in the Department of Agriculture, the Office of Farm Management. These make up three parts of the fragmentary "Atlas of American Agriculture," published between 1918 and 1928 [26]. The main maps are on a generous atlas scale, 1:8,000,000, excellently printed in color by the Geological Survey, on a base that bears a fair representation of relief. The preparation of the Atlas fell in a period in which there was much interest within the Weather Bureau in the statistical expression of the probability of occurrence of given values of the climatic variables. This concern is expressed in the Atlas by the inclusion of numerous maps of frequency and probability as well as of means.

Although a shift in the demands made on the Weather Bureau led to its transfer from the Department of Agriculture to the Department of Commerce in 1940, it contributed much of the essential material to the Yearbook of Agriculture for 1941, *Climate and Man* [27]. Its contributions include not only papers by members of the Bureau, but also a summary of the climate of the United States, presented in both tables and maps, that fills nearly 600 pages. This is the latest general summary of the climate of the country prepared by an official agency.

UNOFFICIAL CLIMATOLOGY

Members of the national weather service have always found outlets for their writings outside official publications. The public, indeed, has normally looked to them for articles and books on weather and climate intended for general information. Such a book was A. W. Greely's *American Weather*, 1888 [28]. It contains the first general set of climatic maps of the United States based on uniform official observations. Historically it is of interest in that it set an example in organization that was largely followed by R. DeC. Ward in his *Climates of the United States,* published a generation later.

Academic institutions, however, bear the heaviest responsibility for scientific work; and in climatology the record of American universities is meager. The older academic tradition is well represented by Elias Loomis (1811-1889) of Yale, who in addition to his many studies in synoptic meteorology compiled the first map of isohyets of the earth [29]. But academic climatology has no continuous tradition in the United States; instead, it is dominated by one figure, Robert DeCourcy Ward (1867-1931). Ward was the first, and for many years the only, professor of climatology in an American university; and none of the few that have arisen later approaches him in national and international prominence.

Ward's scientific preparation was gained at Harvard, under William Morris Davis, and his whole career was spent there, from his appointment as instructor in meteorology (changed to climatology in 1896) to his death. Of his copious writings only his books can be cited here.

Aside from his early manual, *Practical Exercises in Elementary Meteorology,* the first of these is his translation of the first volume of Julius Hann's *Handbuch der Klimatologie* [30]. Ward made no slavish attempt to reproduce Hann's words in English. He divided and rearranged Hann's material to give it a better articulation, omitted passages and inserted new ones to give the English-speaking reader illustrations from places nearer home than Hann provided. He carried out all this adaptation unobtrusively, so that the book reads smoothly. Of Ward's translation it may be said, as of extremely few translations, that even one who reads German easily may prefer the translation to the original.

Some of Hann's ideas, as they appear in the opening pages of this book, can be recognized in Ward's later, original writings. The following are the most important: "Climatology is essentially . . . descriptive in character. . . ." "In climatology . . . those meteorological phenomena become the most important which have the greatest influence upon organic life. The importance of the different climatic elements is therefore determined from an outside standpoint." "Climatology is thus seen to be a branch of knowledge which is in part subordinated to other sciences and to practical ends. . . . This point should be borne in mind when we treat climatology as a science auxiliary to geography."

These ideas articulated closely with certain concepts current at the turn of the century regarding the relations of life, and particularly human life, to nature, which have been subsumed under the term "environmentalism." Ward's teacher, Davis, when he turned from the problems of physical geography, with which he ordinarily dealt, and contemplated mankind on the earth, floundered in environmentalistic quicksand. N. S. Shaler, Davis's teacher, based a good deal of his sincerely humane thought regarding the problems of society on the same derivative of social Darwinism. It occasions little wonder, therefore, that Ward's next work, *Climate, Considered Especially in Relation to Man* [31], has an environmentalistic flavor. It is discursive rather than systematic, in accord with contemporary tendencies in academic geography in the United States rather than with the contemporary state of the physical sciences.

Ward's *Climates of the United States* [32] is a much more substantial work, but retains a good deal of the same tendency. The first group Ward addresses in his preface is "teachers and students of geography." He endeavored "to make the presentation vivid by frequent references to the human relations of the various climatic elements and phenomena." Most of the content of the book, as of *Climate*, had appeared earlier in the form of articles in journals, but in their collected form they constitute a more-tightly organized whole than the earlier work. Though Ward made full use of maps of the climatic elements (indeed, the book could scarcely have been written without the maps recently published in the climatic sections of the *Atlas of American Agriculture*), he made a strong effort to take account of the synoptic aspect of climate. But, like A. J. Henry before him, he was unable to devise any method of articulating weather with climate. A decade earlier [33] he had appealed for the treatment of climate by synoptic units rather than time units, and had indicated by a citation that he had read how Köppen had solved this problem in the early 1870's [34]. Yet neither then nor later did Ward arrive at any better way of discussing weather climatologically than by reproducing graphs of the change in weather during representative sequences of days.

It is remarkable that Ward's *Climates of the United States*, admirably organized and written as it is, offers little or no intellectual stimulation only a little more than a quarter-century after its appearance. The reason is not merely that the revolution in synoptic meteorology that was under way when it was written has been completed. It is rather that the book displays no questioning, no doubt, no consciousness of problems still to be attacked. It has one lasting value: its copious bibliographic footnotes. The student of the history of climatology in the United States scarcely needs to go beyond these footnotes and Ward's bibliographic articles cited in them for a guide to the literature published before the date of the book.

The last large work that bears Ward's name, the section on the United States and Mexico in the Köppen-Geiger *Handbuch* [35], published after Ward's death, contains work by too many hands to be representative of his authorship. Most of what is recognizably his had appeared earlier in his book on the United States. The reader who knows his other work is not astonished to learn that the extensive tabular material and the admirable series of maps this work contains were not compiled and drawn by Ward, but by or under the supervision of C. F. Brooks.

The changing tenor of intellectual life has made another Ward impossible. No one born in the present century could attain a comparable level of learning and literary skill and at the same time accept an essentially verbal climatology only loosely articulated with other aspects of the science of the atmosphere. One must suspect that the popularity of climatology in departments of geography in our universities stemmed largely from Ward's example at Harvard. His books, used as undergraduate textbooks, were "easy," and deliberately in harmony with much instruction in geography. Ward had translated from Hann the statement that "climatology presupposes a knowledge of the most important teachings of meteorology," but his books demanded of their readers little of this knowledge. In many departments of geography, familiarity with the more exacting phases of meteorology came to be looked upon as superfluous, and students were offered courses in climatology without being required to have any such familiarity.

Neglect of the significance of contemporary advances in other phases of the science of the atmosphere for the understanding of climate is characteristic of our climatologic literature, both regional monographs and textbooks, in the third and fourth decades of this century. This shortcoming was in part compensated by a greater willingness to wrestle with masses of numerical data than Ward displayed. But Ward's example provided no guidance toward the discovery of a philosopher's stone that would transmute the base data of observation into the gold of physical generalization.

From the middle 1920's to the present time, much academic instruction in climatology has been focused on the classification of climates and maps of climatic regions based on the classifications used. C. W. Thornthwaite has combined a review of the pertinent literature with a thorough (and adverse) criticism of the classification most frequently used, Köppen's of 1918 [36]. Much earlier, Ward had reviewed for his countrymen various proposed delineations of climatic regions [37], but none of these took root. The wide adoption and imitation of Köppen's classification when it became known in the United States in the 1920's undoubtedly resulted from the abandonment by American academic geography at about this time of its earlier environmentalism in favor of an intense interest in regions. Maps purporting to show the distribution of types of climate fitted easily into the thinking reflected in this new orientation. Much effort was expended on tracing in greater detail the boundaries of climatic regions, especially within the United States, shown on Köppen's small-scale world map, and in modifying his definitions so as to discover better boundaries between these regions than his original ones [38].

Thornthwaite's has been the most powerful voice raised against the uncritical application of Köppen's classification. More than a decade before the date of the article just cited, he had published his own general classification of climates, which grew out of his dissatisfaction with the Köppen classification, in particular with its delineation of the dry climates [39]. The most impressive monument of Thornthwaite's classification is an elaborate atlas issued by the Department of Agriculture [40], which shows the distribution of the types based on Thornthwaite's distinctions of humidity for each calendar year and crop season of a forty-year period. No other document of our climatologic literature presents so effectively as this atlas the fluctuations of the moisture element of climate in the middle regions of North America.

The same dominant concern with temporary or permanent shortage of water marks Thornthwaite's second attempt at a classification of climates, published in 1948 [41], in which the climates are distinguished by the balance, in the course of the year, between precipitation and a quantity he calls "potential evapotranspiration," the amount of water that would be lost in unit time from the surface of the land to the atmosphere if the cover of vegetation were transpiring it at the maximum rate permitted by the prevailing climatic factors, primarily insolation and temperature.

Our most important general treatise since Ward's death is Helmut Landsberg's *Physical Climatology* [42]. It represents a welcome reaction against the trend that Ward established, but is not sufficiently comprehensive to rank with the best works of the past. V. Conrad's *Methods in Climatology* [43] reflects its author's European background rather than the growth of climatology in America. Much of the book is devoted to statistical techniques; but the best statistical practices currently in use here, which will be mentioned later, are far in advance of those Conrad describes.

Since about 1930 a number of our universities and technologic institutions have inaugurated curricula for the technical training of meteorologists. These institutions are the most important centers of investigation of the atmosphere in the country. Their curricula demand a new kind of academic climatology, which has not yet emerged; it must evidently be far different from Ward's or one organized about a classification of climates. One full-sized textbook of climatology [44] has come out of a department of meteorology in a technical institution, but it does not reflect much serious thought on the part of its authors concering the appropriate content of a climatology commensurate in quality with modern meteorology in general.

From one theoretical viewpoint climate signifies the mean state of the atmosphere, a norm with which any instantaneous state can be compared. Landsberg has provided a specimen of that kind of climatology [45], a concise and workmanlike job, the essential feature of which is a series of 29 maps of the earth. The mean state of the atmosphere represents a state of equilibrium with respect to many processes of flux and exchange: flux of heat from low latitudes to high, in equilibrium with the balance of radiation; exchange of radiation among the earth, atmosphere, sun, and space; exchange of angular momentum between the surface of the earth and the atmosphere; and balance of precipitation with evaporation, to name a few. The condition of equilibrium is the starting-point of many investigations of these processes [46].

So far as one can see at present, however, many of the phenomena of motion in the atmosphere that are seen on the weather map and that are responsible for weather and its changes must still in the future, as in the past, be investigated empirically; that is, by synoptic methods. A particularly enlightening use of synoptic methods potentially useful in climatology appears in the section on monthly weather and circulation published in the *Monthly Weather Review* beginning with the issue for January, 1950. Maps of departures of monthly means of temperature and precipitation from long-term means have been published in the *Review* for many years; but only recently have these departures been referred to some more fundamental variable. This variable is the anomaly of mean circulation in the lower middle troposphere [47]. As in most investigations of climate, the month is somewhat too long to be as synoptically homogeneous as is desirable; but even within this unit of time the anomaly of circulation at the 700-millibar level provides more insight into the fluctuations of the familiar aspects of climate than has been available heretofore.

APPLICATIONS OF CLIMATOLOGY

Climatologic observations collected by the procedures standardized a century ago supply answers to the problems posed by science and technology scarcely better than do common sense and the experience of ordinary persons. Blodget pointed out, by comparison of the recorded climates of the western country with those of the Old World, the potential adaptation of these new lands to cultivated crops; but certainly few of the farmers that moved west in the half century following the date of publication of Blodget's great book had ever read it. In the decade in which it was written, however, the climatology of the sea, in the hands of Matthew Fontaine Maury (1806-1873), scored a notable accomplishment by showing the best routes for sailing ships on long voyages [48]. This accomplishment was possible because the phenomena observed, primarily the directions of the winds, were just those that were immediately involved in the technical operations for the sake of which the observations were made. The connections between climate and most operations are much more intricate. The conflicting conclusions that can be drawn from the same observational data are sufficiently illustrated by the disagreement between Glassford and Powell already cited.

The test of the utility of climatologic data in practical affairs is the establishment of quantitative relations between these data and contemplated action. In few instances can these relations be deduced from physical principles; if they are to be determined it must be by empirical procedures, most of which are today called statistical.

The Statistical Treatment of Climatic Data

Aside from the statistical procedures that find their principal use in applied climatology, statistics occupy a central place in the interpretation of the climatic record itself. In the earlier history of American climatology, little attention was paid to the statistical qualities of the record. C. A. Schott, who brought to his climatologic work the habits and methods fixed by long experience in the reduction of observations in another field, provides the one conspicuous exception.

The record of climate has been scrutinized for possible trends or recurrences with time ever since it attained a length of more than a few years. Most collations of observations made for this purpose have aimed at no more than a graphic demonstration, with or without some sort of smoothing of the fluctuations recorded. The vast amount of arithmetic labor expended in the effort to find "hidden" periodicities may be ignored. The study of sequences has led to the recognition of a certain amount of persistence, easily recognized and measured in units of days [49], at some places even in successive years [50]. Interest in the persistence of departures from long-term means was stimulated by the occurrence of successive years of drought in the 1930's [51] and by the general rise of winter temperature in the past 60 or 70 years [52]. It is not to be expected that such persistence as exists can be put to any appreciable use in forecasting, but it poses a problem that insistently demands investigation. How provocative it may be is indicated by H. C. Willett's far-reaching generalizations concerning persistence in climate through periods longer than that spanned by the record of instrumental observation [53]. Although statistical tests for persistence in time series are available [54], they have not been systematically applied for the purpose of determining the coherence or incoherence of the several parts of the United States in the fluctuations of climate. Statistical investigation of the record for this purpose would be a valuable complement to the studies of fluctuations in the general circulation by synoptic methods that have been intensively pursued [47].

Frequency distributions of climatologic observations, or, more often, the return intervals of given values (which are the reciprocals of probability when, as in climatic data, the individual values of a variate pertain to units of time) first became of concern to hydraulic engineers. R. E. Horton wrote in 1913 [55] that he had used expressions based on probability, "in some instances the ordinary Gaussian law and in some instances developing special probability curves, formulas, and diagrams," since about the beginning of this century. Two articles, by Allen Hazen and Thorndike Saville, published in 1916, recommended that the theory of probability be applied to precipitation [56]. Both authors used the normal frequency distribution without questioning its applicability. Hazen's article was accompanied by a map of the United States that showed by isarithms the coefficient of variation (standard deviation expressed as a percentage of the mean) of annual precipitation. This map has often been referred to in the technical literature, even since the appearance of a later map of a closely related quantity, mean deviation expressed as a percentage of mean precipitation [57].

Variation in precipitation has attracted more attention than any other quantity in the statistics of climate. It has usually been conceived as dispersion about the mean, but B. M. Varney, investigating precipitation in California, used instead the change from year to year (variability) [58]. Varney evidently did not know that variability is functionally related to dispersion in a random distribution, and so missed the opportunity of comparing his results with those to be found in a chance sequence. Most authors have been content with such an expression as the one Hazen computed in 1916, though the numerical value of the coefficient of variation evidently depends on the value of the mean, displaying a range of numbers running in a sense opposite to mean precipitation. V. Conrad [59] has attempted to avoid the absurdity involved in the approach of the coefficient of variation to infinity as the mean approaches zero, by fitting an empirical equation to the coefficients computed for a large number of stations. He obtained an expression for the coefficient of variation as a function of the mean, according to which it approaches the value 73 per cent, instead of infinity, as the mean approaches zero. This result is unconvincing, since as the mean approaches zero any measure of variation should also approach zero.

The problem of expressing the dispersion of values of precipitation in units of time about their mean thus remains unsolved. The achievement of a satisfactory measure of dispersion might hasten the abandonment of the often repeated dogma that precipitation is most variable in dry climates; a dogma that has no firmer basis than the distribution over the lands of the coefficient of variation. This measure is appropriate to series in which dispersion both upward and downward from the mean is unlimited, but wholly inappropriate to a quantity such as precipitation, the scale of which is closed in one direction by zero.

The asymmetry of the frequency distributions of some climatic data was recognized rather early, in one of the many articles on the statistical handling of observations that appeared in the *Monthly Weather Review* between 1916 and 1930 [60], but the fitting of asymmetrical curves to such data has never become general. Hydraulic engineers began to work with skewed distributions at about the same time; Pearson's flexible Type III distribution was formally recommended to them in 1924 [61]. Frequency of occurrence of floods has more recently been taken by E. J. Gumbel as a starting point for an improved statistical treatment of extreme values [62], and Arnold Court has applied it, with instructive results, to extremes of temperature and wind speed [63]. Pearson's Type III curve has been used as an expression of the probability of weekly precipitation, in an investigation of drought in Iowa [64].

A minor revolution in the handling of masses of observations has been effected by the introduction of punched cards and machines for sorting them. First applied to climatic data in the United States by the Weather Bureau in the 1930's, the use of punched-card methods was enormously expanded during World War II. A brief account of this growth, with citations of literature, has been given by V. Conrad and L. W. Pollak in the second edition of *Methods in Climatology;* and W. C. Jacobs has described the applications of the procedure made by the Weather Service of the Air Force [65]. The characteristic products of the punched-card technique, aside from the traditional sums and means, are frequency distributions within categories into which observations may be broken down; and in particular frequencies of associated values of different elements, such as temperature or precipitation with direction or speed of wind. One product, to which Jacobs gives appropriate prominence, is a practical synoptic climatology, long asked for but never before realized. A few products of the card-sorting machine published since World War II may be noted: a presentation of observations made at airports that closely resembles the war work described by Jacobs [66], an application of the procedure to agricultural climatology [67], and a bulky compilation of observational data made for a single locality, Washington, D. C. [68].

The record of statistical treatment of climatologic data in the United States is good. If it invites any general conclusion, it is that too few of the persons working in climatology have had a creative knowledge of probability and the statistical procedures founded upon it. Hence too many of the good studies published have remained little more than demonstrations, examples that have not been followed.

AGRICULTURAL CLIMATOLOGY

Until the end of the 19th century American concern with the relation of climate to agriculture was restricted to generalizations regarding the limits of possible cultivation of crops, expressed in terms of precipitation and temperature. In the meantime agricultural settlement had continued, and the available stock of crop plants, including an ever-widening range of varieties of each crop species, had been tried out. The problem conceived by the climatologists of the middle of the century, to discover the possibilities of agricultural development in the parts of the country as yet unsettled, was solved by trial, both by individual farmers and by workers at the State agricultural experiment stations founded, with federal subvention, in the last quarter of the century.

The information that Cleveland Abbe put together in his report on the relations between climate and crops [24] was based mostly on laboratory rather than field experiments, and reflected European rather than American experience. But actual cultivation, whether experimental or commercial, provides quantitative information on crops in the form of yield, and it was with yield that the results of climatologic observations came by preference to be compared.

The earliest procedure for making this comparison was the parallel plotting, on a scale of time, of yield and some climatic factor that might be expected to affect it [69]. At about the turn of the century, however, the techniques of statistical treatment of observational data underwent rapid improvement in England. In 1905 W. N. (later Sir Napier) Shaw published a pioneer effort at correlating yield of wheat with precipitation [70]. It was followed in 1907 by R. H. Hooker's exemplary investigation of a number of crops, which made use of the best contemporary resources of correlation statistics [71]. Four years later the new procedure made its appearance in the United States, in a rather tentative study by J. Warren Smith of the relation of yield of potatoes in Ohio to weather factors, in which Smith used simple linear correlation [72]. This was the first of a long series of studies, of many crops and all available climatic factors, that Smith carried out, in most of which he used linear correlation. His work culminated in his *Agricultural Meteorology*, a most ineptly written and organized book [73]. A paper by Henry Wallace that appeared in the same year as Smith's book [74] represented a great advance over what had been done earlier, especially in the use of multiple correlation and the application of a critical attitude to the problem.

Meanwhile others entered the field opened by Smith. Among them were T. A. Blair, D. A. Seeley, and in particular J. B. Kincer, who succeeded

Smith as chief of the Weather Bureau's Division of Agricultural Meteorology. Kincer's work, which extended over a period of fifteen years, is
noteworthy for its flexibility and ingenuity. All these investigations proceeded from the postulate of one or more critical periods in the annual
cycle of the crop plant, at which it is particularly susceptible to adverse
effects of weather. In them all the resources of Pearsonian statistics were
applied, but with only indifferent success. The latest name to be mentioned is that of W. A. Mattice, who after repeated applications of the
best statistics available took leave in 1931 of this species of investigation,
to which so much labor had been devoted. "The weather in its effect on
agriculture has been scrutinized from afar," he wrote, "as through a long-
range telescope, but very little has been accomplished in pursuing the
microscopic detail necessary for a complete understanding of the underlying principles involved in plant growth. . . . To enable us to know just
how the weather is affecting a crop at any time . . . we need accurate
and comparable data of weather and crop progress, with the details of
various weather phases and of crop developments from planting to harvest
accurately observed and recorded on the ground" [75]. A program of
precisely this sort has been undertaken, with corn, at Iowa State College,
but has been in progress too short a time to produce results as yet.

A new epoch in the statistical treatment of the problem of crops and
climate opened with the publication, in 1921 and 1924, of R. A. Fisher's
masterly study of the yield of wheat on the famous Broadbalk plots at
Rothamsted [76]. The method Fisher devised made it unnecessary to seek
by trial hypothetical critical periods in the annual cycle of the crop plant.
Instead, it provided a continuous measure, in the sense of the familiar coefficient of regression, of the influence of a climatic factor on yield. The
first application of Fisher's method in the United States was apparently
not to crop yields, but to the growth of trees and range grasses [77].
When it finally came to be applied to crops, those who used it were no
longer men of the Weather Bureau, but statisticians of the Department of
Agriculture, to whom the problem was not one in natural science, but
abstractly arithmetic [78]. A later paper emanating from the same group
goes beyond its predecessors by extending the method to measure the
joint effects of temperature and precipitation on yield [79]. It represents
the most original American work to date in the application of the newer
statistics to the relation of climate to the yield of crops.

OTHER APPLICATIONS

Some special uses of climatology are treated in chapters twenty-one,
twenty-two, and twenty-three of this book. The general problem of applied climatology has been so well defined in Landsberg and Jacobs's

admirable survey [80] that it would be presumptuous to try to add anything to what they have said. One of their generalizations, which will be heartily concurred in by all who have attempted the application of climatology to practical problems, may be cited: "It is . . . essential that both parties involved in the climatological analysis of an operation, the climatologist and the *operator*, collaborate." Landsberg has published a demonstration of how the climatologist can go halfway to meet the "operator" by appropriate handling of readily accessible climatologic material [81].

Because the data of standardized observations, whether instituted "to establish and record the climatic conditions" or for the purposes of synoptic meteorology, seldom cast more than a dim light on the problems of applied climatology, many special techniques of observation have been elaborated. These techniques work close to the surfaces of the ground and other objects affected by climate, and record differences within much smaller distances, both along and normal to these surfaces, than do conventional observations, which are designed to minimize such differences. The appropriate term "microclimatology" has been coined to designate investigations using such techniques. W. A. Baum and Arnold Court [82] have reviewed work in this field, with special attention to that done in the United States. Observations on the scale of microclimatology require more elaborate instrumentation than do the conventional ones, and so are expensive, sometimes fantastically so. They are therefore seldom taken except for specific practical purposes. Hence the absence of standardization that Baum and Court deplore; and hence, too, the difficulty of comparing the results obtained by different workers in a field in which comparisons are necessarily difficult. Microclimatology is likely to retain two aspects: on the one hand a complex of instrumental methods, and on the other a body of doctrine that tries to interpret the physics of the layer of air close to the many kinds of surfaces that are exposed to sun, space, and atmosphere. It has already yielded enlightening generalizations concerning the exchange, in nature, of energy and water vapor at those surfaces; and may in the future make possible an illuminating use of the material collected by conventional procedures.

There is no line, but only a gradual transition, between applied climatology and the use of climatologic data by technicians for their own ends. Landsberg and Jacobs cite examples of such use. The literature of engineering, particularly of hydraulics and the technology of heating and ventilating, abounds with further examples. It is along this border that those who accept Landsberg and Jacobs's invitation to "those climatologists who break away from the traditional archivistic or descriptive approach and who are willing to meet the challenge of new frontiers" will find most of their tasks.

PROSPECT

The most impressive contributions being made to climatology in the United States at present are coming from those who are working intensively on the true constituent elements of climate, the constantly changing states of weather recorded on synoptic maps. Their work is welcome after a long period in which the relations between weather and climate received only perfunctory recognition. It is concerned with causes, as William Morris Davis used that word. As for consequences, the other member of the alliterative antithesis Davis was fond of using, it is evident that any investigation must be highly specific, that procedures must be designed with the elaborate attention to detail that has long been familiar in the laboratory. Some aspects of the turnover of energy and water at the surface of the earth, such as the loss of water from the ground that Thornthwaite has made his prime concern, or the exchange of energy between the atmosphere and the surface of the sea that Jacobs [46] has investigated, may be handled by means of conventional observational material. But the relations of organisms to climate present far more difficult problems. F. W. Went [25] has shown what elaborate precautions must be taken if even the simpler reactions of plants to climate are to be defined quantitatively.

It can be foreseen that the solution of any problem in applied climatology in the future must involve special instrumentation and extremely detailed observations through a period of time long enough to sample a good range of weather. Then, in the most fortunate circumstances, ordinary climatologic observations may be used for extrapolation to other places and times. It is not to be expected, however, that circumstances will often be so fortunate. For general observations of weather, primary dependence will be placed on the regular stations established for synoptic purposes. The net of voluntary observing stations, instituted by Joseph Henry more than a century ago, will lose the primary importance it has had so long.

It is to be hoped, however, that the time-honored system of voluntary observers will not disappear. Its disappearance would signify a great cultural loss, a part of the cultural impoverishment that goes with the excessive specialization of our times, in which people are alienated from techniques and modes of thinking not associated with their daily work. Though the scientific significance of his observations may decline, the voluntary observer should remain for his usefulness to the community, especially in the country and the small town, where the sense of neighborhood is not yet lost. The tradition of the scientific amateur, on which Joseph Henry depended for the initiation of a net of observing stations in the early days of the Smithsonian Institution, has never been strong in this country. But to the extent to which it survives in the voluntary observer it should be fostered.

There is a place, too, for a climatology other than the rigorous numerical science with which this chapter has been mainly concerned: a natural history of climate. The rhythm of the seasons echoes deeply in our literature and folk-traditions. In the interpretation of this rhythm, in the communication of insights that are better expressed in words than in equations, the climatologist has a role to play, along with the amateur of plants and animals and the literary artist. These formulations, no less than algebraic ones, can make the dry bones of climatologic tabulations live.

REFERENCES

1. BROWN, H. E. *The Medical Department of the United States Army from 1775 to 1873*, Washington, 1873: 113-116.
2. *Meteorological Register 1822-1825*, Washington, 1826; *ibid 1826-1830* (including also 1822-1825), Philadelphia, 1840; *ibid. 1831-1842*, Washington, 1851; *ibid. 1843-1854*, Washington, 1855; *Statistical Report on the Sickness and Mortality of the Army of the United States . . . from January 1855 to January 1860*, Washington, 1860.
3. FORRY, S. *The Climate of the United States and its Endemic Influences*, New York, 1842 .
4. COFFIN, J. H. *Winds of the Northern Hemisphere*, Smithsonian Contributions to Knowledge, 6, Washington, 1854.
5. BLODGET, L. "On the Distribution of Precipitation in Rain and Snow on the North American Continent," *Proceedings of the American Association for the Advancement of Science*, 7 (1853): 101-108; *idem*. "On the Southeast Monsoon of Texas, the Northers of Texas and the Gulf of Mexico, and the Abnormal Atmospheric Movements of the North American Continent Generally," *ibid.*, 109-118.
6. ———. *Climatology of the United States and of the Temperate Latitudes of the North American Continent*, Philadelphia, 1857.
7. ———. "Agricultural Climatology of the United States, Compared with That of Other Parts of the Globe," in United States Commissioner of Patents, *Report for the Year 1853, Agriculture*, 328-432, Washington, 1854.
8. ———. "Report on the Prominent Features of General Climate in the United States, as Exhibited in the Distribution of Temperature and of Rain and an Explanation of the Illustrative Charts," in *Army Meteorological Register 1843-1854*, Washington, 1855: 681-763.
9. HENRY, J. "Meteorology in its Connection with Agriculture," in United States Commissioner of Patents, *Report for the Year 1855, Agriculture*: 357-374; *ibid. 1856*: 455-492; *ibid. 1857*: 419-506; *ibid. 1858*: 429-493; *ibid. 1859*: 461-524. Reprinted in *Scientific Writings of Joseph Henry*, Smithsonian Miscellaneous Collections, 30, Washington, 1886.
10. *Results of Meteorological Observations, Made under the Direction of the United States Patent Office and the Smithsonian Institution, from the Years 1854 to 1859, Inclusive*, Washington, 1864.
11. COFFIN, J. H. *The Winds of the Globe, or, the Laws of Atmospheric Circulation over the Surface of the Earth*, Smithsonian Contributions to Knowledge, 20, Washington, 1875.

12. SCHOTT, C. A. *Tables and Results of the Precipitation, in Rain and Snow, in the United States* . . . , Smithsonian Contributions to Knowledge, 18, 2, Washington, 1872, 2nd ed., *ibid.*, 24, 2, 1881; *idem. Tables, Distribution, and Variations of the Atmospheric Temperature in the United States, ibid.*, 21, 5, Washington, 1876.

13. GREELY, A. W. *Isothermal Lines of the United States, 1871-1880*, Signal Service Professional Papers, 2, 1881; DUNWOODY, H. H. C., *Charts and Tables Showing the Geographical Distribution of Rainfall in the United States, ibid.*, 9, 1883.

14. GREELY, A. W. *The Rainfall of the Pacific Slope and the Western States and Territories*, 50th Congress, first session, Senate Executive Document, 91, 1888; GREENE, F. *Report on the Interior Wheat Lands of Oregon and Washington Territory, ibid.*, 229, 1888; *idem. The Climate of Oregon and Washington Territory, ibid.*, 282, 1889; GREELY, A. W. *Climate of Nebraska, Particularly in Reference to the Temperature and Rainfall and Their Influence upon the Agricultural Interests of the State*, 51st Congress, first session, Senate Executive Document, 115, 1890; GLASSFORD, W. A. *Rainfall in Arizona and its Effect on Irrigation and Water Storage, with Description of some Favorable Points for Storage Reservoirs*, 51st Congress, first session, Senate Report, 928 (1890): 486-494; GREELY, A. W. *Report on the Climatology of the Arid Regions of the United States, with Reference to Irrigation*, 51st Congress, second session, House Executive Document, 287, 1891.

15. RUSSELL, T. "Depth of Evaporation in the United States," *Monthly Weather Review*, 16 (1888): 235-239.

16. TRANSEAU, E. N. "Forest Centers of Eastern America," *American Naturalist*, 39 (1905): 875-889.

17. *Report of the International Irrigation Congress Held at Los Angeles, California, October, 1893*: 108-113.

18. WHITSON, H. R., and BAKER, O. E. *The Climate of Wisconsin and its Relation to Agriculture*, University of Wisconsin Agricultural Experiment Station, Bulletin 223, 1912.

19. MITCHELL, A. J., and ENSIGN, M. R. *The Climate of Florida*, University of Florida Agricultural Experiment Station, Bulletin 200, 1928 (93-300).

20. FLORA, S. D. *Climate of Kansas*, Kansas State Board of Agriculture, Report 67, 1948.

21. PAGE, J. L. *Climate of Illinois: Summary and Analysis of Long-Time Weather Records*, University of Illinois Agricultural Experiment Station, Bulletin 532, 1949.

22. VISHER, S. S. *Climate of South Dakota*, South Dakota Geological Survey, Bulletin 8, 1918; *idem. The Climate of Kentucky*, Kentucky Geological Survey, series 6, 31, 1929: 81-167.

23. ALEXANDER, W. H. *A Climatological History of Ohio*, Ohio State University Bulletin 28, 1923; VISHER, S. S. *Climate of Indiana*, Indiana University Publications, Science Series 13, 1944.

24. ABBE, C. *A First Report on the Relations Between Climates and Crops*, Weather Bureau Bulletin 36, 1905.

25. WENT, F. W. "The Response of Plants to Climate," *Science*, 112 (1950): 489-494.

26. REED, W. G. *Frost and the Growing Season*, Atlas of American Agriculture, Part II, Section I, 1918; KINCER, J. B. *Precipitation and Humidity, ibid.*, Part II, Section A, 1922; *idem. Temperature, Sunshine, and Wind, ibid.*, Part II, Section B, 1928.

27. United States Department of Agriculture, *Climate and Man*, Yearbook 1941, Washington, 1941.

28. GREELY, A. W. *American Weather*, New York, 1888.

29. LOOMIS, E. "Contributions to Meteorology . . . : Mean Annual Rainfall for Different Countries of the Globe," *American Journal of Science,* 23 (1882): 1-25.

30. HANN, J. *Handbook of Climatology, Part I,* (Translated by R. DeC. WARD), New York, 1903.

31. WARD, R. DeC. *Climate, Considered Especially in Relation to Man,* New York, 1908, 2nd ed., 1918.

32. ———. *Climates of the United States,* New York, 1925.

33. ———. "Suggestions Concerning a More Rational Treatment of Climatology," 8th International Geographical Congress, *Proceedings,* (1904): 277-293.

34. KÖPPEN, W. "Über die Abhängigkeit des klimatischen Charakters der Winde von ihrem Ursprunge," *Repertorium für Meteorologie,* 4, 4 (1874).

35. WARD, R. DeC., and BROOKS, C. F. *The Climates of North America,* in KÖPPEN, W., and GEIGER, R., eds. *Handbuch der Klimatologie,* 2, Part J, Berlin, 1936.

36. THORNTHWAITE, C. W. "Problems in the Classification of Climates," *Geographical Review,* 33 (1943): 233-255.

37. WARD, R. DeC. "The Classification of Climates, Continued," *Bulletin of the American Geographical Society,* 38 (1906): 465-477, and *Climate . . . ,* chapter 3.

38. VAN ROYEN, W. "The Climatic Regions of North America," *Monthly Weather Review,* 55 (1927): 315-319, 410-412; RUSSELL, R. J. "Dry Climates of the United States: I, Climatic Map," *University of California Publications in Geography,* 5, 1 (1931), "II, Frequency of Dry and Desert Years 1901-1920," *ibid.,* 5, 5 (1932); ACKERMAN, E. A., "The Köppen Classification of Climates in North America," *Geographical Review,* 31 (1941): 105-111.

39. THORNTHWAITE, C. W. "The Climates of North America According to a New Classification," *Geographical Review,* 21 (1931): 633-655; *idem.* "The Climates of the Earth," *ibid.,* 23 (1933): 433-440.

40. ———. *Atlas of Climatic Types in the United States, 1900-1939,* United States Department of Agriculture, Miscellaneous Publications 421, 1941.

41. ———. "An Approach toward a Rational Classification of Climate," *Geographical Review,* 38 (1948): 55-94.

42. LANDSBERG, H. *Physical Climatology,* State College (Pennsylvania), 1941.

43. CONRAD, V. *Methods in Climatology,* Cambridge (Massachusetts), 1944, 2nd ed. (with L. W. POLLAK), 1950.

44. HAURWITZ, B., and AUSTIN, J. M. *Climatology,* New York, 1944.

45. LANDSBERG, H. "Climatology," in BERRY, F. A. JR., BOLLAY, E., and BEERS, N. R., eds. *Handbook of Meteorology,* New York, 1945: 927-997.

46. NAMIAS, J., and CLAPP, P. F. "Normal Fields of Convergence and Divergence at the 10,000-foot Level," *Journal of Meteorology,* 3 (1946): 14-22; JACOBS, W. C. "The Energy Acquired by the Atmosphere over the Oceans through Condensation and through Heating from the Sea Surface," *ibid.,* 6 (1949): 266-272.

47. NAMIAS, J. "Evaluation of Monthly Mean Circulation and Weather Patterns," *Transactions of the American Geophysical Union,* 28 (1948): 777-788; MARTIN, D. E., and LEIGHT, W. G. "Objective Temperature Estimates from Mean Circulation Patterns," *Monthly Weather Review,* 77 (1949): 275-283; AUBERT, E. J., and WINSTON, J. S. "A Study of Atmospheric Heat-Sources in the Northern Hemisphere for Monthly Periods," *Journal of Meteorology,* 8 (1951): 111-125.

48. MAURY, M. F. *Explanations and Sailing Directions to Accompany the Wind and Current Charts,* 8th ed., 1, Washington, 1858: 299-302; *ibid.,* 2, Washington, 1859: *passim.*

49. BLAIR, T. A. "The Coefficient of Persistence," *Monthly Weather Review*, 52 (1924): 350; JORGENSON, D. L. "Persistence of Rain and No-Rain Periods During the Winter at San Francisco," *ibid.*, 77 (1949): 303-307.

50. SOLOT, S. B. *Possibility of Long Range Precipitation Forecasting for the Hawaiian Islands*, United States Weather Bureau, Research Paper 28, 1948, SHOWALTER, A. K. "Chance for Successive Dry Years in Southern California," *Monthly Weather Review*, 76 (1948): 221-223.

51. HOYT, J. C. *Droughts of 1930 to 1934*, United States Geological Survey, Water Supply Paper 680, 1936; BLUMENSTOCK, D. I. *Rainfall Characterisics as Related to Soil Erosion*, United States Department of Agriculture, Technical Bulltin 698, 1939; BLUMENSTOCK, G., JR. *Drought in the United States Analyzed by Means of the Theory of Probability*, *ibid.*, 819, 1942.

52. KINCER, J. B. "Our Changing Climate," *Transactions of the American Geophysical Union*, 27 (1946): 342-347.

53. WILLETT, H. C. "Long-Period Fluctuations of the General Circulation of the Atmosphere," *Journal of Meteorology*, 6 (1949): 34-50; LORENZ, E. N. "Seasonal and Irregular Variations of the Northern Hemisphere Sea-Level Pressure Profile," *ibid.*, 8 (1951): 52-59.

54. CLOUGH, H. W. "A Statistical Comparison of Meteorological Data with Data of Random Occurrence," *Monthly Weather Review*, 49 (1921): 124-132; HORTON, R. E. "Group Distribution and Periodicity of Annual Rainfall Amounts," *ibid.*, 51 (1923): 515-521; WOOLARD, E. W. "On the Mean Variability in Random Series," *ibid.*, 53 (1925): 107-111; KENNEDY, R. E. "Analyzing the Degree of Randomness in Weather Data," *Civil Engineering*, 12 (1942): 34-36.

55. HORTON, R. E. "Frequency of Recurrence of Hudson River Floods," in HENRY, A. J. *The Floods of 1913 in the Rivers of the Ohio and Mississippi Valleys*, United States Weather Bureau, Bulletin Z, 1913: 109-112.

56. HAZEN, A. "Variations in Annual Rainfall," *Engineering News*, 75 (1916): 4-5; SAVILLE, T. "Rainfall Data Interpreted by Laws of Probability," *ibid.*, 76 (1916): 1208-1211.

57. National Resources Board, *A Report on National Planning and Public Works*, Washington, 1934.

58. VARNEY, B. M. "Seasonal Precipitation in California and its Variability," *Monthly Weather Review*, 53 (1925): 148-163, 208-218.

59. CONRAD, V. "The Variability of Precipitation," *Monthly Weather Review*, 69 (1941): 5-11.

60. TOLLEY, H. R. "Frequency Curves of Climatic Phenomena," *Monthly Weather Review*, 44 (1916): 634-642.

61. FOSTER, H. A. "Theoretical Frequency Curves and their Application to Engineering Problems," *Transactions of the American Society of Civil Engineers*, 87 (1924): 142-173.

62. GUMBEL, E. J. "The Return Period of Flood Flows," *Annals of Mathematical Statistics*, 12 (1941): 163-190; *idem.* "Floods Estimated by Probability Method," *Engineering News-Record*, 134 (1945): 833-837; *idem.* "Simplified Plotting of Statistical Observations," *Transactions of the American Geophysical Union*, 26 (1945): 69-82.

63. COURT, A. "Temperature Extremes in the United States," *Geographical Review*, 43 (1953): 39-49; *idem.* "Wind Extremes as Design Factors," *Journal of the Franklin Institute*, 256 (1953): 39-56.

64. BARGER, G. L., and THOM, H. C. S. "Evaluation of Drought Hazard," *Agronomy Journal*, 41 (1949): 519-526.

65. JACOBS, W. C. *Wartime Developments in Applied Climatology*, Meteorological Monographs 1, 1947.

66. LAMOUREUX, C. E. "Duration Frequencies of Low Ceilings in Climatology," *Transactions of the American Geophysical Union*, 27 (1946): 27-33.

67. WAKELEY, J. T., and RIGNEY, J. A. "The Use of Frequency Distributions of Weather Factors in Agronomic Practices," *Journal of the American Society of Agronomists*, 39 (1947): 1088-1093.

68. United States Weather Bureau, *The Climatic Handbook for Washington, D. C.*, Technical Paper 8, 1949.

69. SMITH, J. W. "Relation of Precipitation to the Yield of Corn," in United States Department of Agriculture, *Yearbook for 1903*, Washington, 1904: 215-224.

70. SHAW, W. N. "On a Relation Between Autumnal Rainfall and the Yield of Wheat of the Following Year—Preliminary Note," *Proceedings of the Royal Society of London*, 74 (1905): 552-553.

71. HOOKER, R. H. "Correlation of the Weather and Crops," *Journal of the Royal Statistical Society*, 70 (1907): 1-42.

72. SMITH, J. W. "Correlation," *Monthly Weather Review*, 39 (1911): 792-795.

73. ———. *Agricultural Meteorology*, New York, 1920.

74. WALLACE, H. A. "Mathematical Inquiry into the Effect of Weather on Corn Yield in the Eight Corn Belt States," *Monthly Weather Review*, 48 (1920): 439-446.

75. MATTICE, W. A. "The Future of Agricultural Meteorology," *Monthly Weather Review*, 59 (1931): 274-275.

76. FISHER, R. A. "Studies in Crop Variation: an Examination of the Yield of Dressed Grain from Broadbalk," *Journal of Agricultural Science*, 11 (1921): 107-135; *idem.* "The Influence of Rainfall on the Yield of Wheat at Rothamsted," *Philosophical Transactions of the Royal Society of London*, 213 (1924): 89-142.

77. SCHUMACHER, F. X., and MEYER, H. A. "Effect of Climate on Timber-Growth Fluctuations," *Journal of Agricultural Research*, 54 (1936): 79-107; LISTER, P. B., and SCHUMACHER, F. X., "The Influence of Rainfall upon Tuft Area and Height Growth of three Semidesert Range Grasses in Southern Arizona," *ibid.*, 54 (1937): 109-121.

78. DAVIS, F. E., and PALLESEN, J. E. "Effect of the Amount and Distribution of Rainfall and Evaporation during the Growing Season on Yields of Corn and Spring Wheat," *Journal of Agricultural Research*, 60 (1940): 1-24; PALLESEN, J. E., and LAUDE, H. H. *Seasonal Distribution of Rainfall in Relation to Yield of Winter Wheat*, United States Department of Agriculture, Technical Bulletin 761, 1941; DAVIS, F. E., and HARRELL, G. D. *Relation of Weather and its Distribution to Corn Yields, ibid.*, 806, 1942.

79. HENDRICKS, W. A., and SCHOLL, J. C. *The Joint Effects of Temperature and Precipitation on Corn Yields*, North Carolina State College Agricultural and Engineering Experiment Station, Technical Bulletin 74, 1943.

80. LANDSBERG, H. E., and JACOBS, W. C. "Applied Climatology," in American Meteorological Society, *Compendium of Meteorology*, 1951: 976-992.

81. LANDSBERG, H. E. "Use of Climatic Data in Heating and Cooling Design," *Heating, Piping, and Air Conditioning*, 19 (1947): 121-125.

82. BAUM, W. A., and COURT, A. "Research Status and Needs in Microclimatology," *Transactions of the American Geophysical Union*, 30 (1949): 488-493.

GEOMORPHOLOGY[A]

THE DEVELOPMENT OF GEOMORPHOLOGY IN AMERICA
19TH CENTURY ANTECEDENTS
GEOMORPHOLOGY BEFORE 1930
GEOMORPHOLOGY SINCE 1930

CURRENT APPROACHES TO GEOMORPHOLOGY
DESCRIPTIVE GEOMORPHOLOGY
DYNAMIC GEOMORPHOLOGY
Fluvial Processes
Solution Processes
Marine Processes
Glacial Processes
Periglacial Processes
Eolian Processes
Volcanic Processes
APPLIED GEOMORPHOLOGY

THE PROSPECT

A. Prepared by Louis C. Peltier with the assistance of the following scholars who have reviewed the manuscript at various stages: Homer P. Little, Kirtley F. Mather, Hugh Raup, Richard J. Russell, C. F. Stewart Sharpe, R. S. Ullery, and Derwent Whittlesey. Louis O. Quam, in addition to reviewing the whole chapter, provided special assistance in the sections on marine and volcanic processes. The work was sponsored by the Committee on Physical and Bio-Geography, Hoyt Lemons, Chairman.

CHAPTER FIFTEEN

GEOMORPHOLOGY

GEOMORPHOLOGY IS A FIELD of study in which the form of the earth's surface is examined. Studies of the configuration of the earth deal with the continents and ocean basins, the great mountain systems, the broad plains, and with the shapes of hills and valleys. Geomorphology is concerned with the classification, measurement, and description of land-forms, and with the history of the processes that have produced them.

Fifty years ago in the United States most of the geographers were specialists in geomorphology. The outstanding geomorphologists of the first part of the present century in the United States were the teachers of those who later specialized in many other branches of geography. The geographic profession was established in the United States largely through the efforts of scholars such as William Morris Davis, Rollin D. Salisbury, and Wallace W. Atwood, all known first for their teaching and their research studies in geomorphology. There was a time when so many of their students sought to explore the frontiers of geographic thought in other parts of the field that few specialists in geomorphology were left, and this traditional field of interest was kept alive largely through the efforts of Douglas W. Johnson and Kirk Bryan. Now, in recent years, a renewed vitality has become apparent and geomorphology is thriving.

Geomorphology is a borderline field between geography and geology. Its most successful students are those adequately trained in the concepts and methods of both geography and geology. Although geographers are chiefly concerned with the surface characteristics of particular places and geologists are chiefly concerned with the understanding of processes, the two fields are not, and should not be, sharply distinguished along the border area. Geomorphologists will make use of both approaches.

THE DEVELOPMENT OF GEOMORPHOLOGY IN AMERICA

These two approaches were discernible as geomorphology emerged from the 19th century period of naturalists and moved forward into the 20th century period of scientific specialization. One approach was directed toward the analysis and description of landforms; the other examined landforms as indicators of process and therefore as a key to the interpretation of the geologic past.

19TH CENTURY ANTECEDENTS

Geomorphology in America is an outgrowth of the age of naturalists. This age is represented by such philosophically-minded observers as Louis Agassiz [1], Nathaniel Southgate Shaler [2], and Arnold H. Guyot [3], and by such scientific explorers as John Wesley Powell [4], C. E. Dutton [5], and G. K. Gilbert [6]. To be sure these great figures of the 19th century had been preceded by Jedidiah Morse. Morse wrote an encyclopedic kind of descriptive geography which he regarded as "a branch of mixed mathematics [treating] of the nature, figure, and magnitude of the earth; the situation, extent, and appearance of different parts of its surface; its products and inhabitants" [7: 9]. Guyot, on the other hand, insisted that geography should "not only describe, it should compare, it should interpret, it should rise to the how and wherefore of the phenomena which it describes" [3: 21]. Guyot provided a channel whereby the Kantian philosophy of geography was brought to the United States.

Geology in the United States in the 19th century was also strongly influenced by the writings of Charles Lyell [8]. He argued, from the Huttonian doctrine of uniformitarianism, that "the geologist who assents to the truth of these principles will deem it incumbent on him to examine with minute attention all the changes now in progress on the earth, and will regard every fact collected respecting the causes in diurnal action, as affording him a key to the interpretation of some mystery in the archives of remote ages" [8: 166]. Among the geologists of the United States who, following this theme, examined geomorphic phenomena for the light they threw on geologic processes of the past were James D. Dana, Joseph LeConte, and T. C. Chamberlain [9].

GEOMORPHOLOGY BEFORE 1930

Geographical geomorphology in the early part of the 20th century was developed along the lines suggested by Guyot. Geological geomorphology at the same time was developed on principles set down by Lyell. These two points of view are discernible throughout the history of geomorphology in the present century. The kind of geomorphology associated with geography was developed by two outstanding scholars, William Morris Davis and Rollin D. Salisbury. Davis, who taught at Harvard, was a prolific writer and an indefatigable field observer; Salisbury, who taught at Chicago, was an extraordinarily gifted teacher who published very little but whose influence was spread by his many students, notably Wallace W. Atwood.

The thought pattern that William Morris Davis expanded into his philosophy of geomorphology, and which Rollin D. Salisbury, Wallace W. Atwood, Douglas W. Johnson, and others so effectively taught to their students, was an "explanatory description" of landforms. Yet to many of his contemporaries, Davis seemed to preach geography but to do geology. This was in part a result of the great popularity of his physiographic essays, and the lack of appreciation of his other writings [10; 11; 12; 13]. It was perhaps also in part a result of a failure to understand his objectives. He strove to devise a system of descriptive geomorphology that would have intrinsic meaning. The descriptions would, thus, carry implications that could be used in deriving general principles concerning the relations between landforms and the activities of man. Davis' explanatory description of landforms was essentially a natural history of landforms; he hoped to extend his system to include a natural history of man on the earth (chapter three, page 81). Even though he called for an extension and refinement of his descriptive system, however, his writings do not indicate that he considered such descriptions to be the ultimate objectives. Davis' disciples have tended to follow his example more closely in the explanatory description of landforms than in his pleas for geographical integration with other scientific fields.

Davis' cyclical interpretation of landforms and his recognition of the complex history of uplift and erosion of mountains captured the imagination of geologists and geographers alike. It made possible the postulation of geomorphic history from an otherwise meaningless maze of hills and valleys. Many people yielded to the temptation to make such postulates; too often there were those who did not tarry to prove them. Nevertheless, Davis' inspiration led to sound accomplishment. There were the descriptive treatments of the physiography of the United States by Isaiah Bowman [14] and by Nevin M. Fenneman [15], the systematic studies of shorelines by Douglas W. Johnson [16], and the systematic study of soils by Curtis F. Marbut (chapter sixteen) [17].

So powerful was the personality of Davis and so forceful was his defense of the concept he had formulated that few indeed were the geomorphologists, at home or in Europe, who did not adopt his ideas and make application of them in the field. Yet some of his followers were inept in that they applied his concepts too literally, too inflexibly, and in that they failed to carry out the careful and laborious field studies that would have corrected these faults. Furthermore, Davis' system failed to provide for a balanced analysis of all the landforms of an area, for it concentrated attention on erosion surfaces and this led to a corresponding neglect of slopes. In contrast, the analyses of Walther Penck [18] focused attention on slope-development, and offered what appeared to be a challenge to the whole Davis system. For some three decades, Penck's concepts have received scant attention in the United States.

Perhaps inevitably, Davis' retirement led to a reaction. Following the initial surge of activity and enthusiasm in geomorphology between 1905 and 1920, there was a doldrum period of about ten years during which the geographers turned to human geography and the geologists went in search of oil. Many of Davis' concepts became obscured by controversy and unimaginative application. The Davis school was kept alive chiefly by the efforts of Douglas W. Johnson and Kirk Bryan.

GEOMORPHOLOGY SINCE 1930

A strong revival of interest in geomorphology began about 1930. This has been marked by a sharp increase in the number of papers on geomorphic subjects, which reached about one hundred a year, and by an increase in the number of graduate students interested in geomorphology in the universities. There is no evidence that the peak of this rise has yet been reached.

Kirk Bryan, at Harvard, was one of the leaders of this new tide of interest in geomorphology. He effected a combination of geographic and geologic geomorphology in which he borrowed much from Albrecht Penck and his followers. Bryan argued that one could not understand a surface configuration except through an understanding of the surface materials; he also argued that because nearly all existing surfaces came into being during the Pleistocene, explanatory description should not be divorced from Pleistocene stratigraphy. The major achievements of Bryan and his students were in climatic geomorphology, particularly in arid and periglacial regions, and in the application of geomorphology to civil engineering, soils, and archaeology [19; 20; 21; 22; 23; 92].

By 1950, American geomorphology had achieved a reasonably thorough descriptive treatment of the United States, and had formulated many outstanding problems. Several basic concepts had been stated or refined. Among these were the concepts of base level, erosion cycle, available relief, regimen of streams, climatic fluctuations, and polygenetic surfaces. A few American scholars had also examined the ideas of Walther Penck and other European geomorphologists and had subjected to critical evaluation such concepts as the nick point, the retreat of slopes, and the "old at birth" erosion surface (*Primärrumpf*).

CURRENT APPROACHES TO GEOMORPHOLOGY

A new era, dominated by younger men who for the most part are products of the post-1930 period of revival, may be supposed to have started about 1940. The situation reflects the fact that very few persons entered geomorphology between 1920 and 1930. The present active geomorphologists have neither the guidance nor the hindrance of their teachers. With such intellectual freedom in a virile field, one may anticipate the development of several different and divergent approaches. We may expect the entire field of geomorphology and its marginal areas that overlap with other fields to be covered more effectively than would have been done by a more complete adherence to any one school of thought.

Even if some of the broader concepts of geomorphology have been formulated, many concepts remain to be brought forth and tested. Not a single topic has been exhausted. As factual evidence accumulates, new concepts will emerge to provide a frame of reference for detailed studies of small areas or of specific individual topics and to justify the re-examination of many classic areas and problems. The quantitative analysis of landforms for the purpose of determining the factors that govern the dimensions and shape of terrain is one of the lines of detailed investigation by which present genetic theories are being tested and empirical generalizations are being clarified [24].

DESCRIPTIVE GEOMORPHOLOGY

In the field of descriptive geomorphology two current tendencies may be observed. One is the search for a precise descriptive system that can be applied before a geomorphic analysis is completed, and can be used in the analysis of areal relationships between landforms and the occupance of land by man. The other is the search for a broader frame of reference into which a system of explanatory description, based on genetic principles, can be fitted.

The degree of generalization used by Bowman and Fenneman in the works cited previously does not permit the examination of areal relations with the phenomena of human occupance. The reasons for this have been presented in chapters one and two of this book. The categories of landforms useful for chorographic-scale studies leading to the recognition of physiographic provinces are too highly generalized to match with the specific features developed by man: such as fields and farms, roads and settlements, or cities and factories. To be sure, the examples of contour-map interpretation contained in the stimulating volume by Rollin D. Salisbury and Wallace W. Atwood published in 1908 were on a topographic scale [25]. Yet the categories determined on genetic principles overlooked too many slight variations of slope to be applicable to the topographic or detailed, large-scale study of areal relations. The empirical approach to the classification of surface features of the kind suggested by Siegfried Passarge [26], which was vigorously attacked by Davis, also failed to provide a treatment of landforms that could be related in any meaningful way to human occupance. Before 1920 in the United States, however, the field geographers who were experimenting with methods of studying agricultural areas (chapter ten) devised a classification of surface features based on slope or "lay of the land." Categories of this kind were used in the land classification work of the Michigan Land Economic Survey (chapter twenty-four), the Soil Conservation Service, and in refined form by many other geographical surveys, such as those of the Tennessee Valley Authority and the Puerto Rico Rural Land Classification Program. Empirical classifications were also proposed by C. K. Wentworth [27], Guy-Harold Smith [28], and others.

Any consideration of land surfaces that goes beyond empirical description will show that the differences between surfaces are due to differences in the agents or processes that produced them and to the characteristics of the underlying material. A description that gives attention to landforms on a topographic scale (as defined in chapter one) and that is based on categories defined in terms of origin seems most likely to meet the requirements not only of geographers but also of the many others who have found applied geomorphology useful. Notable accomplishments in this direction have already been made in glacial geomorphology with recognition of such glacial landforms as drumlins, kames, moraines, cirques, and other mappable phenomena.

Nevertheless, there remains the need for a broad frame of reference to which such topographic-scale features can be related. In the newly developing field of climatic geomorphology such a frame of reference may be emerging. Here surface features are related, through the formative processes, to the climatic regime under which they have been produced. Soils and biotic phenomena are related to climate. Geomorphology, therefore, can be tied to these other aspects of physical geography and biogeography through the common denominator of climate. The recognition of climatically defined morphogenetic regions does not imply discard or replacement of either the physiographic provinces or the structural regions, for a different rank of regions with a different degree of generalization is treated. It does permit more detailed comparative studies following a rational pattern, and it leads to a clearer recognition of the evidence of a complex climatic history. It provides for a system of analysis of slope categories, of erosional processes, and of soil processes. The system of climatic geomorphology may eventually be applied to topographic detail. A complementary classification which seems likely to become quite useful has been based on slope (or the attitude of the surface), the nature of the underlying material, and the agent that produced the surface [29]. (Compare the similar viewpoint presented in chapter sixteen, and the comments on the future of regional climatology in chapter fourteen.)

Dynamic Geomorphology

Dynamic geomorphology is that part of the field which is concerned with the study of processes. Seven different kinds of processes have been studied: fluvial, solution, marine, glacial, periglacial, eolian, and volcanic.

Of these the greater attention has been given to the study of fluvial and glacial processes.

Fluvial Processes

Fluvial geomorphology has been developed in mountainous areas and in the alluvial valleys of both humid and semi-arid climates. Classic studies in America are G. K. Gilbert's monograph on the Henry Mountains [6] and some of W. M. Davis' early papers on drainage in Pennsylvania and New Jersey [10]. Among the many other works on the fluvial geomorphology of mountainous areas may be mentioned those of Eliot Blackwelder [30], W. W. Atwood and Kirtley F. Mather [31], and W. H. Bradley [32]. Alluvial valleys have been investigated notably by A. C. Lawson [33], Kirk Bryan [19], Richard J. Russell [34], and W. C. Putnam [35]. These and other studies have served to clarify and extend the Davisian concepts of the erosion cycle and the peneplain. Corollary researches have been directed toward an understanding of the development of drainage patterns [10; 13; 36] and of desert pediments [19; 33]. As a result of this work some areas have become classic: the Lower Mississippi Valley, the Colorado Plateau, the Rocky Mountains of Colorado, and the Northern Appalachians. Other promising areas, however, have been largely ignored.

The study of large river floodplains is an aspect of fluvial geomorphology that has aroused increasing interest [20; 34; 35; 37; 38]. The study of alluvial processes is bringing to light new concepts regarding the character of river floodplains and the processes that shape them.

Increased attention has been paid recently to the development of slopes, a field of inquiry that promises much for the future. This field of interest was actively developed by the United States Soil Conservation Service under H. H. Bennett [39], and others have independently focused attention on slope morphology [24; 40; 41].

The major problems faced by students of fluvial geomorphology are concerned with the relationship between fluvial features and other environmental conditions, as, for example, the relation between discharge, stream load, gradient, and channel characteristics. Another concerns the effect of changes of climate, vegetation cover, or sea level on stream-flow. The most pressing need, however, is for a classification of rivers that is related to their characteristics of flow, erosion, and sedimentation.

Solution Processes

Studies of solution geomorphology have been focused on the examination of limestone caverns, following either the approach outlined by W. M. Davis in his monograph on the origin of such caverns [42] or that developed by A. C. Swinnerton [43]. The description of the detailed characteristics of caverns has received much attention, but relatively little has been done toward the classification and interpretation of other solution features or toward the correlation of subterranean features with subaerial phenomena and processes. The few papers on karst surfaces include the review of the work of Jovan Cvijic by E. M. Sanders [44] and investigations of solution landforms in Kentucky by S. N. Dicken [45].

Marine Processes

Marine geomorphology, including the study of shorelines, has developed slowly owing to the meager financial support for research vessels to collect necessary data. G. K. Gilbert's studies of the Lake Bonneville shores [6] and work by N. S. Shaler and F. P. Gulliver, as well as earlier work by Europeans, especially that of Eduard Suess and Elie de Beaumont, formed the conceptual basis for shoreline research by W. M. Davis and Douglas W. Johnson. Johnson's conclusions regarding shore processes and shoreline development, which strongly reflect the views of de Beaumont, Shaler, Gulliver, and Davis, were long thought to be definitive by most American geomorphologists [16].

The principal challenge to Johnson's conclusions regarding emergent and submergent shorelines has come from students of the presumed emergent shorelines of the Atlantic and Gulf Coastal Plain. Quite different interpretations are included in works by N. M. Fenneman [15], Francis P. Shepard [46], Richard J. Russell [47], and others. As modern echo sounding came into general use after World War I, rapid advances were made in the study of the sea bottom. The United States Coast and Geodetic Survey deserves much credit and the gratitude of geomorphologists for the application of precise methods to offshore surveying. The publication of contour maps of the continental shelf and slopes made available for the first time an adequate picture of the true configuration of the submerged borders of the continents. These charts directed attention to the submarine canyons which, although known since James D. Dana's day, had attracted little notice until Francis P. Shepard emphasized the implication of their depth and world-wide occurrence on existing theories of sea-level fluctuations. Many explanations have been advanced for the origin of these canyons. They include hypotheses of subaerial stream erosion during a period of greatly lowered sea level, submarine erosion by turbidity currents, and sapping by submarine artesian springs along the continental slope [48].

Associated with the problems of the development of shorelines as well as of the origin of submarine canyons is the question of eustatic fluctuations of sea level during the glacial epochs. This, too, is an old problem; it was announced as early as 1842, and has been considered from many points of view by many investigators since that time. The most extensive modern study was started about 1910 by R. A. Daly and culminated in his book on the ocean floor, published in 1942 [49]. Local correlations of marine and river terraces with glacial stages have engaged the attention of several scholars, notably Ernst Antevs [50], C. W. Cooke [51], R. F. Flint [54], and R. J. Russell [47].

Another aspect of marine geomorphology is the study of organic reefs. These reefs, built by corals and algae around the shores of islands under appropriate conditions of water temperature, depth, turbidity, circulation, and salinity and in the presence of appropriate species, have the effect of protecting the islands behind them from normal wave erosion. Clipperton Island remains a classic example of a barrier reef; other types of reef are also recognized, such as the atoll, represented by Kwajalein and Wake Islands, the fringing reef, represented in the Hawaiian and Philippine Islands, and the raised reef, represented on Ocean, Nauru, and Samar Islands. The history of these reefs is intimately related to changes in the temperature and relative level of the sea. The rate of reef growth is, among other conditions, closely dependent on the presence of breakers and turbulent water. The relatively simple hypotheses formulated to account for different kinds of reefs by Charles Darwin, J. D. Dana, and William Morris Davis [52] are now being found to require some modification. It now appears that, within the broad range of conditions under which reef-forming organisms can grow, there may exist a variety of detailed, but critical, differences. Reefs may form under conditions of rising or lowering sea level and they may be influenced in their development by marine erosion, mass movement, solution, and wind action. The distribution of reefs and many of the differences observed from island to island are related to the distribution of reef-forming species and to the Pleistocene climatic history of different parts of the oceans. Considerable work leading toward the classification, interpretation, and prediction of the occurrence of organic reefs remains to be done.

The study of shorelines and other aspects of marine geomorphology, long handicapped by the lack of funds for research vessels and technical equipment, is now being rapidly advanced (chapter eighteen). Oceanographic institutions, oil companies, and government agencies have recently embarked on intensive study of all phases of marine science. Especially significant are the oceanographic and geographic research programs of the Office of Naval Research. Although the geomorphology of shorelines and coasts represents a minor part of the total research effort, promising progress is being made in the understanding of the characteristics of shores. If the current level of this research can be maintained, the next decade should see the replacement of many present theories with much more precise concepts based on a new wealth of factual data.

Glacial Processes

Glacial geomorphology is one of the two aspects of dynamic geomorphology in which the most progress has been made. During the past sixty years there have been many studies of the processes and the resulting phenomena of both continental and mountain glaciation. No small part of this stemmed from T. C. Chamberlain [9] and his colleague, R. D. Salisbury, as well as from the studies by William H. Hobbs [53]. A classification of glacial features was developed, based on landforms and stratigraphy, and the formidable task of applying this classification to North America was undertaken [54]. The Upper Mississippi Valley and the Great Lakes area of the United States has become a classic area for the study of glacial phenomena; among the many scholars who have worked in this part of the country, Frank Leverett [55], W. C. Alden, M. M. Leighton [56], G. F. Kay, Earl T. Apfel [57], Paul MacClintock [58], and R. F. Flint [54] have made outstanding contributions.

In the immediate future studies of continental glaciation may well become involved with two major concepts regarding glacial processes: concepts of glacial recession by stagnation, and of separate and distinct glacial stages and sub-stages [59]. One line of investigation may seek to determine the influence on glacial advance and recession of climatic fluctuations and of the nature of the sub-glacial surface. Another line may seek to refine the climatic interpretation of the Pleistocene and may result in a division of the pre-Wisconsin glacial stages into sub-stages. This could lead to a further recognition of buried loess soil and to their interpretation.

Many studies have been made of mountain glaciers. More than a century ago Louis Agassiz presented the basic point of view: "the features of this place look like those observed elsewhere in situations where it is certain they were produced by glaciers; therefore I believe that a glacier was once here, even though none now exists here." Davis' cyclical interpretation of fluvial forms was extended to the postulation of a cycle of erosion in glaciated mountains, in the support of which examples of different stages were reported from many parts of the world. Early in the present century W. D. Johnson (not to be confused with D. W. Johnson) outlined the presumed characteristics of maturity in glaciated mountains [60]. Classic studies of glaciated mountains are those of R. S. Tarr [61], François E. Matthes [62], and W. W. Atwood [63]. Recent work on mountain glaciers and their deposits has dealt with problems of glacier regimen [64; 65], climatology, and glacier mechanics [66; 67]. These studies are closely related to the interpretation of the Little Ice Age of 1650-1850, described by Matthes [68]. The general retreat of American glaciers since 1900 provides an inviting group of problems involving meteorological and hydrological observations in the vicinity of the glaciers, and measurements of the changes in shape and extent of the ice. Meanwhile the problem of correlating the Little Ice Age and earlier sub-stages with features of the land in areas not covered by ice remains to be solved [22].

Periglacial Processes

·Studies of the periglacial geomorphology of the Pleistocene in the United States have only just been started. The whole concept of a periglacial regime has been formulated so recently that its implications have scarcely been touched upon. The more distinctive forms of frost-disturbed soils are now being reported from many places [67; 69; 70]. The less obvious forms will become known in the course of more careful investigations. The frost-produced rubbles, by their association with stream terraces of unglaciated valleys, demonstrate the fact of periglacial alluviation [71]. A similar association with dune sand, ventifact pavement, and loess mantle is leading to the recognition of wind action as an important element of periglacial conditions [53; 70]. Questions which call for immediate attention have to do with the distribution of frost-produced features and periglacial deposits in the United States, and with the variations in size and form of these features from place to place.

The present-day processes and phenomena of periglacial areas have been studied for many years by the Alaska Branch of the United States Geological Survey [72; 73; 74]. This work is being continued by the Alaska Terrain and Permafrost Section of the Geological Survey, by the United States Bureau of Plant Industry, Soils and Agricultural Engineering, and by many individual workers. The *Journal of Geology* for 1949 contains one whole number devoted to periglacial topics. This field is still in the early stages of development, where new observations and new concepts are needed and where the alert student can scarcely avoid making new discoveries.

Eolian Processes

Another undeveloped branch of American geomorphology pertains to eolian processes. Most of the observations on dunes have dealt with sedimentation. Some studies, however, have been made in dune classification [75; 76; 77], in the relationships of dune forms to the supply of sand and to wind velocity, and concerning the nature of the plants associated with the dunes [78]. In Texas, buried soils and multiple generations of sand dunes have been recognized [79].

Studies of loess have advanced much more rapidly. Loesses of the several stages of the Pleistocene have been identified in the United States. Guy D. Smith has shown that the loesses of Illinois are derived chiefly from large alluvial valleys. Thus the regime that produced alluviation also led to loess development. Recent studies by various writers have shown that thin bodies of loess exist throughout northeastern United States, in many cases containing pebbles intermixed by frost-heaving.

The available information concerning eolian deposits in the United States has been compiled on a single map by a committee of the Division of Geology and Geography of the National Research Council, under the leadership of James Thorp and H. T. U. Smith. The need for further field studies in critical areas that have not yet been closely examined is quite apparent from this map.

Volcanic Processes

Volcanism is of interest to volcanologists, petrologists, and geophysicists as well as to geomorphologists. Workers in these other fields have contributed most of the existing knowledge of volcanic landforms. These contributions generally relate to: 1) the mechanism of formation and the lithological characteristics of volcanic rocks; 2) the constructional landforms produced by volcanic eruptions, such as domes, craters, calderas, cones, lava flows, and others; and 3) the destructional landforms resulting from the erosion of volcanic rocks and related deposits.

The 19th century interest in volcanoes is reflected in the works by C. E. Dutton [80], G. F. Becker [81], James D. Dana [82], and especially in I. C. Russell's monograph on the volcanoes of North America [83]. More recent interest in these phenomena and the processes involved is indicated by the writings of G. W. Tyrrell [84], H. T. Stearns [85], W. C. Putnam [86], Howel Williams [87], and others. Their works contain numerous descriptions of constructional volcanic landforms and contribute to the understanding processes and structures that produce volcanic landscapes.

A classical treatment of volcanic geomorphology is exemplified by the work of W. M. Davis who regarded volcanic activity as an "accident" that interfered with the normal cycle of erosion. He and his followers have described rivers, diverted and impounded by volcanic accumulations, which produced new local base levels and in some places resulted in the inversion of the landforms. Davis also dealt with the cycle of erosion on volcanic forms. Further studies of these destructional landscapes have been made by J. S. Diller [88], C. K. Wentworth [89], and N. E. A. Hinds [90]. However, for a thorough, general treatment of volcanic landforms following the precepts of Davis one must consult the work of a New Zealand geomorphologist, C. A. Cotton [91].

APPLIED GEOMORPHOLOGY

Studies in both descriptive and dynamic geomorphology can be undertaken solely for the purpose of bettering our understanding of surface forms and structures; or they may be undertaken because of their application to problems in other fields, as, for example, when they are pursued with a view to unraveling the tangled record of the geologic past. In other parts of this book, various applications of geomorphology to the study of other phenomena of the earth's surface are discussed, for there is scarcely any aspect of geography in which the physical features of the earth are not important. Of special interest, however, are the applications of geomorphology to anthropology and civil engineering.

The application of geomorphology to anthropological problems has taken place largely since 1940. Scholars working on problems of prehistoric anthropology and archaeology had long sought the aid of the geomorphologists in the reconstruction of past environments or for determinations of age, but only during the last two decades or so have the geomorphologists been able to give sufficiently precise answers. Considerable assistance has been given since 1939 in such investigations as those of the Lindenmeier Site [92], the Sandia Cave [93], the Hopi Sites [94], the Boylston Street Fish Weir [95], and of certain Indian sites of Central Mexico [96]. Geomorphologists have also aided in the archaeological surveys of the Lower Mississippi Valley [97].

Civil engineering, by virtue of its preoccupation with design and construction, usually cannot wait on theory but has been forced to push ahead along empirical lines [41]. The engineers have made their own applications of geomorphology, and in doing so have made important contributions to the understanding of various geomorphic processes. Such new understandings have been derived in part from the operation of dynamic models, such as those of the United States Army Engineers Waterways Experiment Station at Vicksburg, Mississippi.

THE PROSPECT

Geomorphology, which overlaps the fields of geology and geography, has perhaps suffered from a relative neglect of the geographic approach. The geographer needs precise, factual information about particular places. What landforms actually exist in a given area? How do they differ? Where are they? What are their distribution patterns? The geomorphologist may concern himself with questions of structure, process, and stage, but the geographer wants specific answers to the questions: what? where? and how much?

These ideas were discussed by Richard J. Russell in his presidential address to the Association of American Geographers in 1948 [98]. He and other writers have argued the need for the development of those aspects of geomorphology that are of particular interest to geographers. Such a geomorphology must be precise in terms both of measurement and location, and it must be free from those genetic speculations that are extraneous to geographical purposes. The geographer needs categories of landforms, defined at a proper degree of generalization (chapter one), and mapped with an attention to detail appropriate to the map scale. He then wants to seek areal relationships with other phenomena by the methods of cartographic analysis.

Geomorphology has entered a new and vigorous period of growth. Its place as a fundamental aspect of the geographic sciences is clearly recognized. It offers to the student approaching it either from the geologic side or the geographic side a variety of interesting problems [99]. These problems may involve the search for general concepts, or they may be concerned with practical applications. Geomorphology may be studied because of the light it throws on geologic processes; or it may be studied for the areal relationships it reveals with other phenomena of the earth's surface.

REFERENCES

1. AGASSIZ, L. "On the Former Existence of Local Glaciers in the White Mountains," *Proceedings of the American Association for the Advancement of Science*, 19 (1870): 161-167.
2. SHALER, N. S. "The Dismal Swamp District," *10th Annual Report*, United States Geological Survey, Washington, 1890: 261-339.
3. GUYOT, A. H. *The Earth and Man: Lectures on Comparative Physical Geography, in its Relation to the History of Mankind*, Boston, 1849; *idem.* "On the Physical Structure and Hypsometry of the Catskill Mountain Region," *American Journal of Science*, 19 (1880): 429-451.
4. DARRAH, W. C. *Powell of the Colorado*, Princeton, 1951.
5. DUTTON, C. E. "Mount Taylor and the Zuni Plateau," *6th Annual Report*, United States Geological Survey, Washington, 1885: 105-198.
6. GILBERT, G. K. *Report on the Geology of the Henry Mountains, Utah*, United States Geographical and Geological Survey of the Rocky Mountain Region, Washington, 1877; *idem. Lake Bonneville*, United States Geological Survey, Monograph 1, Washington, 1890.
7. MORSE, J. *The American Geography*, . . . , Boston, 1789; *idem. The American Universal Geography*, . . . , Boston, 1793.
8. LYELL, C. *Principles of Geology: Being an Inquiry How Far the Former Changes of the Earth's Surface are Referable to Causes Now in Operation*, London, 1830-1843.
9. CHAMBERLAIN, T. C. "Preliminary Paper on the Terminal Moraine of the Second Glacial Epoch," *3rd Annual Report*, United States Geological Survey, Washington, 1883; *idem.* "A Proposed System of Chronologic Cartography on a Physiographic Basis," *Bulletin of the Geological Society of America*, 2 (1891): 541-544; *idem.* "Proposed Genetic Classification of Pleistocene Glacial Formations," *Journal of Geology*, 2 (1894): 517-538.
10. DALY, R. A. "Biographical Memoir of William Morris Davis, 1850-1934," National Academy of Sciences, *Biographical Memoirs*, 23 (1945): 263-303.
11. DAVIS, W. M. *Geographical Essays*, Boston, 1909 (a collection of 26 papers).
12. ———. "The Colorado Front Range, A Study in Physiographic Presentation," *Annals of the Association of American Geographers*, 1 (1911): 21-84.
13. ———. "The Principles of Geographic Description," *Annals of the Association of American Geographers*, 5 (1915): 61-105.

14. BOWMAN, I. *Forest Physiography: Physiography of the United States and Principles of Soils in Relation to Forestry*, New York, 1911.

15. FENNEMAN, N. M. *Physiography of Western United States*, New York, 1931; *idem. Physiography of Eastern United States*, New York, 1938.

16. JOHNSON, D. W. *Shore Processes and Shoreline Development*, New York, 1919; *idem. New England-Acadian Shoreline*, New York, 1925; *idem. Stream Sculpture on the Atlantic Slope*, New York, 1931.

17. MARBUT, C. F., and others. *Soils of the United States*, United States Department of Agriculture, Bureau of Soils, Bulletin 96, Washington, 1913.

18. PENCK, W. *Die Morphologische Analyse*, Stuttgart, 1924; *idem. Morphological Analysis of Land Forms: A Contribution to Physical Geology*, (translated by H. CZECH, and K. C. BOSWELL), New York, 1953.

19. BRYAN, K. "Erosion and Sedimentation in the Papago Country, Arizona, with a Sketch of the Geology," United States Geological Survey, *Bulletin* 730 (1922): 19-90.

20. ———. *Geology and Ground Water Resources of the Sacramento Valley, California*, United States Geological Survey, Water Supply Paper 495, Washington, 1923.

21. ———. *Geology of Reservoirs and Dam Sites*, United States Geological Survey, Water Supply Paper 597, Washington, 1929.

22. ———. "Cryopedology: The Study of Frozen Ground and Intensive Frost-Action with Suggestions on Nomenclature," *American Journal of Science*, 244 (1946): 622-642.

23. BRYAN, K., and ALBRITTON, C. C. "Soil Phenomena as Evidence of Climatic Changes," *American Journal of Science*, 241 (1943): 469-490.

24. STRAHLER, A. N. "Equilibrium Theory of Erosional Slopes Approached by Frequency Distribution Analysis," *American Journal of Science*, 248 (1950): 673-696, 800-814.

25. SALISBURY, R. D., and ATWOOD, W. W. *Interpretation of Topographic Maps*, United States Geological Survey, Professional Paper 60, Washington, 1908.

26. PASSARGE, S. *Grundlagen der Landschaftskunde*, Hamburg, 1919-1920.

27. WENTWORTH, C. K. "A Simplified Method of Determining the Average Slope of Land Surfaces," *American Journal of Science*, 20 (1930): 184-194.

28. SMITH, G.-H. "The Relative Relief of Ohio," *Geographical Review*, 25 (1935): 272-284.

29. VEATCH, J. O. *Soils and Land of Michigan*, East Lansing (Michigan), 1953.

30. BLACKWELDER, E. "Post-Cretaceous History of the Mountains of Central Western Wyoming," *Journal of Geology*, 23 (1915): 97-117, 193-217, 307-340.

31. ATWOOD, W. W., and MATHER, K. F. *Physiography and Quaternary Geology of the San Juan Mountains, Colorado*, United States Geological Survey, Professional Paper 166, Washington, 1932.

32. BRADLEY, W. H. *Geomorphology of the North Flank of the Uinta Mountains, Utah*, United States Geological Survey, Professional Paper 185, Washington, 1936.

33. LAWSON, A. C. "The Epigene Profiles of the Desert," *University of California Publications in Geology*, 9 (1915): 23-48.

34. RUSSELL, R. J. "Physiography of the Lower Mississippi River Delta," in *Reports on the Geology of Plaquemines and St. Bernard Parishes*, Louisiana Department of Conservation, Geology Bulletin 8 (1936): 3-193.

35. PUTNAM, W. C. "Geomorphology of the Ventura Region, California," *Bulletin of the Geological Society of America*, 53 (1942): 691-754.

36. CAMPBELL, M. R. "Drainage Modifications and Their Interpretation," *Journal of Geology*, 4 (1896): 567-581, 657-678.

37. FISK, H. N. *Geological Investigation of the Alluvial Valley of the Lower Mississippi River*, Mississippi River Commission, Vicksburg, 1947.

38. Brown, C. B. *The Control of Reservoir Silting*, United States Department of Agriculture, Miscellaneous Publication 521, Washington, 1944.
39. Bennett, H. H. "The Quantitative Study of Erosion Techniques and Some Preliminary Results," *Geographical Review*, 23 (1933): 423-432.
40. Sharpe, C. F. S. *Landslides and Related Phenomena*, New York, 1938.
41. Horton, R. E. "Erosional Development of Streams and their Drainage Basins; Hydrophysical Approach to Quantitative Morphology," *Bulletin of the Geological Society of America*, 56 (1945): 275-370.
42. Davis, W. M. "Origin of Limestone Caverns," *Bulletin of the Geological Society of America*, 41 (1930): 475-628.
43. Swinnerton, A. C. "Origin of Limestone Caverns," *Bulletin of the Geological Society of America*, 43 (1932): 663-693.
44. Sanders, E. M. "The Cycle of Erosion in a Karst Region (after Cvijic)," *Geographical Review*, 11 (1921): 593-604.
45. Dicken, S. N. "Kentucky Karst Landscapes," *Journal of Geology*, 43 (1935): 708-728.
46. Shepard, F. P. *Submarine Geology*, New York, 1948; *idem.* "Revised Nomenclature for Depositional Coastal Features," *Bulletin of the American Association of Petroleum Geologists*, 36 (1952): 1902-1912.
47. Russell, R. J. "Coastal Advance and Retreat in Louisiana," *Proceedings of the 19th International Geological Congress*, Algiers, 1952.
48. Shepard, F. P. "Mass Movements in Submarine Canyon Heads," *Transactions of the American Geophysical Union*, 32 (1951): 405-418.
49. Daly, R. A. *The Floor of the Ocean*, Chapel Hill (North Carolina), 1942.
50. Antevs, E. "Quaternary Marine Terraces in Non-Glaciated Regions and Changes of Level of Sea and Land," *American Journal of Science*, 17 (1929): 35-49.
51. Cooke, C. W. *Geology of the Coastal Plain of South Carolina*, United States Geological Survey, Bulletin 867, Washington, 1936.
52. Davis, W. M. *The Coral Reef Problem*, American Geographical Society, Special Publication no. 9, New York, 1928.
53. Hobbs, W. H. "The Cycle of Mountain Glaciation," *Geographical Journal*, 35 (1910): 146-163, 268-284; *idem. Characteristics of Existing Glaciers*, New York, 1911; *idem.* "Loess, Pebble Bands, and Boulders from Glacial Outwash of the Greenland Continental Glacier," *Journal of Geology*, 39 (1931): 381-385.
54. Flint, R. F., and others. *Glacial Map of North America*, Geological Society of America, Special Paper 60, Washington, 1945; Flint, R. F. *Glacial Geology and the Pleistocene Epoch*, New York, 1947; *idem.* "Pleistocene Drainage Diversions in South Dakota," *Geografiska Annaler*, (1949): 56-74.
55. Leverett, F. *The Illinois Glacial Lobe*, United States Geological Survey, Monograph 38, Washington, 1899; Leverett, F., and Taylor, F. B. *The Pleistocene of Indiana and Michigan and the History of the Great Lakes*, United States Geological Survey, Monograph 53, Washington, 1915; Leverett, F. *Quaternary Geology of Minnesota and Parts of Adjacent States*, United States Geological Survey, Professional Paper 161, Washington, 1932.
56. Alden, W. C., and Leighton, M. M. "The Iowan Drift, A Review of the Evidences of the Iowan Stage of Glaciation," *Iowa Geological Survey*, 26 (1915): 49-212; Alden, W. C. *The Quaternary Geology of Southeastern Wisconsin with a Chapter on the Older Rock Formations*, United States Geological Survey, Professional Paper 106, Washington, 1918.
57. Kay, G. F., and Apfel, E. T. "The Pre-Illinois Pleistocene Geology of Iowa," *Iowa Geological Survey, Annual Report*, 34 (1929): 1-304.
58. MacClintock, P., and Apfel, E. T. "Correlation of the Drifts of the Salamanca Re-entrant, New York," *Bulletin of the Geological Society of America*, 35 (1944): 1143-1164.

59. FLINT, R. F. "Glacial Geology," in *Geology 1888-1938*, Geological Society of America, 50th Anniversary Volume, Washington, 1941.

60. JOHNSON, W. D. "The Profile of Maturity in Alpine Glacial Erosion," *Journal of Geology*, 12 (1904): 569-578.

61. TARR, R. S. "Glaciers and Glaciation in Alaska," *Science*, 12 (1912): 241-258.

62. MATTHES, F. E. "Glacial Sculpture of the Bighorn Mountains, Wyoming," *21st Annual Report*, United States Geological Survey, 1900: 167-190; *idem. Geological History of the Yosemite Valley*, United States Geological Survey, Professional Paper 160, Washington, 1930.

63. ATWOOD, W. W. *Glaciation of the Uinta and Wasatch Mountains*, United States Geological Survey, Professional Paper 61, Washington, 1909.

64. DYSON, J. L. "Shrinkage of Sperry and Grinnell Glaciers, Glacier National Park, Montana," *Geographical Review*, 38 (1948): 96-103.

65. FIELD, W. O. JR., and MILLER, M. M. "The Juneau Ice Field Research Project," *Geographical Review*, 40 (1950): 179-190; FIELD, W. O. JR., and HEUSSER, C. J. "Glaciers—Historians of Climate," *ibid.*, 42 (1952): 337-345.

66. DEMOREST, M. H. "Glacier Regimens and Ice Movement within Glaciers," *American Journal of Science*, 240 (1942): 29-66.

67. SHARP, R. P. "Ep-Archean and Ep-Algonkian Erosion Surfaces, Grand Canyon, Arizona," *Bulletin of the Geological Society of America*, 51 (1940):1235-1270; *idem*. "Periglacial Involutions in Northeastern Illinois," *Journal of Geology*, 50 (1942): 113-133; *idem*. "The Constitution of Valley Glaciers," *Journal of Glaciology*, 1 (1948): 182-189.

68. MATTHES, F. E. "Rebirth of the Glaciers of the Sierra Nevada during Late Post-Pleistocene Time," *Bulletin of the Geological Society of America*, 52 (1941): 2030.

69. DENNY, C. S. "Periglacial Phenomena in Southern Connecticut," *American Journal of Science*, 32 (1936): 322-342.

70. SMITH, H. T. U. "Aerial Photographs in Geomorphic Studies," *Journal of Geomorphology*, 4 (1941): 171-205; *idem*. "Periglacial Features in the Driftless Area of Southern Wisconsin," *Journal of Geology*, 57 (1949): 196-215; *idem*. "Physical Effects of Pleistocene Climatic Changes in Non-Glaciated Areas: Eolian Phenomena, Frost Action and Stream Terracing," *Bulletin of the Geological Society of America*, 60 (1949): 1485-1516.

71. PELTIER, L. C. *Pleistocene Terraces of the Susquehanna River, Pennsylvania*, Pennsylvania Geological Survey, Bulletin G-23, 1949; *idem*. "The Geographic Cycle in Periglacial Regions as it Is Related to Climatic Geomorphology," *Annals of the Association of American Geographers*, 40 (1950): 214-236.

72. LEFFINGWELL, E. DE K. *The Channing River Region, North Alaska*, United States Geological Survey, Professional Paper 109, Washington, 1919.

73. CAIRNES, D. D. "Differential Erosion and Equiplanation in Portions of Yukon and Alaska," *Bulletin of the Geological Society of America*, 23 (1912): 333-348.

74. EAKIN, H. M. *The Yukon-Koyukuk Region, Alaska*, United States Geological Survey, Bulletin 631, Washington, 1916.

75. MELTON, F. A. "A Tentative Classification of Sand Dunes; Its Application to Dune History in the Southern High Plains," *Journal of Geology*, 48 (1940): 113-174.

76. BAGNOLD, R. A. *The Physics of Blown Sand and Desert Dunes*, New York, 1942.

77. RUSSELL, R. J. "Landforms of San Gorgonio Pass, Southern California," *University of California Publications in Geography*, 6 (1932): 23-121.

78. HACK, J. T. "Dunes of the Western Navajo Country," *Geographical Review*, 31 (1941): 240-263.

79. HUFFINGTON, R. M., and ALBRITTON, C. C. "Quaternary Sands on the Southern High Plains of Western Texas," *American Journal of Science*, 239 (1941): 325-338.

80. DUTTON, C. E. "Hawaiian Volcanoes," *4th Annual Report*, United States Geological Survey, Washington, 1884: 75-219.
81. BECKER, G. F. "The Geometrical Forms of Volcanic Cones," *American Journal of Science*, 30 (1885): 283-293.
82. DANA, J. D. *Characteristics of Volcanoes, with Contributions of Facts and Principles from the Hawaiian Islands*, New York, 1890.
83. RUSSELL, I. C. *Volcanoes of North America*, New York, 1897.
84. TYRRELL, G. W. *Volcanoes*, New York, 1932
85. STEARNS, H. T. "Volcanism in Mud Lake Area, Idaho," *American Journal of Science*, 11 (1926): 353-363; *idem. Geology and Ground Water Resources of the Island of Oahu, Hawaii*, Hawaii Division of Hydrography, Bulletin 1, Honolulu, 1935; *idem.* "Origin of Haleadala Crater," *Bulletin of the Geological Society of America*, 53 (1942): 1-14.
86. PUTNAM, W. C. "The Mono Craters, California," *Geographical Review*, 28 (1938): 68-82.
87. WILLIAMS, H. "The History and Character of Volcanic Domes," *University of California Publications, Bulletin of the Department of Geological Sciences*, 21 (1932): 51-146; *idem.* "Calderas and their Origin," *ibid.* 25 (1941): 239-346; *idem.* "Pliocene Volcanoes of the Navajo-Hopi Country," *Bulletin of the Geological Society of America*, 47 (1936): 111-171.
88. DILLER, J. S. "Did Crater Lake, Oregon, Originate by a Volcanic Subsidence or an Explosive Eruption?" *Journal of Geology*, 31 (1923): 226-277.
89. WENTWORTH, C. K. "Principles of Stream Erosion in Hawaii," *Journal of Geology*, 36 (1928): 385-410.
90. HINDS, N. E. A. "The Relative Ages of Hawaiian Landscapes," *University of California, Bulletin of the Department of Geological Sciences*, 20 (1931): 143-260.
91. COTTON, C. A. *Volcanoes as Landscape Forms*, Wellington (New Zealand), 1944.
92. BRYAN, K., and RAY, L. L. *Geological Antiquity of the Lindenmeier Site, Colorado*, Smithsonian Miscellaneous Collections, 99, no.2, Washington, 1940.
93. HIBBEN, F. C. *Evidences of Early Occupation in Sandia Cave, New Mexico, and Other Sites in the Sandia-Manzano Region*, Smithsonian Miscellaneous Collections, 99, 1941.
94. HACK, J. T. *The Changing Physical Environment of the Hopi Indians of Arizona*, Papers of the Peabody Museum, 35, Cambridge (Massachusetts), 1942.
95. JOHNSON, F., ed. *The Boylston Street Fish Weir*, Robert S. Peabody Foundation Papers, 4, Andover (Massachusetts), 1949.
96. COOKE, S. F. "Soil Erosion and Population in Central Mexico," *Ibero-Americana*, 34 (1949): 1-86.
97. PHILLIPS, P., FORD, J. A., and GRIFFITH, J. B. *Archaeological Surveys in the Lower Mississippi Alluvial Valley, 1940-1947*, Papers of the Peabody Museum, 25, Cambridge (Massachusetts), 1951.
98. RUSSELL, R. J. "Geographical Geomorphology," *Annals of the Association of American Geographers*, 39 (1949): 1-11.
99. BRYAN, K. "Physiography," in *Geology, 1888-1938*, Geological Society of America, 50th Anniversary Volume, Washington, 1941: 1-15; *idem.* "The Place of Geomorphology in the Geographic Sciences," *Annals of the Association of American Geographers*, 40 (1950): 196-208.

THE GEOGRAPHIC STUDY OF SOILS[A]

CONCEPTS CONCERNINC SOIL CLASSIFICATION
EARLY SOIL CONCEPTS IN AMERICA
COFFEY'S APPROACH TO SOIL STUDY
THE RUSSIAN SCHOOL OF SOIL SCIENCE
MARBUT AND THE NATURAL CLASSIFICATION OF SOILS
CURRENT CONCEPTS OF SOIL CLASSIFICATION

THE CONCEPT OF USE POTENTIAL
KEYS OF USE POTENTIAL
RELATING SOIL PERFORMANCE TO SOIL CLASSIFICATION

THE GEOGRAPHIC INVENTORY OF THE SOIL RESOURCE
THE SOIL SURVEY
THE PROBLEM OF SCALE
RECONNAISSANCE MAPPING

A. Original draft by Carleton P. Barnes and sponsored by the Committee on Physical and Bio-Geography, Hoyt Lemons, Chairman.

CHAPTER SIXTEEN

THE GEOGRAPHIC STUDY OF SOILS

THE WORD SOIL is commonly used in two different senses. The civil engineer concerned with problems of soil mechanics, or the geologist concerned with the processes operating to shape the surface of the earth may think of soil as the unconsolidated mantle of rock waste, the regolith. On the other hand, the agronomist, the ecologist, and the specialist in soil science think of a soil as a natural body, in which inorganic and organic materials are intimately combined. Although soil in both senses could be examined by geographic method, this chapter deals only with soil in the second sense.

Soils can be studied in two quite different ways. They can be, and for many years have been, analyzed by laboratory methods for the purpose of identifying their chemical and some of their physical properties. The procedure is to collect samples in the field and then to test the samples in the laboratory. Yet to estimate the potentialities of a soil by chemical analysis is no more possible than to estimate the characteristics of a river by testing a sample of its water. The geographic method of studying soils requires the identification of kinds of soil and the mapping of the areal spread of these types.

Like all geographical study, this involves generalization. In the introduction to this volume the fact was stated that no two points on the face of the earth are identical; and this statement applies as much to soil as to any other geographical phenomenon. Yet to study each separate point on the earth is obviously impossible. The geographer is faced with the problem of finding useful generalizations, of devising classifications in which each category is defined by specific criteria.

Soils may be classified for the purpose of grouping together those kinds that have similar characteristics, or that were formed by similar processes; or they may be classified for the purpose of grouping together those kinds that have similar use potentials. Proposed classifications are to be judged on the basis of the illumination they throw on one or another of these problems. Over the years in which scholars have been interested in soils and have accumulated information about them, the concepts regarding methods of selecting categories have been sharpened and the classifications have become more realistic.

A soil, as the soil geographer now sees it, is not a single feature. Any individual soil type possesses a number of different characteristics which together determine its significance. The importance of any one characteristic depends on the others with which it is associated. Thus to know that a soil has a loam texture is important; but if it is not known also whether the soil has a high water table, a shallow depth over an impervious layer, or a steep slope, no adequate estimate of its nature or its utility can be made. Therefore soil classifications which ignore these attributes and are based on only one item of the total complex are not suitable for geographic study. Failure to grasp this fundamental principle has been the cause of more misleading characterizations of soil than almost any other single thing.

CONCEPTS CONCERNING SOIL CLASSIFICATION

To paraphrase a remark of Sherlock Holmes, the soil is something which every one sees but few observe. Farmers, especially, have worked so intimately with the soil that it is not always easy to convince them of the need for a wider and more careful observation of what they see every day. The farmer may speak of black land, gumbo, hummock land, clay; on the limited area of a single farm such designations may be satisfactory. Classifications of soil by geologists based on the age of the rock from which the parent material was derived are scarcely better. Yet even today many persons are content to speak of Silurian soils, crystalline soils, or residual, alluvial, and colluvial soils. The fact is that soils have been observed and mapped out-of-doors (in contrast to the laboratory) for only a relatively short time.

EARLY SOIL CONCEPTS IN AMERICA

Some of the earliest ideas about soils in the United States were derived from field studies by geologists. In the late 19th century N. S. Shaler recognized that soil was more than regolith, and lamented the lack of systematic study of soils in relation to geology, chemistry, and botany [1]. E. W. Hilgard was another of the early pioneers in America. In 1906 he published the results of many years of field study, pointing out that in the semi-arid West the soils contained lime regardless of the nature of the underlying rock [2]. M. Whitney summed up the early concepts regarding soils in 1909 [3].

The real advance of soil science, however, was dependent on the systematic observation of soils in the field. The United States government instituted the first soil survey in 1899, and in the first decade of the 20th century an important collection of new data regarding the relation of soils to climate, vegetation, drainage, landforms, lithology, and agricultural utilization became available as a consequence.

Coffey's Approach to Soil Study

In 1912 the Bureau of Soils published the doctoral dissertation of George N. Coffey [4]. In this publication Coffey advanced the idea of a soil classification based on all the important characteristics of the soils themselves, rather than one based on any single feature. He suggested a broad grouping of the soils of the United States based on this principle. This work was the first step toward the treatment of soils as a phenomenon separate from the underlying geology.

The Russian School of Soil Science

Meanwhile, the Russians were far in advance of their professional colleagues in other parts of the world. Observing the soils of the great plains of Eastern Europe crossed by broad belts of contrasting climates, the Russians early came to recognize that soils could be defined each with a definite form and structure of its own, and that these soils were associated with particular combinations of climate, vegetation, slope, parent material, and age. In 1870 Dokuchaiev established a school of soil science based on this new concept. In 1895 Sibirtsev published a genetic soil classification in which, for the first time, the distinction between zonal, intrazonal, and azonal soils was made. Zonal soils, as the term has come to be used, reflect the influence of the pervading factors of soil genesis, namely climate, and vegetation; intrazonal soils have more or less well-developed soil characteristics that reflect the dominant influence of some local factor of slope or parent material; azonal soils are without well-developed soil characteristics [5: 980-981].

These Russian ideas had little influence in America until after K. D. Glinka had translated them into German in 1914 [6]. Glinka's classification subdivided the zonal soils according to the amount of moisture reaching the surface horizons.

Marbut and the Natural Classification of Soils

Curtis F. Marbut introduced the Russian concepts of soil classification into the United States through his translation of Glinka's German version [7]. Many of the Russian concepts applied well in the United States because both countries contain somewhat similar great soil groups. In Western Europe, on the other hand, there are few areas where the contrasts between light-colored forest soils and dark-colored prairie and steppe soils could be clearly observed. Here most of the soils reflect local differences of parent material and slope, as is also the case in the hilly parts of the eastern United States.

When Marbut attempted to apply the Russian concepts in North America he reached the conclusion that the Russians had placed undue emphasis on climate. Marbut was seeking a natural classification of soils that is, a system of categories based on the inherent characteristics of the soil itself. Soil mapping in the United States brought out clearly the close relationship of the underlying soil to the vegetation cover. The soils that had developed under tall grass were markedly different from those that had developed under a forest, and the two groups were separated by sharp boundaries that could scarcely be ascribed directly to climate.

Marbut also recognized the time factor in soil development. Influenced by the teachings of William Morris Davis especially concerning the sequence of stages in the development of landforms (chapter fifteen), he attempted to apply the concept of youth, maturity, and old age to soils. Materials freshly deposited by water, wind, or ice express primarily the differences in parent material, whereas materials that have been exposed to an environment over relatively long periods have received the stamp of that environment. An old soil, according to Marbut's concept, was one in which further zonal development could not be anticipated. This is an idea closely parallel to that of the climax in plant ecology (chapter nineteen), or that of the peneplain in geomorphology. Modern soil scientists, like modern plant ecologists, are not finding the concept of the climax useful.

There are four basic factors involved in soil formation: parent material, slope, vegetation, and climate [8]. In the case of any one soil one of these factors may be important out of all proportion to the others, yet all must be kept in mind in any natural classification. The operation of these four factors may vary not only from place to place but also from time to time. That climates and the phenomena related to climate have changed during recent geologic times has been recognized by the geomorphologists [9] who, as was noted in the preceding chapter, are currently devoting much attention to the evidences and results of these changes. The soil development, too, reflects such changes, and in many places, especially those subjected to recent glaciation or rapid loess accumulation, an insufficient time has elapsed since the establishment of the present conditions of climate and vegetation cover to permit the development of mature soils.

CURRENT CONCEPTS OF SOIL CLASSIFICATION

Considerable progress has been made during the last decade or so toward the formulation of an illuminating system of soil classification in which the categories are defined and distinguished in terms of their own natural characteristics. The need for classifications at different ranks or degrees of generalization has been recognized. Yet present knowledge is incomplete, for soils are still described from time to time in new areas which do not fit into any of the well-established categories, even those of the highest rank. Present American theory and practice in soil classification are well summarized in a symposium published in *Soil Science* for February, 1949 [10]. It is not necessary to repeat here the principles on which soil classifications now rest, or the fundamentals of the American system now accepted as standard in this country.

It is necessary, however, to point out the need for a natural classification of soils rather than a classification based on potential use. In the latter kind of system a soil might be described as a "third-class soil for cultivated crops." But what does such a description reveal? Is it suitable for alfalfa? Does it need lime? Must it be leveled before irrigation? Will it drain quickly after a rain? Can it be used as a highway sub-grade? Will it afford a source of gravel for road construction? Not one of these questions is answered from the description of it as a third-class soil for cultivated crops. Furthermore, as the techniques of agriculture change, it may not remain a third-class soil. The identification of the soil in a natural classification, however, should disclose information about its character that permits the ready answering of all these questions.

The need for taxonomy in soils is analogous to the need for it in botany. Knowing that a tree is a honey locust tells us its characteristics and utility, and the performance of its wood. Similarly, knowing that a soil is a Tama silt loam, tells us its characteristics, utility, and behavior. Lacking a natural classification system it would be necessary to test each separate plant, or the soil at virtually every spot, to estimate its utility.

But taxonomy alone is not enough. Obviously the classification by itself does not include enough information to predict the capabilities of soils in particular places and for particular purposes. To do this it is necessary to prepare keys of use potentials, which requires knowledge gained by experience or by experiment regarding the behavior of each category of soils. Again, this may be illustrated by a botanical analogy. It may be possible to identify a Virginia pine and to place it in a taxonomic system. But until some one has attempted to make pulp and paper from this tree, its utility as pulpwood is not known. Once such tests have been made with samples of Virginia pine, and the wood has been found suitable for the manufacture of paper, then it is possible to predict with some confidence that this species has good pulping potentialities wherever it is found. So it is with soils. Furthermore, if the technique of paper manufacture is changed, the basic classification is not rendered worthless; only the key to use-potential must be reexamined.

The geographic study of soils, therefore, involves two phases. First is the definition of categories of soil at different degrees of generalization and the mapping of these categories. Second is the development of keys to indicate the use-potential of the defined categories of soil.

THE CONCEPT OF USE-POTENTIAL

The concept of use-potential involves two ideas. In the first place, an estimate of the utility of a soil should be based on the whole complex of characteristics by which the category is defined, not on any one feature by itself. In the second place, the use-potential must be recognized as a temporary matter, for with each change in technology or in such societal elements as economic conditions, forms of land tenure, or tax systems, it may need to be reconsidered and perhaps recalculated.

With a map of soils identified as units of a natural classification system, it is possible to delimit a number of single-feature regions, each showing the distribution pattern of some one soil characteristic affecting land use, crops, farm practices, or engineering practices. For example, all soils having poor drainage can easily be distinguished from soils having good drainage. All soils having gentle slopes can be separated from those with steeper slopes. All soils with root-hindering layers can be outlined, as can those with bedrock at slight depth.

But the significance of such single-feature maps is questionable, for there is a considerable possibility of error in judging a soil on the basis of one feature alone. Rather it is the combined effect of the whole complex of features that determines soil capacity.

The map of soils provides the basis for recording the results of experience and experiment with different kinds of use. There is need for keys related to each soil type which define productivity for particular crops, response to irrigation or drainage, erosion hazard under specified management, or behavior under different kinds of structures. Such keys take many forms and can deal with many varied uses and practices, engineering as well as agricultural. Land classification, in the sense of land-suitability prediction, can be derived from use-potential keys related to the soil categories of the basic classification. To attempt land classification directly without a soil map based on the natural characteristics of the soil can bring only temporary and uncertain results.

KEYS OF USE-POTENTIAL

The problem is to find a way to prepare these keys so systematically that workers who need to appraise the suitability of soils for specific purposes can do so from the soil map and the accompanying data without much research and interpretation on their own part. One device recently coming into use is the tabular summary of soil properties as an appendix to the map. An example of such a key, related to the soil map of Tama County, Iowa, is given in Table I. This kind of key contains the results of observations and interpretations made in the field by the soil mapper, and greatly facilitates the use and understanding of the soil map on the part of workers who need geographic information on soil properties and

TABLE I

TABLE I

SUPPLEMENT TO SOIL MAP OF TAMA COUNTY, IOWA, SHOWING PRINCIPAL CHARACTERISTICS OF THE SOILS

Soil (type, phase, complex, or land type)	Map symbol	Surface		Subsoil permeability	Surface soil consistence	Subsoil consistence	Probable erosion hazard[1]
		Situation	Slope percent				
remer silt loam	Bl	Slight depressions in terraces (nearly level)	. . . .	Moderate to moderately slow	Friable	Firm	None
remer silty clay loam	Bs	"	. . . .	Moderately slow	"	"	"
uckner sandy loam	Bu	Terraces (undulating)	0-5	Very rapid	Loose when dry; friable when moist	Loose	Slight
arrington loam	Cl	Uplands	3-8	Moderate	Friable	Friable	Moderate
roded gently rolling phase	Cle	"	8-12	"	"	"	Serious
arrington silt loam	Ca	"	3-8	"	"	"	Moderate
roded gently rolling phase	Cae	"	8-12	"	"	"	Serious
roded rolling phase	Cao	"	12-16	"	"	Firm when moist; slightly plastic when wet	Very serious

Under row crop cultivation without special practices

behavior. It seems likely that such keys accompanying soil maps will come into general use and that the number of properties and behavior characteristics to be portrayed will be expanded as experience and research disclose new facts about the different soil types.

RELATING SOIL PERFORMANCE TO SOIL CLASSIFICATION

Gradually the importance of this concept of the use-potential, and of the procedure of relating the keys to the taxonomic system, is gaining acceptance. The soil map provides the framework on which the facts may be organized and applied geographically from the continuously increasing body of information regarding the performance of soils. The analogy of the Virginia pine and the pulpwood helps to illustrate the point. Suppose the wood which is tested for pulping quality were simply wood of an unknown and unidentified species. Of what value would be the findings? None at all until the tree species had been identified. The same is true for soils. If an experiment is conducted at a particular spot, how generally applicable will the results be? There is no assurance that the results will apply to any place except the particular one where the experiment was conducted, until the soil is specifically identified. Thereafter the results can be applied to that soil type wherever it is found.

This means that an important part of the geographic study of soils consists of relating the results of experience to the kinds of soil on which the experience was obtained. The results can then be keyed to the soil type, and information of great prediction-value built up. Crop yields are one of the well-known kinds of performance data related to soils that have been collected and used in predictions of use-potential. It may be pointed out further that the application of soil science to engineering is developing rapidly and that this procedure provides a means for relating the results of physical tests to soil categories, thus greatly expanding the meaning and utility of the results [10: 159-161].

THE GEOGRAPHIC INVENTORY OF THE SOIL RESOURCE

The inventory and appraisal of resources is a broad field to which geographic methods can profitably be applied. One aspect of this work involves the examination of the soil resource. The kind of inventory most widely known to non-geographers is based on the counting of things; but a geographic inventory provides not a count but a map: it portrays the facts not in arithmetic terms but in geometric terms.

Applied to soils, as to other geographic phenomena, a count is of little significance. It is possible, of course, to add up the number of acres of this or that soil type, possessing this or that use-potential. Yet any one who has mapped soils in the field is aware of the complex intermingling of individual small bodies of different, sometimes sharply contrasting, kinds of soil. Small bodies of deep, level, fertile soil, interspersed through a matrix of rough, stony land, may add up to a considerable total acreage and yet be low in total value. A smaller total acreage of such soil concentrated in a single compact area would have far greater importance in a resource inventory.

The Soil Survey

Approximately half of the area of continental United States has been covered by soil maps since 1899 [11]. During this period the soil mapping has been done in increasing detail and with more and more effective categories. Some of the older maps have today much less utility than the newer ones. Yet even the older ones permit some attempt at the development of keys of use-potential, not in so much detail, to be sure, as is possible with the latest surveys. The current methods of procedure are described by Charles E. Kellogg [12]. Much of this work has been done by the Soil Survey of the United States Department of Agriculture, and by State agricultural experiment stations and agricultural colleges.

The Problem of Scale

As a result of these many decades of mapping experience, the problem of scale has emerged as a major question demanding solution. Field maps in farming sections of the United States have commonly been made on air photographs at a scale of about 1/15,840. But such large-scale maps are too big to publish. They are printed on a scale of 1/31,680, with some loss of detail through the combination of small areas into larger and more highly generalized ones. The published maps lose so much detail in this process that they are not generally considered suitable for farm planning, which must be based on the larger-scale field sheets. If the maps are thus to be generalized for publication, could they not be generalized much more, and still show enough for uses not requiring full detail, thereby saving much expense? How should soil categories be defined at the smaller scales? Can sample areas be selected for detailed, topographic-scale mapping without covering the whole country in such detail? These questions need careful consideration, with the needs of different groups of map-users in mind.

When soil mapping is done at scales too small to show the specific soil types as they are arranged on the ordinary farm, there are two methods of procedure. The older one, is to identify the soil type which occupies the largest percentage of an area and to show the whole area as occupied by this predominant type. This method disregards many soil types occupying small areas, even if these small areas are uniformly scattered. It

generalizes in terms of predominance and thereby disregards the existence
of less extensive bodies of other soil types. For example, in the area shown
in Figure 14, generalization in terms of predominance would be accomplished by showing the Clyde soil as a narrow ribbon in the valley, and
by lumping together the Floyd, Carrington, and Coggon types, designating the whole area as occupied by the one with the greatest spread.

The more modern method of generalizing on smaller-scale maps makes
use of the principle of association. Fortunately soil types commonly occur
in patterns that are repeated again and again. Two or more types will be
found in association in more or less characteristic proportions and landscape positions. The whole of an area possessing a characteristic pattern of
associated soils is now called a soil association. The block diagram (Fig.14)
shows part of an area mapped as a Carrington-Clyde Association, where
the Floyd type commonly occupies the gentler slopes, the Carrington and
Coggon types are on the somewhat steeper valley slopes, and the Clyde
occupies the poorly drained swales. So common is this grouping of contrasted soil types that the map which shows the areas where they are associated has a much greater value for the preparation of use-potential
keys, than does the map of the predominant type.

The principle of association, which geographers as a whole derived
from the experience of the soil geographers, is now proposed for land-use
mapping, or the mapping of other geographic phenomena at what have

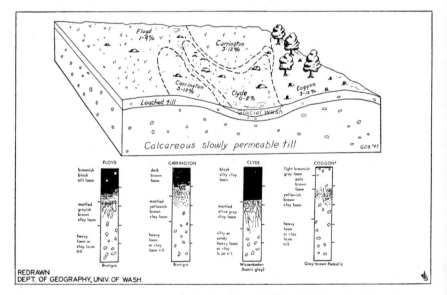

Fig. 14. Components of the Carrington-Clyde Soil Association.

been called the chorographic scales. Also from the soil geographers has come new insight regarding the methods of selecting sample areas for detailed mapping. The purpose of the sample studies is to reveal the nature and arrangement of the components of an association. The more uniform the make-up of an association, the fewer are the sample studies needed to bring out its essential characteristics. If the method of mapping at chorographic scales in terms of associations, supported by detailed samples at whatever intervals are required, can be perfected, the rapidity and effectiveness of the resource inventory will be greatly increased.

RECONNAISSANCE MAPPING

There are large areas of the world for which no soil maps at any scale exist. Can any worth-while soil surveys be made of such areas on a reconnaissance basis?

The first step in such a reconnaissance survey would be to make the best possible schematic soil-association map, on a scale such as 1/1,000,000. The mapper would utilize all the evidence available, such as existing soil maps, geological maps, contour maps, maps of vegetation, climatic maps, and the like. In view of the known fact that soils owe much of their character to the rocks, climate, vegetation, and slope, if the geographical arrangement of these things is known, fair approximations of the soil character can be formulated. These hypothetical soil associations are checked by direct observation wherever possible.

The second step is to select sample areas representative of each of the broad soil associations, and to map the soil types in detail, fitting them into the standard classification as far as possible. At the same time all possible information about the productivity of the different mapped soil types, or of their response to various systems of management, would be accumulated by the usual methods of field study. Investigation of each sample area would, in effect, give an understanding of the pattern of soils throughout the whole soil association of which it is representative.

The value of such a reconnaissance with its schematic map of soil associations, supported by detailed samples, would vary with the competence of the worker and the complexity of the area under investigation. Assuming reasonable competence and the use of acceptable field techniques, the resulting map would at least provide a tentative frame of reference into which agricultural research scientists, farm advisors, or land-use planners could insert the results of their experience in specific small areas.

The full utility of soil classification of any kind, whether detailed or generalized, will not be realized until the ability to identify soils is one of the working skills of farm advisors, and of those engineers whose problems are related to kinds of soil. The accurate mapping of representative sections in each of the important soil associations by soil scientists experienced in soil mapping procedures, is essential if the meaning of the more general soil categories is to be widely understood. But even wealthy countries cannot afford to map the soil everywhere in the detail needed for intensive use, as on the site of an individual building. The geographer who specializes in the study of soil is in a position to enlarge the importance of his contribution through the development of better methods of defining significant broad categories to permit more rapid and less costly coverage.

REFERENCES

1. SHALER, N. S. "The Origin and Nature of Soils," in *U. S. Geological Survey, Twelfth Annual Report, Part I,* Washington, 1891: 213-345.
2. HILGARD, E. W. *Soils, their Formation, Properties, Composition, and Relation to Climate and Plant Growth in the Humid and Arid Regions,* New York, 1906.
3. WHITNEY, M. *Soils of the United States,* U. S. Department of Agriculture, Bureau of Soils, Bulletin 55, Washington, 1909.
4. COFFEY, G. N. *A Study of the Soils of the United States,* U. S. Department of Agriculture, Bureau of Soils, Bulletin 85, Washington, 1912.
5. BALDWIN, M., KELLOGG, C. E., and THORP, J. "Soil Classification," in *Soils and Men,* U. S. Department of Agriculture Yearbook, 1938: 979-1001.
6. GLINKA, K. D. *Die Typen der Bodenbildung, Ihre Klassifikation und geographische Verbreitung,* Berlin, 1914.
7. GLINKA, K. D. *The Great Soil Groups of the World and their Development,* Translated from the German by C. F. MARBUT, Ann Arbor (Michigan), 1927.
8. JENNY, H. *Foundations of Soil Science,* New York, 1946.
9. BRYAN, K., and ALBRITTON, C. C. JR. "Soil Phenomena as Evidence of Climatic Changes," *American Journal of Science,* 241 (1943): 469-490.
10. KELLOGG, C. E., CLINE, M. G., and others. *Soil Science,* 67 (1949): 77-260.
11. MARBUT, C. F. "Soils of the United States," Part III in *Atlas of American Agriculture,* U. S. Department of Agriculture, Washington, 1935.
12. KELLOGG, C. E. *The Soils that Support Us,* New York, 1941; *idem. Soil Survey Manual,* U. S. Department of Agriculture, Misc. Publ. 274, 2nd ed., Washington, 1951.

THE GEOGRAPHIC STUDY
OF WATER ON THE LAND[A]

WATER FORMS AND THEIR IMPACT ON THE LAND

BASIC WATER DATA AND THEIR GEOGRAPHIC USE
PRECIPITATION
STREAMFLOW
LAKES
GROUND WATER
SNOWFALL
EVAPORATION AND TRANSPIRATION

AREAL SYNTHESIS OF WATER FORMS

THE PROSPECT

A Original draft by Peveril Meigs III, sponsored by the Committee on Physical and Bio-Geography, Hoyt Lemons, chairman.

CHAPTER SEVENTEEN

THE GEOGRAPHIC STUDY
OF WATER ON THE LAND

OF THE VARIOUS elements that make up the geographic scene, water probably has been studied the least systematically by geographers, although their total contribution to this field has been considerable. Only the first phase of the hydrologic cycle—precipitation, and to a lesser extent other aspects of hydrometeorology—has received full geographic treatment and mapping. The land phases of the cycle have until recently received only sporadic attention from geographers. The land phases, as subdivided by the International Association of Scientific Hydrology, include: potamology (the study of rivers), limnology (the study of lakes), cryology (the study of ice phenomena), and geohydrology (the study of subsurface waters). The morphology of the waters of the lands has been particularly slighted. Even when the water element is included in the description of an area it is likely to be given a rather summary treatment, rarely a systematic analysis of all its phases.

The bulk of American study of hydrology (defined by a leading hydrologist as "that branch of physical geography dealing with the waters of the earth" [1: 1]) has been conducted to meet urgent economic needs. The geologists, meteorologists, and engineers making these studies have been concerned primarily with applied hydrology: estimating the amounts, dependability, and behavior of water as a resource; devising ways of putting the water where it is needed at the time when it is needed; and trying to keep it within bounds during floods. But geographers have had a creditable share in developing the theoretical concepts, the field studies, and the broad plans of action, and, increasingly, in working out special applications.

Although pioneer work had gone on in most phases of hydrology in the last quarter of the 19th century, the main impulses to hydrologic study came after the turn of the century. The reclamation era, stimulated by passage of the Reclamation Act in 1902, led to a pressing demand for more information on water needs, resources, and techniques of development. The first American textbook on hydrology [2] appeared two years later, in time to be used by the early reclamationists. It was followed by a later edition and by the comprehensive works of A. F. Meyer [3], O. E. Meinzer [4], and R. K. Linsley[1]. The symposium edited by Meinzer, a geohydrologist, is especially useful to geographers, treating the elements of the hydrologic cycle with a broad areal approach, and with well-selected bibliographies. The work by Linsley, a hydrometeorologist, is an indispensable key to modern developments, although designed primarily as an applied technical text and handbook.

The creation by the American Geophysical Union in 1930 of the first American hydrological organization, its Section on Hydrology, with vigorous research committees, gave a great impulse to scientific development and coherence in the field of hydrology. In five of its twelve research committees, geographers have played a leading role: glaciers (François Matthes, William O. Field, Jr.); snow (J. E. Church, P. E. Church); evaporation and transpiration (C. W. Thornthwaite); land erosion (W. C. Lowdermilk); and lakes (P. E. Church). The scope of the work in hydrology is suggested by the names of the additional committees: precipitation; infiltration; physics of soil moisture; chemistry of soil moisture; runoff; dynamics of streams; and groundwater. The *Transactions of the American Geophysical Union* promptly became the chief medium of publication for studies in theoretical and applied hydrology. Not the least contribution of the Union has been its annual bibliographies of hydrology, culminating in the publication in 1952 of a combined annotated bibliography of 4,500 works [5]: which suggests the hopelessness of touching more than a few highlights in this present brief chapter.

Great natural hydrologic disasters such as the Mississippi flood of 1927, the Ohio River flood a few years later, and the droughts of the Great Plains in the 1930's, have led to upsurges of hydrological activity by governmental agencies; and the economic depression of the early 30's, with its pressure for putting the unemployed to work, gave added stimulus to the demand for vast reclamation and water-development works from the Tennessee to the Sacramento. By about 1935 there began the proliferation of hydrologic research that has continued to the present.

WATER FORMS AND THEIR IMPACT ON THE LAND

A partial explanation of the early neglect of the study of water forms by geographers is to be found in their preoccupation with geomorphology at the beginning of the century (chapter fifteen). So all-consuming has been the concern with elucidating the history of landforms that running water, moving ice, and lake waves were of interest chiefly as agents of erosion, not as significant elements of the landscape in their own right. It is true that water bodies, even as forms of their own, do not exist in a vacuum but in a sheath of land that must be considered a part of their character.

As the century opened, the stage for geomorphologic emphasis had been set by the alluring geographical cycle of William Morris Davis [6], in which running water was the touchstone to an understanding of land forms. Persistent exploration of this theme followed, to a large extent adhering to the "classical" and often unrealistic concepts laid down by Davis. In recent decades, emphasis has shifted away from physical towards economic and cultural geography. For such studies of water features and their works as are being made, Richard J. Russell has issued a clarion call for fresh examination untrammelled by *a priori* systems, for objective descriptions and evaluations of forms and relationships observed in the field [7].

This morphologic approach increasingly has pervaded geographic research under the influence of Carl Sauer and others since the 1920's. Russell's own work exemplifies the fruitfulness of this philosophy. By the process of recording and considering all discernible forms, rather than only those that fit a theoretical pattern, he has modified profoundly our concept of the character and history of deltas and other forms of river deposition, and has given concrete examples of a morphologic classification of water forms themselves. One of the most comprehensive of his many contributions in geographical hydrology is a report on the physiography of the lower Mississippi River delta [8]. Here he describes and classifies the water elements of the landscape: the distributary patterns, the levee-flank lakes, the round lakes and other features of the marshes, and sheds new light on the origin and growth of these features and related landforms. He punctures, with solid observation, the classical theory of the importance of downstream migration of meanders, demolishes the thesis that obstructions initiate meanders, proves that crevasses are more likely to be self-sealing than self-perpetuating, and points out errors into which two-dimensional thinking had led the classicists.

In recent years hydrological studies with a bearing on the geomorphic work of streams have emphasized particularly the hydraulics of stream flow, sedimentation, and soil erosion. For the most part these studies have been designed to help solve specific problems of water management or control. Some of the most significant, however, have been theoretical, like John Leighly's pioneering investigations of stream turbulence and its three-dimensional effects upon transport and deposition of sediments and upon the morphology of meanders [9]. Recent studies by L. B. Leopold reverse the widely-held belief that streams have their highest velocities in their upper reaches; he shows that actually the Mississippi greatly increases its velocity downstream: depth and smoothness of channel here more than compensate for the greater slope upstream [10].

Recent study of sedimentation—that is, of the load carried by streams—has been spurred by the realization that the life of reservoirs and consequently the annual cost of reclamation projects are seriously affected by sediment contributions. There has also been continuing interest in the behavior and control of sedimentation in navigable streams. Basic knowledge has been obtained both of geomorphic principles of sedimentation and regional applications, especially by the Corps of Engineers and the Mississippi River Commission [11; 12].

Finally, the very origin of sedimentation, soil erosion, involving the action of slope wash and streams, has been studied theoretically and with a view towards minimization by many federal agencies and individuals. The Soil Conservation Service of the Department of Agriculture, under the leadership of H. H. Bennett, the Tennessee Valley Authority, and the Forest Service are among the more active agencies in this field. Among the research themes that have engaged the attention of geographers has been the question of causes of renewed gullying and arroyo-trenching in the arid West. C. W. Thornthwaite, C. F. S. Sharpe, and E. F. Dosch made an exhaustive study of this process as affecting a particular area in Arizona [13]. They concluded that human activity rather than climatic change was primarily responsible.

The geographic study of glaciers, like that of streams, has been concentrated largely upon the physiographic effects of the ice, though the morphology of the glaciers themselves has received increasing attention. Many individual glaciers and groups of glaciers have been studied during the past fifty years, starting in 1900 with F. E. Matthes' report on the glacial landforms of the Bighorn Mountains [14]. The American Geographical Society has played an important part in promoting original studies of glaciers, most recently with its Juneau Ice Field research project led by William O. Field, Jr. [15; 16]. In recent years the Arctic Institute of North America has channelled funds into studies of glaciers as well as of other Arctic phenomena. An expedition to the Baffin Island glaciers has been one of its latest research projects, as reported in the pages of *Arctic*. The American Geophysical Union Committee on Glaciers, long headed by Matthes and later by Field, has stimulated discussion and publication of glacial studies.

The above-cited studies and scores of technical engineering reports have begun to produce the basis for a genuine morphology not only of water-created landforms but also of the water bodies themselves. The coordination of principles and the systematic analysis of water-related forms remain largely to be done.

BASIC WATER DATA AND THEIR GEOGRAPHIC USE

Unlike contour maps and soil maps, for which quantitative data can be obtained on the basis of observations in the field made at one time, quantitative generalizations about water features must be based upon years of continuing painstaking measurements, because of the fluctuating character of these forms. Devices and techniques for measuring precipitation, stream flow, and the water equivalent of snow on the ground must be developed first, and then the uniform application of these techniques arranged. Although individuals can develop the techniques, only the government has sufficient resources and popular support to maintain on a non-profit basis a network of long-term observations. It is to the government statistics that geographers must turn for the raw materials of their quantitative studies of the regional distribution of water, seasonal fluctuations, abnormal fluctuations (floods and droughts), and long-time trends.

PRECIPITATION

The present status of basic water statistics in the United States is presented in a report by J. R. Mahoney [17] and by a chapter and series of maps in the report of the President's Water Resources Policy Commission [18: Vol. 1, 336-364]. For two elements only, precipitation and streamflow, have fairly adequate water data been collected. By 1900 a sizable network of precipitation gauges was in operation by the Weather Bureau, following earlier, more limited observations by the Signal Corps. By 1950 the Weather Bureau had more than 10,000 regular and cooperative stations measuring precipitation. On the basis of these observations, it now is possible to obtain estimates of the total amount of water reaching the surface of the United States and each of its main regions, and to make maps of various phases of precipitation distribution, of which may be mentioned those published in 1922 in the *Atlas of American Agriculture* under the supervision of O. E. Baker [19], and the numerous studies by S. S. Visher.

For some areas, notably mountains, many additional stations are needed to provide a complete estimate of precipitation. One of the most thorough attempts to develop methods of mapping precipitation in mountainous areas with few stations was made by W. C. Spreen [20], who found that such factors as exposure and orientation are more important than elevation for this purpose.

The results of regular precipitation measurements are published by the Weather Bureau, notably in *Climatological Data* for basic statistics, and in the *Weekly Weather and Crop Bulletin* for current summaries and maps of drought or surplus water conditions.

STREAMFLOW

Only scattered measurements of streamflow had been made by 1900, but systematic and continuous measurements, made by the Geological Survey directly or through the States, were begun in 1900, and the basic network of gauging stations was established during the period 1910 to 1940 for the most part. By 1950 observations were being taken regularly at about 6,000 points, and a comprehensive map of average annual runoff had been published [21]. For most of the small streams and even some sizable tributaries of the large ones we still have no measurements of flow. Runoff data are published and summarized by the Geological Survey annually in *Water Supply Papers,* monthly in the *Water Resources Review.* Included in the former series are special reports of individual floods and general summaries of floods [22].

LAKES

Limnology in the United States has been concerned primarily with the biologic aspects of lakes although occasional studies have been made of other of their aspects. Systematic continuous observation has been made of lake levels of the larger lakes by the Geological Survey (published in annual runoff reports) and, of navigable lakes, by the Corps of Engineers. Few of the smaller lakes are measured, although Henry Thoreau demonstrated a century ago even without instruments the interesting character and results, if not the causes, of seasonal and long-term changes in the level of Walden Pond [23: 156-157]. While there have been several studies of individual lakes or groups of lakes (notably the Great Lakes and the Pleistocene lakes of the Great Basin), there is a lack of broad works in which the lake types of the country as a whole have been classified and mapped. The small study by I. C. Russell in 1895 [25], considering the origin and characteristics of a few of the outstanding lakes of the country, seems to have had no successors. As with rivers, geographers have been more concerned with physiographic effects of past and present lakes than with lakes as features of the landscape.

Of late years, attention has been concentrated upon evaporation from the surface of artificial lakes, as affecting the efficiency of water storage. Although it was known that increased evaporation caused by reservoirs sometimes exceeds the amount of usable water gained by storage [26], measurement of the precise amount of water lost presented great difficulties. In 1952 for the first time a thoroughly adequate evaporation study was completed of a lake (Lake Hefner). It was determined that a knowledge of lake temperature and the standard meteorologic data of a nearby weather station are adequate for making close estimates of actual evaporation [27]. Incidental to the main purpose of the study, a contribution was made to our knowledge of lake morphology.

GROUND WATER

Ground water, comparable with runoff in economic significance, lies at the very core of the interest of the Geological Survey, and it is to the Survey that we must turn for most of the available quantitative data on this subject. Such data on the ground water of the United States are fragmentary, despite the Survey's program of well measurement. Adequate field studies have been made primarily where there are pressing problems, especially where the water table has been falling dangerously. Elsewhere there is little specific knowledge of the movement and source of underground water. Geographically, ground water commonly has escaped attention, as it does not constitute nor create surface forms of the earth except where it emerges as springs. Its powerful effect upon surface cultural features, especially in the arid West, is of course fully recognized.

Our knowledge of ground water depends more upon intensive areal surveys than upon data accumulated over the years. The *Water Supply Papers* of the Geological Survey are the main source of information and present the results both of areal surveys and of nation-wide generalizations. Since early in the century when the series was started with papers by I. C. Russell and others on ground-water conditions in volcanic areas of the West and of the Hawaiian Islands, new reports have been issued unceasingly. Highlights in summaries of ground-water knowledge are important works of O. E. Meinzer [4; 28; 29; 30] appearing between 1923 and 1942, with generalized regional maps; and, of most recent date, the volume of case histories on ground water compiled and digested by H. E. Thomas for the Conservation Foundation [31], and the regional summary by C. L. McGuinness [32].

Certain occurrences of ground water, amounting to underground rivers rather than to "normal" slowly seeping conduit movements, have attracted attention in the limestone areas of the East and the volcanic areas of the West. Another factor receiving increasing attention with its growing economic seriousness has been the encroachment of sea water in bodies of ground water near the coast [33; 34]. Altogether, there is no shortage of distributional aspects of ground water for the geographer's attention, and one geographer recently has made a regional compilation of ground-water studies [35].

SNOWFALL

In the field of cryology, only for snowfall have basic continuous measurements been made for long periods by the Weather Bureau. Measurement of the water equivalent of snow on the ground, a highly important factor for estimating potential spring and summer runoff from the Western mountains, is of comparatively recent date. Even though the invention of the weight snow sampler by J. E. Church in 1909 introduced a rapid method of sampling snow, the mushrooming of Western snow courses did not occur until the decade between 1930 and 1940. By 1950 there were more than 1,000 snow courses in operation in the West, coordinated by the Soil Conservation Service. In the East, the Eastern Snow Conference has maintained snow courses in New York and New England since 1940, and the Weather Bureau publishes the results annually [18: Vol. 1, 340]. J. E. Church, the father of snow surveying, gives an account of it in its early days [36].

Outstanding intensive field research on the whole complex of climatic, terrain, vegetation, and hydrologic phenomena associated with snow cover in the mountains is being published by the Corps of Engineers from studies of experimental basins in the Rockies, the Sierra Nevada, and the Cascades. The purpose of these studies has been to estimate the rate and the amount of the snowmelt contribution to runoff [37]. In this research, the energy-budget concept has proved useful in reconstructing hour-to-hour variations in stream flow produced by changes in the meteorological elements. Work in the opposite direction, namely the effect of snow cover on the atmosphere, carried on by D. H. Miller, has thrown new light on the characteristics of mountain climates. He finds, for example, that in the high Sierra the daytime temperature of the air rises high above the melting point of snow as a result of the solar energy absorbed by trees and released as heat.

Very helpful in cryological research are two major bibliographies. One is the *Arctic Bibliography* prepared under the direction of the Arctic Institute of North America, consisting of 20,000 annotated thoroughly indexed items [38]. The other is the annotated bibliography, now numbering about 7,000 items and steadily growing, of the Snow, Ice, and Permafrost Research Establishment sponsored by the Corps of Engineers.

EVAPORATION AND TRANSPIRATION

Evaporation data, obtained from a representative but inadequate network of standard pans, have been summarized by the Weather Bureau [39]. Much work has been done to devise more sensitive methods of estimating evaporation, utilizing measurements of air turbulence [40], but the pan-method is still standard, and the studies at Lake Hefner [27] suggest that it may be possible to work out usable correlations between pan and lake evaporation. If the transpiration of plants be considered, as it must in a comprehensive system of water budgeting, the picture becomes still more complicated. Quantitative measurement of evapotranspiration has been carried out experimentally for many years, by the plant geographer H. L. Shantz, for example [41] and by others. Methods of measuring evapotranspiration on a basis nearly approaching that of natural conditions have been devised by Thornthwaite [42], and a few experimental stations have been set up, but no regular measurements of this element are made by governmental agencies. The relation between evapotranspiration and the rest of the hydrologic cycle has received much attention during the past decade, both experimentally and by the simple system of subtracting runoff from precipitation figures for a given area. The importance of evapotranspiration is indicated by an estimate for the United States that the total precipitation amounts to about 30 inches, the runoff to about 8½ inches, leaving an evapotranspiration loss of 21½ inches [21: 12]. An ingenious and useful geographic application of potential evapotranspiration, calculated primarily from temperature data and mapped for the United States, was used by Thornthwaite to work out regions of differing moisture adequacy [43]. The necessity for reducing every possible source of water waste in the dry West has led to study of phreatophytes, or plants that draw upon the ground water [44; 45]. The effect of vegetation as a conserver of runoff on slopes is another aspect of the relation of plants to the hydrologic cycle.

As plant geography impinges upon hydrology, so also does soil science. Infiltration, an important initial process in the production of ground water, is dependent upon the size of the soil pores, and soils must be classified upon a special basis where their hydrologic work is considered [46; 47].

AREAL SYNTHESIS OF WATER FORMS ..

More and more the elements of hydrology are being combined in complete studies of stream basins and larger areas. The professional hydrologists have devoted their attention increasingly to analyzing precipitation and runoff with a view to forecasting stream hydrographs (curves showing changes in water flow) that result from various amounts of precipitation under various basin conditions [48]. Such studies are essential for the efficient management of multiple-purpose interrelated dams and other water conservation works involving a constantly-changing balance of choices between the demands of navigation, flood control, irrigation, power generation, and recreation. There is a voluminous literature on this phase of hydrology, which for the most part is of greater interest to engineers than to geographers. It does, however, shed light on the character of hydrologic features, and to that extent can aid the geographer in classifying water features. Furthermore, those studies that attempt to give a complete picture of the terrain, vegetation, and other factors affecting runoff constitute specialized geographic studies in themselves. An early summary of quantitative aspects of basin analysis was given by R. E. Horton, who emphasized methods of classifying drainage-density and slope [49]. Basin studies for runoff analysis now give consideration to a large number of factors and are applied to snow as well as rainfall. A recent example of an areal synthesis is the study of Castle Creek Basin in the

Sierra Nevada by David H. Miller, in which he discusses the significance of slope, the curvature of surfaces, the soils, the rocks, the intensity of insolation, and the vegetation [50].

Geographers have also devoted themselves to still broader areal syntheses. Some studies of this kind have been made in connection with the development of specific reclamation projects, for example, Grand Coulee or the Central Valley of California. Geographers such as E. N. Torbert in the Columbia Basin and John Abrahamson in the Missouri Basin have played leading parts in the practical development of plans by the Bureau of Reclamation. Other areal studies are of a more academic type, aimed primarily at making an intelligible synthesis of the diverse facts bearing upon water and its potentialities [51; 52; 53]. One deals with the whole gamut of areal hydrology in relation to a specific drought [54].

An outstanding series of areal syntheses of water problems for ten important river basins was assembled by E. A. Ackerman [18: Vol. 2] as part of the work of the President's Water Resources Policy Commission. One of the directors of the Commission, Gilbert F. White, had served earlier as principal geographer of the National Resources Planning Board during the drought and depression years of the 30's. Yet another geographer, Harlan H. Barrows, served the National Resources Board as a member of the Water Resources Committee. These geographers did their part in developing one of the major policies of the Board, that of basin-wide syntheses [55]. A valuable study of water problems of the arid portions of North America, with a bibliography of 424 items, has recently been published by Linsley [56]. Peveril Meigs summarized some water problems of the United States in a paper published in 1952 [57].

THE PROSPECT

In spite of the many contributions of geographers to the study of water and especially to the formulation of water policies, a vast amount of detailed work remains to be done. There is need for the development of a system of categories of water features; and when such a classification has been established, the mapping of the distribution of the different forms of water throughout the United States must be undertaken.

There is a wealth of statistical data on water. Statistics have been accumulated through long years of continuous observation by government agencies, giving information concerning precipitation, stream flow, snow depth and water equivalent, lake levels, and evaporation. Other agencies have provided supplementary data on the basis of shorter, intensive periods of field study, such as the materials brought together for the Muskingum Valley by the Soil Conservation Service.

All these data await use by geographers for geographical purposes. River basins can be examined for the purpose of gaining a picture of their normal conditions; and in relation to these normals the eccentricities of drought and flood can be measured. The synthesis in specific drainage basins of all relevant aspects of water availability and potential water use can yield important results. The relations of vegetation and crops to water as an element of the environment offers a wide range of practical problems in the solution of which the geographic point of view can be useful.

REFERENCES

1. LINSLEY, R. K. JR., KOHLER, M. A., and PAULUS, J. L. H. *Applied Hydrology*, New York, 1949.
2. MEAD, D. W. *Notes on Hydrology*, Chicago, 1904.
3. MEYER, A. F. *The Elements of Hydrology*, New York, 1919.
4. MEINZER, O. E., ed. *Hydrology, Physics of the Earth IX*, National Research Council, New York, 1942.
5. LINSLEY, R. K., ed. *Annotated Bibliography on Hydrology, 1941-1950, United States and Canada*, United States Federal Inter-Agency River Basin Committee, Notes on Hydrologic Activities, No. 5, 1952.
6. DAVIS, W. M. "The Geographical Cycle," *Geographical Journal*, 14 (1899): 481-504.
7. RUSSELL, R. J. "Geographical Geomorphology," *Annals of the Association of American Geographers*, 39 (1949): 1-11.
8. RUSSELL, R. J., and others. *Lower Mississippi River Delta*, Department of Conservation, Louisiana Geological Survey, Geological Bulletin No. 8, New Orleans, 1936; *idem.* "Louisiana Stream Patterns," *Bulletin of the Association of Petroleum Geologists*, 23 (1939): 1199-1227.
9. LEIGHLY, J. B. *Toward a Theory of the Morphological Significance of Turbulence in the Flow of Water in Streams*, University of California Publications in Geography, 6 (1932): 1-22; *idem.* "Turbulence and the Transportation of Rock Debris by Streams," *Geographical Review*, 24 (1934): 453-464.
10. LEOPOLD, L. B., and MADDOCK, T. *The Hydraulic Geometry of Stream Channels and Some Physiographic Implications*, United States Geological Survey, Professional Paper 252, Washington, 1953; *idem.* "Downstream Change of Velocity in Rivers," *American Journal of Science*, 251 (1953): 606-624.
11. FRIEDKIN, J. F. *Meandering of Alluvial Rivers*, United States Waterways Experiment Station, Vicksburg (Mississippi), 1945.
12. FISK, H. N. *Fine-Grained Alluvial Deposits and their Effects on Mississippi River Activity*, 2 vols., Corps of Engineers, Washington, 1947.
13. THORNTHWAITE, C. W., SHARPE, C. F. S., and DOSCH, E. F. *Climate and Accelerated Erosion in the Arid and Semi-Arid Southwest, with Special Reference to the Polacca Wash Drainage Basin, Arizona*, United States Department of Agriculture, Technical Bulletin 808, Washington, 1942.
14. MATTHES, F. E. *Glacial Sculpture of the Bighorn Mountains, Wyoming*, United States Geological Survey, 21st Annual Report, Part 2, (1900): 167-190.
15. FIELD, W. O. JR., and MILLER, M. M. "The Juneau Ice Field Research Project," *Geographical Review*, 40 (1950): 179-190.
16. LAWRENCE, D. B. "Glacier Fluctuations for Six Centuries in Southeastern Alaska and Its Relation to Solar Activity," *Geographical Review*, 40 (1950): 191-223.
17. *A Comprehensive Basic-Data Program for Water Resources*, 81st Congress, Second Session, House Document 706, Part 2, Washington, 1950.

18. The President's Water Resources Policy Commission, *A Water Policy for the American People* (Vol. 1); *Ten Rivers in America's Future* (Vol. 2); *Water Resources Law* (Vol. 3), Washington, 1950.

19. KINCER, J. B. *Climate: Precipitation and Humidity*, Washington, 1922, in *Atlas of American Agriculture*, United States Department of Agriculture, Washington, 1936.

20. SPREEN, W. C. "A Determination of the Effect of Topography upon Precipitation," *Transactions of the American Geophysical Union*, 28 (1947): 285-290.

21. LANGBEIN, W. B., and others. *Annual Runoff in the United States*, United States Geological Survey Circular 52, Washington, 1949.

22. JARVIS, C. S., and others. *Floods in the United States: Magnitude and Frequency*, United States Geological Survey, Water-Supply Paper 771, 1936.

23. THOREAU, H. D. *Walden*, Penguin Books Edition, New York, 1943 (first published 1854).

24. PETTIS, C. R., and HICKMAN, H. C. "Hydrology of the Great Lakes–a Symposium," *American Society of Civil Engineers, Proceedings*, 65 (1939): 584-606.

25. RUSSELL, I. C. *Lakes of North America*, New York, 1895.

26. LINSLEY, R. K. "The Hydrologic Cycle and Its Relation to Meteorology: River Forecasting," in *Compendium of Meteorology*, American Meteorological Society, Boston, 1951: 1048-1054.

27. *Water-Loss Investigations: Vol. 1–Lake Hefner Studies, Technical Report*, (Collaborating Agencies: Department of the Navy, Bureau of Ships, Navy Electronics Laboratory; Department of the Interior, Bureau of Reclamation, Geological Survey; Department of Commerce, Weather Bureau), United States Geological Survey, Circular 229, Washington, 1952.

28. MEINZER, O. E. *Outline of Ground-Water Hydrology*, United States Geological Survey, Water-Supply Paper 494, 1923.

29. ———. *Large Springs in the United States*, United States Geological Survey, Water-Supply Paper 557, 1927.

30. ———. *Ground-Water in the United States*, United States Geological Survey, Water-Supply Paper 836-D, 1939.

31. THOMAS, H. E. *The Conservation of Ground Water*, New York, 1951.

32. McGUINNESS, C. L. *The Water Situation in the United States with Special Reference to Ground Water*, United States Geological Survey, Circular 114, Washington, 1951.

33. TODD, D. K. "Sea-Water Intrusion in Coastal Aquifers," *Transactions of the American Geophysical Union*, 34 (1953): 749-754.

34. GREGOR, H. F. "The Southern California Water Problem in the Oxnard Area," *Geographical Review*, 42 (1952): 16-36.

35. BARTZ, P. McB. *Ground Water in California: The Present State of Our Knowledge*, Giannini Foundation Ground Water Studies 1, University of California, College of Agriculture, Berkeley, 1949.

36. CHURCH, J. E. "Snow Surveying–Its Principles and Possibilities," *Geographical Review*, 23 (1933): 529-563.

37. *Synopsis of Snow Investigations and Plans for Fiscal Year 1954*, Corps of Engineers, North-Pacific Division, Portland (Oregon), 1953.

38. *Arctic Bibliography*, 3 Vols., Department of Defense, Washington, 1953.

39. *Mean Monthly and Annual Evaporation from the Free Water Surface for the United States, Canada, Alaska, Hawaii and Puerto Rico*, United States Weather Bureau, Technical Paper 13, Washington, 1951.

40. THORNTHWAITE, C. W., and HOLZMAN, B. *Measurement of Evaporation from Land and Water Surfaces*, United States Department of Agriculture, Technical Bulletin 817, 1942.

41. SHANTZ, H. L., and PIEMEISEL, L. N. "The Water Requirements of Plants at Akron, Colo.," *Journal of Agricultural Research*, 34 (1927): 1093-1190.

42. SANDERSON, M., "Measuring Potential Evapotranspiration at Norman Wells, 1949," *Geographical Review*, 40 (1950): 636-645.
43. THORNTHWAITE, C. W. "An Approach Towards a Rational Classification of Climate," *Geographical Review*, 38 (1948): 55-94.
44. GATEWOOD, J. S., and others. *Use of Water by Bottom-Land Vegetation in Lower Safford Valley, Arizona,* United States Geological Survey, Water-Supply Paper 1103, Washington, 1950.
45. ROBINSON, T. W., and others. "Symposium on Phreatophytes," *Transactions of the American Geophysical Union,* 33 (1952): 57-80.
46. HORTON, R. E. "The Role of Infiltration in the Hydrologic Cycle," *Transactions of the American Geophysical Union,* 14 (1933): 446-460.
47. FREE, G. R., BROWNING, G. M., and MUSGRAVE, G. W. *Relative Infiltration and Related Physical Characteristics of Certain Soils,* United States Department of Agriculture, Technical Bulletin 792, Washington, 1940.
48. FOSTER, E. E. *Rainfall and Runoff,* New York, 1949.
49. HORTON, R. E. "Drainage Basin Characteristics," *Transactions of the American Geophysical Union,* 13 (1932): 350-361.
50. *Terrain Characteristics, Central Sierra Snow Laboratory Basin,* Cooperative Snow Investigations, Technical Report 4A, Corps of Engineers, South Pacific Division, San Francisco, 1951.
51. MEIGS, P. "Water Planning in the Great Central Valley, California," *Geographical Review,* 29 (1939): 252-273.
52. MASON, C. Y. "Water in the Laramie Region of Wyoming," *Economic Geography,* 15 (1939): 271-282.
53. SCHWENDEMAN, J. R. "Water Supply Geography of the Basin of the Red River of the North in the United States," *Economic Geography,* 20 (1944): 153-188, 258-285.
54. *The Drought in Southwestern United States as of October 1951,* United States Department of the Interior, Washington, 1951.
55. BARROWS, H. H. "A National Plan and Policy for the Control and Use of Water Resources," in COLBY, C. C., ed. *Geographic Aspects of International Relations,* Harris Foundation Lectures, 1937, Chicago, 1938: 99-123.
56. LINSLEY, R. K. "Report on the Hydrologic Problems of the Arid and Semi-Arid Areas of the United States and Canada," *Reviews of Research on Arid-Zone Hydrology* (UNESCO), Paris, 1953: 128-152.
57. MEIGS, P. "Water Problems in the United States," *Geographical Review,* 42 (1952): 346-366.

THE GEOGRAPHIC STUDY OF THE OCEANS[A]

THE PHYSICAL BASIS OF OCEANOGRAPHY
CURRENTS IN THE SURFACE LAYER
WIND, SEA, SWELL, AND SURF
WATER MASSES AND UNDERWATER CURRENTS
ECOLOGICAL ASPECTS

THE APPLICATION OF THE REGIONAL CONCEPT
THE WORK OF GERHARD SCHOTT
A NEW APPROACH TO THE REGIONS OF THE OCEANS
THE OCEANIC REGIONS
THE CONTINENTAL REGIONS

APPLICATIONS OF OCEANOGRAPHY

A. Original draft of the section on the physical basis of oceanography by C. J. Burke; original draft of the section on the regional aspects of the oceans by Francis E. Elliott. The chapter was sponsored by the Committee on Physical and Bio-Geography, Hoyt Lemons, chairman.

CHAPTER EIGHTEEN

THE GEOGRAPHIC STUDY OF THE OCEANS

MOST OF THE PROFESSIONAL geographers of the United States are landlubbers. American geography, almost without exception, has dealt with the phenomena and processes of land areas. To be sure, the climatologists have recognized the great importance of oceanic conditions as affecting the climates of the lands, and have extended their climatic maps over the oceans. The synoptic weather maps for whole hemispheres cover land and water alike. Yet few professional geographers would list oceanography as their topical or an ocean as their areal specialties.

There is, however, no logical justification for the restriction of geography to the study of land areas. There is a geography of the oceans almost as broad in its physical and biotic ramifications as the geography of the lands. Many of the concepts and methods of geographic research presented in the several chapters of this book are fully as applicable to the ocean areas of the world as to the land areas. Yet the geographic point of view has not been widely adopted by students of the oceans.

There are two chief reasons for this. In the first place, the oceanographers include among their numbers very few persons who have contact with modern geography. Most of them are physicists, meteorologists, chemists, geologists, and biologists. In the second place, oceanography has been chiefly advanced during the past three decades or so by the more and more precise understanding of the physical processes at work in the ocean areas. From the descriptive writings on the ocean by Elisée Reclus [1] to the highly technical treatise of H. U. Sverdrup, M. W. Johnson, and R. H. Fleming [2] the understanding of oceanic processes shows a notable advance. Oceanography rests on a firm mathematical basis for theoretical research. But the detailed observations in particular parts of the oceans, where the interaction of unsystematically associated phenomena can be observed, are few. The number of vessels adequately equipped to carry on oceanographic observations and measurements at all times of the year and in the open sea can be counted on the fingers of one hand.

Modern oceanography faces an unusual problem. New kinds of instruments, made possible by advances in applied physics, now permit a greatly increased precision of measurement of oceanic phenomena. The theoretical grasp of the processes of physical oceanography is well advanced. Yet there is a notable lack of detailed observations, and an almost complete absence of any reports on the study of small oceanic areas where all the various processes could be examined in their interaction. Marine biology is also a well-advanced field, especially with respect to commercial fisheries; but here also there is a lack of information regarding the interaction of physical, biotic, and economic processes in particular places. The data needed for the application of geographic concepts and methods are not available. Here is an inviting prospect for the unusual kind of scholar who can combine an understanding of physical theory with the geographer's insistence on direct observation in specific locations.

THE PHYSICAL BASIS OF OCEANOGRAPHY

In the oceans as well as on the land physical, biotic, and cultural processes operate in complex and unsystematic association to produce the phenomena observable at any one time. The study of these processes and their related phenomena can be approached from the point of view of the student of process, such as the physical oceanographer or the marine biologist, or it can be approached from the point of view of the geographer, stressing the characteristics of particular areas and the modifications of the processes resulting from the particular associations of phenomena in specific locations. The greatest progress toward an understanding of the oceans would presumably result from the successful combination of these approaches.

Ocean water can be characterized by such physical and chemical properties as temperature, pressure, density, conductivity, salinity, and chlorinity. At any given time, and at any given spot in the ocean, the water has unique values for these physical and chemical variables, for even in the oceans no two points on the earth are identical. The physical oceanographer, facing this situation, works toward a complete theoretical description of the distribution of the properties of sea water and of the temporal changes in these distributions. He considers the ocean as a bounded system in which changes are effected by external processes at the boundaries of the system and by internal processes which depend only on the properties of the system.

The approach of the physical oceanographer can be illustrated by a specific problem. Suppose that the distribution of temperature throughout the ocean were known exactly at a certain time in the past and that it was desired to calculate the distribution after a given time interval. Knowledge of the following factors would be required:

a. The radiation absorbed and emitted at the sea surface;
b. The heat interchange with the atmosphere due to the convection of sensible heat;

c. The interchange of mechanical energy with the atmosphere;

d. The evaporation and condensation of water vapor at the sea surface;

e. The convection of heat at the ocean bottom;

f. The gain or loss of heat at the mouths of rivers due to the influx of water with its own characteristic temperature;

g. The transformation of mechanical energy into heat due to internal friction in the ocean water;

h. The transformation of mechanical energy into heat due to bottom friction;

i. Heating due to chemical and biological processes within the ocean;

j. The extent of turbulent mixing within the ocean and its effect on the distribution of temperature;

k. The extent of transformation of heat into mechanical energy of thermal currents.

No oceanographer has, of course, attempted a complete solution of this kind, but these are the processes he would consider if he did. He recognizes that the effects of some are negligible, and that a close approximation of the complete picture can be had from a consideration only of the more important factors. Yet the theoretical picture is a complete one.

The geographic approach is quite different. Instead of seeking as complete a description as possible of a specific process, such as the changes in the distribution of temperature in the ocean, the geographer seeks as complete a description as possible of the associated properties of specific areas. He proceeds 1) to seek criteria by which masses of ocean water can be distinguished; 2) to map the areas occupied by water masses that are homogeneous in terms of these criteria; and 3) to discover the areal relations between different kinds of regional systems. The physical oceanographer seeks ultimately a complete theoretical picture of the ocean on a scale of 1/1; the geographer seeks to arrive at generalizations concerning the associated characteristics of ocean areas, always at scales sufficiently reduced to permit cartographic analysis. When the latter approach is guided by the theoretical understandings of the former, and the theoretical understandings of the former are illuminated and extended by the effective generalizations of the latter, the greatest over-all progress toward an understanding of the oceans will have been made.

There are several aspects of oceanography in which the concepts developed by the physical oceanographers are especially important. Four of these will be considered here.

CURRENTS IN THE SURFACE LAYER

The amount of observational data on surface currents in the oceans provides a more nearly adequate picture of this aspect of oceanography than is provided by the data on other ocean characteristics. The comparison of the maps of ocean currents with those of average wind directions reveals a close correspondence of patterns. Can it be said, then, that ocean currents are a simple and direct effect of the wind stress? Such an interpretation has been assumed in the past, but is no longer tenable in the light of modern theory.

A steady current cannot exist at the surface or at any depth below the surface unless the isobaric surface at the same depth is sloping. A steady surface current can only be produced by a slope of the free surface. The earth's rotation causes moving bodies on the earth to curve to the right in the Northern Hemisphere and to the left in the Southern Hemisphere. Surface water currents curve in these directions until they move at right angles to the slope of the free water surface. Therefore, in the Northern Hemisphere the ocean surface is high to the right of a current and low to the left.

It follows that the existence of permanent ocean currents requires a permanent non-homogeneous distribution of mass in the ocean basins. It is possible that this non-homogeneous distribution of mass was brought about and is maintained by thermo-haline effects in a manner similar to that outlined by Bjerknes and his collaborators for the atmosphere [3]. It is also possible that the mass distribution was brought about and is maintained by the stress of winds, or by some combination of the two. Recent investigations indicate that perhaps the wind stress alone may be sufficient to account for the observed effects of this mass distribution [4; 5; 6; 7]. But it is not the wind stress alone which produces the current; rather it is a secondary effect resulting from the transportation and accumulation of water to the right of the wind in the Northern Hemisphere and to the left of it in the Southern Hemisphere. It is the vorticity of the stress field rather than the simple stress that is important. The significance of this can be illustrated by a simple example.

Consider a rectangular area in mid-ocean extending over several degrees of latitude and longitude in the Northern Hemisphere. The winds over this area are westerly, strong to the north and progressively weaker toward the south. The strong winds to the north, acting by direct stress on the ocean surface, would cause the transport of large quantities of water into the area (since water moved along by the wind tends to swing to the right of the wind in the Northern Hemisphere); the weaker winds to the south would cause a smaller transport of water out of the area. Under these circumstances, mass must accumulate inside the area, running like a ridge from west to east. Such a slope of the free water surface would then give rise to two currents: one would flow toward the east on the northern side of the area; and another would flow toward the west on the southern side of the area. Both would have been produced by a westerly wind.[B]

WIND, SEA, SWELL, AND SURF

The importance of surf conditions for landing operations in World War II gave impetus to extensive research on short-period surface waves. Results of this research are summarized by H. U. Sverdrup and W. H. Munk [8; 9]. Qualitative observations indicate that these waves are formed by wind stress; hence, in principle, it should be possible to predict the wave characteristics whenever the wind field is known.

In the theoretical approach, the energy imparted to the water by the wind is computed. This energy gives a motion to the water particles which is described by the wave equation. The energy received may be divided into two parts, associated respectively with the height and period of the waves. It is then possible to compute the height and period of the waves in the generating area in terms of the direction and duration of the wind. The results are in good agreement with available observations of waves and wind in generating areas, an agreement brought about, to be sure, by the somewhat arbitrary division of the energy into two parts. The wave height that is calculated is approximately the average height of the highest third of the waves present (known as the significant waves) and the period is that associated with these waves.

After the waves have been generated they leave the generating area and move as a train. The wave front (a region of sharp transition from low to high waves) moves with a characteristic group velocity. During this movement the height of the significant waves decreases and the period associated with them increases. This phenomenon is handled by means of an interchange of the energies associated with height and period. From a knowledge of the wave characteristics in the generating area and the distance over which they must travel, it is possible to compute the height, period, and arrival time of the resulting swell at a distant point.

Sverdrup has re-examined the increase in period and decrease in height of the significant waves during the travel from the generating area [10]. He assumes a spectrum of impulsively generated waves in the generating area and shows that, on the basis of conservation of energy, one would expect the height of the waves at the wave front to decrease, but not so rapidly as in the first statement of this theory. He then introduces air resistance as an additional factor and obtains results in general agreement with the earlier theory but which show that the first statement was an

B. Note that it is customary to name a wind for the direction from which it comes; and it is also customary to name an ocean current for the direction toward which it goes.

oversimplification. In his new formulation of the theory, the character-istics of the decay process and movement are not completely determined by the height and period of the waves as they leave the generating area. When the effect of air resistance is considered the theory predicts that short-period waves will die out first, in agreement with observations that swell is longer and smoother than waves in the generating area.

Interest in surf conditions has stimulated research on the transforma-tions of waves as they near the shore. The refraction effects of the bottom, longshore currents due to water transported by the waves, and the move-ment of sand on both shore and bottom due to wave action and longshore currents have received attention [11; 12].

Many interesting and important theoretical problems remain in the study of wave generation, modification, and decay since the theory re-mains quite tenuous at several points. Yet the general system has enjoyed considerable success in the prediction of wave phenomena from weather data.

WATER MASSES AND UNDERWATER CURRENTS

A mass of water can be identified by its physical properties and chemi-cal content. The most important variables used in this identification are temperature and salinity. Rather uniform geophysical processes tend to produce systematic differences in the water characteristic of various oceanic regions. In regions of high evaporation, the water tends to have a high salinity. In warm regions, water temperature is characteristically high. In cold regions, especially where there is a large influx of fresh water from the mouths of rivers, water of low temperatures and low salinity is formed. In the Gulf Stream system, water of high salinity is cooled as it moves northward, forming a water mass characterized by high salinity and fairly low temperature.

As a water mass moves from its place of origin its temperature and salinity are gradually modified, yet the water mass can be traced as it moves from its source. In general, a water mass cannot be identified by a single value of temperature and salinity since these quantities are subject to variation in space and time. Rather, the identification is by means of a characteristic temperature-salinity curve. The possibility of tracing the movement of water masses through measurements of temperature and salinity enables the oceanographer to construct the patterns of deep-water flow. Currents in the ocean depths have not been accessible to direct ob-servation, but measurements of temperature and salinity can be made at any depth. The detailed results of such analyses are presented by Sverdrup, Johnson, and Fleming [1].

ECOLOGICAL ASPECTS

As water masses move, their properties are modified very slowly. The conditions which give rise to the formation of characteristic masses are relatively uniform from year to year. Over large areas in the ocean, the only changes in physical conditions that can occur, aside from the surface layers, are due to the influx of water masses formed elsewhere in the ocean. The deep water remains remarkably uniform over long periods of time. Even in the surface layers, accessible to the effects of solar radiation, evaporation, and interchange of energy with the atmosphere, changes tend to be rather slow because of the physical properties of sea water. Its high heat capacity and a high latent heat of evaporation tend to keep changes in temperature small and gradual.

In this connection, it should be noted that ocean-living animals have not developed the complex temperature regulating mechanisms necessary to land animals. In the highly stable deep-water environments, marine organisms become closely attuned to the prevailing mean conditions so that slight environmental differences from place to place are associated with recognizably different organisms. An interesting phenomenon resulting from slow diurnal changes in the amount of light at intermediate depths is a diurnal vertical movement of organisms to preserve optimal conditions. The upper layers and areas near shore where conditions are less uniform are inhabited by more mobile and adaptive organisms. It is possible to divide the oceans into more or less clearly defined biotic regions, each occupied by organisms of specific types with little overlap in the forms of life.

THE APPLICATION OF THE REGIONAL CONCEPT

The geographical study of the oceans involves the use of the regional concept and the regional method as defined in chapter two of this book. The understandings of the processes involved, such as those described above, provide a basis for the selection of significant criteria: that is, criteria which are relevant to the causes and consequences of differences from place to place in the oceans. The purpose is to define oceanic regions, homogeneous in terms of the criteria by which they are defined. So long as observational data are meager, only very general regions can be identified; the more detailed and less highly generalized examination of particular parts of the ocean, such as the Norwegian Sea, cannot, be undertaken without additional information of the kind that could be plotted on large-scale maps.

The popular division of the marine waters of the earth into oceans, seas, bights, gulfs, bays, and others, is no more meaningful in terms of regional study than is the similar division of the lands into continents. There is need for the application of the regional method to the study of likenesses and differences in the oceans through the formulation of criteria, and the examination of areal relations.

THE WORK OF GERHARD SCHOTT

Students of the oceans are indebted to the monumental works of Gerhard Schott [13; 14]. In his earlier works Schott gathered together a vast amount of oceanographic data and presented them in the form of maps and text. In 1935 he first published a suggested division of the Pacific and Indian Oceans into natural regions; and in a separate paper in 1936 extended his regional system to the Atlantic. His divisions of the oceans are based on "a synthesis of the most important characteristics and processes in the body of water and in the air above it." Unfortunately Schott does not define his regions in terms of specific criteria, nor does he specify what are the most important characteristics and processes. Because his regions are designated as subdivisions of the oceans, the analogous regions in different oceans are not brought together. To use regions of this sort as a framework for general description may be satisfactory, but it is impossible to use them for the kind of analysis of accordant and discordant relationships described earlier in this book as an essential part of the regional method.

A NEW APPROACH TO THE REGIONS OF THE OCEANS[c]

In seeking a division of the world's oceans into regions homogeneous in terms of specific criteria, it is necessary to deal with the lower layers of the atmosphere and the upper layers of the ocean. Generally the upper layer of the ocean extends in depth to the most prominent break in temperature. In winter this surface layer in the middle latitudes is about 300 feet deep; it grows shallower toward the equator and toward the continental shores; it deepens to about 400 feet in the polar areas. In summer the depth is considerably reduced, especially in the middle latitudes, because the heating of the surface water produces greater stability [15].

In arriving at the regional divisions of this zone between the atmosphere and the hydrosphere, maps of seven different phenomena were examined and their patterns matched. These maps were:

a. Mean winter temperature of the surface water;
b. Mean annual variation of surface water temperature;
c. Average surface salinity;
d. Major surface currents and major wind belts;

---

c. The regional divisions presented here were defined by Francis E. Elliott (U.S. Navy Hydrographic Office).

e. Average winter limits of the Polar Ice Pack;

f. Mean annual variation of air temperature;

g. Mean annual precipitation.

When the patterns shown on these seven maps are examined for their areal relations they are found to accord in a broader pattern of multiple-feature regions. The single-feature boundaries are bunched in boundary girdles around the areas of homogeneity. It is found, moreover, that the two most important single features of the oceans are temperature of the water and salinity, for these two coincide closely with the boundaries on the other five maps. The resulting regions are shown on figure 15. Eight kinds of regions are recognized, two of them subdivided on the basis of high and low salinity.

The patterns of oceanic regions are arranged by latitude and by position with respect to the large-land masses. The general latitudinal arrangement is clear, especially in those parts of the oceans that are far from the continents. Near the land, on the other hand, the regions lie generally parallel to the coast. The sea has an ameliorating influence making for small seasonal contrasts, while the land has the opposite effect, bringing about marked seasonal contrasts. Applying this concept, familiar in climatology, the oceanic regions can be divided into those with little seasonal variation (A through D), and those with marked seasonal variation (E through H).

It may be helpful to imagine for a few moments what the regions would be on the earth if the surfaces were entirely covered by water. Eliminating interference by land, with all its implications such as differential heating and cooling, or deflection of currents, we would have only three types of surface water arranged in latitudinal bands: the low-latitude waters, the middle-latitude waters, and the high-latitude waters. They would owe their differentiations primarily to the varied effects of the sun. As in the hypothetical planetary-wind belts of a homogeneous globe, the circulation would be westward in the low latitudes, eastward in the middle latitudes, and westward again in the high latitudes.

This idealized pattern is still clearly recognizable in the Types A, B, and C shown in figure 15. The latter two types demonstrate this almost perfectly in the Southern Hemisphere, where the continents do not reach beyond the higher middle latitudes. All other types owe their existence, shape, and location to the influence of land masses. Type D is grouped with the oceanic types because its physical characteristics are not a direct result of continental influence in the sense this term is used by climatologists (that is, the implications of differential heating and cooling by land and sea); it owes its being to the fact that the land masses by their mere existence force the circulation equatorwards, which in turn, through the deflective effect of the earth's rotation, causes upwelling and changes in physical characteristics.

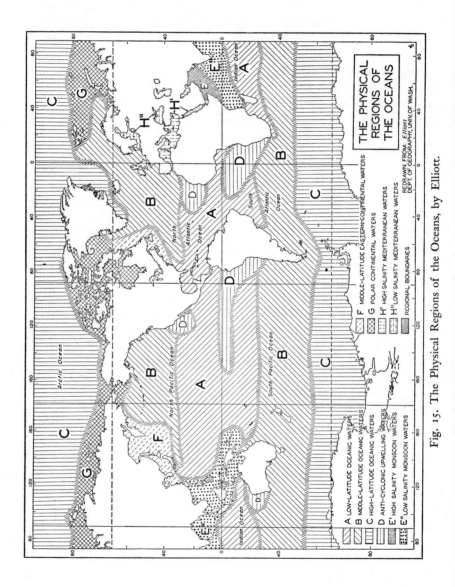

Fig. 15. The Physical Regions of the Oceans, by Elliott.

In general, one may say that the oceanic group, not being under continental influence, is characterized by comparatively marked regional uniformity. Seasonal variation of phenomena is small. The continental group, on the other hand, is under strong continental influence, and is characterized by marked regional variability, and great seasonal variation.

THE OCEANIC REGIONS

Type A—Low-Latitude Oceanic Waters

Type A occupies a position on both sides of the equator, with its poleward boundaries following approximately the mean stand of the longitudinal axes of the subtropical high-pressure cells. There is no continental influence. Surface temperatures are high and annual variation of temperatures is very small. Excess of evaporation over precipitation causes high surface salinities except in the doldrums. The major surface currents generally set westward, converging from the east and diverging in the west, except for the comparatively narrow Equatorial Countercurrents, which are directly contrary. Most of the region is within the belts of Trade Winds. Annual variations of air temperatures are very small. Precipitation (P) is very heavy in the doldrums, exceeding evaporation (E), with a decrease in the eastern parts of the regions and poleward.

Winter temperatures	20°C to 25°C
Annual variation of temperature	Less than 5°C
Average salinities	35 0/oo to 37 0/oo
Annual variation of air temperature	Less than 5°C
Precipitation-evaporation balance	E exceeds P

Type B—Middle-Latitude Oceanic Waters

Type B occupies the middle latitudes between the poleward limits of Type A and the equatorward limits of the Polar Ice Packs in winter. The difference in configuration of the land masses and the comparatively wide connection of the North Atlantic with the Polar Basin accounts for the fact that this region extends much farther north in the North Atlantic than in the North Pacific. This type shows the greatest regional variability of all oceanic types. There is no continental influence, temperatures show a fairly large latitudinal range, and seasonal variations of temperature are the widest for the oceanic group. Salinities are close to the average of the world ocean. The major currents of the regions are generally within the belts of Westerlies. Annual variation of air temperature is intermediate, and precipitation is heavy and fairly evenly distributed throughout the area, exceeding evaporation.

Winter temperatures	5°C to 20°C
Annual variation of temperature	About 10°C
Average salinities	About 35 0/oo
Annual variation of air temperature	About 10°C
Precipitation-evaporation balance	P exceeds E

Type C—High-Latitude Oceanic Waters

Type C occupies a position in the high latitudes. Its equatorward limits where it borders Type B are generally determined by the winter ice. For the most part Type C is covered by ice throughout the year. There is no continental influence except possibly in the immediate vicinity of the Antarctic Continent. Surface-water temperatures are always near the freezing point of sea water, and annual variations are small. Salinities are low. The surface-water circulation and the wind circulation are clockwise around the poles. Air temperatures are generally below freezing, and the annual variation of air temperatures probably large. Because of the low air temperatures and the resultant low capacity of the air to hold moisture, the annual amount of precipitation is very small but exceeds evaporation.

Winter temperatures	About -2°C
Annual variation of temperature	Less than 5°C

Average salinities   28 0/oo to 32 0/oo
Annual variation of air temperature   Up to 40°C
Precipitation-evaporation balance   P exceeds E

Type D—Anticyclonic Upwelling Waters

Type D occurs only on the west side of large continents and is set off from the adjacent low-latitude oceanic waters by lower temperatures, somewhat wider variation of temperatures, and lower salinities. The reason for this phenomenon is that the winds in the eastern sector of the subtropical high-pressure cells blow equatorward, parallel to the coast, deflecting surface waters seaward, thereby causing upwelling from intermediate depth along the coasts of the continents. These waters are relatively cold and only moderately saline. The effects of the upwelling are particularly pronounced immediately offshore, where temperatures and salinities are exceptionally low throughout the year. There is no continental influence and variations throughout the year are small. Air temperatures are also rather low for the latitude and comparatively uniform throughout the year. Rainfall is very scanty; this low amount of rainfall is the result of the stabilization of the air mass over the cold water surface. However, this is one case in which small rainfall does not correspond to high salinities. Probably the main reason is that the surface waters are driven by the winds at such a rate that there is always present cool, low-salinity water from intermediate depth. Evaporation exceeds precipitation.

Winter temperatures   15°C to 20°C
Annual variation of temperature   About 8°C
Average salinities   About 35 0/oo
Annual variation of air temperature   About 5°C
Precipitation-evaporation balance   E exceeds P

THE CONTINENTAL REGIONS

Type E—Monsoon Waters

Type E is a modification of the low-latitude oceanic waters by continental influence. The type is restricted to the Indian Ocean and the western Pacific. Annual variation of temperature is quite small; however, currents, winds, salinities, and precipitation show marked seasonal change because of the monsoon winds, which cause a reversal of the direction of the surface currents with the seasons. Furthermore, these winds are relatively dry in the northern part of the Arabian Sea and humid in the rest of the region, a fact that makes for a marked difference in salinity. For this reason this type is subdivided into High-Salinity Monsoon Waters, Type E', and Low-Salinity Monsoon Waters, Type E". The dry winds over subtype E', bring about very low precipitation, strong evaporation, and high salinities. In Subtype E", salinities are considerably lower because of the monsoon rains and high run-off (R) from the land. The currents set generally northwestward with the summer monsoon and southeastward with the winter monsoon. The annual variation of air temperature is relatively high near the coast.

Type E:
Winter temperatures   20°C to 25°C
Annual variation of temperature   Less than 5°C
Average salinities   30 0/oo to 37 0/oo
Annual variation of air temperature   Up to 10°C
Subtype E':
Average salinities   About 37 0/oo
Precipitation-evaporation balance   E exceeds P
Subtype E":
Average salinities   30 0/oo to 35 0/oo
Precipitation-runoff-evaporation balance   P and R exceed E

Type F—Middle Latitude Continental Waters

Type F is restricted to the Northern Hemisphere and is distinguished from the adjacent Type B by strong continental influence and by the inflow of cold waters from the higher latitudes. It is located only adjacent to Asia and North America, continents in the middle latitudes of the Southern Hemisphere being too small to exert such a strong influence. This continental influence is imposed upon the sea by strong outflows of cold continental air masses during winter producing low winter temperatures and very wide annual variations of temperature. The boundaries correspond with the average positions of the polar front. Comparatively low salinities are caused by heavy run-off from the land, great excess of precipitation over evaporation, and inflow of low-salinity polar continental waters. The major currents set equatorward. The winds shift from a winter northwest to a summer south off Asia, and from a winter northwest to a summer southwest off North America. The annual variation of air temperature is very large and precipitation is abundant.

Winter temperatures	About -1°C to 5°C
Annual variation of temperature	15°C to 25°C
Average salinities	30 0/oo to 33 0/oo
Annual variation of air temperature	15°C to 35°C
Precipitation-evaporation balance	P exceeds E

Type G—Polar Continental Waters

Continental influence also sets off Type G from the adjacent Type C. This type occurs only in the Northern Hemisphere. The seaward boundary of this region follows the winter limits of fast ice, which is defined as sea ice frozen to the coast during a single winter and not in horizontal motion. The outstanding feature of this region is that it is generally frozen over in winter and open in summer. Winter temperatures are obviously near the freezing point of sea water and annual variations of water temperature are small. Salinities are even lower than in Type C, and show a very wide range, with the lowest values near the mouths of the large rivers. Currents flow eastward or counterclockwise. Winds shift with the seasons, blowing generally seaward in winter and landward in summer. Annual variations of air temperature are fairly large and precipitation is small.

Winter temperatures	About -1°C
Annual variation of temperature	Up to 7°C
Average salinities	From brackish to 28 0/oo
Annual variation of air temperature	About 10°C
Precipitation-runoff-evaporation balance	P and R exceed E

Type H—Mediterranean Waters

Type H varies greatly from all others. Continental influence is extremely strong because the individual units are almost landlocked and water exchange with the world oceans is very slow. It has been estimated, for instance, that it would take about 2500 years for complete renewal of the water below 30 meters in the Black Sea [2]. Therefore, despite the fact that they are all connected with waters of the oceanic type, their environment is so dominated by the surrounding land masses that they show very little similarity to the neighboring oceanic waters. Each unit shows a strong individuality but nevertheless there are certain characteristics which permit grouping them into one major type. These common peculiarities are: almost landlocked position, extremely strong continental influence, wide range and variation of temperatures, wide range of salinities, wide variation of air temperatures, great range of precipitation, and generally counterclockwise surface circulation. Despite these wide ranges in the individual criteria, it is quite possible to bring order into this apparent chaos, by dividing this group into two subtypes: The High-Salinity Mediterranean Waters, H', characterized by excess of evaporation over precipitation, represented by the Mediterranean Sea, the Red Sea, and the Persian Gulf; and the Low-Salinity Mediterranean Waters, H", characterized by excess of precipitation and run-off over evaporation, represented by the Black Sea and the Baltic Sea. In the case of the former, evaporation probably exceeds precipitation; however, the run-off of rivers like the Danube, the Dnieper, and the Don, is so heavy that the water is still very dilute.

Type H:

Winter temperatures	0°C to 25°C
Annual variation of temperature	8°C to 20°C
Average salinities	Brackish to 41 o/oo
Annual variation of air temperature	10°C to 20°C

Subtype H':

Winter temperatures	15°C to 25°C
Annual variation of temperatures	About 10°C '
Average salinities	37 o/oo to 41 o/oo
Precipitation-evaporation balance	E exceeds P

Subtype H":

Winter temperatures	0°C to 5°C
Annual variation of temperatures	15°C to 20°C
Average salinities	Brackish to 33 o/oo
Precipitation-evaporation-runoff balance	P and R exceed E

APPLICATIONS OF OCEANOGRAPHY

Ocean regions of the kind described here represent only a preliminary application of the geographic approach. In a hierarchy of ranks, of the kind proposed in chapter two, such a system of oceanic regions corresponds to the realms, the regional divisions of the lands having the highest degree of generalization. When such very general regions are the result of the aggregation of more detailed regions they are found to be more illuminating than is the case when realms are defined without knowledge of component detail. Yet the information about the oceans remains scanty: there is a dearth of observed data regarding the total association of characteristics in particular parts of the oceans; and the oceanographic vessels equipped to carry out such observations are still few. The great advance of physical oceanography with its insights regarding the operation of physical processes is not matched by geographical knowledge of the oceans.

There are, however, a variety of practical problems of importance to industries, shore communities, military and naval establishments, and others to the solution of which technically trained oceanographers could contribute. Four such practical problems are offered as examples.

1. Many seaside communities have set up sewage disposal plants which utilize the ocean as a dumping ground. These communities frequently discover that the sewage is returning to pollute their beaches. In some cases, an oceanographer could have predicted this and could also, at a small added cost to the community, have arranged the sewage disposal so as to avoid the resultant beach pollution.

2. At many points along our coasts, breakwaters have been erected for various purposes. The breakwaters disturb the pattern of currents along the shore, and over a period of years, there is a filling of sand on one side of the breakwater and a cutting away of the beach accompanied by considerable damage to valuable property on the other. This result could invariably be predicted by a physical oceanographer. If such a person were consulted during the planning of a project of this kind, he could usually place the breakwater in a position where property damage could be minimized.

3. Various industries build structures offshore, sometimes in water of considerable depth. Here, an oceanographer can estimate the maximum stresses to which the structure will be exposed, providing information of value to the engineer.

4. The fishing industries notice seasonal migrations of various fish. Occasionally the fish migrate out of season and some time elapses before they are again found. On the hypothesis that those migrations are determined by anomalies in the physical properties of the water, oceanographic study of the problem might well be worthwhile. One applied research project of this kind is already being carried on.

Students of the oceans could contribute in important ways to problems such as these. Unfortunately not many people are aware of the existence of a considerable amount of technical "know-how" available for the solution of perplexing problems involving marine water. In addition there is a wide field for the application of geographical methods to the study of the oceans. Important progress in the field would result if the two approaches could be combined and if the dependence of one upon the other could be widely appreciated.

REFERENCES

1. RECLUS, E. *The Ocean, Atmosphere and Life*, New York, 1873.
2. SVERDRUP, H. U., JOHNSON, M. W., and FLEMING, R. H. *The Oceans; Their Physics, Chemistry, and General Biology*, New York, 1942.
3. BJERKNES, V., BJERKNES, J., SOLBERG, H., and BERGERON, T. *Physikalische Hydrodynamik*, Berlin, 1933.
4. SVERDRUP, H. U. "Wind-Driven Currents in a Baroclinic Ocean, with Application to the Equatorial Currents of the Eastern Pacific," *Proceedings of the National Academy of Science*, 33 (1947): 318-326.
5. REID, R. O. "The Equatorial Currents of the Eastern Pacific as Maintained by the Stress of the Wind," *Journal of Marine Research*, 7 (1948): 74-99.
6. STOMMEL, H. "The Westward Intensification of Wind-Driven Ocean Currents," *Transactions of the American Geophysical Union*, 29 (1948): 202-206.
7. MUNK, W. H. "On the Wind-Driven Ocean Circulation," *Journal of Meteorology*, 7 (1950): 79-93.

8. SVERDRUP, H. U., and MUNK, W. H. *Wind, Sea, and Swell: Theory of Relations for Forecasting,* United States Hydrographic Office, Technical Bulletin in Oceanography, No. 1., Washington, 1946.

9. SVERDRUP, H. U., and MUNK, W. H. "Empirical and Theoretical Relation between Wind, Sea, and Swell," *Transactions of the American Geophysical Union,* 27 (1946): 823-827.

10. SVERDRUP, H. U. "Period Increase of Ocean Swell," *Transactions of the American Geophysical Union,* 28 (1947): 407-417.

11. MUNK, W. H., and TRAYLOR, M. A. "Refraction of Ocean Waves," *Journal of Geology,* 55 (1947): 1-26.

12. ARTHUR, R. S. "Refraction of Water Waves by Islands and Shoals with Circular Bottom Contours," *Transactions of the American Geophysical Union,* 27 (1946): 168-177.

13. SCHOTT, G. *Geographie des Atlantischen Ozeans,* Hamburg, 1912; 2nd ed. 1936; 3rd ed. 1942; 4th ed. 1944; *idem. Geographie des Indischen und Stillen Ozeans,* Hamburg, 1935; *idem.* "Die Aufteilung der Drei Ozeane in Natürliche Regionen," *Petermann's Mitteilungen,* 82 (1936): 165-170; 218-222.

14. JAMES, P. E. "The Geography of the Oceans: A Review of the Work of Gerhard Schott," *Geographical Review,* 26 (1936): 664-669.

15. ISELIN, C. O'D. "Preliminary Report on Long-Period Variations in the Transport of the Gulf Stream System," *Papers in Physical Oceanography and Meteorology,* 8 (1940): 1-40.

PLANT GEOGRAPHY[A]

FLORISTIC PLANT GEOGRAPHY

ECOLOGICAL PLANT GEOGRAPHY

THE STUDY OF VEGETATION

VEGETATION MAPS

PROSPECT

[A]. By A. W. Küchler.

CHAPTER NINETEEN

PLANT GEOGRAPHY

Plants GIVE a landscape much of its character and therefore have long been recognized as geographical phenomena of fundamental significance. As a result, plant geography, or phytogeography as it is sometimes called, developed as a field of study, investigating the geographical distribution of plants on the earth.

There are various approaches to the study of plant geography. Many botanists have emphasized the geographic distribution of species or other taxonomic units. This is called the floristic approach or floristic plant geography. It is of importance because the inherent characteristics of the species are largely responsible for the extent and the location of the area occupied by a species.

In the process of spreading over the globe, plants meet with many obstacles such as adverse climates and soils, oceans, mountain ranges, deserts, and others. A given species occurs, therefore, where conditions permit its survival. This implies that the geographic distribution of plants is also the result of the relations between species and their environment. The study of plant distribution as related to environmental conditions is termed ecological plant geography. It is important to be aware of two considerations: 1) that the inherent characteristics of the species and the circumstances of the environment jointly control the spread of plant life over the earth; and 2) that both the species and the environment undergo a continuous process of change.

The distribution of plants can also be examined in terms of plant communities. Instead of considering the distribution of individual species, plants are recognized as occurring in characteristic groupings, such as a meadow, a type of forest, a beach, a bog, or a desert community. A plant community has a quality of its own, with generally homogeneous appearance, within limits, and with a characteristic floristic composition. These plant communities, ranging in size from less than an acre to many square miles, carpet the landscape in ever-changing patterns. The sum total of all these plant communities is called vegetation. To many geographers plant geography is synonymous with the study of vegetation.

Geographers, characteristically, record on maps their observations regarding patterns of distribution, and the maps, in turn, are used for the study of areal relations. In plant geography, vegetation maps are receiving more and more attention. Such maps have triple value: as inventories of plant resources; as indicators of associated environmental conditions; and as a basis for planning for more desirable forms of land use.

This chapter considers these various aspects of plant geography as they have developed in the United States.

FLORISTIC PLANT GEOGRAPHY

The basic problem of floristic plant geography is to establish the area that a species occupies and to explain the extent and location of this area. Once such areas have been established they, in turn, may suggest solutions of related problems. In view of the continuous evolution of plants and of the environment it is usually impossible to account for the area a species occupies by examining present conditions only. The historical aspects of plant geography form, therefore, an essential part of all areal investigations, even though they may lead to much speculation.

To illustrate: M. L. Fernald [1] introduced a theory on conservative versus aggressive species, and as a sort of corollary, his nunatak theory. The latter, without doubt, is better known among American geographers. After detailed studies of the areas of many species Fernald boldly proclaimed that the North American extent of the Pleistocene glaciation had been overestimated and that many areas protruded through the ice sheet like islands (known as nunataks) serving as refuges for a host of plants. The evidence, however, was contradictory. Plants seemed to have survived in areas which were certainly glaciated, although it was not always clear which particular advance of the ice sheet was involved. The investigations of Fr. Marie-Victorin [2] seemed to prove that the peculiar distribution of some species in eastern North America is little related to the glacial processes, if at all. However, Fernald's ideas did lead to the re-examination of many broader generalizations, and it is now clear that nunataks must have been more common than was realized earlier. Present nunataks in Greenland bear considerable plant life, and it may be assumed that this was also true in Pleistocene times. The nunatak theory has resulted in much valuable research that would have remained undone without Fernald's provocative speculations.

An altogether different approach to historical problems was used by Hugh Raup [3] whose investigations tested Hulten's [4] theories of equiformal progressive areas of present plant species. His positive and very stimulating results might well have been much deeper and more far-reaching, had more detailed data on the geographic distribution of species been at his disposal. His work clearly points up the need for accurate distribution maps if this research is to be continued and refined.

In the meantime the geographic distribution of species became an object of study for geneticists. The outstanding works of J. Clausen, D. D. Keck, and W. M. Hiesey [5], and of W. H. Camp [6; 7] provide a new approach to plant geography. The genetic methods make it possible to follow an evolving species through both time and space and to find an explanation for distributional patterns which must otherwise remain baffling.

Perhaps it was Stanley Cain among all contemporary plant geographers who rendered geography the most signal service, through publication of his *Foundation of Plant Geography* [8]. The book is limited almost entirely to floristic plant geography, but it is precisely the floristic aspects of plant geography that geographers know least about, and Cain supplied them with a detailed and comprehensive summary of the latest developments. He presents the basic features that affect the distribution of species through time and space; hence his strong section on paleoecology; hence also his excellent chapters on areography which deal with the areas that species occupy, and, as this is an especially important aspect of plant geography that had not been adequately presented in America, geographers may welcome this part of Cain's book above all. Cain closes his book with two sections on the relation between plant geography and genetics. As most geographers have little or no background in genetics, and in view of the growing significance of genetical work in phytogeography, Cain's contribution is a very real one.

A significant development is taking place at the present time with the application of the carbon-14 method. The decay of radioactive carbon-14 can be used to establish the age of organic matter, and in post-Pleistocene times enough material has been deposited to permit numerous studies in many places. These investigations reveal that in North America the length of the period between the last glaciation and the present may have been less than had been assumed only a few years ago. The effect of such results on our theories of species migration may be far-reaching, and plant geographers everywhere must follow this development with serious attention.

We are now witnessing a vigorous evolution of floristic plant geography in several directions, and it is impossible at the present time to foresee where these exciting developments may lead.

ECOLOGICAL PLANT GEOGRAPHY

No aspect of plant geography has received more attention in the United States than ecological plant geography. This is the investigation of the relations between the geographic distribution of plants and environmental influences, and it is here that Americans have made their greatest and most numerous contributions. Ecology is one of the several cornerstones of plant geography, and, in the United States, it has been given a greater breadth than in Europe or elsewhere, in part because Americans often and mistakenly equated it with plant geography. In two journals, *Ecology* and *Ecological Monographs*, a wealth of phytogeographical information has been brought together, and, although the other branches of plant geography are now rapidly developing, it is still true as it has been for the last six decades that American efforts in the field of plant geography are largely focused on ecological considerations.

Perhaps it was C. H. Merriam [9] who made the first major American contribution to ecological plant geography when he proposed his "life zones." He divided North America into climatic zones each of which is characterized by certain biota. This was a fertile idea and found wide acceptance; it has been studied and applied primarily in the eleven western states where the bold relief and the resulting altitudinal zonation of vegetation facilitated the recognition of the life zones. A great refinement of Merriam's theories lies in A. D. Hopkins' [10] use of phenological data to establish the zones, but very few ecologists have ever availed themselves of the possibilities which Hopkins offered.

Early American ecology evolved under the leadership of F. E. Clements and H. C. Cowles. Clements' [11; 12; 13] dynamic approach to ecology proved immensely stimulating. He reintroduced the concept of succession, developed and refined it, and classified the vegetation on the basis of succession. Clements early introduced the concept of climax vegetation, in which a plant community has passed through all phases of its succession and reached a state of equilibrium with the climate of the region. Quite logically, his ecological investigations led to the concept of indicators, which he greatly developed. Later on, he became interested in relicts and studied vegetation from this point of view. There can be no question about the enormous impetus that ecology all over the globe received from Clements' vivid imagination.

But Clements never accepted the possibility of an edaphic climax and apparently was not adequately informed regarding environmental conditions in high or low latitudes. His ideas of the climax concept were tested throughout the world, and it is becoming increasingly evident that the climax is often too elusive to be useful and that the succession leading to it may be lacking. Raup's observations in the Far North led to the conclusion that succession as presented by Clements requires a certain stability of the environment through long periods, so that the changes can proceed as postulated. The manner in which lakes gradually fall victim to an ever-increasing vegetation, or the vegetation changes under the influence of the erosional cycle, all require a certain stability of climatic conditions. But Raup finds that this stability is often wanting and that Clements' theories are less useful in higher latitudes. According to Forrest Shreve

[14] the theory of succession, as usually formulated, does not apply to the desert.

H. C. Cowles [15] in the meantime developed physiographic ecology, observing the inevitable changes of the vegetation as the landforms are changed by the work of water and wind. The basic ideas of dynamic ecology had been imported from Europe where they had never been applied. But when W. M. Davis developed his theory of the erosion cycle (chapter fifteen), it was Cowles who recognized the correlation between the dynamics of landforms and those of vegetation.

In spite of the powerful influence of Clements and Cowles, it was Forrest Shreve who, more than any other American, saved ecological plant geography from the much narrower confines of pure ecology. His clear geographic reasoning and his insight into the nature and meaning of regions placed his works among the most important ever to appear in America in the field of plant geography. The value of his early work in the eastern United States and of his discussion of tropical problems was quickly recognized, but it was to the North American desert that he devoted many of his most fruitful years. Although Shreve is classified professionally as a botanist, geographers fully appreciated the quality of his work and elected him vice-president of the Association of American Geographers and later president of the Pacific Coast Geographers.

The environmental factors weighed heavily on the minds of the ecologists, and all efforts were devoted to grasping the meaning of each individual factor. One learned to speak of the climatic, edaphic, or biotic environment and studied climate, soil, and biota in isolation. As each of these factors is actually a whole complex of factors that can be investigated separately, the research possibilities grew to vast dimensions.

The attempts to explain the distribution of vegetation on the basis of a single environmental complex, climate, reached their culmination in the painstaking researches of B. E. Livingston and F. Shreve [16]. The amount of thought and energy spent on this monumental work has never been matched in any similar or related project, and at the end the authors had the rare courage to recognize that such attempts must needs be futile; but so far, the great lesson that they taught has been ignored by nearly everybody.

Ecology was eventually divided into autecology, which deals with the environmental relations of individual plants, and synecology, which studies the environmental relations of plant communities. Our present autecological knowledge was recently summarized by R. F. Daubemmire [17], while H. J. Oosting [18] did a similar service to synecology. Ecological research continues unabated but it seems that plant ecology has somehow reached its growth limits. It is proving its usefulness in applied fields (agronomy, forestry, range research, and entomology), but it has distinct theoretical limitations and new ideas are extremely rare. Ultimately, ecology will be forced to regenerate itself or be dissolved in physiology, pedology, or climatology.

THE STUDY OF VEGETATION

In the minds of most American geographers the study of plant geography has come to be more or less synonymous with the study of vegetation. This is as narrow an approach as that of the botanist who sees little more in plant geography than the geographic distribution of species. But the viewpoint of the geographer is more acceptable than it appears at first sight. For the study of vegetation implies not only the investigation of the most significant aspect of the landscape with which the geographer deals; it implies also a comprehensiveness of approach that is not equalled in any other branch of plant geography. Vegetation consists of plant communities, and these in turn consist of individual species and life forms that live together. In cities the vegetation has been almost entirely removed, but even the largest cities are no more than tiny dots on the globe; and outside the cities the vegetation spreads its meaningful cover over the continents and islands and extends far out into the sea. Nothing can help a geographer to appreciate the character of a region as well as the vegetation. Hence, he speaks about vegetation when he thinks of plant geography and in his textbooks (if he refers to plant geography at all) he rarely mentions the floristic aspects previously outlined; he limits himself almost exclusively to vegetation.

The significance of vegetation to geographers rests not only on esthetic qualities which have been ignored almost completely, and quite unjustly so, but above all on its very far-reaching indicator value. Individual species have long been known for their use as indicators of the environment, but in most cases they do not compare for effectiveness with plant communities as indicators. The area of a species is determined by different controls in different places, and while a species may well dominate a community and justly be called the index species of that community, this same species may and usually does also occur elsewhere, often far removed from the formation in which it finds its optimum conditions of growth and survival. On the other hand, a plant community with its characteristic floristic composition and its particular set of life forms is not only a sure indicator of the coarser environmental and historical aspects of the landscapes, but it unerringly points out the more subtle and hidden characteristics as well.

A thorough knowledge of plant communities has great practical value. In forestry, it is leading toward a complete re-evaluation of the conventional outlook. In pasture management, in land classification for land use planning, in industrial planning, and in numerous other connections, the study of plant communities is proving its worth. Already there is such a specialist as a "vegetationist" whose main activity it is to advise commercial and governmental organizations with regard to the management of plant communities. This is an immensely promising field of investigation which can be profitably expanded by enterprising and imaginative students of vegetation.

In America the study of vegetation is nearly as old as the study of plant geography itself. One of the first significant publications, by R. Pound and F. E. Clements, discussed the phytogeography of Nebraska [19]. The last part of this monograph is largely a vegetational study. Since then there has been a steady stream of vegetation studies, mostly with a strong ecological slant, scattered throughout the botanical literature.

At the same time, historical studies of vegetation have not been neglected. Many geographers know the work of Hugh Raup [20] on New England. Lucy Braun [21; 22; 23] wrote a series of papers on the evolution of some of the forests of the Appalachian Mountains. A great stimulus to this kind of work came with the introduction of pollen analysis. This method, largely developed in Sweden, is used most successfully in areas of Pleistocene glaciation. The principle is simple: with the discovery that pollen grains of many species are preserved in bogs (under exclusion of oxygen), it became largely a matter of taking cores from bog deposits and observing the pollen accumulations in the various layers. Presumably the pollen comes from plants in the vicinity of the bog and is blown in by the wind. The sources of error are considerable; but by correlating a very large number of tests it becomes possible to obtain a rather accurate picture of the vegetation at various times. This not only permits conclusions as to the past composition of the vegetation but gives direct clues to the past migration of plants and indirect ones to the climatology of the past. H. P. Hansen [24] has done much work of this sort, and his monograph on the vegetation of the Pacific Northwest is a fine example of what can be done. Potzger, Sears, Cain, and many others have carried on similar researches in the eastern part of the United States.

The need for an over-all presentation of the vegetation of large areas was recognized early in Europe, and at the behest of a German publishing firm, John Harshberger [25] wrote his massive treatise on the vegetation of North America. This great piece of work is much more realistic than some of its reviewers would indicate, especially as one can get a rather true impression of the vegetation as it actually occurs in the many parts of this continent free from those speculations about what the vegetation ought to be on the basis of some successional scheme of the sort that one finds so often in later papers. Certainly, no one has been able so far to produce a superior work covering the whole continent.

A classic of American plant geography was the "Vegetational History of the Middle West" by H. A Gleason [26] in which he foretold the story that pollen analysis is now laboriously working out. Also, Gleason's individualistic association hypothesis is an original approach and a rebellion against both Clements and Braun.

Recently Braun [27] published a comprehensive treatment of the deciduous forests of eastern North America. Although more limited in scope than Harshberger's work, Braun nevertheless covers a vast area and incorporates the ideas of the last four decades. Because there seems to be neither the time nor the inclination in American geographical circles to obtain a thorough botanical training, few geographers possess enough background to devote themselves to detailed vegetational studies. Therefore a book like that by Braun is doubly valuable. Here we find in bold strokes a picture of the vegetation of a very large territory, both in its present setting and in its historical development.

Recently Frank Egler [28; 29; 30], after examining the enormous amount of material that has accumulated, began to question the very concepts with which American students of vegetation had become so accustomed to work. The philosophic depth of his work has placed Egler among the foremost thinkers in the field of plant geography. Simultaneously with his theoretical work he has carried on important experiments: for example, his concept of semi-natural self-perpetuating plant communities has become valuable to hydro-electric power companies. Possibly Egler's research will lead to a re-evaluation of criteria, methods, and goals in plant geography.

VEGETATION MAPS

Vegetation maps assume special significance for geographers who wish to do research in plant geography. Unfortunately, vegetation maps have often been neglected, and it is common enough to find regional descriptions of vegetation without a vegetation map. Progress in this country has lagged behind that of Europe. The United States has no agency for mapping vegetation, such as the *Service de la Végétation de la France* of the French, or the *Zentralstelle für Vegetations-Kartierung* in Germany.

However, America is by no means without its distinctive achievements. At the beginning of the second decade of this century, enough material had accumulated to permit Harshberger [25] to prepare a vegetation map of North America (1:40,000,000). It cannot be overlooked here that Europeans had preceded Harshberger by several years, and especially Schimper's map [31] of the world had become widely known.

But progress was slow. In 1917 Shreve [32] presented his new vegetation map of the United States to the readers of the *Geographical Review*. Although of small scale and highly generalized, this map revealed an unusually clear insight into the nature and the problems of American vegetation.

The maps of Africa (1:10,000,000) by Homer Shantz [33] and of the United States (1:8,000,000) by Shantz and Rafael Zon [34] have a standard of accuracy and clarity all the more remarkable because until then source material was sketchy. Although these two maps are now three decades old, American geographers are, generally speaking, more familiar with them than with any other vegetation maps that have been published since, regardless of region, scale, or quality.

In addition to the various vegetation maps which have been published as a result of scientific investigations and for scientific use, there have appeared especially during the last two decades a considerable number of maps portraying specific economically important features. A fine example is the Forest Type Map of Oregon [35] which shows forest types in terms significant to the lumberman. For this purpose, great emphasis is placed on the forest types with merchantable timber, especially if their species produce valuable wood, such as the Douglas fir (*Pseudotsuga taxifolia*) or the western yellow pine (*Pinus ponderosa*). The importance of such trees is indicated by several height, size, and age classes. Less valuable trees are mentioned in a more cursory manner, and where no trees occur, everything is shown under the heading of "non-forest land," regardless of whether the vegetation of such areas is grass, sagebrush, or is absent, as in urban areas. This one-sided approach may perhaps surprise the scientific investigator but the purpose of the map has nevertheless been fulfilled.

Attention should be directed, also, to two other recent developments in vegetation mapping. The work of A. E. Wieslander [36; 37] in the Forest Service has resulted in the development of a genuinely systematized approach to the problem. Although limited to use in California, his method can be expanded to apply to other parts of the country. Wieslander's colored vegetation type maps (1:62,500 and 1:125,000) and vegetation-soil maps (1:31,680) represent today the greatest achievement in vegetation mapping in the United States. His techniques are now employed commercially, which emphasizes their practical value. And A. W. Küchler [38; 39] developed a technique of mapping the physiognomy of vegetation. He reasoned that in the study of vegetation, physiognomy can be quite as significant as the floristic composition, especially for non-botanists. His method makes it possible to map the physiognomy of vegetation in any part of the world at any scale.

American vegetation mapping was slow to start. But the interest in vegetation maps has at last been aroused, and Küchler [40; 41] has recently summarized some of the theoretical and practical aspects of such maps. Several good mapping techniques have been developed, and the use of aerial photographs renders vegetation mapping simpler, more accurate and less expensive. The available information on American vegetation and the existing mapping techniques are now adequate to serve as a basis for preparing more detailed vegetation maps of all parts of this country.

PROSPECT

For geographers, vegetation will long remain the most absorbing aspect of biogeography, with vegetation maps as the most valuable tool. Recent years have witnessed an extension of American plant geography, made possible by a growing interest in the subject and a greater output of research papers.

During the first decades of this century, the trend in plant geography was to accumulate information. The latest trend, as yet only just beginning, differs essentially from the earlier one in that the work changes from amassing to coordinating data. Already, Cain [8] has fused a variety of phytogeographic ideas and has thus made a major contribution to modern plant geography.

Now that plant geography is more readily approached from several sides at once, one may anticipate the formulation of new concepts. Whether plant geography will ever have its Darwin remains to be seen. At last, the way to some essentially new "Origin of Vegetation" is open to those who are able to bring the various aspects of plant geography into focus as they proceed on their road to discovery.

REFERENCES

1. FERNALD, M. L. "Persistence of Plants in Unglaciated Areas of Boreal America," *American Academy of Arts and Sciences, Memoirs*, 15 (1925): 241-342.
2. MARIE-VICTORIN, FR. *Phytogeographical Problems of Eastern Canada*, Contributions of the Institute of Botany, University of Montreal, no. 38, Montreal, 1938.
3. RAUP, H. M. "The Botany of Southwestern Mackenzie," *Sargentia*, 6 (1947): 1-275.
4. HULTEN, E. *Outline of the History of Arctic and Boreal Biota During the Quaternary Period*, Stockholm, 1937.
5. CLAUSEN, J., KECK, D. D., and HIESEY, W. M. *Experimental Studies on the Nature of Species*, Carnegie Institution Publication, no. 520, Washington, 1940.
6. CAMP, W. H. "Studies in the Ericales," *Bulletin of the Torrey Botanical Club*, 68 (1941): 531-551.
7. ———. "Distribution Patterns in Modern Plants and the Problems of Ancient Dispersals," *Ecological Monographs*, 17 (1947): 159-183.

8. CAIN, S. A. *Foundations of Plant Geography*, New York, 1944.
9. MERRIAM, C. H. *Life Zones and Crop Zones of the United States*, United States Department of Agriculture, Biological Survey Bulletin 10, Washington, 1898.
10. HOPKINS, A. D. *Bioclimatics*, United States Department of Agriculture, Miscellaneous Publications, no. 280, Washington, 1938.
11. CLEMENTS, F. E. *Plant Succession*, Carnegie Institution Publication no. 242, Washington, 1916.
12. ———. *Plant Indicators*, Carnegie Institution Publication no. 290, Washington, 1920.
13. ———. "The Relict Method in Dynamic Ecology," *Journal of Ecology*, 22, (1941): 39-68.
14. SHREVE, F. "The Desert Vegetation of North America," *Botanical Review*, 8 (1942): 195-246.
15. COWLES, H. C. "The Physiographic Ecology of Chicago and Vicinity," *Botanical Gazette*, 31 (1901): 73-108, 145-182.
16. LIVINGSTON, B. E., and SHREVE, F. *The Distribution of Vegetation in the United States, as Related to Climatic Conditions*, Carnegie Institution Publication, no. 284, Washington, 1921.
17. DAUBENMIRE, R. F. *Plants and Environment*, New York, 1947.
18. OOSTING, H. J. *The Study of Plant Communities*, San Francisco, 1948.
19. POUND, R., and CLEMENTS, F. E. *The Phytogeography of Nebraska*, Lincoln (Nebraska), 1898.
20. RAUP, H. M. "Recent Changes of Climate and Vegetation in Southern New England and Adjacent New York," *Journal of the Arnold Arboretum*, 18 (1937): 79-117.
21. BRAUN, E. L. "Glacial and Postglacial Plant Migrations Indicated by Relic Colonies of Southern Ohio," *Ecology*, 9 (1928): 284-302.
22. ———. "Forests of the Illinoian Till Plain of Southwestern Ohio," *Ecological Monographs*, 6 (1936): 89-149.
23. ———. "Some Relationships of the Flora of the Cumberland Plateau and Cumberland Mountains in Kentucky," *Rhodora*, 39 (1937): 193-208.
24. HANSEN, H. P. "Postglacial Forest Succession, Climate, and Chronology in the Pacific Northwest," *Transactions of the American Philosophical Society*, N. S. 37 (1947): 1-130.
25. HARSHBERGER, J. W. *Phytogeographic Survey of North America*, Leipzig, 1911.
26. GLEASON, H. A. "The Vegetational History of the Middle West," *Annals of the Association of American Geographers*, 12 (1922): 39-85.
27. BRAUN, E. L. *Deciduous Forests of Eastern North America*, Philadelphia, 1950.
28. EGLER, F. E. "Vegetation as an Object of Study," *Philosophy of Science*, 9 (1942): 245-260.
29. ———. "Arid Southeast Oahu Vegetation," *Ecological Monographs*, 17 (1947): 383-435.
30. ———. "A Commentary on American Plant Ecology." *Ecology*, 32 (1951): 673-694.
31. SCHIMPER, A. F. W. *Pflanzengeographie auf Physiologischer Grundlage*, Jena, 1898.
32. SHREVE, F. "A Map of the Vegetation of the United States," *Geographical Review*, 3 (1917): 119-125; map, 1:9,600,000.
33. SHANTZ, H. L., and MARBUT, C. F. *The Vegetation and Soils of Africa*, American Geographical Society, Research Series, no. 13, New York, 1923; map of vegetation, 1:10,000,000.
34. SHANTZ, H. L., and ZON, R. "Natural Vegetation," in the *Atlas of American Agriculture*, United States Department of Agriculture, Washington, 1924.

35. *Forest Type Map of Oregon*, United States Forest Service, Portland (Oregon), 1936.

36. WIESLANDER, A. E. *Vegetation Types of California*, Map series 1:62,500, United States Forest Service, Berkeley (California), irregularly since 1937.

37. ————. *Timber-Stand and Vegetation-soil Maps of California*, United States Forest Service, San Francisco, 1949.

38. KÜCHLER, A. W. "A Physiognomic Classification of Vegetation," *Annals of the Association of American Geographers*, 39 (1949): 201-210.

39. ————. "Die Physiognomische Kartierung der Vegetation," *Petermanns Mitteilungen*, 94 (1950): 1-6.

40. ————. "The Relation Between Classifying and Mapping Vegetation," *Ecology*, 32 (1951): 275-283.

41. ————. "Some Uses of Vegetation Maps," *Ecology*, 34 (1953): 629-636.

A. Original draft by L. C. Stuart and sponsored by the Committee on Physical and Bio-Geography, Hoyt Lemons, Chairman.

CHAPTER TWENTY

ANIMAL GEOGRAPHY

ANIMAL GEOGRAPHY, or zoogeography, is concerned with animals below the level of acculturized man. Zoogeographers approach their subject both regionally and taxonomically. Although some are primarily interested in distinguishing and understanding the differences from place to place on the earth in terms of animal populations, and others are primarily interested in the biotic processes involved, both groups consider themselves equally to be zoogeographers. Most zoogeographers were originally trained as systematic zoologists, paleontologists, or ecologists, and perhaps for this reason there has been little interest in the formulation of unified concepts concerning the nature of the field and its relations to bordering fields. This inventory of work in zoogeography, therefore, is presented in terms of what has been done by scholars who call themselves zoogeographers, whether or not such work is always in accord with the view of the field of geography as a whole as presented and developed in the first three chapters of this book.

Zoogeographers and phytogeographers may be grouped together as biogeographers because they deal with essentially the same biotic processes. Both recognize that the explanation of biogeographic patterns involves two sets of factors: those that are intrinsic to the organism and fundamentally genetic; and those that are extrinsic to the organism and fundamentally environmental. Biogeographers study the mutual relations between plants, animals, and environments, both past and present; they accept the concept of environmental determinism (in the geographic, not the biologic, sense) as fundamental to their approach. Plants and animals lack the capacity consciously to alter their environment; but animals, unlike plants, possess inherent powers of movement and may thus seek environments that are congenial to them. For this reason the methods of the zoogeographers and the phytogeographers must differ. Careful consideration of the nature of biogeography has recently been offered in the writings of Pierre Dansereau [1] and of W. C. Allee and K. P. Schmidt [2].

THE DEVELOPMENT OF ZOOGEOGRAPHY

The history of zoogeography may be divided into three general periods. The first of these lasted for a century after Linnaeus had laid the groundwork for systematic zoology and was characterized by descriptions of the distribution of various animal groups. These descriptions were drawn together in the first attempts to define zoogeographic regions.

The second period began with the announcement of the Darwin-Wallace concept of organic evolution through natural selection. This formulation led not only to a refinement of regional description but also to efforts to explain zoogeographic patterns in terms of evolutionary history.

The third period, which began about 1900, was marked by increasingly detailed attention to the environment, by the more precise consideration of the mutual relations between organism and environment, and by the development of new concepts that were injected into systematic zoology by the rediscovery and application of Mendelian principles. American zoogeography, though rooted in Europe, shared in only the last two of these periods.

European Roots

Although van der Hoeven (1864) [3] was credited with presenting the first clear distinction between geographical zoology, the study of the distribution of systematic units, and zoological geography, the study of faunal regions, both approaches had been utilized by zoogeographers before his time. Zoological geography, however, requires a considerable knowledge of the paleontological record and of the processes of differentiation, neither of which field was well developed before 1900. The greater part of the effort during the first two periods of zoogeographic history was devoted to the study of patterns of animal distribution [4].

The first soundly-constituted system of faunal regions was proposed by P. L. Sclater in 1858 [5]. His six major zoogeographic divisions of the world were based on avifauna. In the terms suggested in chapter two for regions of the highest rank, Sclater's regions would be called realms, as, indeed, they were called by some workers. These animal realms were given additional support by the studies of men like A. R. Wallace [6] and R. Lydekker [7], and, with certain minor modifications as suggested by T. H. Huxley [8] and A. Heilprin [9], they are reasonably acceptable even today.

During the second period of zoogeography, European scholars sought to explain the distribution patterns of animals. These early attempts at explanation, however, were handicapped by an insufficient knowledge of the basic factors previously noted. The zoogeographers of that period were divided into two groups: 1) those who believed that zoogeographic patterns could be interpreted only through postulated Gondwana Lands and Atlantises as land bridges (R. F. Scharff [10], A. E. Ortman [11]); and 2) those who insisted on the permanency of continental masses and ocean basins (Wallace [6]). This controversy continues today, although the idea of land bridges has been replaced by the Wegener hypothesis of continental drift.

The ecological approach to the study of animal geography also made its first appearance in Europe during the 19th century. In 1881, K. Semper published a work in which he viewed the organism in terms of its adjustment to the environment [12]. He attempted no regional classification. The plant geographers, A. F. W. Schimper [13] and E. Warming [14], blazed the trail in ecological classification, but it was some years before zoologists essayed regional studies from this point of view.

ZOOGEOGRAPHY IN AMERICA

During the latter part of the 19th century zoogeographers in the United States, influenced by their European colleagues, became interested in the patterns of animal distribution. At first they were concerned with the definition of world-wide realms [15], especially with the refinement of such realms on a continental scale. In America this phase of zoogeographic study culminated in the formulation, by Charles H. Merriam, of the "life zone hypothesis" [16]. Merriam based his life zones largely on temperature, and attempted to demonstrate the existence of accordant areal relations between these zones and the genera of plants and animals.

The zoogeographers of the early 20th century, like their colleagues in geomorphology (chapter fifteen), were not content with the description of observed patterns of distribution: the urge to seek explanations, as expressed by Arnold Guyot with reference to landforms, was also felt by students of animal geography. Furthermore, the basic tools that made possible a more effective interpretation of observed patterns were being rapidly forged. The paleontological record was being expanded and filled out; the elements of the environment were being analyzed by improved methods; and in zoology the "new systematics," summarized by Huxley [17], provided, for the first time, a sound basis for the study of animal distributions. The 20th century was marked by the focus of attention on details, on studies that traced the evolution and movement of animals through both space and time. Especially significant were the studies of A. G. Ruthven [18] and W. L. Tower [19] dealing with minor systematic categories, the garter snakes and a certain genus of beetles, respectively.

As work of this kind progressed it became evident that the histories of many different groups of animals were remarkably similar. From data gathered in these detailed studies it became possible to formulate generalizations regarding the centers of origin of the larger systematic groups and even of animal associations, or faunas. W. D. Mathew was able to explain the geographical pattern of larger mammalian groups in terms of their evolution and movements [20], and C. C. Adams was able to trace the post-glacial dispersal of plants and animals in North America [21]. It became clear that the explanation of observed patterns of distribution must be sought through the historical study of faunas.

Since 1900, the ecological approach to this field has been increasingly developed in America. Such scholars as H. C. Cowles [22], C. C. Adams [23], and V. E. Shelford [24] focused their attention on the relationships between the organism and its environment. A large number of ecological studies, in both Europe and America, provided the data on which to base such general works as those of F. E. Clements and V. E. Shelford [25], and R. Hesse, W. C. Allee, and K. P. Schmidt [2].

PRESENT-DAY ZOOGEOGRAPHY

There are, then, three different approaches to the study of zoogeography in the modern period. These are: 1) the regional approach; 2) the historical approach; and 3) the ecological approach. These three approaches are not completely separable, for the individual scholar may make use of all three in the study of a zoogeographic problem, perhaps stressing one or another because of his own background of training and experience, or interest.

THREE APPROACHES

The regional approach to zoogeography, first used effectively about a century ago, is still actively cultivated. The earlier efforts to define world realms on the basis of homologous faunas have been succeeded by studies in which there has been a greater and greater attention to detail. Merriam's life zones represent one of the first attempts to subdivide the realms into smaller units. Now Merriam's hypothesis has largely been superseded by L. R. Dice's biotic-province hypothesis [26]. That Dice's provinces provide for a more effective definition of biotic regions than have the earlier attempts along these lines is due to his recognition of other relevant environmental factors than temperature alone, and to the availability of new concepts and new data regarding the systematic classification of animals and regarding animal distribution. His provinces have been widely adopted by North American zoologists, who have extended the system of classification southward into Central America [26; 27; 28; 29].

As experience has been gained with the application of different systems of regions, zoogeographers have come to realize the futility of seeking any regional classification that fits all groups of animals. A zoogeographic region is defined on the basis of the existence within it of a more or less homologous association of animals. The utility of animal realms has come to be doubted, for such regions based on highly generalized criteria overlook too many relevant details. For example, a physical or ecological barrier that restricts the range of one group of organisms, such as mammals, may serve as a pathway for the dispersal of another group, such as the earthworms.

The principal tool of the modern regional zoogeographer is a thorough knowledge of systematics—that is, of the systems of animal categories. Only when the student has an understanding of the nature and range of variation of the organism under consideration can he determine its geographic extent. Systematics, moreover, views the organism as dynamic, a concept that leads to a realization that regions themselves are only momentary stages in a series of stages. Some zoogeographers still confine their efforts to static descriptions of existing geographic patterns, but the majority seek to inject the time sequence into their analyses, a phase of zoogeography that may be referred to as "historical" or "faunal" (compare the concepts developed in chapters one, two, and three).

This new historical approach to the study of faunas reveals that the existence of a more or less homologous association of animals in an area may be the result of complex processes. Although the several groups of animals that are associated in an area may today have approximately the same geographic range, it can often be shown that they have been derived from quite different sources. Thus E. R. Dunn has shown that the amphibian and reptile fauna of the Americas is composed of four major elements, each with a very different history [30]. Working on a small area in Alta Verapaz in Guatemala, L. C. Stuart has shown that even when examined in detail a homologous fauna contains three of the four elements listed by Dunn [31]. In studies of this kind, the zoogeographer is called upon to handle data drawn from such fields as stratigraphy, paleoclimatology, and paleontology. Animal groups are traced through both time and space as they evolved, moved, and were subjected to varied environmental conditions affecting them.

The third approach to modern zoogeographical studies, the ecological approach, is quite different from the first two. Regional zoogeography, as the term is used by some zoogeographers, is the study of areas characterized by associations of animals with homologous ranges. The patterns of zoogeographic regions defined by such studies are interpreted by the historical approach. Ecological zoogeography seeks to analyze faunas or taxonomic units in terms of common or equivalent adaptations to a specific category of environment. In such studies the regional approach may or may not be utilized. Hierarchical arrangements of regions include such units as the *biome*, a broad area of homologous response to a climax, and the *biotope*, a very specific environment of small areal spread, such as a moss-bank [2; 32]. A biotic province, of the kind defined by Dice, must be a continuous area; the same biome or the same biotope may be found in widely-separated areas. For example, the tropical forest is a biome which is found in different continents with wet-tropical areas. The species of animals occupying the East Indian forests are very different from those found in a similar environment in Madagascar. Yet the animals inhabiting the two areas are similar in their responses to environmental conditions. Thus the flying lemur of Madagascar and the flying squirrel of the East Indies are volant types that have responded to the arboreal habitat in the same way, yet phylogenetically the two are only remotely related.

The tools of ecological zoogeography, in addition to the understanding of the earth sciences demanded for environmental analysis, include physiology and functional morphology. Workers in this field, also, were quick to realize that it is often necessary to approach their problems through controlled experimentation [24].

These three approaches to modern zoogeography cannot be viewed as necessarily distinct and separate. If the zoogeographer is to achieve anything approaching a complete understanding of the distribution of animals on the earth, all three approaches must be employed. Regional zoogeography will supply the basic pattern of arrangement; historical zoogeography will show how this pattern was developed; and ecological zoogeography will explain how it is sustained in terms of animal-environment relations. As already explained all three approaches are often used in the study of a single problem, in which case they become supplementary aspects of one approach.

The Interpretation of Zoogeographic Data

In attempting to throw light on the areal relations of changing organisms and of a changing environment, the zoogeographer is faced with the necessity of using data derived from many disciplines. He cannot be a specialist in all the diverse fields with which he must deal. He is primarily concerned with the synthesis of concepts and data derived from a variety of fields, and he must be sufficiently grounded in each to be able to estimate the validity and significance of the conclusions reached by scholars in them.

The history of zoogeography suggests that many of the failures experienced in the attempts to interpret animal distributions have resulted less from the inadequacy of available data than from the inability of the individual scholar to interpret these data. The failure of the "bridge-building" school of historical zoogeographers is typical. The transoceanic connections they postulated were described without due consideration of the data from geology and geophysics, or of the negative zoogeographic evidence. W. D. Mathew was successful where others had failed in explaining the discontinuities observed, by postulating a center of origin in the north and a southward dispersal at least of the higher and more recent animal groups [20].

THE PROSPECT

Zoogeographers have only started to attack their problems with modern methods [33]. They are learning that regions representing too high a degree of generalization cannot be expected to illuminate the areal relations of animals and their environments; they are learning that environments or faunas distinguished on the basis of a single criterion selected as an indicator are not the same as the total environment or the total fauna. They are learning, as geographers have learned in other parts of the field, that illuminating ideas of wide significance are often derived from the detailed examination of small areas. They have, as a group, been too prone to accept deductive procedure. Specifically, they are faced with the problem of relating the smaller areal units of the regional hierarchy with the larger and more highly-generalized units.

It has been pointed out that zoogeographic research has been carried on chiefly by systematic zoologists, paleontologists, and ecologists, and the character of their individual backgrounds of training and experience has been reflected in the diverse approaches to zoogeography. Perhaps if scholars trained in geography and thoroughly grounded in the methods of regional study were as adequately trained in systematic zoology and paleontology, they would be in a position to render important service to zoogeography. In the ecological approach, species and faunas were examined for the purpose of bringing to light their responses to environment and of describing the environment-produced modifications of their morphology or physiology; or environments were defined for the purpose of identifying the particular response produced among the animal inhabitants. Here is a field in which the regional method, as developed at the beginning of this book, might yield valuable new insights.

It is unfortunate that the aims and methods of zoogeography have never been clearly formulated. This discipline promises to have many applications to related fields, but these applications cannot be made effectively until the basic concepts and procedures of zoogeography are formulated and accepted. Certainly there are many problems of climatic change, or of changes in landforms and drainage, that have been illuminated by data from zoogeography [34]. In its ecological aspects, zoogeography can render important service to wildlife management; and it is indispensable in the field of medical geography, the subject of the next chapter, where attention is focused on host-parasite relations or on factors relating to the distribution of disease vectors.

REFERENCES

1. DANSEREAU, P. "The Scope of Biogeography and its Integrative Levels," *Révue Canadienne de Biologie,* 10 (1951): 8-32.
2. ALLEE, W. C., and SCHMIDT, K. P. *Ecological Animal Geography,* 2nd ed. New York, 1951 (1st. ed., 1937, by HESSE, R., ALLEE, W. C., and SCHMIDT, K. P.).
3. VAN DER HOEVEN, J. *Philosophia Zoologica,* Leiden, 1864.
4. MURRAY, A. *The Geographical Distribution of Animals,* London, 1866.
5. SCLATER, P. L. "On the General Geographical Distribution of the Members of the Class *Aves,*" *Proceedings of the Linnaean Society,* (1858): 130-145.
6. WALLACE, A. R. *The Geographical Distribution of Animals,* New York, 1876.
7. LYDEKKER, R. *A Geographical History of Mammals,* Cambridge, 1896.
8. HUXLEY, T. H. "On the Classification and Distribution of the *Alectoromorphae* and *Heteromorphae,*" *Proceedings of the Zoological Society of London,* (1868): 294-319.
9. HEILPRIN, A. *The Geographical and Geological Distribution of Animals,* London, 1887.
10. SCHARFF, R. F. *Distribution and Origin of Life in America,* London, 1911.
11. ORTMAN, A. E. "The Geographical Distribution of Fresh-Water Decapods and its Bearing upon Ancient Geography," *Proceedings of the American Philosophical Society,* 41 (1902): 267-400.
12. SEMPER, K. *Animal Life,* New York, 1881.
13. SCHIMPER, A. F. W. *Plant Geography upon a Physiological Basis,* (translated by W. R. FISHER), Oxford, 1903.
14. WARMING, E. *Oecology of Plants,* (translated by P. GROOM and I. B. BALFOUR), Oxford, 1909.
15. ALLEN, J. A. "The Geographical Distribution of Mammalia," *Bulletin of the United States Geological Survey,* 4 (1878): 313-377.
16. MERRIAM, C. H. "Laws of Temperature Control of the Geographic Distribution of Terrestrial Animals and Plants," *National Geographic Magazine,* 6 (1894): 229-238.
17. HUXLEY, J. *The New Systematics,* Oxford, 1940.
18. RUTHVEN, A. G. "Variations and Genetic Relations of the Garter Snakes," *Bulletin of the United States National Museum,* 61 (1908): 1-201.
19. TOWER, W. L. *An Investigation of Evolution in Chrysomelid Beetles of the Genus Leptinotarsa,* Carnegie Institution, Publication 48, Washington, 1906.
20. MATTHEW, W. D. "Climate and Evolution," *New York Academy of Science,* 24 (1915): 171-318.
21. ADAMS, C. C. "The Post-Glacial Dispersal of the North American Biota," *Eighth International Geographical Congress,* 1904: 623-637.
22. COWLES, H. C. "The Physiographic Ecology of Chicago and Vicinity," *Botanical Gazette,* 31 (1901): 73-108, 145-182.
23. ADAMS, C. C. "Isle Royale as a Biotic Environment," in *An Ecological Survey of Isle Royale, Lake Superior,* Report of the Michigan Geological Survey, 1908: 1-53.
24. SHELFORD, V. E. "Physiological Animal Geography," *Journal of Morphology,* 22 (1911): 551-618.
25. CLEMENTS, F. E., and SHELFORD, V. E. *Bio-Ecology,* New York, 1939.
26. DICE, L. R. *The Biotic Provinces of North America,* Ann Arbor (Michigan), 1943.

27. GOLDMAN, E. A., and MOORE, R. T. "The Biotic Provinces of Mexico," *Journal of Mammology*, 26 (1945): 347-360.

28. SMITH, H. M. "Herpetogeny in Mexico and Guatemala," *Annals of the Association of American Geographers*, 39 (1949): 219-238.

29. STUART, L. C. "Taxonomic and Geographic Comments on Guatemalan Salamanders of the Genus *Oedipus*," *Miscellaneous Publications of the Museum of Zoology, University of Michigan*, 56 (1943): 1-33.

30. DUNN, E. R. "The Herpetological Fauna of the Americas," *Copeia*, (1931): 106-119.

31. STUART, L. C. "A Geographic Study of the Herpetofauna of Alta Verapaz, Guatemala," *Contributions from the Laboratory of Vertebrate Biology, University of Michigan*, 45 (1950): 1-77.

32. SHELFORD, V. E. "Basic Principles of the Classification of Communities and Habitats and the Use of Terms," *Ecology*, 13 (1932): 105-120.

33. MURPHY, R. C. "Animal Geography: A Review," *Geographical Review*, 28 (1938): 140-144.

34. DEEVEY, E. S. JR. "Biogeography of the Pleistocene," *Bulletin of the Geological Society of America*, 60 (1949): 1315-1416.

MEDICAL GEOGRAPHY[A]

PATHOGENS AND GEOGENS

COMMUNICABLE AND NON-COMMUNICABLE DISEASES
The Pathogens of Communicable Diseases
The Pathogens of Non-Communicable Diseases

AMERICAN CONTRIBUTIONS
Malaria Control
Pellagra
Fluorine and Dental Caries
Tularemia
Rickettsial Diseases

PROBLEMS AND METHODS

PROSPECT

A. Original draft prepared by Jacques M. May.

CHAPTER TWENTY-ONE

MEDICAL GEOGRAPHY

THERE IS NEED today for a better understanding of the scope and nature of medical geography and for an evaluation of its techniques and potentialities. The field has too long been identified with an aggregate of loose impressions and shaky hypotheses concerning relationships between pathological phenomena and their environment. Yet such relationships unquestionably exist, and when studied with discrimination may shed light on the causes, occurrences, and effects of disease.

The earliest physicians knew little of the cause of diseases beyond the fact that certain ones seemed to be found in certain localities only; and even today, there are a number of diseases about which not much more than this is understood (as reflected in such locational names as Bunyanwera fever, Wumba fever, Semliki Forest fever). With the development of medical knowledge the influence of the environment was soon recognized and continued to be regarded as important until the Pastorian discoveries turned attention to the study of pathogenic organisms. Interest in the environment revived when subsequent research made it clear that these organisms are themselves profoundly affected by environmental circumstances [1:128].

The following is suggested, in the light of current thought, as a working definition: "Parasitology, epidemiology, and medical entomology take into consideration special aspects of the relationship between disease and environment. Medical geography professes to make its principal objective the study of the areal distribution of disease and its relationship to the environment."

Hitherto this field has been developed almost exclusively by the medical profession with a view to the treatment, eradication, or control of disease. Consequently the emphasis has been predominantly upon the study of the environment, both natural and human, as causing and conditioning pathological phenomena. The geographer, however, holds medical geography as comparable with the other so-called topical branches of his field. He thus conceives of it as dealing potentially not only with the influence of environment upon pathological phenomena but also with disease as an element of the complex of associated phenomena in particular places, and as a factor in the relationships of one area to another. The geographer is concerned no less with the effects than with the causes of disease, effects that are often powerful, indeed, upon the circumstances of human life. Attention is called to this point of view in order to link the present discussion with the thought of American geographers as expressed elsewhere in this volume, and also to suggest an important, but as yet little-explored, avenue of research. So little has it been explored, however, that from here on the treatment will be confined within the conceptual framework of that part of the field of medical geography in which the most substantial results are now being achieved in the United States.

PATHOGENS AND GEOGENS

The first question that demands answer in the study of medical geography is "who has what diseases and where?" Until these fundamental facts are known, the attempt to correlate environmental with pathological circumstances is like navigation without a chart. The initial and most urgent task of the medical geographer would therefore seem to be the mapping of the distribution throughout the world of the occurrence of manifested and potential diseases and also of the pathological factors, which, when combined in certain pathological complexes, cause each disease. Not until these factors have been mapped is the way opened for the discovery of correlations that may exist between the pathological complexes and environmental factors. To emphasize the distinction here, two terms have been coined and are gaining currency: *pathogen*, meaning a pathological factor, and *geogen*, meaning an environmental factor known or believed to be correlated with a disease or its pathogens [2]. The geogens, of course, comprise factors not only of the physical and biotic, but also of the human, environment.

In the attempt to establish such correlations, statistical techniques are employed wherever possible; otherwise less refined methods of assessment must be used. In both procedures, extreme caution is needed to avoid the frequent error of mistaking a coincidence of A with B as necessary evidence that A is the cause of B, or vice versa. Despite such pitfalls, however, present-day techniques seem to be catching up with the ambitions of our predecessors. Before too long it may be possible to discern the real, but concealed, picture of many a disease. To reach this goal, the following approach is suggested:

1) sample studies of populations to determine the extent of exposure to different diseases;

2) analysis of the influence of each environmental factor, or geogen, (insofar as this influence may be isolated), upon the various pathological factors, or pathogens;

3) synthetic area studies to show how groups of diseases are fostered by the combined action of different geogens.

In most studies in medical geography attention is focused mainly upon the pathological factors, which are made to form the frame of reference, and the environmental factors, or geogens, are fitted into the frame only insofar as they are known or believed to influence the pathogens. For this reason, and also because their number is legion, no comment seems needed here on individual geogens. The following list, however, may give an idea of their nature [2:10].

1) PHYSICAL
   CLIMATE
       Latitude
       Rainfall and humidity
       Temperature
       Barometric pressure
       Sunshine and cloudiness
       Wind direction and velocity
       Radiation
       Static electricity
       Ionization
   RELIEF
   SOILS
   HYDROGRAPHY
   TERRESTRIAL MAGNETISM

2) HUMAN OR SOCIETAL
   POPULATION DISTRIBUTION AND DENSITY
   STANDARD OF LIVING
       Housing
       Diet
       Clothing
       Sanitation
       Income
   TRANSPORTATION AND COMMUNICATIONS
   RELIGIOUS CUSTOMS AND SUPERSTITIONS
   DRUG ADDICTIONS

3) BIOLOGICAL
   VEGETABLE LIFE
   ANIMAL LIFE, ON EARTH AND IN WATER
   PARASITISM, HUMAN AND ANIMAL
   PREVALENT DISEASES
   DOMINANT BLOOD GROUPS

COMMUNICABLE AND NON-COMMUNICABLE DISEASES

Disease is the term we apply to the signs of suffering of living tissues (pathology). For the purpose of medical geography it is convenient to divide human diseases in two groups:

1) Communicable diseases, occurring as the result of the aggression of other animal or vegetable organisms;

2) Non-communicable diseases, occurring without the aggression of other animal or vegetable organisms. This group can be divided into: a) diseases due to tissue changes which are essentially reversible by spontaneous physiological adjustments (these should be studied together with the problems of physiological climatology considered in the next chapter); and b) diseases due to changes which are essentially nonreversible.

THE PATHOGENS OF COMMUNICABLE DISEASES

The pathogens of communicable diseases may be tentatively classified as: causative agents, vectors, intermediate hosts, reservoirs, and man himself.

The causative (or infective) agent is the immediate cause of the disease, as for example, a virus, a fungus, a bacterium, a rickettsia, a spirochete, a protozoon, a metazoon. In certain diseases, such as epidemic meningitis, tuberculosis, typhoid fever, cholera, tetanus, and many others, the agent is introduced directly into the human body by inhalation, ingestion, or through wounds or abrasions. In other diseases it is carried to and introduced into the human body by a vector, usually an arthropod. Also in some diseases causative agents can develop in their larval stages only in an intermediate host, such as a mollusk, fish, or mammal, including man. Vertebrates other than man may serve as a habitat, or reservoir, of the causative agents of certain diseases, thus helping to support their life cycles and increasing their quantity in nature and hence the chances of epidemics. Finally, man himself tops the list of pathogens. He is obviously an element in the pathological complex sometimes as an intermediate host, or as the definitive host. His reaction to the environment when he is healthy determines his reaction to it when he is sick. The study of the pathogen "normal man" in connection with the environment is the field of physiological climatology.

These biological processes suggest a classification of communicable diseases according to whether they represent two-factor or multiple-factor complexes. This facilitates their study from a geographical point of view. Each of the factors in each complex, including man himself, is closely linked to the environment. The occurrence of disease at a given spot is the result of a combination at the most favorable time of geographical circumstances that bring together causative agents, vectors, intermediate hosts, reservoirs, and man.

While cholera has been observed all over the world in epidemic forms, the endemic foci from which the epidemics spread appear to be limited to certain areas in India. Here the permanent occurrence of endemic cholera is correlated with low-lying terrain in which there are bodies of water well sheltered from both rain and sun and rich in organic material and salts. Epidemics seem to break out under conditions of high absolute humidity, of failure of the rains, and of high temperature, and when great crowds of susceptibles gather. The relationship between cholera outbreaks and religious pilgrimages is now well established. The onset of the rains may in some places dilute the concentration of germs in bodies of water, thus apparently playing a part in bringing an epidemic to an end. But in most cases heavy rainfall, especially after a prolonged drought, may first increase the number of cases of cholera, because the rivers receive a mass of infection as a result of the downpours.

A correlation between the character of the soil and the occurrence of hookworm in the United States was first emphasized by C. W. Stiles in 1903. D. L. Augustine and W. G. Smillie, quoting Stiles in 1926, wrote: "As soon as I entered the sandy areas uncinariasis was found. As soon as I left the sand, local foci of the infection disappeared" [3: 36]. Augustine and Smillie made a detailed study of the different soil types of Alabama to determine their effectiveness in producing hookworm larvae. Sandy soil was observed to give a high yield, whereas clay soils gave widely varying yields, never reaching a percentage of effective hatching of eggs higher than 12. It was thus demonstrated that the disease was related closely to the texture of the soil. In a further study [4] Augustine found that no hookworm larvae could be isolated from polluted soils from December 27th to March 3rd, thus showing that the winter frosts in Alabama prevented the disease from developing to the infective stage. Dry periods during the spring were also found to check its development entirely.

Conversely, the type of soil that favors the development of the eggs of *Ancylostoma duodenale* is hostile to *Ascaris lumbricoides*. An inverse proportion of infestation by both worms in males and females is usually found in Egypt, where men work in the fields and women stay around the home. In Italy, however, the women work in the fields and men go into the towns to market the produce; hence, women bear the brunt of hookworm infestation and men get the ascarids. Cultural factors, such as less soil pollution and wearing of shoes (in the case of *A. duodenale*), may change the picture. On the other hand, infection of the soil may be increased if an unabsorbtive subsoil makes the drainage system inadequate [2:22].

E. C. Faust in his treatise on *Human Helminthology* pointed out that on the island of Vitilevu, in the Fijis, a mountain range prevents the rains from reaching the northwestern side of the island [5: 36]. Hookworm infestations on the rainy side involve 90 percent of the native population, whereas on the dry side only 38 percent are affected. The moisture factor also affects the distribution of strongyloidosis, the free-living larvae of the parasite being very sensitive to drought.

Some of the helminthic infections require an intermediate host, such as snails, fish, and mammals, all of which are also closely dependent on physical surroundings. As opposed to the free-living larvae of the hookworm, those of the ascarids are tougher and withstand more drought, cold, and sun. However, the eggs that do not survive direct sunlight embryonate favorably on hard clay soil, whereas the hookworm larvae require sandy humus.

It seems quite clear that ever since man has studied medical geography there has been a belief that certain insects cause diseases. An allusion in the Bible to the corruption of Egypt by a swarm of flies is perhaps the first recorded statement of medical entomology. In 1587 de Souza definitely asserted that yaws was translated from a sick to a healthy person by the suction of flies. In 1848 Josiah C. Nott of New Orleans expressed the view that mosquitoes produce malaria and yellow fever, an opinion shared by Beauperthuy, a contemporary French physician of the West Indies. In 1895 Bruce discovered the role of the tsetse fly in the sleeping sickness of cattle (Nagana) and in 1897 Ronald Ross announced that he had found forms of the malarial parasite in two mosquitoes [6].

At the same time a number of investigators suspected and established the role of mosquitoes in the transmission not only of malaria but also of other diseases. One of the outstanding field experiments in epidemiology was accomplished by the United States Yellow Fever Commission in 1900 on the island of Cuba under the leadership of Walter Reed [7] with the help of Carrol, Lazear, and Agramonte. At the same time L. O. Howard demonstrated that malaria in the Catskill Mountains could be controlled by using kerosene. Thus in the last years of the 19th century the foundations of disease control through knowledge of the contribution of geographical factors to pathological complexes were definitely laid.

Marston Bates [8] has made a masterly study of the various factors of the environment as they affect mosquitoes and has stressed the importance of their habits, time and distance of flight, resting places, longevity, and seasonal distribution; and also the factors influencing the sexual behavior of the various species and their food habits. The mechanism of biting, egg development, and larval development (which is affected in turn by temperature, light, movement of water, surface characteristics, gases dissolved, pH, organic materials, and salts), all play roles in the development of the larva and hence in the distribution of the diseases.

Flies, ticks, and mites are also important vectors of disease, causing such scourges as African trypanosomiasis, Rock Mountain spotted fever, and scrub typhus respectively. Their dependence upon the physical environment and particularly their sensitivity to weather changes hold the key to the occurrence of the diseases mentioned.

Dependence of the pathological complex upon the presence of an intermediate host occurs chiefly in helminthic complexes. A typical example can be found in the study of the schistosomiases. Here the importance of the environment has been demonstrated in Egypt, where changes in irrigation methods caused a change in the pathological picture of schistosomiasis. The intermediate host required for the development of *Schistosoma haematobium* is *Bulinus truncatus,* which is not commonly found under conditions of basin irrigation. When perennial irrigation was adopted in southern Egypt, as it had been earlier in the Delta, *S. haematobium* spread to upper Egypt. In the United States schistosome dermatitis has been studied by S. Jarcho and A. Van Burkalow, W. W. Cort, D. G. Mac-Farlane, F. F. Ferguson, and many others. Snails of the family *Planorbidae, Lymnaeidae,* and *Physiodae* serve as intermediate hosts for the schistosomes responsible for "swimmer's itch." Physical factors such as the nature of the lake bottom, chemical composition of the water, exposure to wind, and so on, are influential in localizing the snail and hence the disease [9].

THE PATHOGENS OF NON-COMMUNICABLE DISEASES

Much the same line of thought may be pursued in the study of noncommunicable pathological changes. The pathogens here are represented by the various physiological functions. How do the liver functions behave under different climatological circumstances? How do heat, temperature, and altitude modify the creation and destruction of red cells, the excretion of urea nitrogen, the metabolism of water, the behavior of the blood electrolytes? Once these factors of what might be called "anthropological physiology" are better known, a better understanding may be reached of the geographical pattern of distribution of degenerative diseases, such as goiter, hardening of the arteries, arthritis, and cancer.

The American Geographical Society is now engaged in studies of the geographical pathology of cancer and has established contacts with key workers and observers throughout the world. With the same objective in view, the National Cancer Institute in Bethesda, Maryland, has set up a unit of geographical pathology, which has the cooperation of pathologists, epidemiologists, statisticians, nutritionists, and endocrinologists. In 1950 the Council for the Coordination of International Congresses of Medical

Sciences (CCICMS) conducted at Oxford a symposium in which such questions were discussed.

Workers in different countries have noticed considerable variations in the prevalence of cancer by sites, types, histological definition, and age groups, and under various living conditions. It is reported that pulmonary cancer does not exist in Iceland and Korea, that pancreatic cancer is common in Finland, that gastric cancer is not as common in England as on the continent of Europe. But not all these observations are authenticated by statistical evidence.

H. F. Dorn reported that in the United States cancer was common among all age groups of the population but that the morbidity rates for all forms of cancer and for some specific primary sites vary rather widely. He observed that the incidence rate for all forms of cancer is nearly 50 percent higher among whites living in the south then among those living in the north, which he explained largely in terms of the higher incidence of skin cancer and of cancer of the buccal cavity, especially lip cancer among persons living in the south. Although it is true that even in the most advanced countries statistical data give only a grossly approximate idea of the prevalence of cancer, and that in most places in the world where they would be needed to support a scientific judgment, statistical data on cancer distribution and demography are virtually nonexistent, there is enough presumptive evidence of the influence of environmental factors on the occurrence of this disease to justify further studies. Thus, among the facts as gathered in 1950 at the Oxford symposium were the following:

1) Primary cancer of the liver is relatively more common in Africa, Indo-China, India, Malaya, Indonesia, and the Philippine Islands than anywhere else in the world.

2) Cancer of the cervix of the uterus is relatively uncommon in Jewish women.

3) Cancer of the stomach is relatively infrequent in Javanese, African negroes, and the indigenous people of French North Africa.

4) Cancer of the base of the tongue is relatively frequent in certain communities in India.

5) Cancer of the scalp is frequent in inhabitants of French North Africa, of the skin of the trunk in the inhabitants of India, of the scalp and skin of the legs in African negroes and Indonesians.

6) A relatively large number of cancers of the pancreas has been reported from the Mulago Hospital at Kampala, Uganda.

7) In Indonesia, India, and Indo-China an unusual number of malignant tumors of the cervical lymph nodes has been reported.

Although present knowledge is insufficient for one to state with finality that certain factors, exclusive of others, are instrumental in governing the occurrence of cancer cases, there are enough indications to show which of these factors may be of particular importance.

a) Infestation of parasites. The carcinogenetic properties of certain helminths were reported long ago. W. Kouwenaar told the Oxford symposium that in Indonesia and India scars of yaws seem to be correlated with skin cancer in significant numbers. Wherever *S. haematobium* is common in Egypt, Tunisia, Morocco, cancer of the bladder seems to be prevalent.

b) Common usage of irritant substances seems to have an influence on certain forms of cancer. Some of these substances may be connected with certain occupations (for example, scrotal cancer was common among chimney sweeps); others are correlated with local usages and culture. There seems to be a correlation between the occurrence of oral cancer and the chewing of betel nuts; according to Joyeux this is found in Indo-China only when the betel nut is taken with lime, mainly provided by burnt sea shells. Stewart remarks that in certain parts of the United States physicians have been impressed by the development of cancer in the parts of the mouth where tobacco chewers keep their cud. Furthermore, there is a correlation between heavy, persistent smoking and cancer of the lungs.

c) Nutrition also seems to have an important relation to the occurrence of cancer, although how is not yet clear. In certain types of starvation there appears to be an increase in estrogenic hormones, which may in turn result in gynecomastia and also in precancerous or cancerous conditions. Bonne, van Veen, and Tjokronegoro [10] produced primary cancer of the liver in rats submitted to a diet of polished rice, thiamin, cod liver oil, carrots, and butter yellow in olive or coconut oil. However, it must be kept in mind that no definite proof can be established that these factors have any revelance in humans until studies are made in various countries on comparable populations. There is no doubt that, in countries where life expectancy at birth does not exceed 35 to 40 years, the cancer-age groups are practically non-existent.

AMERICAN CONTRIBUTIONS

An inventory of our knowledge in the field of medical geography demonstrates that none of the discoveries described were made by scientists who considered themselves medical geographers. Our stockpile of medico-geographical information has been accumulated from a number of allied sciences: bacteriology, biology, entomology, and others. Among institutions in America which have promoted research in these fields are the Rockefeller Foundation, the United States Public Health Service, the United States Army, and the United States Navy, which have gathered an impressive body of medical intelligence basic to the study of medical geography. American research has made history in the fields of hookworm, yellow fever, and malaria.

American interest in hookworm disease started after the Spanish-American war when B. K. Ashford [11] found the larvae of *A. duodenale* in great numbers in the feces of patients in Puerto Rico. Following Ashford's paper, C. W. Stiles [12] of the United States Department of Agriculture, described in 1902 a new indigenous species of *Ancylostoma* (*N. Americanus*), which he claimed to be widespread in the United States. In 1905 a hookworm campaign on a limited scale was begun in the southern States by the United States Public Health Service. Finally the Rockefeller Sanitary Commission for the Eradication of Hookworm Disease was organized on October 26, 1909, and was endowed with a gift of $5,000,000 by John D. Rockefeller, Sr.

American interest in yellow fever, perhaps the most dreadful scourge in the Western Hemisphere, was awakened early. The crowning event was the sending of the Walter Reed Commission to Habana in 1900 to protect the lives of American soldiers, who were dying in great number of the dreaded disease. Their discovery of the role of *Aedes aegypti* (then called *Culex fasciatus*) in the transmission of the disease laid the groundwork for several control campaigns which were to have dramatic repercussions on the health of mankind. Most famous among these were the Habana, New Orleans, and the Panama Canal Zone campaigns.

The Habana campaign, based on the natural history of the mosquito, was entrusted to Colonel Gorgas, who succeeded in completely eradicating yellow fever from Habana. Similar successful results were obtained in New Orleans in 1905.

The Panama Canal Zone campaign demonstrated that a knowledge of the domestic characteristics of *A. Aegypti* (which may breed in houses in the smallest collections of water left in tubs, barrels, kerosene tins, and bottles, or in crab holes, tree cavities, and roadside ditches) was essential to the eradication of the disease. On this knowledge was based the work of the Yellow Fever Service of the Rockefeller Foundation's International Health Division and what F. L. Soper has described as the golden age of achievement in the eradication of yellow fever.

By 1925 it was believed that the Western Hemisphere had been freed of the yellow fever danger, but in 1926 an outbreak occurred in northeastern Brazil, and the way was open for new research which eventually brought about the discovery of jungle yellow fever. At the end of 1949 the Rockefeller Foundation concluded 27 years of cooperation with the Yellow Fever Control Program of Brazil [13]. A 15-year investigation of the geographical distribution of immunity to yellow fever among the primates of Brazil has established the fact that jungle yellow fever has prevailed practically everywhere in Brazil during the past decade and a half.

MALARIA CONTROL

Since 1916 the International Health Division of the Rockefeller Foundation has assisted various countries in antimalaria schemes, working hand in hand with the respective governments, notably in Argentina, Puerto Rico, Nicaragua, and Brazil. In 1924 and 1925 a similar helping hand was held out to Italy, Poland, Palestine, and the Philippines.

The anti-*gambiae* campaign jointly developed by the Rockefeller Foundation and the Brazilian government is a good example of applied medical geography and is a monumental contribution to the science. The arrival of *Anopheles gambiae* in Brazil in 1930 was followed by serious outbreaks of malaria in 1930 and 1931. The first organized campaign in 1931 apparently resulted in the eradication of *A. gambiae* from Natal, its port of entry, but not until after it had found a footing in the interior of Rio Grande do Norte. After a quiescent period, terrific outbreaks of the disease with high mortality rates occurred in 1938. To meet this emergency the malaria service of the northeast was organized, supported jointly by the Brazilian Ministry of Health and the Rockefeller Foundation. This service showed to what extent biological information and cartographic and geographical techniques could be combined to achieve success.

A. gambiae is one of the most dangerous malaria vectors because of its almost complete affinity for human blood. The breeding requirements of *gambiae* had first to be ascertained. They were found to be in a climate characterized by absence of frost, and a range of temperature of less than 40°F. [14: 10], and in small shallow pools, either clear or muddy, free of vegetation, exposed to sunlight and close to human dwellings. One of the first activities of the new anti-malaria service was to establish a cartographic unit to map the "operative regions" and keep close track of the progress and retreat of the enemy. Air maps of certain sections were made and the necessary ground work, developing and printing of the photographs and preparation of the final maps, was done on the spot. The technique of eradication was established after a close study of the microclimates that govern the life of the vector. It was demonstrated that the complete eradication of the carrier species was feasible and at the time more effective than cruder anti-mosquito measures.

PELLAGRA

In 1913 the United States Bureau of Public Health started an investigation into the cause of pellagra. J. Goldberger, head of the mission, demonstrated the correlation between the disease and dietetic and standard of living factors, thus paving the way for the present-day conception of the disease.

FLUORINE AND DENTAL CARIES

Although a number of references to the relationship between the fluorine content of water supplies and dental caries can be found in the literature of the last quarter of the nineteenth century, systematic study of this correlation dates from the publication of *Dental Cosmos* by G. V. Black and F. S. McKay in 1916. These studies have been continued chiefly by officers of the United States Public Health Service, among whom the names of H. Trendley Dean, F. A. Arnold, and P. Jay Elvoa must be cited. More recently Anastasia Van Burkalow of the American Geographical Society published a series of maps showing the distribution of fluorine in United States water supplies [15].

TULAREMIA

Tularemia has also been chiefly studied in the United States. G. W. McCoy discovered in a ground squirrel (*Citellus beecheyi*) a disease characterized by lesions resembling those of plague. Later, in Tulare County, California, G. W. McCoy and C. W. Chapin described the causative agent, a bacterium, and named it *Tularense*. The first infection in man was recognized by Wherry and Lamb in 1914. Further work has been done by officers of the United States Public Health Service on the numerous reservoirs found in the American fauna, on such vectors as the deerfly (*Chrysops viscalis*), the rabbit louse (*Haemodinus ventricosus*), and on a number of ticks acting both as reservoirs and vectors.

RICKETTSIAL DISEASES

Rickettsial diseases and the medical geography of some of them have been made the subject of significant American studies. The United States Public Health Service has established a laboratory in Montana where work has been done on ticks known to be natural vectors of Rocky Mountain spotted fever. In the United States, Rocky Mountain spotted fever has been reported from 43 out of the 48 States: Maine, Vermont, Connecticut, Rhode Island, and Michigan are the exceptions. The geographical distribution of various carrier ticks has been established—notably of *Dermacentor andersoni* (the Rocky Mountain wood tick), *D. variabilis, Amblyomma americanum,* and *Haemaphysalis leprosis palustris* (the rabbit tick)—by such workers as Cooley, Kohls [16], Breman, Bishopp, Trembley, Parker, and Steinhaus. Rickettsial pox was described for the first time in New York City by Huebner, Stamp, and Armstrong. It is carried by a mite (*A. sanguineus*) the locale of which seems to be mainly cities, such as New York, Boston, Philadelphia, Indianapolis, and others.

PROBLEMS AND METHODS

The individual stones brought together to build the existing edifice of medical geography have been gathered from various branches of the medical sciences. Each of these component sciences has its own methodology, which might perhaps lend itself to the study of medical geography. Is there, however, such a thing as a distinctive method applicable to the systematic acquisition of knowledge in medical geography? How does one think in medical geography and how does one acquire knowledge?

The first problem is to find a means whereby accurate data can be collected. Unfortunately, our knowledge of the distribution of disease is now chiefly based on reports made by various individual physicians to governmental health agencies, on reports of hospitals, on published papers dealing with individual cases, and on a small number of field surveys. These give at best an incomplete picture of disease occurrence and no picture at all of diseases which may have occurred in the past but were not reported. The vast majority of pathological cases are never scientifically observed. Some are observed, but not recognized. Some are recognized, but not recorded. Fewer still are both accurately diagnosed and efficiently reported.

Aware of this condition, the United States Public Health Service has made several attempts to investigate some sections of the United States population on a broader basis. A. L. Chapman has recommended multiple screening, performing a battery of tests simultaneously on a given section of the population, and has drawn attention to the hygienic value and low cost of the method. These tests reveal incipient or concealed diseases. The program is designed to reduce the cost of public health expenditures, rather than to promote medico-geographical research.

Harold F. Dorn has conducted a master survey of cancer in ten cities of the United States in order to give a picture of cancer occurrence; also, under the auspices of the United States Public Health Service, a national health survey has been undertaken to discover prevalent diseases among large numbers of families. This survey has been described by G. St. John Perrott and collaborators [17].

For the purposes of medical geography the recording of actual diseases is indispensable, but hardly less needed are methods of detecting the traces left in the individual by non-clinical forms of diseases, the footprints of disease. Until such information is available one cannot hope for an accurrate picture of the prevalence of diseases in a given region and to proceed, on the basis of this picture, to the study of their geographical correlations.

To this end the American Geographical Society has recommended a regional investigation in which data on the prevalence of certain diseases can be brought to light. The Society's method is based on the principle of sample studies of populations. These samples would be carefully chosen so as to be representative both of normal types and of deviations from them. Details of the statistical problem cannot be given here. Briefly, it is estimated that the study of a unit sample of 3,000 would give information on 1,000,000 people for the most common diseases under consideration. The margin of error would be small where the prevalence of the diseases is great.

To simplify the technique, the persons investigated would be examined at five different desks: a general information desk, a clinical desk, a blood desk, an X-ray desk, and a stool center. The data thus secured would disclose the more important diseases to which the sample group of subjects had been exposed and also the areal distribution of these diseases within the area inhabited by the population represented. Such information, plotted on maps, would permit further study of the correlations between the diseases and environmental factors.

PROSPECT

This survey suggests that problems of gathering information and of presenting the essential facts in an orderly manner are of paramount importance to the future of research in medical geography. This requisite information pertains to the health as well as to the diseases of man. Knowledge of the normal physiological character of the populations under study is indispensable to a sound understanding of the occurrence of pathological conditions. We must know what the "normal man" is in different regions before we can recognize when, and where, and why he is sick.

No less essential is the cartographic representation of such information as can be mapped pertaining to the normal man in the physiological sense and to the diseases that afflict him. The effective study of correlations awaits the solution of these informational problems.

Recognition of causal connections between environmental and pathological phenomena through the study of their correlations might procede in two steps, the first analytical, breaking down the most prevalent pathological phenomena into their pathogenic complexes and determining their relationships to the environmental factors, and the second synthetic, the building up of composite pictures of the combined operation of pathogenic and environmental factors.

Techniques for establishing correlations between pathogenic and environmental factors and for interpreting their meaning need much further development and improvement. Some of the pertinent factors lend themselves to quantitative measurement or enumeration and to statistical methods of correlation. Others may be correlated by the comparison of distributions as shown on maps. Still others may be assessed and correlated in qualitative terms only on the basis of descriptive statements. Also pertinent correlations can be studied in the laboratory by entomological and bacteriological techniques. Testing of the behaviour of certain pathological complexes of physiological functions under artificially established climatic or other conditions is a method with a short past, but, perhaps, with great promise. The future of medical geography depends upon the judicious use of all of these methods often in combination, and of other methods yet to be developed.

In concluding this exposé it may not be too presumptuous to hazard a prediction. Once a global picture of the proportionate prevalence of certain pathological phenomena is outlined, it may be found that the same pathological phenomena always occur under similar environmental conditions and also that the occurrence of certain diseases precludes that of others in the same areas. The latter might be the case, for example, in a hypothetical region where cow-pox was so universally prevalent as to eliminate the possibility of small-pox. This line of thought also suggests that the foods produced by certain soils, the diets based upon these foods, and the cultural characteristics that have contributed to the diets, may permit or forbid the prevalence of such non-communicable diseases as goiter, cancer, and hardening of the arteries. In other words, it may be found that, just as men, plants, and animals live together in communities, so also do pathological phenomena occur in communities or associations.

Should this prediction come true, a great advance would be made in the study of medical geography as a means of understanding and mastering disease. Meanwhile the concept of pathological phenomena occurring in characteristic associations might well provide a useful working hypothesis in the further conduct of researches in this field.

REFERENCES

1. MAY, J. M. "The Geography of Pathology," *Scientific Monthly*, 72 (1951): 128-131.
2. ———. "Medical Geography: Its Methods and Objectives," *Geographical Review*, 40 (1950): 9-41.
3. AUGUSTINE, D. L., and SMILLIE, W. G. "The Relation of the Type of Soils of Alabama to the Distribution of Hookworm Disease," *American Journal of Hygiene.* 6 (1926), March supplement: 36-62.
4. AUGUSTINE, D. L. "Studies and Observations on Soil Infestation with Hookworm in Southern Alabama from October 1923 to September 1924," *American Journal of Hygiene*, 6 (1926), March supplement: 63-79.
5. FAUST, E. C. *Human Helminthology* (3rd Ed.), Philadelphia, 1949.

6. The following references are taken from HERMS, W. B. *Medical Entomology* (3rd ed.), New York, 1939: 7, Notes 2, 6, 7;
   DE SOUZA, G. S. *Tratado descriptivo do Brasil em 1587*, Rio de Janeiro, 1587;
   NOTT, J. C. "On the Origin of Yellow Fever," *New Orleans Medical and Surgical Journal*, 4 (1848): 563-601;
   BEAUPERTHUY, L. D., "Transmission of Yellow Fever and Other Diseases by Mosquito," *Gazeta oficial de Cumaná*, Año 4, no. 57, May 23, 1854.

7. REED, W. "The Etiology of Yellow Fever," *Philadelphia Medical Journal*, 6 (1900): 790-796.

8. BATES, M. *The Natural History of Mosquitoes*, New York, 1949.

9. JARCHO, S., and VAN BURKALOW, A. "A Geographical Study of 'Swimmers' Itch' in the United States and Canada," *Geographical Review*, 42 (1952): 212-226.

10. BONNE, C., VAN VEEN, A. G., and TJOKRONEGRO, S. "Over het experimenteele botergeel carcinoom van de level," *Geneeskundig Tijdschrift van Ned. Indie*, 81 (1941): 2448.

11. ASHFORD, B. K., KING, W. W., and BUTIERREX IGARAVIDEZ, P. *Report of the Committee for the Study and Treatment of "Anemia" in Puerto Rico*, San Juan, 1904.

12. STILES, C. W. *Report on the Prevalence and Geographical Distribution of Hookworm Disease in the United States*, Hygienic Laboratory U. S. Public Health and Marine Hospital Service, Bulletin No. 10, 1903.

13. The Rockefeller Foundation, International Health Division, *Annual Report*, 1949: 77.

14. SOPER, F. L., and WILSON, D. B. *Anopheles Gambiae in Brazil 1930-1940*. The Rockefeller Foundation, New York, 1943: 10.

15. VAN BURKALOW, A. "Fluorine in United States Water Supplies: Pilot Project for the Atlas of Diseases," *Geographical Review*, 36 (1946): 177-193.

16. KOHLS, G. M. "Vectors of Rickettsial Diseases," *The Rickettsial Diseases of Man*, Publication of the American Association for the Advancement of Science, 1948: 83-96.

17. PERROTT, G. ST. J., TIBBITTS, C., and BRITTEN, R. H. "The National Health Survey: Scope and Method of the Nation-wide Canvass of Sickness in Relation to its Social and Economic Setting," *Public Health Reports*, 54 (1939): 1663-1687.

PHYSIOLOGICAL CLIMATOLOGY[A]

STEPS IN THE STUDY OF CLIMATIC EFFECTS
THE SEARCH FOR NATURAL ASSOCIATIONS
BASAL METABOLISM
THE PHYSIOLOGICAL SIGNIFICANCE OF ATMOSPHERIC CONDITIONS
SYSTEMATIC STUDY OF HUMAN REACTIONS TO CLIMATIC STRESS

PRACTICAL APPLICATIONS
CLOTHING
HOUSING
APPLICATIONS TO ANIMAL INDUSTRY

APPLICATION OF THE REGIONAL CONCEPT

BEYOND THE REGIONAL CONCEPT

A. Original draft by D. H. K. Lee, sponsored by the Committee on Physical and Bio-Geography, Hoyt Lemons, Chairman.

CHAPTER TWENTY-TWO

PHYSIOLOGICAL CLIMATOLOGY

Pₕᵧₛᵢₒₗₒ𝒈ᵢcₐₗ not...

P HYSIOLOGICAL CLIMATOLOGY is a field of study that is peripheral to the central theme of geography. It is concerned with the effect of atmospheric conditions on the functioning of the normal human body, and of the bodies of other warm-blooded mammals. A large amount of new evidence has been brought together during the past two decades regarding the effect of climate on man, evidence derived largely from laboratory methods. The factors involved have been identified and isolated, and the processes have been observed under controlled conditions.

The effect of climate on man has received much attention from writers of geography during the past fifty years or more. In fact, some of the most imaginative and stimulating works deal with one form or another of climatic determinism, and these works still enjoy a wide popularity. It seems, however, that all these earlier works must be re-examined in the light of the new understandings of process provided by the studies in physiological climatology.

Contributions to this new field have come from climatologists, physicists, physiologists, and geographers. Workers in the United States and in Canada have taken the lead in bringing together these diverse ideas to form a unified approach. The problems of World War II, which called for new kinds of clothing and housing to provide protection from a variety of extreme environments, gave a great impetus to the scientific study of environmental stress. The impetus of war needs has been continued in the postwar period.

The studies of physiological climatology center on the heat balance of the warm-blooded animal. Quantitative measures are sought to define the effect of thermal conditions on this heat balance and the adaptations made by the normal body to meet these conditions. Thermal stress can be calculated for given conditions of activity, clothing, temperature, humidity, air movement, and radiation. The assessment of the physiological significance of thermal stress remains uncertain. Important practical applications have been made to problems of clothing, housing, and animal production. But the geographical distribution of the climatic elements that produce thermal stress is not yet well known, nor have the various processes and relationships isolated in the laboratory yet been examined in their natural associations. The application of the new knowledge of physiological climatology to the enlargement of geographic understanding remains to be accomplished.

STEPS IN THE STUDY OF CLIMATIC EFFECTS

Interest in the effect of atmospheric conditions on human health and comfort is not new. Observations and speculations on this subject date back at least to Hippocrates and doubtlessly were made again and again long before his time. Some speculations on the correlation of human behavior with climate, representing the work of a scholar who derived his data from books rather than from out-of-doors observation, came from the pen of the English historian H. T. Buckle. Most of these early writings were inspired by imagination and preconception rather than by careful observation. Yet it is interesting to note that one of the earliest reports based on a close look at the facts came from Benjamin Franklin, who, in 1758, wrote his friend Lining an account of his observations of the cooling effects of evaporation and stated his conviction that evaporation played a prominent part in protecting man against the ill-effects of heat [1]. Not long after this C. Blagden reported on certain spectacular and well-designed experiments carried out in London [2]. He set out all the evidence a modern student could require regarding the part played by evaporation in sensible temperature, yet he did not make the obvious deduction. It was from Georgia and South Carolina that H. Ellis [3], and J. Lining [4] wrote circumstantial accounts of the effect of heat, with a praiseworthy collection of quantitative observations in support. In spite of these and a few other 19th century approaches, it was not until 1904 that J. S. Haldane published his classic observations of the effects of heat and humidity in the Cornish tin mines [5]. A study by J. Leferre in 1911 for the first time offered a thorough exposition of the quantitative physical approach to the measurement of body heat [6].

THE SEARCH FOR NATURAL ASSOCIATIONS

Early in the present century the interest of professional geographers in the effect of climate on man led to a search for natural associations of certain kinds of climatic phenomena with supposed human responses. The ideas of this period are summarized in the early works of Robert DeCourcy Ward (cited in the chapter on climatology). But it was Ellsworth Huntington whose keen imagination and facile pen blazed new trails in the search for the effects of climate on man [7].

Huntington's method was to seek for and report on natural associations of climatic conditions and human behavior. The associations he examined involved both time and space: he sought evidence of natural events that were followed or accompanied by specific kinds of behavior; and he sought accordant areal relations between categories of climate and categories of societal activity. But when Huntington began his studies in 1907, there were no data of the kind he needed, no large body of carefully recorded and tested observations. The physiologists and medical men to whom Huntington might have turned for guidance on biological matters were still preoccupied with other tasks or insufficiently advanced in their understanding of natural phenomena to be of much assistance. All too often phenomena that are really not at all comparable were combined, as in Huntington's map of "civilization." All too often there was an unconscious selection of data that conformed to theory and a rejection of contradictory evidence. Statements such as the following were taken as principles (from the *Principles of Human Geography*): "The geographical distribution of health and energy depends on climate and weather more than on any other single factor. The well-known contrast between the energetic people of the most progressive parts of the Temperate Zone and the inert inhabitants of the Tropics and even of intermediate regions such as Persia, is largely due to climate." To be sure, in his last work (*Mainsprings of Civilization* [7]) he gave first place to diet, but this is usually overlooked. So keen is man's interest in his own welfare, and so readable was Huntington's prose, that his writings had an immediate popular appeal, an appeal that is still evident. Administrators, sociologists, geographers, and even medical men read his works with avidity and absorbed his ideas uncritically.

But the world of scholarship has moved on. The support of this or that concept cannot be made a matter of faith or loyalty or emotion: concepts in any field of scholarship must be subjected to tests and verified or abandoned. Huntington's brilliant generalizations, covering such a wide range of relationships to climate, are worth reading for two reasons: first, they are thought-provoking and not all of them have been disproved; and second, as a demonstration of effective presentation they are unequalled. But all must now be examined in the light of newly acquired knowledge.

BASAL METABOLISM

In the meanwhile the physiologists had turned their attention to studies of the production of energy by the oxidative process, which is fundamental to most living forms. This had narrowed down to a measurement of so-called basal metabolism, the rate of energy production by the resting animal. To investigate the state of basal metabolism under various climatic conditions was an obvious line of inquiry. Considerable data from different climatic regions were accumulated.

The value of the concept seems uncertain. It now appears that measurements made under different climatic conditions may not be strictly comparable. Also the actual metabolic rate of natural activity seems to be much more relevant in problems of adjustment to thermal stress than is the basal rate; and here the problem is greatly complicated by the cultural and nutritional habits of the people being investigated. Furthermore, the word "metabolism" seems to have acquired an innate virtue in the minds of some who are interested in human climatology but are not sufficiently versed in physiology. Such are apt to speak of metabolism as a good thing, something of which the more you have the better. Yet as a general proposition such a stand is not defensible. As with most physiological processes there is a range that is suitable to the particular individual, and rates above or below the limits of this range are not suitable. The limits of the range vary greatly from person to person. If there is such a thing as climatic stimulus, and if it does not have desirable effects, it is most certainly not expressed merely through an increase in metabolism.

The Physiological Significance of Atmospheric Conditions

Probably the most dramatic contribution by workers in the United States to the progress of studies of the effect of atmospheric conditions on human physiology was the development of the concept of effective temperature. This concept was formulated by a group of investigators working for the American Society of Heating and Ventilating Engineers (ASHVE). That human comfort and thermal stress depend not on air temperature alone, but on the combination of temperature, humidity, and air movement, has long been appreciated. Attempts to provide an instrument for the measurement of these three elements together had been made but without much success. The ASHVE group made use of the human body as the measuring instrument. They defined combinations of dry-bulb temperature, wet-bulb temperature, and air movement that resulted in similar sensations of comfort or discomfort [8]. On the basis of these results they drew up a nomogram from which can be read off the temperature of a theoretical atmosphere, still and saturate, which would have the same effect on comfort as any observed atmosphere of which the temperature, humidity, and movement are known. This they called the effective temperature. This gave to those concerned with indoor conditions a measure for assessing the combined effect on man of the three most important variables affecting thermal balance and comfort. Its use out-of-doors is restricted by the exclusion of all but the simplest corrections for radiation exchange, and by the condition that the person shall either be clad in a medium weight suit or stripped to the waist [9].

In the meanwhile C-E. Winslow and others at the Pierce Laboratory of Hygiene at New Haven studied the partition of heat exchange between the human body and its environment among the channels of radiation: conduction-convection, and evaporation-convection. This led them to two concepts of importance in theoretical considerations of heat exchange and thermal stress: the concept of percent wetted area and that of the operative temperature. While the skin may sometimes be completely wet, it is never dry. At any one time, if the numerous small areas of water film on the skin are added together they would constitute some important fraction of the total skin area. This gives a physical measure of what is meant by percent wetted area [10], and enters into the calculations of heat exchange. The second concept provided a formula for the measurement of complex factors. For any situation in which the walls and air are at different temperatures, there is another situation with walls and air at the same temperature which would result in the same exchange of heat with the naked human body. The common temperature of this second situation is, briefly, the operative temperature of the first [11].

The Pierce Laboratory work dealt primarily with the naked man [12], but practical considerations called for quantitative methods of dealing, not only with clothed man, but with the significance of the clothing itself. In 1941 the concepts used in the Pierce Laboratory were extended into this field with the definition of the "clo" unit as the unit of insulation [13], and its application to the determination of clothing requirements for men at different degrees of activity in different thermal environments [14]. The effect of clothing on heat exchange between body and environment was studied in terms of radiation and evaporation-convection.

The application of these ideas to the systematic treatment of clothing problems was discussed at a conference held under the auspices of the Climatology Section (Office of the Quartermaster General, United States Army) in 1944 on "The Principles of Environmental Stress on Soldiers." As a result of this conference, interest has been focused on the search for an index of climatic stress and strain that would be more fundamental in its development and wider in its application than the effective temperature index or other current indices used for the conjoint assessment of thermal factors in the environment. The difficulty lies, not so much in calculating the applied thermal stress, as in determining the physiological significance of a given stress, and in deciding the extent to which relative contributions by the stress-producing components affect that significance.

Systematic Study of Human Reactions to Climatic Stress

While the generalists have been seeking the touchstone that will reveal the true significance of any set of environmental conditions, systematic physiologists have been going about their normal business of describing in greater detail the specific responses of the human body to various environmental situations, and in working out the mechanisms responsible for such responses. The Harvard Fatigue Laboratory served as the focal point in the 1930's, but the central figure in later investigations in the Western Hemisphere has been H. C. Bazett, whose death on his way to the International Physiological Congress in 1950 put an end to a most stimulating career. By virtue of personal example and constant effort he brought American, Canadian, British, and even Australian investigation during World War II to some semblance of coordination, and prompted an understanding of problems among all these workers which might otherwise never have developed [15].

The war-time and postwar studies of human reaction to climatic stress are numerous. Outstanding are the investigations of reactions to hot-dry climates carried out by E. F. Adolph [16], and of warm-humid climates by S. Robinson [14]. Research projects of great importance were also carried out at two army research laboratories: the Armored Medical Research Laboratory, and the Quartermaster Climatic Research Laboratory. In recent years the beginnings of organized field tests out-of-doors have been made. Various centers in Alaska and Canada have been utilized by service units for the study of cold conditions; Death Valley in California was the scene of hot-weather studies under the supervision of Hoyt Lemons and under the technical direction of D. H. K. Lee. These field studies have contributed not only to the testing of materiel and the observation of human reactions, but also to the development of new or improved field techniques.

PRACTICAL APPLICATIONS

There is no doubt that in the field of physiological climatology the demands of practical problems have greatly increased the effort expended in research and accelerated the pace of development. Some of these practical problems have had to do with military needs, but there are many applications outside the military, especially with regard to clothing and housing.

Clothing and housing are cultural matters. Inherited customs, tenaciously maintained, offer an obstacle to the spread of new ideas based on the findings of research scholars. It may well be that the necktie, for example, is more of a barrier to human comfort under hot weather conditions than any physiological processes within the normal body. In the United States, however, changes in scientific knowledge and technology are more rapidly reflected in such elements as clothing and housing than they are in most other parts of the world. It is becoming more and more necessary to understand just what requirements are imposed on clothing and housing by different climatic conditions, and within just what margins individual caprice or cultural lag may be safely indulged.

CLOTHING

In relation to clothing, there is only one comprehensive book available [14], that of L. H. Newburgh, which gives the principles and the methods of applying the principles to a variety of problems. It is possible, for instance, to estimate the amount of insulation that clothing would have to provide in a given cold environment to maintain comfort in a man carrying out a specified degree of activity; or, conversely, to estimate the maximum activity that would be permissable for a man wearing specified clothing in a given hot environment. It is possible, also, to determine the relative contributions of yarn, weave, thickness, and design to insulation or to the impedance to evaporation. There are numerous refinements that can be made, and data based on field observations in a variety of places need to be collected before the principles can be applied over a satisfactorily wide range of problems; but the methods of application are known, and the need is recognized.

Paul A. Siple was largely responsible for the interest taken in this approach to clothing problems by the Quartermaster Corps of the United States Army. To him belongs the credit for showing how basic clothing requirements for an area could be determined from a knowledge of the climatic conditions, and for the formulation of these requirements in the "clothing almanac." The work initiated by Siple was continued and expanded in postwar years by Hoyt Lemons and a staff of geographers and climatologists in the Office of the Quartermaster General. In this postwar work, attention was given to the coordination of climatic and other conditions of the physical environment with human requirements in the determination of clothing needs [17].

HOUSING

There is still no comprehensive book on housing comparable to that cited for clothing. The principles, however, have been formulated and presented in several publications. D. H. K. Lee has brought together the basic concepts regarding hot-weather housing in a monograph discussing many aspects of physiological climatology as related to tropical living [18]. The monograph contains maps showing the distribution of certain categories of climate that are relevant to housing problems. In still another form the principles are applied to different parts of the United States. In roughly alternate numbers of the *Bulletin of the American Institute of Architects*, starting in September 1949, there have appeared tables under the title of "Climatic Control Project." Each set of tables deals with a particular part of the country. For a central point within the area treated, the frequency of the occurrence, by months, of selected values for temperature, solar radiation, cloud cover, precipitation, relative humidity, vapor pressure, wind force, and other climatic elements are given. Diagrams indicate the extent to which zones about the central point might be expected to differ from it. Then the significance of the distribution of each variable for each of several key features of housing is given in tabloid form. Sponsored by the magazine *House Beautiful*, and directed by a board of well-known persons (R. Linton, H. Landsberg, L. P. Herrington, W. A. Taylor, D. Coman, and M. L. Colean), it illustrates graphically and forcibly what can be done by any country which possesses adequate climatic data and the means for processing them.

Research conducted by the American Society of Heating and Ventilating Engineers and by Herrington and his colleagues at the Pierce Laboratory of Hygiene, combined with studies in other countries, have been largely responsible for the information upon which to base the recommendations for the adjustment of housing to climate. In the United States the Bureau of Standards serves in somewhat the same way as the government-sponsored Experimental Building Stations in England, Australia, and South Africa. The latter are equipped for laboratory investigation, and also participate in the testing of ideas and practices through government-financed housing schemes.

Applications to Animal Industry

While consideration of the role played by climate in the general field of animal ecology rightly belongs in zoogeography, the biophysical approach which has advanced our understanding of man's reactions to climatic stress is now being applied to problems of animal production, and the course of this development may well be traced here.

Although it has been known for a long time that domestic animals of highly selected mid-latitude breeds often deteriorate when introduced into sub-tropical and tropical climates, the causes have remained conjectural and corrective measures a matter of marked dispute. The most cursory analysis of the problem indicates that at least three major groups of factors call for consideration in such transfers of livestock: climatic, nutritional, and infective. Veterinary medicine was not slow in recognizing the tremendous improvement brought to human medicine about the turn of the century by the study of infective pathology. The significance of climatic factors for the transmission and development of many types of infective disease was similarly appreciated, and led to the early development of a climatic pathology of domestic animals. In the case of nutritional factors, however, attention seems to have been concentrated largely upon mid-latitude conditions, and knowledge has remained relatively incomplete regarding what may be involved in introducing to the nutritional circumstances of the tropics animals which have been highly selected over a hundred years or more for the relatively stable and lush pastures of mid-latitude regions. This gap in our understanding is gradually being closed, however, as the techniques of nutritional investigation, in which American workers have played no small part, are applied to problems in various tropical and sub-tropical regions.

Questions of direct climatic effects, however, have been left largely unexplored, some adopting the position that such direct effects are relatively unimportant, others accepting the view that they are of major consequence but inescapable. For the latter, there are two alternative policies. The first is to cease importing mid-latitude animals, and to concentrate upon breeding desirable qualities in native stock. The second is to try to concentrate both the desirable production qualities of the mid-latitude animals and the tolerance qualities of the native animals by careful cross-breeding.

To either of these policies the biophysical approach can offer assistance and suggestions. That certain individual animals have a superior tolerance to heat stress is fairly easily demonstrated. That types or breeds vary in their mean or characteristic tolerance has been established both in the laboratory and in the field. But just why one individual, type, or breed is more tolerant than another has not been satisfactorily worked out. There seems no good reason why this should not be done. It appears reasonable to suggest that breeders would be able to direct their programs, select their breeding stock, and perhaps even discover superior genotypes if they could recognize in their individual animals those characteristics, anatomical or physiological, which determine degrees of climatic tolerance. Guesses have been made, and most of those items which would be suggested by a study of the basic physical processes involved have been named. However, little attempt has been made to determine systematically the relative importance of such things as variable blood-flow through the skin, extent of the surface area, insulative properties of the coat, density and functional capacity of the sweat glands, respiratory cooling, degree of activity, mechanical efficiency of activity, and endocrine control of metabolism.

The central idea, however, has been accepted by such influential organizations as the Bureau of Dairy Industry (United States Department of Agriculture) and the Food and Agriculture Organization of the United Nations. The former has a program of investigation in hand, which will apply this line of inquiry to dairy cattle. The latter has made it one of the major items at two international conferences on animal production, one at Lucknow in India [19], and one at Turrialba in Costa Rica [20]. Through the latter organization studies in animal climatology in both the field and the laboratory have been encouraged in different parts of the world, and provision made for facilitating the exchange of information between the workers themselves, and between the scientists and animal producers.

The outstanding American pioneers in this field are S. Brody and A. O. Rhoad. From a study of the energetic efficiency of stock—the energy value of the meat, milk, or work produced by the animal in comparison to the energy value of the food consumed—Brody naturally passed to considerations of the influence of climatic factors thereon. His book [21] is the only text available and will long remain a valuable reference. His climatic laboratories at the University of Missouri are the most complete of those devoted to animal climatology. Rhoad has emphasized field work, and is best known for his "Iberia Heat Tolerance Test" [22], which provides an approximate method for determining the tolerance of a number of animals under field conditions. With developing interest other centers have started work, notably the University of California Agricultural Experiment Station at Davis, where Kleiber's basic work on metabolism has been extended by his colleagues to questions of partitional calorimetry under different environmental conditions. Through the interest of the Bureau of Dairy Industry, other institutions are being b·ought into collaboration on both the physiological and the climatic aspects, especially in the southeastern states.

APPLICATION OF THE REGIONAL CONCEPT

Geographic thought tends to group phenomena by regions, whether defined by the phenomena themselves, or by their relationship with other phenomena already regionally demarcated. The recognition of a general regional pattern of vegetation led Köppen, and later Thornthwaite, to develop classifications of climate which could be correlated with the vegetation regions; and it might be thought that similar associations would have provided a stimulus to the formulation of a regional classification of climate in so far as it is areally related to man and other animals. That such classifications for animal climatology have not developed beyond the simpler pragmatic stage is probably due to the fact that the association of man and other animals with climate has been examined, not so much by the ecologist, as by the physiologist, who is interested more in processes than in the distribution of species, even when the species is man.

The physiologist feels that a rational, as opposed to an empirical, classification of climate in terms of its significance to man could be sought only for those effects which have a definable and quantitative relationship to measurable climatic factors. In the present state of our knowledge, this would be permissible only in respect to the heat balance of the warm-blooded animal. For other processes, such as behavior, infectious disease, productivity, cultural development, he feels that the antecedents are so complex, measurements so difficult, and distribution determined by so many specialized factors, that quantitative classifications are out of the question.

Even within the limits of heat balance, the physiologist has pursued a somewhat different approach from that adopted, for example, in plant geography. He sees the thermal balance of the body affected by four simultaneously acting climatic elements: air temperature, vapor pressure, air movement, and the radiative state of the surroundings. These are separate physical entities, but they all operate through their effects upon one bodily process, loss of heat to the environment. They are thus interchangeable, and a situation can only be characterized when they are all simultaneously known. Most physiological thought has tended, therefore, to concentrate on establishing an index of composite thermal stress, or, better still, an index of resultant thermal strain. The difficulties of this approach have been mentioned previously [23].

Even if an acceptable index of thermal strain were established, it is doubtful whether a worker with the point of view of a physiologist would be happy about plotting its distribution on a map. The physiologist sees variations in response to climatic conditions or variations in the stress-producing climatic elements as continuing processes, so fluid that to map their patterns of arrangement on the earth at any one time might be highly misleading. Yet attempts to make maps based on the human reaction to climatic conditions have been made, in some cases with the assistance, or at least the acquiescance, of the physiologists. An excellent example is the atlas of clothing zones used by the Quartermaster Corps [17].

BEYOND THE REGIONAL CONCEPT

Of more far-reaching importance, however, is the integration of knowledge concerning the reactions of man and other animals to climatic stress with that concerning other interactions which have an important bearing upon the history, present condition, or potentialities of an area, especially with reference to human occupance. The physiologist may well feel that his competence and even interest stops when the discussion moves on from the physiological reaction to the social significance of the reaction, and in this he may even be accompanied by the physiological climatologist. It is here that what may be the crucial epistemological problem of our age is encountered. It is comparatively simple to break a problem up into component parts, for each of which a pre-existing discipline of thought exists. And even where one does not exist, it is not long in our highly scientific and technical culture until a discipline is developed to meet a need. Deductive thought has always been simpler than inductive, and scientists usually prefer to probe more and more deeply on a narrowing front than to range over a wide field in which thought must be synoptic. But practical demands require that synoptic judgment be made, that the specific answers to specific studies be somehow integrated to give unified answers to the original unit problem. Older scientists sometimes discourage the younger from adopting interests in which energy may be dissipated, and are apt to be suspicious of those who reason in a field in which conclusions are so difficult of verification.

The dangers are real, but integration must be effected. It is much more desirable that the integration be made by those who are familiar with the detailed analysis, than by the uninformed, the superficial, or the irresponsible. How is this to be done? To take the case in point, how are the facts and ideas deriving from geographical, physiological, anthropological, economic, sociological, political, and agricultural investigation to be put together in a form which can be used by administrators, politicians, and statesmen who have to make the judgments and take the direct action which will be in accordance with the best interests of the people of an area? To rely upon the emergence of extraordinarily gifted leaders is both unrealistic and escapist, though such men would provide significant impetus when and where they chanced to occur. A mode of thought and machinery for its realization are required. From our schools through to the highest deliberations of the land, the methods, checks, and limitations of synoptic thought should be taught, encouraged, and practiced. Teachers, tutors, project leaders, administrators, and executives can set the stage, being successively producers, audience, and critics until sublimation of the actors to the performance becomes the accepted creed. This can be done; but it requires faith, insight, and patience. Only when it is accomplished will it be possible for questions of occupance to be answered as such and not with unrelated and conflicting recommendations from a bevy of ruggedly independent scientific groups.

REFERENCES

1. FRANKLIN, B. "Letter from Benjamin Franklin to John Lining, New York, April 14, 1757," and "Letter from Benjamin Franklin to John Lining, London, June 17, 1758," in The Works of Benjamin Franklin, Vol. VI: 203-210, Boston, 1938; idem. "IIᵉ Lettre de M. Franklin au Docteur Lining sur le refraichissement produit par l'evaporation," Observations sur la Physique, 2 (1774): 453-457.
2. BLAGDEN, C. "Experiments and Observations in a Heated Room," Philosophical Transactions of the Royal Society of London, 65 (1775): 111-123, 484-494.
3. ELLIS, H. "An Account of the Heat of the Weather in Georgia," Philosophical Transactions of the Royal Society of London, 50 (1758): 754-756.
4. LINING, J. "Concerning the Weather in South Carolina; with Abstracts of the Tables of his Meteorological Observations in Charles-Town," Philosophical Transactions of the Royal Society of London, 45 (1748): 336-344.

5. HALDANE, J. S. *Report to the Secretary of State for the Home Department on the Health of Cornish Miners*, London, 1904.
6. LEFERRE, J. *Chaleur Animale et Bioenergetique*, Paris, 1911.
7. HUNTINGTON, E. *Civilization and Climate*, New Haven, 1915; *idem. Principles of Human Geography*, New York, 1922; *idem. Mainsprings of Civilization*, New York, 1945.
8. HOUGHTEN, F. C., and YAGLOU, C. P. "Determining Lines of Equal Comfort," *Transactions of the American Society of Heating and Ventilating Engineers*, 29 (1923): 163-175; *idem.* "Determination of the Comfort Zone," *ibid.*, 361-384; *idem.* "Cooling Effect on Human Beings Produced by Various Air Velocities," *ibid.*, 30 (1924): 193-212.
9. YAGLOU, C. P., and MILLER, W. E. "Effective Temperature with Clothing," *Transactions of the American Society of Heating and Ventilating Engineers*, 31 (1925): 89-99; *idem*, "The Comfort Zone for Men At Rest and Stripped to the Waist," *ibid.*, 33 (1927): 165-179.
10. GAGGE, A. P. "A New Physiological Variable Associated with Sensible and Insensible Perspiration," *American Journal of Physiology*, 120 (1937): 277-287.
11. ———. "Standard Operative Temperature, A Generalized Temperature Scale Applicable to Direct and Partitional Calorimetry," *American Journal of Physiology*, 131 (1940): 93-103.
12. WINSLOW, C-E. A., and HERRINGTON, L. P. *Temperature and Human Life*, Princeton, 1949.
13. GAGGE, A. P., BURTON, A. C., and BAZETT, H. C. "A Practical System of Units for the Description of the Heat Exchange of Man with his Environment," *Science*, 94 (1941): 428-430.
14. NEWBURGH, L. H., ed. *Physiology of Heat Regulation and the Science of Clothing*, Philadelphia, 1949.
15. BAZETT, H. C. "Physiological Responses to Heat," *Physiological Review*, 7 (1927): 531-599.
16. ADOLPH, E. F. *Physiology of Man in the Desert*, New York, 1947.
17. LEE, D. H. K., and LEMONS, H. "Clothing for Global Man," *Geographical Review*, 39 (1949): 181-213.
18. LEE, D. H. K. *Physiological Objectives in Hot Weather Housing*, United States Housing and Home Finance Agency, Washington, 1953.
19. PHILLIPS, R. W., ed. *Improving Livestock under Tropical and Sub-Tropical Conditions*, United Nations, Food and Agriculture Organization, Development Paper no. 6, July, 1950.
20. ———. *Inter-American Meeting on Livestock Production*, United Nations, Food and Agriculture Organization, Development Paper no. 8, December, 1950.
21. BRODY, S. *Bioenergetics and Growth, with Special Reference to the Efficiency Complex in Domestic Animals*, New York, 1945.
22. RHOAD, A. O. "The Iberia Heat-Tolerance Test for Cattle," *Tropical Agriculture*, 21 (1944): 162-164.
23. LEE, D. H. K. "Physiology as a Guide to Combatting Tropical Stress," *New England Journal of Medicine*, 243 (1950): 723-730.

MILITARY GEOGRAPHY [A]

THE GEOGRAPHIC BASIS OF MILITARY SCIENCE
SCOPE OF MILITARY GEOGRAPHY
THE TERRAIN BASE OF ARMY OPERATIONS
GEOGRAPHICAL INFORMATION FOR AIR FORCE AND NAVY
APPLICATIONS OF THE REGIONAL METHOD

MILITARY GEOGRAPHY IN TWO WORLD WARS
GEOGRAPHERS IN WORLD WAR II
APPLICATIONS OF CARTOGRAPHY TO MILITARY GEOGRAPHY

MILITARY GEOGRAPHY SINCE WORLD WAR II
GEOGRAPHIC INTELLIGENCE
RESEARCH AND DEVELOPMENT

[A]. Original draft by Joseph A. Russell (Chairman), Alfred W. Booth, and Sidman P. Poole.

CHAPTER TWENTY-THREE

MILITARY GEOGRAPHY

MILITARY GEOGRAPHY involves the whole range of geographic research as it is applied to military problems. Military activities of all kinds are applied in particular places, and the success of any given undertaking depends in considerable measure on the flexibility with which military principles are adjusted to the natural and cultural conditions existing in specific strategic and tactical situations. Thus it is impossible to separate a military operation from the geographic conditions which make up the area of conflict; likewise in modern warfare the space relationships of the theater of operations with all other parts of the world have economic and political, and therefore, strategic, significance.

Inevitably the field of study in which attention is focused on the actual associations of phenomena and on the actual operation of processes which characterize specific parts of the earth, and which differentiate one part from another, is of basic importance to all military planning and operations. The impact on military affairs of physical and biotic features—landforms, underlying rock, soils, vegetation, drainage, water supply, animals, vectors of disease, and the like—must be assessed by the military scientist. Climatic differences from place to place affect men and equipment and their employment; seasonal and other short-term weather sequences are vital considerations. Nor can a military campaign be planned without reference to the elements of human geography: the patterns of population and settlement, the roads, railroads, and telecommunications, the structures, the patterns of town and rural land use. In addition to the material features of the landscape there are also the non-material features: the demographic characteristics of the population, the educational levels, religions, languages, government structure, and many others. All of these have an impact on military affairs and advance knowledge of the varying combinations of these features is indispensable to a successful military plan.

Furthermore, all the foregoing elements must be appraised in terms of the total environment. Rarely does any one element completely dominate military operations, unless it is in a denying fashion. Rather it is the total complex of elements as they are associated in particular places which has military significance. It is almost axiomatic to say that if a military force is

485

properly manned and properly equipped, adequately supplied and thoroughly trained, it can operate successfully in virtually any natural and cultural environment. However, the personnel, equipment, and supply requirements of a military force are a function of the kind of area in which that force is to operate. Military geographers supply planners and commanders with an evaluated description of areas so that an appraisal can be made of the significance of the total environment with regard to military operations.

In a sense every officer is called upon to make applications of geographic knowledge, whether in detail as in a tactical situation or to large areas as in a strategic plan. All too often a lack of familiarity with the concepts and procedures of geographic analysis handicap otherwise well-trained officers in the performance of this function.

Work in military geography can be done best by persons who have had both geographic and military training. Unfortunately this is a rare combination. A few professional geographers have long held commissions in the reserve, and a few professional soldiers have a keen appreciation for and competence in geographic procedures. It has been suggested that the necessary supply of "terrain" specialists must come from the reserve where both military specialty and civilian professional occupation effectively can be combined. In addition the need for at least some training in geographic concepts and methods for all officers and for advanced training for a few regular officers has been recognized by the Army and the Air Force. This may be because the training geographers receive in synthesizing and coordinating the elements of many other scientific disciplines tends to keep analysis of areas and processes well balanced and without undue stress or bias on any one element.

THE GEOGRAPHIC BASIS OF MILITARY SCIENCE

Geography finds its chief applications in the military fields of intelligence, and of research and development. The regional approach has been found most effective in intelligence studies and such applications call for an understanding of the theory of the region, the regional concept, and for appreciation of the possibilities of the regional method of revealing meaningful areal relations (as presented in chapter two). The topical approach has been applied with great success to certain research and development problems, such as providing the environmental knowledge basic to the design and testing of materiel and equipment.

SCOPE OF MILITARY GEOGRAPHY

The conduct of total war is a task of almost infinite complexity calling for the coordinated application of all kinds of special skills. Reduced to the simplest terms these tasks involve: 1) the design, development, and production of war materiel and equipment and the supply of these in adequate quantities at the right time and place; 2) the recruiting and training of men to employ this materiel and equipment efficiently; 3) the planning and execution of military operations; and 4) the military government of occupied areas. All of these tasks require a knowledge of, and an ability to adjust to, the geographic conditions of particular places.

When war is conceived as total war, the geography of war is the geography of the earth. To be sure certain parts of the earth at any one time are of greater immediate importance than other parts; yet no place is so remote or so insignificant that it can safely be neglected. To gather, analyze, and present in usable form all possible relevant geographic information about all parts of the earth is an undertaking of impossible magnitude. Such an undertaking would be futile, moreover, because of the changing significance of geographic data. With each change in the techniques of warfare, the elements of the environment must be re-evaluated. Under these circumstances there is need for: 1) certain basic geographic information regarding all parts of the earth to provide a strategic frame; 2) detailed and comprehensive studies of critical areas; 3) exhaustive topical research on specific environmental conditions that affect the design, development, and employment of men, materiel, or military techniques; and 4) maintenance of a corps of professionally competent people, including area specialists, ready to undertake studies when and where they are needed.

For the planning of global warfare at the strategic level certain broad geographic understandings are needed. In the popular vocabulary such broad understandings are included in the term geopolitics. As explained in the chapter on political geography, however, professional geographers tend to avoid this term because of its connotations. Geopolitics was used by the Germans as a propaganda vehicle, as a pseudo-scientific rationalization of political policies (chapter seven). The scholarly appraisal of the significance of the major features of world geography in terms of global strategy, however, is an important undertaking calling for the highest degree of competence.

The general appraisal of an actual or potential theater of war also calls for a high degree of competence. The conduct of military operations differs considerably from one theater to another, and any one theater has many special characteristics which require special planning, call for special kinds of equipment, pose special supply problems, or demand the use of unusual tactical procedures. Some such special characteristics have long been recognized by military scientists; but the existence of many other such characteristics is only now being appreciated and studied.

THE TERRAIN BASE OF ARMY OPERATIONS

Terrain can be looked at broadly on a continental basis as a part of the strategic picture. There are, for example, certain historic military gateways where the surface features, the lines of communication, and the absence of forests permit easy passage through broad natural barriers. These gateways have been used as routes of military advance or retreat: for example the Moravian Gate, the Gap of Belfort, the Pear Tree Pass, the Khyber Pass, or, in a somewhat different sense, the Shenandoah Valley.

But terrain can also be looked at on a topographic scale, as the stage setting to which tactical plans must be adjusted. The details of the terrain, such as hills, valleys, streams, marshes, roads, fences, buildings, or forests, offer unique opportunities or pose special problems to a military force. Every army officer who has planned or commanded a military operation, every soldier who has sought the protection of a wall, a ditch, a hill, or a tree in the face of enemy fire, knows something of the varied associations of features that make up a terrain, and probably has wished that he knew more about what to expect to find just beyond the range of his vision. He has wanted to know how to appraise an area in terms of possible lines of advance or retreat, of possible strong points, of possible ambush or camouflage, of passability for his vehicles, of protection from missiles or from flank attack; or perhaps in terms of the availability of food, water, and shelter. There have been notable examples to demonstrate that the lack of such terrain knowledge may turn the tide of battle, as in the case of the sunken road at Waterloo. The tactical importance of terrain features is exemplified by the military axiom, "take the high ground and hold it." But it is the unexpected obstacle, which advance geographic knowledge could have warned against, that is most dangerous: the hedgerows of Normandy, the coral reefs of the South Pacific, the soft volcanic sands of Iwo Jima.

GEOGRAPHICAL INFORMATION FOR AIR FORCE AND NAVY

The Air Force and the Navy also carry on operations in the planning and execution of which geographic knowledge is of fundamental importance. The interest of the Air Force in weather and climate is primary and obvious. Yet air operations also depend on knowledge of terrain, including elements added by man. Such an undertaking as the construction of an airfield calls not only for knowledge of airstrip foundation, drainage, and other physical matters, but also of the availability of labor, and the characteristics of the local population. Strategic bombing operations are planned on the basis of an analysis of urban patterns, structures, diurnal labor movements, and many other problems included in the general fields of urban geography, transportation geography, and industrial geography.

The techniques of photo-interpretation are indispensable in bomb-damage assessment, as well as in most other intelligence activities. Knowledge of terrain patterns, including the patterns of settlement, are of importance to airmen who are off their course and must find some way of establishing their position.

The needs of the Navy are also numerous. The nature of shores where landing operations may be required is a field to which geographers have contributed. The Navy needs information concerning ports and the facilities they contain. But in its own special medium, the Navy needs geographic knowledge concerning the oceans, of the kind suggested in the chapter on the study of the oceans. Military oceanography and military geography are companion fields.

APPLICATIONS OF THE REGIONAL METHOD

The regional concept and the regional method, as presented in the introduction and in chapter two, are fundamental to military geography. At various scales and degrees of generalization, areas homogeneous in terms relevant to military problems are defined and mapped. The value of such regional systems for operational planning or for the design of materiel and equipment depends in large measure on the pertinence of the criteria adopted for the definition of regions. The analysis of areal relations among different regional systems by the technique of matching maps is a difficult one to demonstrate; yet such analyses applied to the geographic elements relevant to military problems offer the best way of determining with effective detail the significance of the total environment.

In studies of military geography, as in other aspects of geography, a considerable part of the work is done by persons who are not professional geographers. Workers in many other fields contribute to the analysis of particular areas: geologists, meteorologists, specialists in soils, hydrologists, plant ecologists, engineers, economists, and many others. As pointed out in the introduction, when these specialists in other fields discuss the phenomena and processes as they are associated in particular places on the earth they are writing geography. In these circumstances the role of the professional geographer is three-fold: 1) to establish standards for geographic analysis and presentation; 2) to devise new and more effective methods of research; and 3), as a specialist in a particular area, to provide expertness in the coordination of studies dealing with the different aspects of that area. However, because military requirements do not usually include discussion of processes leading to the presence of landscape features, but rather demand collated descriptions of the features as they co-exist in areas the geographer is ideally equipped to prepare the final reports.

MILITARY GEOGRAPHY IN TWO WORLD WARS

In two world wars geographers in the United States have served either as officers or as civilians employed in military agencies. During World War I the role of the military geographer was not so well recognized as it was in World War II. In the trench warfare of the World War I European Theater, there was a considerable demand for geologists and geomorphologists and for experts in groundwater occurrence and fluctuations. Geographers contributed to these needs and were used in considerable number in officer-training programs. They were also used in the study of boundary problems which followed the conclusion of the war. The one outstanding monograph by an American scholar dealing with military geography resulted from experience in World War I. It remains even now a classic in this field, although its conclusions regarding the military significance of terrain features are now quite out-of-date as a result of the change from warfare on foot or on horseback to warfare in tanks, trucks, and airplanes.[B]

Geographers in World War II

At the outbreak of World War II the basic geographic knowledge regarding the various theaters of war was seriously deficient. In the months just before Pearl Harbor professional geographers who were also reserve officers were called to duty, primarily in the Military Intelligence Service (G-2) of the Army, which early recognized the need for world-wide "terrain" information. Along with many officers who were not geographers by training they attempted to fill the gaps hurriedly, and chiefly from documentary sources. The Research and Analysis Branch of the newly formed Office of Strategic Services called a number of professional geographers to Washington, some as civilian workers, some assigned as reserve officers. In the Washington agencies, and later in all the theaters of war, geographers labored to make up for the lack of attention to such work in the decades after World War I. But the appalling lack of geographic understanding in high places is illustrated by the clothing and equipment issue of the Army Quartermaster. In 1941, there were three standard issues: *Temperate, Torrid, and Frigid*, with boundaries along the latitude lines!

As the war proceeded, and as large sums of money were made available to the agencies contributing to the war effort, a major mobilization of geographic man-power was accomplished, and the magnitude of the geographic work completed during the war years dwarfed all previous professional efforts. Geographical work went on in the Army, the Navy, and the (then) Army Air Corps. Geographical work was also carried on in

B. Johnson, D. W. *Battlefields of the World War*, American Geographical Society, Research Series, No. 3, New York, 1921.

many other government agencies: in the Office of Strategic Services, in the Board on Economic Warfare, the War Production Board, the Board on Geographic Names, and many others. Geography was a basic course in training programs which the universities carried on for the armed forces. Any one, even with only a beginning of graduate study, was pressed into service as a professional geographer.

Most of the geographical war effort was devoted to the preparation of area intelligence reports bearing on military, economic, or administrative problems. The reports varied in topical coverage, in scale, scope, and format. In some cases there were specific problems to answer; in other cases the requirement was for a presentation of the basic facts and relationships in an area for use in strategic planning. In planning for operations, emphasis was placed on descriptions of landforms, soils, vegetation, drainage, and on evaluations of cities, roads, railroads, and other elements of the circulation pattern, for the military geographer was assigned the primary function of informing the policy planners regarding the possible routes to objectives and the environmental hazards which would be encountered along these routes. However, other aspects of the areas were not neglected. Handbooks were prepared for the actual and many potential invasion areas.

Perhaps the finest examples of wartime area reports were the Joint Army and Navy Intelligence Studies (JANIS). To be sure these volumes were compiled mostly from documentary sources; but the range of topics covered, the variety of sources tapped, and the high quality of the writing and the cartographic work placed them among the major geographic achievements of recent decades.

The JANIS reports were produced under the supervision of geographers. More than half of each of these reports was written by geographers, but many chapters were contributed by workers in other fields. The economic, social, and political conditions were described by specialists in these matters. Geologists contributed sections on geological topics of engineering interest. The soil specialists developed classifications of soil meaningful in military terms. The sections on climate were prepared by meteorologists. Professional geographers, however, developed the basic outline of the JANIS studies, and for each report a geographer was the director and coordinator of the members of the research team. The final editing of the studies was done by a geographer.

These wartime studies, involving the application of many topical specialties to the examination of particular areas, provided much valuable experience in the organization of inter-disciplinary area research. The specialists in the Latin American area had made a start in this direction before 1941; but the war gave new impetus and new breadth to such joint studies based on the coordinated efforts of scholars in different disciplines. Many social scientists believe that the focus of attention on the problems of particular areas may provide the integration so widely recognized as lacking among the various social-science disciplines. Furthermore, specialists in strategic intelligence recognize the need for a variety of topical studies organized on an area basis.[c] As was to be expected, the large organized group efforts terminated with the end of the national emergency. The geographers, who use the materials of a variety of fields, are the professional group most interested in going forward with inter-disciplinary studies for purposes other than purely military.

In addition to the studies of areas which were prepared for planning purposes, special types of areal or topical studies were produced to assist Technical Service personnel in the development of ordnance, clothing, and other materiel which required adjustment to environmental conditions. Of particular significance were climatic interpretations upon which to base the issue of clothing, medical supplies, shelter equipment, food, and other items.[D] A particularly useful set of nearly 100 monthly climatic-zone maps, by continents, produced by the geographers and climatologists of the Office of the Quartermaster General, provided a means for estimating the characteristics of unknown areas by showing their analogy with familiar areas and climates. Appraisals were made of vegetation and animal life as potential sources of food in a survival situation.

The monumental results produced by geographers during and since World War II, in conjunction with scientists of related fields, is of necessity hidden by security classification. This collection of data in the form of maps and reports, represents perhaps the most voluminous work by American geographers in the present generation. It is important to the advance of geography in the United States, and the world as a whole, that these classified compendia become available for ready reference. It is therefore important that declassification of these works be accomplished as soon as safety permits, and that surplus copies be stored, rather than destroyed, so that eventually wide dissemination may be made to universities and public libraries. The postwar distribution of large quantities of war surplus maps to libraries selected as repositories was a most welcome boon to geo-

c. The lessons of the wartime experience have been brought together and published in several sources. Chief among these may be listed: Committee on Training and Standards in the Geographic Profession (National Research Council), "Lessons from the War-Time Experience for Improving Graduate Training for Geographic Research," *Annals of the Association of American Geographers*, 36 (1946): 195-214; report on a symposium on "Geographers in the National Defense Program," *The Professional Geographer*, 7 (1948); and KENT, S. *Strategic Intelligence for American World Policy*, Princeton, 1949.

D. LEE, D. K., and LEMONS, H. "Clothing for Global Man," *Geographical Review*, 29 (1949): 181-213.

graphers. In the event of future national emergencies that require the large scale service of American geographers, it is important that the past records be available in peacetime if only for the identification of the gaps that need to be filled. To make these works public may, in fact, outweigh the advantage of keeping them indefinitely in security protection, where the weakness as well as the strength of the documents is hidden from those who must be called on later to bring them up to date.

APPLICATIONS OF CARTOGRAPHY TO MILITARY GEOGRAPHY

The geographer's special experience in selecting, analyzing, and presenting data on particular areas is based in part at least on his training in and use of the cartographic techniques. The applications of cartography in wartime military geography were numerous and, in many cases, quite new. Never before had so much financial support been available for the development of new cartographic procedures and for the publication of expensive maps in color. As developed in the several geographical agencies of the government, cartography progressed along four lines: 1) the compilation of information on maps, and the presentation of this information on published maps as a part of the program of studies in military geography; 2) map intelligence; 3) place-name intelligence; and 4) terrain modeling.

These developments in the field of cartography are discussed at length in the chapter on cartography. It should be emphasized here that many of the new ideas in this field stemmed from the needs and experience of wartime. For example, faced with the lack of certain important map series, a map procurement program was initiated in February 1942, which became virtually a combat operation, for in some cases teams of map procurement officers actually preceded combat units in the field. Map intelligence reports on map reliability and content became a regular and valuable part of military geographic studies.

Another aspect of military geography and its meaning for cartographic development had to do with the studies of place names. It became obvious early in the war that there was need for a standardization of place names. They presented a major problem because the best maps existing for certain theaters of war were in alphabets other than the Latin. Geographers with linguistic ability played a leading role in these place-name studies in the military mapping agencies. The Board on Geographic Names of the Department of the Interior, which was given final responsibility for such decisions, was called upon to increase its personnel enormously and quickly in order to meet this new wartime demand.

MILITARY GEOGRAPHY SINCE WORLD WAR II

Since World War II, studies in military geography have been continued. Many of the geographers who were employed in wartime agencies either as civilians or as officers returned to their civilian posts. However, some remained in the several agencies active in the field, and large numbers of new geographers, many with military experience as well as graduate training in geography, were employed.

GEOGRAPHIC INTELLIGENCE

During the postwar years the geographic research directed toward intelligence requirements was in large measure a continuation of the same procedures developed during the war. The same broad objectives were served, although postwar developments in weapons and equipment have required a re-evaluation of preconceived ideas of the impact of specific environmental elements on military affairs.

RESEARCH AND DEVELOPMENT

Geographic studies in the field of research and development, as distinguished from intelligence, were started during the war, but have been considerably expanded in the postwar period. The studies of extreme and unfamiliar environments were undertaken to note the effects of these environments on men, equipment, and materiel. There was need for the development of new machines, new lubricants, new methods of upkeep to operate efficiently in extreme cold, or in extreme heat, or on steep slopes. The studies in physiological climatology related to clothing and housing have been described (chapter twenty-two).

Research and development has to do with the search for better equipment, better materiel, better machines and weapons. It touches upon intelligence concerning specific areas only with respect to the methods of gathering, analyzing, and presenting geographic knowledge. Generally the approach to research and development is topical rather than regional; yet, of course, topical within regions. The environmental elements are singled out, and their effects studied both singly and as parts of total environments. The objectives are to identify and examine the impact of critical elements of an area. More emphasis is placed on the understanding of processes than on the understanding of total environments in particular places. Emphasis is also placed on studies leading to prediction of future conditions, chiefly natural rather than cultural. This emphasis is related to the fact that developmental programs are concerned with supplying materiel for use x years from now, and in any part of the world. The materiel must be suitable to conditions at whatever time and in whatever place it is required. Therefore only those landscape features that can be foreseen with reasonable accuracy can be studied meaningfully. Thus development research projects comprise a limited selection of studies that promise long-term value in the evolution of new weapons or supply systems, or are projects that seek a better method of procuring and of analyzing geographic intelligence that may subsequently be required.

It is understandable that much of this developmental work is concentrated on extreme environmental conditions, represented by very hot or very cold, very wet or very dry climates, very deep snow or very rugged and high terrain, which, if encountered for any extended period of time, either singly or in combination, can impair seriously the performance of machines and humans. Once the specific qualities of such environments have been isolated and carefully measured, there still remains the enormous task of determining their relative importance in the over-all military picture as reflected in their areal distribution and probability of occurrence in other parts of the world. This is, in some instances, a problem in statistical analysis to which climatologists are devoting their attention.

In similar manner, attention is being devoted by the Corps of Engineers and Ordnance departments to a careful examination of other elements of the environment, for instance, soils and their bearing strength in relation to trafficability, and to design of vehicle suspension gear. These examples point to the desirability of having ultimately an analysis of the physical environment of world regions in terms of the critical elements or combinations of elements that impede or preclude satisfactory equipment and human performance. If this cannot be accomplished for all regions, the process may be developed at least to the point of enabling prediction, with more accuracy than is now possible, of the probable effectiveness of all principal types of military gear in terms of environmental conditions. It will be apparent that competent geographers with special interests in soil science, physiography, plant ecology, and climatology of various world regions can be immensely useful in a program of this nature, which would represent the ultimate in military geography.

CHAPTER OUTLINE

FIELD TECHNIQUES[A]

FIELD STUDY

PRELIMINARY PROCEDURES OF FIELD STUDY
The Statement of Objectives
The Search for Documentary Materials
The Selection of Base Maps or Air Photographs
The Reconnaissance
The Formulation of a Field Plan

RECORDING DIRECT FIELD OBSERVATIONS
Field Mapping
  Scale
  The Traverse
  Mapping on Base Maps
  Mapping on Air Photographs
Techniques of Recording on Maps and Photographs
  Mapping Single Phenomena
  Mapping Associations of Phenomena
Mapping Phenomena of Spatial Interchange
Note Taking
  Notes as Primary Recording Techniques
  Notes as Adjuncts to Field Mapping
  Sketching, Sketch Maps, and Field Diagrams
Photographs as Recording Devices

THE INTERVIEW OF INFORMANTS
Interviewing by Questionnaire
Interviewing by Informal Conversation

APTITUDES

A. Original draft by Charles M. Davis, including material submitted by Fred W. Foster, G. Donald Hudson, Wellington D. Jones, Preston E. James, Richard F. Logan, and Robert S. Platt.

496

CHAPTER TWENTY-FOUR

FIELD TECHNIQUES

CERTAIN RESEARCH procedures are common to all the various branches of geography discussed in the preceding chapters. There are four sources of factual information: 1) documents, such as maps, ground photographs, statistics, and written materials; 2) air photographs; 3) direct observation; and 4) interviews with informants. And there are four ways of analyzing factual information for the purpose of identifying and measuring areal, functional, or causal relations, each requiring the use of symbols: 1) analysis by expository methods, using word symbols; 2) analysis by statistical methods, using mathematical symbols; 3) analysis by cartographic methods, using map symbols; and 4) analysis by photo-interpretation methods, using photo-interpretation keys.

Most geographical research problems require the use of all of these methods of gathering and analyzing data. To be sure, there are some in which the statistical sources are fundamental, as in some phases of economic geography, in population geography, or in climatology; and there are some in which circumstances dictate an exclusive use of written materials or of photographs. Yet such restriction always imposes a certain handicap on geographic research: the enumeration areas within which statistics are summarized (chapter four) can best be evaluated by direct field observation; photo-interpretation keys can best be drawn up by ground study; and all written and air-photographic materials can be used most effectively by scholars who are familiar with the areas portrayed. Direct contact with the area in which a problem is located is one of the hallmarks of sound geographic research.

The last three chapters of this book deal with certain distinctively geographic methods. Chapter twenty-six discusses various aspects of geographical cartography, including the use of maps for purposes of analysis and presentation, and the evaluation of maps as documentary sources. Chapter twenty-five treats the relatively new field of air-photo interpretation. The present chapter has to do with the procedures and techniques of field study. For training in the methods of evaluating the written record, which is especially important in historical geography (chapter three) the geographer may well turn to the historian; for training in the statistical method the geographer works with the economist, the sociologist, or the statistician; and for contact with several other techniques of research, especially those of the interview, the geographer strengthens his background by studying with the anthropologist, the political scientist, the social psychologist, and others.

FIELD STUDY

By the nature of the problems he studies, the geographer is led to think of the field as his laboratory. The focus of attention on particular places and on the complex associations of phenomena and processes that differentiate one place from another means that the geographer must emerge from the seclusion of the laboratory, the library, the archive, or the office. Field study, as the term is used here, refers to the collection of information, the formulation of meaningful categories of regions, the development of hypotheses of cause-and-effect relations through direct contact with the phenomena and processes in the area where a problem is located. Unless a scholar possesses an aptitude for field study there are many aspects of geography in which he should not attempt to make a career.

The reason for this emphasis on field study is simple. In geographic research, as in the research of other scholarly disciplines, much of the analytic work and all of the presentation of results is done in terms of symbols. The symbols may be words, or numbers, or shades of gray on a photograph, or the points, lines, and bounded areas on a map. In any discipline the worker attempts to come as close as possible to the actual phenomena for which his symbols stand, but in some disciplines it is never possible to come directly into such contact. The geographer can confront and deal directly with his materials in the field: only in the field can he make the initial transfer from phenomenon to symbol. Normally he needs this fundamental experience in order to understand the meaning and limitation of his symbols. Only out-of-doors can he absorb from personal experience the fact, underlying the regional concept, that no two points on the face of the earth are identical.

Field study is not limited to the recording of direct observations of fact. All geographic research requires the generalization of point to point differences, the selection of criteria for defining homogeneous areas, the identification of areal associations. The resulting map of regional patterns is a generalization intellectually conceived and justified by the illumination it casts on the factors of a problem. In the field, in the presence of the complex phenomena which are being generalized, the scholar is best able to select those items relevant to his problem and safely and consciously to decide which items are irrelevant. In the field he is best able to formulate, test, and revise his ideas concerning functional connections and sequences of cause and effect. The map record of a field study, then, is not a precise portrayal of locations, like the results of a geodetic survey; it is a scholar's interpretation of the meaning and significance of areal differentiation in terms of the phenomena relevant to a problem.

A distinctive feature of professional geography as practised in the United States during the past four decades has been the attention to the field study of small areas. The techniques for such large-scale work were first formulated by a group of geographers at the University of Chicago in the decade between 1910 and 1920. One of the earliest published outlines describing field techniques for use in the study of agricultural areas was by Wellington D. Jones and Carl O. Sauer [1]. Annual field conferences, in which geographers from several Mid-Western universities participated, were held during the 1920's and 1930's. At these conferences field techniques were proposed and tested. As suggested in the chapter on agricultural geography (chapter ten), the concepts and methods derived from these discussions in the field have had a profound and lasting influence on the course of American geography. Some of the techniques described in this chapter were first proposed at that time, and have been revised and improved by further testing and by application during the subsequent years.

PRELIMINARY PROCEDURES OF FIELD STUDY

The procedures of field study must necessarily be adapted to the peculiar conditions of each problem and each area. Nevertheless, it is possible to suggest, more or less in sequence, some of the preliminary steps of a field investigation and some of the questions of method that must ordinarily be answered. These preliminary steps include: 1) the statement of objectives; 2) the search for documentary materials both before and during work in the field; 3) the selection of base maps or air photographs; 4) the reconnaissance; and 5) the formulation of a field plan.

THE STATEMENT OF OBJECTIVES

The first step in effective research is the formulation of a clear statement of the objectives. Many years ago William Morris Davis, attacking the then common procedure of encyclopedic description, pointed out that a person who looked at things in general would either see nothing coherently, or would see only those things illuminated by his preconceptions. The idea of looking at an area "to see what it is like," or of going into the field with a blank mind to search for problems, is generally unacceptable. Most people who claim to have blank minds are simply not conscious of their preconceptions.

The kinds of problems or purposes of geographic research that have been effectively pursued or that should be pursued have been discussed at length in the preceding chapters of this book. From the point of view of method, it is important to select a problem involving the use of concepts and procedures that lie within the range of competence of the individual scholar, and to formulate a clear statement of the problem at the beginning. This statement may be revised and focused more sharply as an investigation procedes. But only when the purpose is clearly in mind can the geographer working among the complex associations of phenomena in an area distinguish between the relevant and the irrelevant, or effectively select the criteria by which he defines his categories of regions.

THE SEARCH FOR DOCUMENTARY MATERIALS

The first steps in a field study are taken at home, in the office, or in the library. From documentary sources the geographer gains an initial familiarity with an area through the writings of other people. He may wish to plot census data on maps, or to transfer to maps of a uniform scale various features that have been published on a variety of maps at different scales. This work is preliminary and is done for the purpose of constructing a general frame of reference and of identifying the unknowns.

Documentary research involves two principal operations: locating materials and evaluating them for the purposes in mind. The first of these is bibliographical in nature. The worker starts with general bibliographies of published writings, maps, photographs, or statistics, such as are included in the *Bibliographie Géographique*, or in such areal bibliographies as the *Handbook of Latin American Studies*. A valuable guide to bibliographical sources is contained in *Aids to Geographical Research* by J. K. Wright and E. T. Platt [2]. The geographer makes use also of the research collections of such geographical repositories as the Library of Congress or the library of the American Geographical Society of New York.

Once located, the documentary sources must be evaluated. For the specific purposes of a particular problem much documentary material concerning the area in which the problem is located proves to be of marginal utility. The methods of evaluating map sources are discussed in chapter twenty-six. The student should be wary of accepting statistical data at face value, for in many cases such data are lacking in precision or in comparability, or may be found to refer to areas that are not exactly defined.

The search for documentary material does not end when the geographer departs for the field. There are many documents of great value to be found in local archives or libraries, or in the possession of the inhabitants of the area being studied. Problems in historical geography can often be solved by the discovery of old maps, manuscripts, newspaper files, or old photographs.

The Selection of Base Maps or Air Photographs

Insofar as a geographic research undertaking involves direct observations in the field, one of the first steps is the selection of an appropriate means for recording such observations. It is still common in many countries for geographers to record their field observations almost exclusively in the form of written notes. But notes, valuable as they often are, are peculiarly lacking in geometric precision and are not easily checked and evaluated by another person. One of the most important heritages of modern professional geography in America are the techniques of recording direct observations in the form of points, lines, and bounded areas plotted on base maps or on air photographs. These techniques are discussed at length in the next section of this chapter. Assuming, then, that the geographer will follow what has come to be the accepted procedure among the geographers of the United States and will record his observations on maps in the field, it is necessary in advance to select and prepare base maps or photographs.

For most parts of the world some kind of base map is available, at least on a scale of 1/1,000,000, and in many cases on much larger scales. Where no base map of suitable scale exists the geographer may have to make a rough map of his own. Generally, however, he has neither the time, equipment, nor training to make a precise base map. In unexplored areas a geographical expedition may run a traverse and tie it to a few positions determined astronomically. The construction of topographic maps with precision instruments, using the methods of photogrammetry to superimpose the base data (coasts, rivers, roads, settlements, elevations) on a grid of geodetically determined points and lines, is the job of an engineer, not a geographer. The geographer is one of the chief users of such maps, and he must select the base that most nearly meets his requirements for the plotting of the information relevant to his problem.

Usually the best available base map is not on the scale demanded by the needs of a particular study. In the United States the topographic maps are generally on a scale of 1/62,500, about an inch to the mile. The modern topographic maps, however, are made with sufficient accuracy so that they can be enlarged to almost any desired scale. But these enlargements will not carry details of pattern corresponding with the enlarged scale. Enlargement by photography produces lens error that increases from the optical center; if this equals one-quarter of an inch at the scale of two inches to a mile, the roads along the margins of the photograph will be 660 feet too long in each mile.

Base maps are seldom perfect instruments. They contain technical errors and depict conditions at the time they were made, which may be many years previous to their use. The original data from which they were constructed have been generalized; two bends in a road may have been amalgamated into one either by a cartographer or a highway engineer. The geographer will readily recognize discrepancies between what he sees and what is on the map, but the failure of a traverse to close on a base map may be caused either by his own inaccuracy or some fault of the base map, perhaps both.

Quite a different type of base map is provided by a vertical air photograph. In addition to furnishing a large amount of base data, it gives a unique view of the terrain and a resolution of certain distribution patterns that may be of value to the field worker and could be obtained by ground techniques only with much greater effort. Although airphotos and mosaics made from them are superior to ordinary base maps for most geographic field purposes, they too have their deficiencies, including matters of availability and cost. The patterns recorded by the camera may not be the ones desired by the investigator and may actually obscure those sought. The use of airphotos in field mapping, however, is discussed later in this chapter.

THE RECONNAISSANCE

The first step a geographer needs to take in the field is the general reconnaissance of his area. A reconnaissance is a rapid survey of the whole or major parts of an area for the purpose of gaining an over-all view before the start of detailed study. During the reconnaissance the field worker checks the information he has received from previous documentary study; he identifies the larger regional divisions, the critical boundary zones, and the lines of circulation; he makes a preliminary determination of the categories he will use; he sharpens the statement of the problem he is investigating.

The techniques suitable for reconnaissance are those that will cover the maximum area with whatever degree of detail is required by the nature of the study. It is usually desirable to record observations on maps, but the degree of generalization is great, the scale of the map small, the criteria used for the definition of categories quite different from those to be used in detailed work. Automobile traverses along roads, train traverses along railroad lines [3], even traverses by boat and airplane have been made [4]. Sometimes sample areas are mapped in detail on an experimental basis, and random interviews with informants are made.

In a strict sense reconnaissance is not intended to produce observations of direct value to the objectives. Its important results are contained in the plan of operations which establishes categories of phenomena, scales for field operations, and techniques for the collection and recording of the materials. Properly, a reconnaissance is preliminary to some other operation. Published field studies of superficial or fragmentary nature have been entitled reconnaissance studies, but such a designation is more apologetic than accurate. Reconnaissance is not a substitute for other methods but a different procedure which has its own specific place and utility. It requires more experience than other methods, since it implies that the investigator already possesses a keen appreciation and understanding of the complexities of areal distributions and relationships gained from intensive and detailed field observation. For this reason reconnaissance is often misused and the results obtained are open to criticism. The apparent ease of this procedure appeals to those who have not previously prepared themselves for its proper use by an apprenticeship in detailed field observation and who mistake the results of reconnaissance for those of more detailed study.

The reconnaissance is especially important where a complete field survey is to be undertaken by a corps of field workers. The field assistants are divided into teams each of which covers a part of the area. The teams are supplied with detailed instructions defining the categories of phenomena to be plotted on the maps, the degree of generalization to be used in mapping, and the items of information to be gathered from informants. Such instructions are drawn up on the basis of the reconnaissance by the director of the project [5]. If a large area is to be covered on maps of intermediate scale and intensive topographic coverage is to be limited to sample areas, the selection of such samples is made on the reconnaissance [6]. On the reconnaissance, too, the specific techniques to be used in gathering and analyzing information are determined.

The Formulation of a Field Plan

The field plan is a schedule for the most effective use of the available time in gathering and analyzing the information relevant to the objective of the study. To draw up such a plan much experience is needed, for the beginner may either underestimate the time necessary for the completion of a field job, or he may not realize the flexibility necessary to follow up ideas or information that are uncovered in the course of the work. Each kind of problem requires its own peculiar balance between complete field mapping, sample-area mapping, and other procedures. Some kinds of information, also, can best be gathered by direct observation, but other kinds require the interview of informants. The field plan sets a proper balance between the time spent in these two techniques.

RECORDING DIRECT FIELD OBSERVATIONS

A central problem of field study is that of selecting a suitable technique for the recording of direct observations. There are three basic ways of doing this: 1) plotting categories of phenomena on maps or on air photographs; 2) writing descriptive notes; and 3) taking ground photographs or low-oblique photographs from the air. There was a time when note-taking was the technique most commonly used, and this is still the best technique for certain purposes. Notes describe what the observer sees: if he writes vividly and observes carefully, his notebook offers a means for preserving what he finds out about an area. The trouble with note-taking as a basic technique is the difficulty if not impossibility of recording areal spread with any precision and of measuring the degree of correspondence between two phenomena. For the precise recording of phenomena which occupy areas and for the measurement of degrees of correspondence there is nothing to compare with the field map. Today most experienced field workers record observations primarily on maps, and make use of notes and photographs to supplement the map.

FIELD MAPPING

It is in field mapping that the geographer makes his primary application of the regional method as discussed in chapter two. He selects the criteria by which he defines categories of phenomena. These categories are the elements of a system of classification, a regional system. Only on maps of very large scale can the geographer map in detail such specific items as individual houses or roads. Usually he represents these point or line phenomena with out-of-scale symbols, and on this framework of points and lines he plots the spread of areas that are homogeneous in terms of his criteria. So important is the technique he uses for this that it is recommended that some mention of procedure be included in every published report on a field study.

Scale

The scale of the field map is a fundamental consideration. Obviously the larger the scale the smaller the homogeneous areas that can be plotted on the map. The smaller the spread of homogeneous areas the less they need be generalized. The term "topographic" is used in this book to describe scales on which it is possible to plot specific features of the human occupance, such as the individual fields on a farm; but there is wide range in the topographic scales.

The selection of a minimum scale for a field map depends on how much space on the piece of paper is required for recording a symbol. An artist with a sharp colored pencil could record color symbols the size of a pinhead. But except for certain kinds of mapping, the use of color in the field is not recommended. Often it is necessary to record a written symbol, and some symbols, as explained in a later section, are fractions with several digits in both numerator and denominator. As an arbitrary working definition, some field men agree that the minimum space required for such a symbol would be a square quarter inch. Granted that this minimum space is flexible, it may nevertheless serve to illustrate the point that map scales must be coordinated with the areal spread of the defined categories. The smallest areas on the ground that can be outlined and identified with written symbols on a field map at different scales are:

Topographic	Chorographic
Scale of 1/63,360—40 acres	Scale of 1/1,000,000—4 miles square
Scale of 1/31,680—10 acres	Scale of 1/500,000 —2 miles square
Scale of 1/15,840— 2.5 acres	

If, therefore, the objective of a field study requires the mapping of phenomena no smaller than 40 acres in extent, a field map can be used with a scale of one inch to the mile; but if the smallest phenomena to be mapped are only 2.5 acres in area, the field map must be at least four inches to the mile. Conversely on a scale of 1/500,000, no phenomena that occupy less than an area of two miles square can be plotted to scale.

It is not always necessary to select a map scale on the basis of the smallest item that must be shown on the map. Very small items such as houses, mine openings, or wells would require very large scales to plot with correct dimensions, with the result that the map would have to be drawn on a very large piece of paper. Such items can be shown with out-of-scale symbols representing point or line phenomena.

The Traverse

Three techniques are used by geographers in field mapping: traverse mapping, mapping on base maps, and mapping on air photographs. If no base is available at the required scale, the geographer may have to make his own rough traverse.

Traverse mapping consists of carrying a continuous line of position from one known point, the point of departure, either back to the point of departure or forward to another known point. The traverse is the main element of control within the field map after the mapper has left the point of departure. From this line of position he extends the mapping laterally by intersection and other means. If the line of position is carried in whatever direction the needs of the moment indicate, the process is referred to as a random traverse; if it is carried in a regular predetermined pattern it is a planned traverse. A series of traverses across an area may provide a complete coverage; if they do not, the construction of a map along a zone on either side of a traverse line is called strip mapping.

There are two ways of making traverses. One method makes use of the compass and protractor. Directional azimuths are read from a sight compass and plotted into a cross-sectioned notebook in which the map is constructed. The other method requires the use of a plane table: the azimuths are plotted directly along the sighting alidade and the table is oriented by an attached compass. The compass-protractor method is well-suited for rough or heavily forested country, or for use in places difficult of access. Plane-table mapping, however, is normally faster and more accurate.

The measuring of distance in both methods of traverse mapping may be done in a variety of ways. Pacing is generally the most practical but it is also very tedious. Stadia readings or a surveyor's tape may be used in rocky areas or along lake shores where a constant pace cannot be maintained. In reconnaissance surveys distance of a sort may be computed by timing the more or less constant speed of an airplane, a boat, or even a horse. Where roads are sufficiently straight, a base line can be measured along them by using the odometer on an automobile. Another method of traverse mapping is done by plane-table triangulation wherein the traverse is carried forward from a carefully measured base line by angular intersections. This is particularly adapted to rough terrain where the constant variation between the sloping and the horizontal distances makes pacing difficult and inaccurate.

The locations of point positions away from the line of traverse may be determined by intersection or by resection on two or more known points along the traverse line. With practice, the mapper may develop his sense of distance perception to such a degree that he can estimate with reasonable accuracy. The factor of error in estimation precludes such estimates for distances on the traverse line; its chief application is in mapping laterally from the line. In areas originally laid out by the General Land Office survey, field and property lines will show a marked correspondence to divisions of the mile into halves, quarters, and eighths, thus providing a means for making estimates.

Traverse mapping is a basic type of pattern delineation. Its principles should be part of every field worker's training. Some of the procedures of traverse mapping, such as distance and direction keeping and accurate estimating, are required in almost any other kind of field mapping; for this reason alone, experience in traverse mapping is an extremely important part of a field geographer's education. Even with adequate base maps or airphotos, a geographer is constantly required to make rough maps at large scales to record the details of things too small to be depicted at the field scale.

Mapping on Base Maps

Where reasonably reliable base maps are available, the geographer may avoid much tedious work. The utility of base maps for reconnaissance, for recording point and line phenomena, for strip mapping, and for sampling is obvious. The most commonly available base maps in the United States are the topographic maps and planimetric sheets of the Geological Survey, but many other public agencies produce maps from which base data may be prepared; notably the soil-survey maps, post-office maps, sectional aeronautical charts, highway-survey maps, and national forest maps. The best place to locate base maps is within the field area itself, where the local public agencies are likely to have or to know about all existing base information and also to possess locally produced or corrected maps. County and city officers and local offices of national government bureaus may have not only the base maps, but also at times other information sought.

The advantages gained by use of a base map are speed in operation and accuracy of location. The field worker is relieved of the necessity of developing the pattern of roads and of other features for orientation. One may move with greater freedom within the mapping area, if it is always possible to pick up points of known location. Lines can be run without closing; a traverse may be interrupted and resumed along a line of control, such as a road shown on the base. Because a base map contains a large amount of locational data, it may be used in a variety of ways within the limits of accuracy required for the ordinary geographical field study. Commonly roads form the traverse bases, and an automobile the distance-keeping function (Fig. 16). This combination is especially effective in those parts of the United States where the General Land Office pattern has resulted in straight roads at regular intervals. Very satisfactory intersection can be done with a telescopic aledaide within a distance of one-third of a mile by using a half mile of straight road for a base line. Various adaptations of the base map-automobile combination have been used in field mapping. A plane table set up on a light truck or even on the top of a passenger automobile, may be oriented by a backsight along the road even if the compass is affected by the automobile. In such mapping careful note is made of prominent features that might serve as resection points: silos, windmills, and similar point phenomena are good checks, whereas hilltops and the corners of woodlots and of fields are poor, because from some other sight angle they may be difficult to identify. High-powered field glasses make such identification easier.

The techniques of using the automobile-base map combination have been developed for many kinds of boundary-finding problems. If the road net is close enough, very little foot mapping may be necessary. This method is excellent for establishing areas of homogeneity for sampling purposes, for delineating land types (Fig. 16), for network traverses, and for locating cultural features associated with the road system.

Base maps have been used in combination with airplanes both for reconnaissance [4] and for actual mapping, and also with boats for mapping swamp areas [7]. In preparing depth contours of lakes a base map

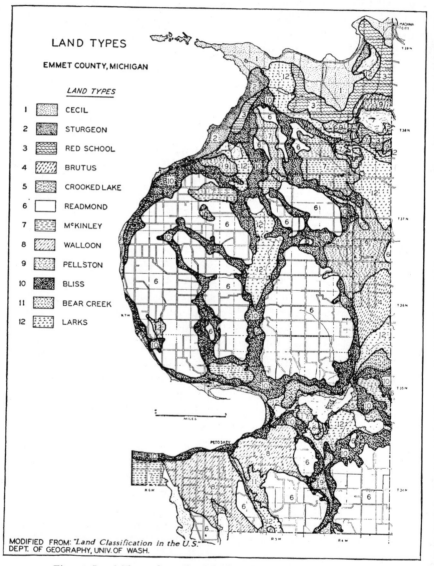

LAND TYPES

EMMET COUNTY, MICHIGAN

LAND TYPES

1		CECIL
2		STURGEON
3		RED SCHOOL
4		BRUTUS
5		CROOKED LAKE
6		READMOND
7		McKINLEY
8		WALLOON
9		PELLSTON
10		BLISS
11		BEAR CREEK
12		LARKS

MODIFIED FROM: "Land Classification in the U.S."
DEPT. OF GEOGRAPHY, UNIV. OF WASH.

Fig. 16. Land Types from the Michigan Land Economic Survey.

of the lake outline is commonly constructed and boat courses laid out between recognizable shore features; along these the position is established by intersection from a third point each time a sounding is made.

Urban mapping differs from other cover mapping in scale, categories, and the occasional representation of the vertical dimension. It is commonly done on large scales on base maps available from municipal departments or on the atlases used by insurance companies. Airphotos and mosaics make excellent bases [8]. Their detailed character make the distance and direction keeping relatively minor problems; any street corner on an urban map becomes a closing point, so orientation is readily maintained [9].

The most common type of urban mapping done by geographers is the delineation of functional zones in cities [10]. This is accomplished by direct observation of the various kinds of utilization, which are identified by keys and recorded on the base map, the pattern of city blocks being used as a framework (Fig. 17). A careful and quantitative definition of categories is required in this type of mapping. The recording of the vertical dimension to give volumetric significance is relatively much more important in urban than in rural mapping. This dimension may be readily recorded by appropriate key symbols in the field but is difficult to show in the finished cartographic presentation. It may be represented by a series of maps, each showing utilization at a separate story or level, or by cross-sections [11].

Many of the maps necessary to urban study are made from documentary rather than from field sources. Maps and cartograms of traffic and pedestrian movements are actually based on sampling techniques, in which statistics are collected at selected points and plotted upon base maps of the city.

Mapping on Air Photographs

The rapid development of aerial photography after World War I brought to field geography a most important new tool. Many of the patterns of phenomena that geographers had laboriously delimited on the ground could now be readily studied as revealed on vertical airphotos. The much larger amount of base data shown on an airphoto than on an ordinary base map almost eliminated the problem of constantly keeping location by traverse methods. Mapping land use or cover phenomena on an airphoto has become largely a matter of identifying on the photograph features previously recognized on the ground by direct observation and of placing symbols in the appropriate areas. The boundaries appear on the photograph. Other phenomena, however, such as soils or land types, are not so clearly or directly shown, although the wealth of easily identified base data on a photograph makes the plotting of soil boundaries relatively easy. There are also many non-material features that a geographer may wish to plot on his field map, such as trade areas, ethnic distribution, or even micro-climatic observations. Again, the airphoto may provide a convenient base for the plotting of such material.

The various ways in which direct field observations can be recorded on airphotos were worked out during the late 20's and early 30's. In 1929, K. C. McMurry made pace-traverse strips across Isle Royale in Lake Superior and by transferring the data to airphotos completed a cover map of the island. In 1934 the geographers of the Tennessee Valley Authority adapted multiple-phenomena techniques to airphotos and plotted information directly on the prints [12]. The Michigan Land Economic Survey continued its single-phenomenon mapping by the use of air mosaics and automobiles [13].

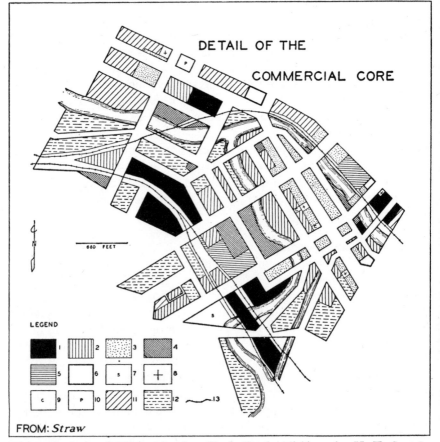

Fig. 17. Example of Urban Mapping (Battle Creek, Michigan, by H. T. Straw); 1 is manufacturing, 3 is commercial core, 5 is land devoted to transportation, etc.

The use of vertical airphotos as mapping bases is significantly different from their interpretation as documentary sources. In mapping, the observer recognizes a complex of phenomena on the ground as forming an identifiable pattern of tone density on the photograph and marks its limits as the boundaries of the complex; he can observe the phenomena, and the photograph discloses their distribution. The main purpose in photo interpretation is to determine what phenomena are represented by the tone patterns of the photograph (chapter twenty-five). In field practice the two procedures are blended: the mapper often interprets the photograph where the pattern boundaries are beyond the range of his vision, and photo interpretation without some ground observation is likely to be incomplete and unsatisfactory.

In no phase of field operations are airphotos of more value than in reconnaissance; they enable the field man to observe beyond the range of his vision, to trace critical boundaries, and to select sample areas with accuracy and dispatch. For such purposes mosaics are preferable to separate prints. The Photo Index sheets furnished at small cost by several government agencies in the United States are suitable for reconnaissance in areas of a few hundred square miles.

To the individual geographer engaged in the study of a field problem the disadvantages of using airphotos lie chiefly in their cost and in the difficulty of obtaining suitable prints. There is airphoto coverage for most parts of the United States, but even at the low unit cost of the separate prints any extensive areal coverage represents a substantial aggregate expense. In direct mapping, the scale of the print is necessarily the field scale, but this may be altered by photographic enlargement or reduction. Where a reasonably good base map is available, vertical or near-vertical photographs by hand cameras from light airplanes following C. H. MacFadden's technique for obliques [14] would seem to be a satisfactory way of obtaining photographs (Fig. 26).

TECHNIQUES OF RECORDING ON MAPS AND PHOTOGRAPHS

Geographers make use of many different techniques for recording direct field observations on maps or photographs. The different techniques, however, fall into two groups. Some field workers prefer to map one feature at a time, using several bases for the separate plotting of associated phenomena. Other geographers prefer to make use of the fractional-code system for the plotting of associations, using one base. In either case the end results are the same: the several maps of individual phenomena can be matched for the identification of associations and for the measurement of the degree of correspondence between them; the multiple-feature maps can be taken apart by tracing the spread of one phenomenon at a time.

Mapping Single Phenomena

Where the geographer prefers the single-phenomenon technique he may use a single air photograph as a base by placing transparencies over it and plotting on these overlays the areal spread of different features. In this way he might map land use on one overlay, soil types on another, slope on another, vegetation on another, and so on.

He may, of course, find that his categories for any one phenomenon, such as vegetation cover, are made up of multiple aspects of that phenomenon. For example, the field workers on the Michigan Land Economic Survey, using the single-phenomenon technique, recognized categories of forest cover that involved a record of the species, their order of dominance, and the size and density of growth of the trees. They recorded this information by means of a complex code symbol. Thus, for an area occupied by white, norway, and jack pine, the symbol might read as follows:

$$\frac{\text{W-N-J}}{\text{12-18}' : \text{1-3}''}$$

The upper part of the symbol indicates the tree associations; the lower part, that there is a thin stand of trees (indicated by the single accent mark) 12 to 18 inches in diameter and a thicker stand of trees (shown by the three accent marks) only 1 to 3 inches in diameter [15].

The mapping of single phenomena has both advantages and disadvantages. Since elaborate symbols are usually not necessary, the map-space required for writing the symbol can be smaller than where fractional codes are entered. Areal differences occupying less than a square quarter inch can be plotted to scale (Figs. 18 and 19). Each phenomenon can be observed as a separate pattern, and thus, the human tendency avoided to see coincidences of association where actually two phenomena only correspond (see chapter one). The disadvantages include the number of separate overlays that must be handled in the field, the danger of overlooking associations of phenomena, and the difficulty of maintaining the same degree of generalization among several different kinds of phenomena that are later to be matched.

Mapping Associations of Phenomena

The field conferences of geographers of the American Middle West held during the years after World War I, at which many of the field techniques of geographic research were first formulated, focused attention on the possibility of mapping associations of phenomena and therefore of measuring the degree of association. This represented a major step forward in geographic research methods. The note-taking technique, previously current, led field observers into the common error of asserting the existence of coincident areal relations that were not proved by actual maps of areal spread. As pointed out previously (chapters one and two) much of the field support of the concept of environmental determinism was derived from this procedure and from the failure of field observers to bring back precise maps of areal relationships. The difficulty of making precise maps without the aid of air photography was great. When the techniques for large-scale field mapping were worked out and applied, geographical field study entered a new phase.

The Mid-Western conferences stimulated the development of techniques not only for identifying associations of phenomena by direct field observation but also for recording them in measurable form. On the field map an area was identified and outlined in which a particular kind of land use and a particular combination of slope, soil, and drainage were associated. Wherever any one of the elements of an association changed a boundary line was drawn and a new association recognized. Theoretically, at least, the several associated features were mapped just as they would have been mapped on separate overlays, the only difference being that they were all plotted on one sheet and identified by a complex fractional code, a technique first suggested by Charles C. Colby. The possibility that the boundaries of associated phenomena might be made to coincide where

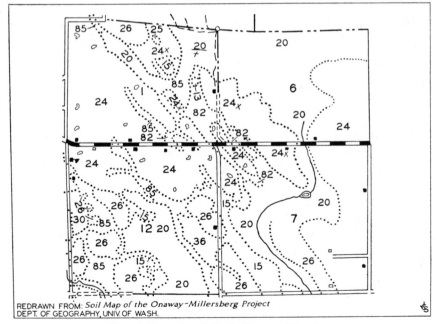

REDRAWN FROM: *Soil Map of the Onaway-Millersberg Project*
DEPT. OF GEOGRAPHY, UNIV. OF WASH.

Fig. 18. Example of Single-Feature Mapping.

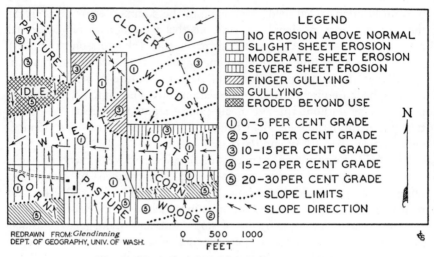

Fig. 19. Example of Multiple-Feature Mapping.

they actually are very nearly but not quite coincident, is off-set by the focus of attention on associations. The boundary difficulty is not simply a result of inattention on the part of the field observer. Because of the space-requirement of a square quarter inch on the paper for the recording of a fractional-code symbol, there are numerous small areas within which differences must be overlooked. For example, a soil boundary may cut across the corner of a field of wheat, leaving an area of less than an acre as a separate association. On the map of associations this has to be combined with bordering areas because it is too small to show; on the map of single phenomena, where soils are mapped on one overlay, crops on another, the small discordance of soil and crop at a field corner is not lost in the mapping.

The first critical evaluation of the technique of mapping associations was made by Vernor C. Finch [16]. In a study of Montfort he devised a fractional-code symbol which distinguished six features: above the line, three separate groups of digits identified the major kind of land use (tilled land, permanent grass, timber, idle), the specific kind of crop or other use, and the condition of the crop; below the line three groups of digits identified categories of slope, soil type, and drainage condition (Fig. 20). The resulting map is complex, but its component elements can be separated and shown as individual maps. Finch's conclusion that the results were not worth the expense (in terms of time and personnel) are modified by the later development of labor-saving procedures and by the demonstration of the practical utility of such detailed maps in studies of agricultural or urban land use.

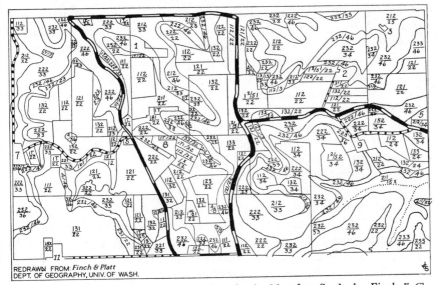

Fig. 20. Use of Fractional Code Technique in the Montfort Study by Finch [16].

A further development of the technique of mapping associations was carried out by the geographers employed by the Tennessee Valley Authority. Under the direction of G. Donald Hudson, starting in 1934, an inventory and appraisal of land conditions and land uses was made as a background for the planning program of the Authority. The first project was a detailed cross-section strip map of the Upper Tennessee Valley comparable in scale to the Montfort study, but with phenomena, criteria, and categories relevant to the needs of the TVA. Air photographs were used as bases for the plotting of field data, probably for the first time in any large geographic survey. As a result, categories of associations were defined, scales of work adopted, and the procedure was applied to a large area. Hudson called this the "unit-area method" [17]. This designation is a satisfactory synonym for what is called here the "association method," provided it be recognized that a unit area is a unit only by definition and not in fact. As pointed out previously, there is no such thing as a homogeneous area that is not divisible into smaller areas if a problem demands such subdivision.

The fractional-code system was applied to a number of problems of the TVA. A survey of the whole area was carried out rapidly on airphotos at a scale of about 1/24,000, using automobiles for the transport of teams of workers. The minimum areas mapped occupied 200 acres (Fig. 21). The codes used included five items above the line (major land use, crop, field size, amount of idle land, and farmstead quality) and seven items below the line (slope, drainage, erosion, stoniness, rock outcrop, soil depth, and fertility). It was then possible in the field to scan each fractional notation in combination with landscape features not recorded, and to arrive at an over-all estimate of the meaning of the fraction in terms of use. The observer recorded a short fraction: above the line he rated the effectiveness of the current land use on a scale of 1 to 5; and below the line he indicated the quality of the physical conditions of the land for agriculture also on a scale of 1 to 5. As a final step, the field observer appraised the severity of the problems of agricultural use, and recorded this appraisal on a scale of I to V, as indicated by a Roman numeral.

A more recent large land-survey program to employ the technique of mapping associations of phenomena and identifying these with a fractional-code symbol is the Rural Land Classification Program of Puerto Rico [18]. The mapping has been done by teams equipped with jeeps, and using airphotos. Owing to the small fields and miniature soil differences on this crowded and mountainous island, the work has been done on a larger scale than that of previous surveys. It is on the scale of 1/10,000, with areal differences as small as a fraction of an acre mapped. The work was under the direction of Clarence F. Jones.

MAPPING PHENOMENA OF SPATIAL INTERCHANGE

Field mapping, as discussed in the preceding sections of this chapter, is concerned with the recording of information concerning the patterns and

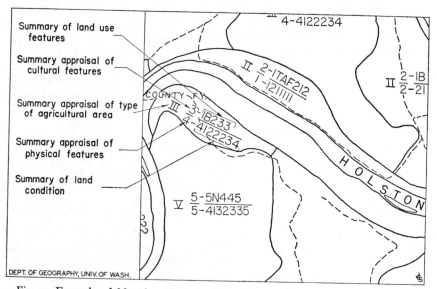

Fig. 21. Example of Mapping Technique Employed in the Work of the TVA.

areal associations of phenomena. These are the phenomena that, when reduced to map-size, appear as patterns of points, lines, or areas.

But geographers are also concerned with the phenomena of motion and functional association. Chapter thirteen deals at length with the problems of measuring spatial interchange. The nodal region, as defined in chapter two, is an area that is homogeneous in terms of organization and function. The map expression of such phenomena is not a portrayal of areal likenesses and differences or of static areal associations: rather it is a portrayal of the pattern of circulation and functional association and it shows points of focus, lines of flow, and limits of influence. Attention is directed to areal connections and separations rather than to the static patterns of uniform regions.

Phenomena of this sort are observed, measured, and mapped in the field. They are less likely to be directly visible, however, than are the features of landscape morphology. Movement and spatial interchange must be observed over a period of time, and this usually requires the work of an established data-gathering agency whether public or private. The individual geographer may make sample counts of the phenomena of movement to supplement statistical information, but, as pointed out in chapter thirteen, he is especially dependent on data systematically gathered by an agency organized for this purpose.

Similarly the functional associations that involve the patterns of circulation of the inhabitants of an area cannot usually be observed directly.

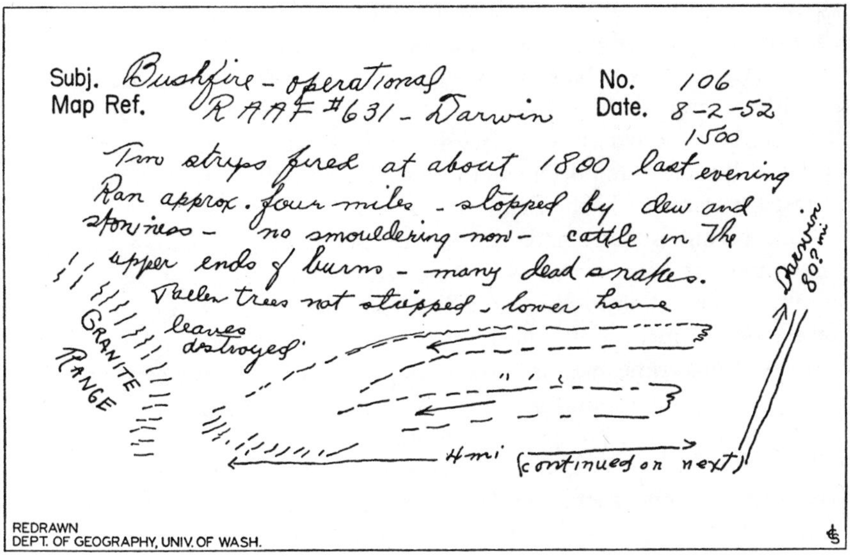

Fig. 22. Example of a Field Note.

The field geographer turns to the techniques of the interview to gather the information he then plots on his map.

NOTE TAKING

Memory cannot be relied upon to hold for long either the details of fact and relationship as observed in the field or the ideas simultaneously formulated regarding what is observed. Facts should be made matters of record as soon after they are observed as possible, or preferably while they are being observed, and the same principle applies to hypothetical identifications of areal associations and hypotheses regarding the operation of processes. Notes serve not only as a means of bringing information back to the office, but also of carrying it from place to place in the field, and thus permit comparisons to be made between earlier and later observations. Notes as here defined, include any kind of written or sketched record other than those recorded on maps or airphotos; they include diagrams, cross sections, and sketches.

Notes as Primary Recording Techniques

Some kinds of information are better recorded by notes than by mapping. It is difficult to express cartographically changes or movements among the phenomena of an area. Seasonal changes and sequences of agricultural practice, and observations of pedestrian or traffic movement at a specified location need to be described, and also many intangible qualities, such as those that lend to some Ohio towns a "New England aspect." In recording the characteristics especially of the higher ranks of compages, notes offer a more flexible technique of recording than do maps. Where it is necessary to work at such small scales that it is not possible to record direct observations on maps because the largest area that can be seen at any one place is too small to plot to scale, notes become of primary importance. In this case the final interpretations are phrased tentatively in the field.

Note-taking is not standardized. For the individual geographer who must interpret his own notes it is necessary only that they recall to his mind the field scene and his immediate thoughts about it (Fig. 22). Some field workers make only the briefest notes on the spot and enlarge on these as soon as time and circumstances permit. In team projects, however, a certain amount of uniformity, and therefore standardization, must be employed to secure comparable results. It is not likely that geographers will ever adopt such rigid and formalized noting systems as those employed by surveyors and engineers. The great diversity of objectives and the variety of techniques, together with the differences of individual interest probably make any standardized system undesirable.

Nevertheless, two minimum requirements in note-taking may be mentioned. Every note taken in the field ought to be keyed to a map, even if it is only a very sketchy road map. Every note ought to be dated.

Notes as Adjuncts to Field Mapping

The map imposes certain limitations on expression inherent in its scale and in the categories of phenomena mapped. Notes, on the other hand, provide an unlimited opportunity for expanding upon the factual and conjectural aspects of any situation under investigation. They permit the field worker to describe, to interpret, and to speculate at any length necessary for his future information. Notes also clarify matters obscured or eliminated by the limitations of the scale. Furthermore, the need for proceeding systematically with the collection of data required by the mapping objectives precludes the development of many less important elements in map form.

Notes of summation are made from time to time, most effectively at the close of a particular phase of investigation when the observations are currently in mind and the situation can readily be reviewed in the field if necessary. They bring a freshness and renewed validity to the manipulation of the field data.

Sketching, Sketch Maps, and Field Diagrams

In field operations many observations are best recorded by pictorial notes. Particularly, these include observations with respect to physiographic features too large or obscure for photography. The techniques by which such items are recorded include sketches, sketch maps, and field diagrams.

Field diagrams are graphic representations of generalizations arrived at by accumulating much experience in direct observation in the field. For geographers of half a century ago, particularly the physiographers, field sketching was a primary field technique. It has now been largely replaced by photography. Sketching enables the field worker to portray selected features of interest and to omit irrelevant detail, to alter the scale in parts of the sketch, and to annotate significant facts. The elements of the technique are not difficult to master [19; 20], but artistry is also important, especially if field sketches are to be published (Fig. 23).

Sketch maps are field records of distributions in which relative rather than exact scale representation is sufficient. They may be made by traverse techniques with distances and directions either estimated or only roughly measured. Sketch maps are designed for later study and are not

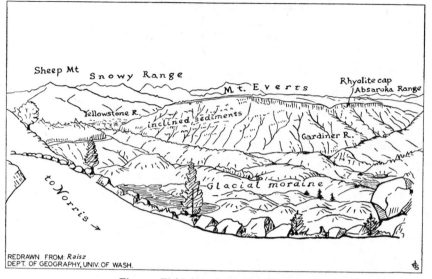

Fig. 23. Field Sketch by Erwin Raisz.

commonly published. They preserve details that cannot be shown on the field scale or recorded by a simple note (Fig. 24).

Field diagrams are graphic representations of generalizations arrived at from repeated observations and are intended to record characteristic occurrences or relationships, rather than specific instances of phenomena. They may be in either map or cross-section form (Fig. 25); as they deal with generalizations, the scale of presentation, either horizontal or vertical, may be distorted for better effect. Field diagrams are summary devices of great value in both reconnaissance and in the finished presentation of field studies.

Photographs as Recording Devices[B]

The camera is a useful tool for recording certain kinds of phenomena in the field. The photograph is non-selective, and in this way serves as a good check on the subjective interpretation of the facts and conditions of an area as portrayed in the symbolism of the map. The photograph permits the checking of impressions months or years later. It can also be used to convey to the mind of another person the visible components of a landscape form which is generalized as a category by the field observer. For example, a geographer working in an arid area might identify an alluvial fan as a category of landform. One clear photograph would do more to recreate the image of an alluvial fan than would pages of words.

B. Original draft of this section prepared by William G. Byron.

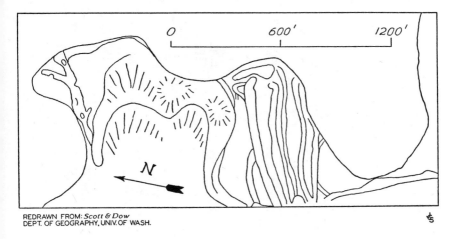

REDRAWN FROM: *Scott & Dow*
DEPT. OF GEOGRAPHY, UNIV. OF WASH.

Fig. 24. Sketch Map of Sand Dunes.

To take good pictures requires skill and a technical knowledge of photography. The field worker should understand the advantages and limitations of different kinds of films and cameras, and should have had experience in the use of his own equipment. A light meter is essential. The amount of additional equipment depends on the competence of the individual. Good photography also involves an understanding of the principles of composition: of the need for balance in a picture between close objects and more distant ones, and of the need for contrasts of light and shadow.

Good field photography for geographical purposes also involves a careful selection of subjects. Some pictures are taken to illustrate details, such as human types, tools, ways of doing things, kinds of plants or associations of plants, even soil profiles. Pictures are always more interesting if some one is doing something in the foreground. On the other hand, pictures for purely professional purposes are usually best when taken at a medium distance. This may be defined as the distance at which associations of phenomena may readily be identified in the photograph. When the subject is too close, associations are not easy to see; when the subject is too distant, associations that may be clear to the eye in the field are too indistinct to appear in the photograph. For example, if it is desired to show the characteristic position of a ranch near the apex of an alluvial fan in dry country, the photograph must be taken from a sufficient distance to show the whole fan. Care should be taken to see that the subject is truly

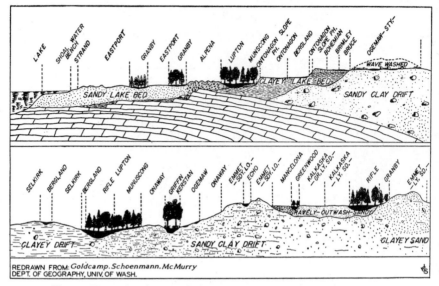

Fig. 25. Example of Cross-Section Drawing Used as a Field Note.

representative; or else a number of photographs should be taken to show the range of individual differences within a class of association.

Good geographic pictures should be able to withstand certain critical challenges. Does the picture document a geographical idea regarding a characteristic phenomenon or association, or is it unique and random? Can the picture be defended as a representative sample? Do unrelated objects obscure the main subject? Is the picture properly composed; the exposure technically correct? Can the view be precisely located on a map?

In every case a note should be written giving information about the picture. Data to be recorded in notes include the date and time of day, the compass direction, and a careful statement of the features or relationships the picture is intended to illustrate, perhaps even with a field sketch to bring out the relevant items.

In recent years technical advances have made new photographic methods possible. Color, infra-red, and ultra-violet photography offer opportunities not yet explored. Better black and white film make it possible to photograph subjects that could be captured only indistinctly by older equipment. Low-level, oblique air views taken from small planes with ordinary cameras (Fig. 26) provide opportunity for the illustration of associations of phenomena in places where such illustrations could not have been made from the ground [14]. The recent perfection of an instantaneously developing camera should simplify the use of ground photography as a recording procedure. If a field worker can take a picture, develop it on the spot, and dry it well enough to write on it an interpretation of the features it shows, he will have both a written and a visual record of great value.

THE INTERVIEW OF INFORMANTS

Another method of gathering information in the field is by interviewing informants. An informant is a person who possesses specific factual knowledge, or who has ideas or opinions worth recording. The use of informants as a source of information is a procedure widely followed by geographers as well as by anthropologists, sociologists, and others.

Information is secured from informants in two chief ways. One of these is through the questionnaire, which may be either a standard, printed set of questions asked of all the inhabitants of an area, as in a census, or a less formal but nevertheless standard set of questions asked of a representative sample of the population. The other method of securing information is by means of a properly directed conversation with an individual. Information gathered by informal conversation is usually supplementary to that gathered by other means.

INTERVIEWING BY QUESTIONNAIRE

There are four important steps in the use of a questionnaire. First, the questions are formulated as clearly and simply as possible. Second, the questions are asked of a sample group of informants, and on the basis of this sample are evaluated and revised. Third, the questionnaire is circulated. Finally the results are compiled and evaluated. Of these steps, only the second and third must be done in the field.

The first step, the framing of simple questions, is not so easy as it may appear to one who has never used this technique. If the questionnaire is to be filled in by the informant alone, with the investigator absent, it is essential to phrase questions that bring in the desired information and that cannot be misinterpreted. For example, the question "how much land do you cultivate?" may be interpreted to mean the acreage of the farm, the acreage that is tilled with or without rotation pasture, or the acreage in clean-cultivated crops as contrasted to crops that are not cultivated. Questionnaires addressed to individuals are most effective if they require only such information as is readily available. If much effort, or much statistical compilation, is required to complete a questionnaire, dependable returns may ordinarily be expected only from public officials, or from organizations that have some interest in public relations [21]. Very good results were obtained, for example, from a mail questionnaire asking county assessors in Michigan for the number of recreational establishments in their districts [22].

Questionnaires can be used not only to ascertain facts but also to measure attitudes. To achieve comparable results and avoid emotional overtones, a scale of attitudes should be provided. For example, if information regarding the prestige of various kinds of occupations in a society is needed, a questionnaire might be made up as follows:

"How would you feel if a person in any one of the following groups should ask to marry your daughter? Mark a cross in the appropriate column for each group."

Table I

	Welcome with Enthusiasm	Accept Willingly	Accept but Not Happily	Refuse
Physician				
Retail merchant				
Army officer				
Porter				
Farm laborer				
Lawyer				

If the questions in a questionnaire can be phrased so that interest is aroused, and if a report on the results is promised, an incentive for answering is developed. For the methods of arousing interest and willing participation, and for discussions of other problems of making and administering questionnaires, the geographer might turn to reports based on wide experience in such matters in other disciplines [23].

The second step—that of evaluating the questions before they are circulated—is of vital importance. In almost every case this second preliminary step brings to light necessary revisions. The most surprising difficulties arise from local differences in word usage, perhaps even within parts of the area being covered. In order to secure results that justify the time and effort expended, it is necessary to be certain not only that a question can be understood, but also that it cannot be misunderstood.

There are two principal methods of administering a questionnaire. The first of these is by mailing it to the respondents, who fill in the required information and return it. The other is through an interviewer who records the facts as stated by the respondent. The second produces better results; it assures a more complete response and avoids errors of misinterpretation by the respondent. Mailed questionnaires addressed to public bureaus and officials, large business organizations, and schools commonly bring satisfactory responses; when addressed to individuals who have little interest in the information being sought, they are likely to produce indifferent results.

The administration of a questionnaire through an interviewer combines the techniques of the questionnaire with that of the conversation. This is the means by which the United States Census is taken each decade. Effective questionnaires, administered through interviews, ask for relatively few, simple facts and figures of a kind involving the respondent personally and on which he is an authority. Facts about himself, his activities, his possessions, his personal history, and matters on which he is qualified to give an informed opinion are readily obtained. An excellent example is found in the work of Hans Kurath [24] and his associates in determining linguistic patterns by interview-questionnaires. The questionnaire completed through an interview is the most versatile of the informant techniques because it allows exact quantities to be recorded and additional information to be obtained (Table II). In some instances, by careful preparation, the validity of the quantitative material may be checked by other questions. A large amount of the public opinion gathering in the United States is done by the interview-questionnaire technique and an extensive bibliography on its methodology is available [25].

INTERVIEWING BY INFORMAL CONVERSATION

Interviews by informal conversation are undertaken for quite different purposes than are those in which the questionnaire technique is employed. Here the information usually desired either applies to a specific kind of occupation, or is derived from some especially well-informed individual. Interviews of this kind may be used to obtain specific factual data, but more often they are used to gather qualitative information, such as the seasonal variations of farm practice, or individual preferences for competing market centers.

An interview of this sort is a conversation that the interviewer carries on as informally as possible with the informant. The interviewer must fix firmly in his memory in advance the main items of information he desires to get, for it is bad practice to make use of a note book. Note-taking during a conversation tends to reduce the spontaneity of the responses. The interviewer must again and again bring the conversation back to the questions he wants answered, but he should do so without making it apparent. Interviewing by conversation requires considerable experience before satisfactory results, at all commensurate with the time involved, can be obtained, for the beginner is almost certain to phrase questions that bring answers either desired or expected by the interviewer.

Table II

Form of questionnaire used in the Rural Land Classification Program of Puerto Rico. Reference [18]

QUESTIONNAIRE FOR CATTLE FINCA
UCT-3   NORTH COAST OF PUERTO RICO

Owner's Name ———————————— Barrio ————————————————
                                 Municipio ———————————————
                                 Unit No. ————————————————
                                 Photograph No. ——————————
                                 Topographic Map ———————————
                                 Number of Map ——————————

General Farm

1. a. Number of years in cattle business?——— b. Number of years on this farm? ———

2. a. What other major items are produced on this or other farms in addition to dairy products?————

3. a. Number of dairy barns?——— b. Number of cattle stalls?——— c. Cement floor?——— d. Dirt floor?——— e. When built?———

4. a. Number of other barns?——— b. When built?——— c. Number of shed and equipment storage buildings?———

5. a. Number of family residences on farm?——— b. Number of men actively employed in the cattle and dairy business?———

6. a. Number of kilometers of improved road on farms?——— b. Unimproved road?———

7. Machinery and equipment (indicate the number). a. None——— b. Truck——— c. Tractor——— d. Pick-up Truck——— e. Wagons and Carts——— f. Milking machines——— g. Feed grinder———

8. Electricity on farm? a. Yes——— b. No———

9. Source of electricity? a. Public power——— b. Private generator——— c. Other ———

Disease and Sanitation

10. Number of diseased cattle in the past five years?———

Disease	Year	Number of Cows	Deaths from this Disease	Veterinary Attended? Yes	No
1.					
2.					
3.					
4.					

Information obtained from informants is an ancient source of geographical field data. Since early Greek times, if not long before, geographers learned how people lived by asking them questions. It seems probable that more of the information included in current geographical writings is derived from informants than from direct field observation. Yet the techniques used are not often described. Formal descriptions and experiment with the techniques of the interview are found in the literature of anthropology or sociology but not of geography.

APTITUDES

A very large part of modern geographic research involves study out-of-doors in the field. There are certain exceptions to this. A research undertaking in climatology, for example, might depend for its basic information on statistical and cartographic sources and might be carried out entirely by statistical and cartographic techniques. Yet any study in micro-climatology requires field observation. Some work in economic geography based on statistical data, can be completed without field study. For practical reasons much of the contribution by geographers to military geography, especially that completed in wartime, has had to be made from office work only. But these exceptions are few; and, even where a problem is examined wholly in the office, the most effective workers are those who have had field experience that enables them the better to interpret the mapped data they derive from documentary sources.

An important aptitude test for prospective field geographers, therefore, might involve some measurement of those qualities of mind and body that lead a person to enjoy field work. What are those qualities? They have never been defined in exact terms; although some general understanding of the desirable qualities, in addition to intellectual capacity, that are expected in graduate students is shared by most directors of university training centers. A prospective geographer should not be burdened with strong prejudices and preconceptions. He should be adjustable: he should be sufficiently at home to work effectively in a wilderness or in a strange city or among strange people whose customs differ from those of his own society. Without forming value judgments, he should be able to accept people and conditions of life different from those with which he is familiar. Furthermore, he should be able to meet and mix with all kinds of people with whom he must come in contact, for, unlike the geologist, the geographer cannot ignore the inhabitants of an area where he is working. He should not be gripped with the zeal of a reformer or missionary, for then his studies are directed toward a preconceived goal. Rather he should be willing to play the part of observer, recorder, interpreter, or clarifier of complex issues. But above all, he should be able to carry on his professional work in all kinds of surroundings, with living conditions that may range from the merely strange to the definitely uncomfortable. If unpleasant surroundings impair the effectiveness of his work as a field observer, or if he carries out his field studies without enjoyment, the professional quality of his work will suffer, no matter how superior a person he may be in his intellectual capacity.

These rather vague qualities of mind, and the physical stamina which makes carrying them easy, are the ingredients of what we may call geographical aptitude. Any one considering field geography as a professional career would do well, in the absence of specific aptitude tests, to measure himself against these requirements and to ask whether he would be more at home in an office, a library, or a laboratory, than out-of-doors.

REFERENCES

1. Jones, W. D., and Sauer, C. O. "Outline for Field Work in Geography," *Bulletin of the American Geographical Society*, 47 (1915): 520-525.
2. Wright, J. K., and Platt, E. T. *Aids to Geographical Research*, American Geographical Society Research Series no. 22, New York, 1947.
3. Colby, C. C. "The Railway Traverse as an Aid in Reconnaissance," *Annals of the Association of American Geographers*, 23 (1933): 157-164.
4. Platt, R. S. "An Air Traverse of Central America," *Annals of the Association of American Geographers*, 24 (1934): 29-39.
5. Hudson, G. D. "Methods Employed by Geographers in Regional Surveys," *Economic Geography*, 12 (1936): 98-104.
6. James, P. E. "The Blackstone Valley: A Study in Chorography in Southern New England," *Annals of the Association of American Geographers*, 19 (1929): 67-109.
7. Debenham, F. "The Bangweulu Swamps of Central Africa," *Geographical Review*, 37 (1947): 351-368.
8. Russell, J. A., Foster, F. W., and McMurry, K. C. "Some Applications of Aerial Photographs to Geographic Inventory," *Papers of the Michigan Academy of Science, Arts and Letters*, 29 (1943): 315-341.
9. Branch, M. C. *Aerial Photography in Urban Planning and Research*, Cambridge, 1948.
10. Jones, W. D. "Field Mapping of Residential Areas in Metropolitan Chicago," *Annals of the Association of American Geographers*, 21 (1931): 207-214.
11. Parkins, A. E. "Profiles of the Retail Business Section of Nashville, Tenn., and their Interpretation," *Annals of the Association of American Geographers*, 20 (1930): 164-175.
12. Hudson, G. D. *The Rural Land Classification Program*, Tennessee Valley Authority, Washington, 1935.
13. Foster, Z. C. "The Use of Aerial Photographs in the Michigan Land Economic Survey," *American Soil Survey Association*, Bulletin 13 (1932): 86-88.
14. MacFadden, C. H. "Some Preliminary Notes on the Use of the Light Airplane and the 35 mm. Camera in Geographic Field Research," *Annals of the Association of American Geographers*, 39 (1949): 188-200.
15. Schoenmann, L. R. "Description of Field Methods Followed by the Michigan Land Economic Survey," *American Soil Survey Association*, Bulletin 4 (1923): 44-52.
16. Finch, V. C. "Geographic Surveying," *Geographic Society of Chicago*, Bulletin 9 (1933): 3-11; *idem.* "Montfort, A Study in Landscape Types in Southwestern Wisconsin," *ibid.*: 15-44.

17. HUDSON, G. D. "The Unit Area Method of Land Classification," *Annals of the Association of American Geographers*, 26 (1936): 99-112.
18. Northwestern University, Department of Geography, *The Rural Land Classification Program of Puerto Rico*, Northwestern University Studies in Geography, no. 1, Evanston (Illinois), 1952.
19. RAISZ, E. *General Cartography*, (2nd Ed.), New York, 1948.
20. SWEENEY, W. C. *Sketching Methods*, Army War College, Washington, 1917.
21. MARTS, M. E. *An Experiment in the Measurement of the Indirect Benefits of Irrigation, Payette, Idaho*, Bureau of Reclamation, Boise (Idaho), 1950.
22. HEDRICK, W. O. *Recreational Use of Northern Michigan Cut-Over Lands*, Michigan State College, Agricultural Experiment Station, Bulletin 247, East Lansing (Michigan), 1934.
23. BENNETT, A. S. "Some Aspects of Preparing Questionnaires," *Journal of Marketing*, 10 (1945): 175-179.
24. KURATH, H. *A Word Geography of the Eastern United States*, Ann Arbor (Michigan), 1949.
25. United States Civil Service Commission, *Interviewing: A Selected List of References*, Library of Congress, Washington, 1945.

[A] Original draft by Hibberd V. B. Kline, Jr.

CHAPTER TWENTY-FIVE

THE INTERPRETATION OF AIR PHOTOGRAPHS

AIR PHOTOGRAPHY brings to geographic research an important new tool. Until men could fly, much of the areal differentiation of the face of the earth could be perceived only dimly beyond its more obvious outlines. The case of the archaeological sites that were first discovered from the air is well-known. There are many other kinds of patterns on the earth's surface that are not at all apparent to men on the ground but that stand out sharply when viewed from above. Observation from the air, as suggested in the previous chapter, has proved to be a valuable new technique of field investigation, especially in the reconnaissance phase. But air photography provides for an additional step beyond such direct observation; for the first time in the history of geographic research it is possible to bring back a visual record of what one sees from the air for unhurried examination in the office.

Like other tools, air photographs have their own limitations as well as advantages. These are inherent not only in the nature of air photographs but also in the effectiveness with which they are used. Effective use depends on an understanding of certain essential characteristics of the photographs themselves. Interpretation procedures involve the employment of some simple instruments and manipulations of the photographs. Fundamentally, however, the photo-interpreter must possess a thorough substantive knowledge of some branch of geography, for only then will he know what to look for. The selection of the relevant from the irrelevant is a matter of trained judgment: the interpretation of photographs is fully as dependent on professional competence in some aspect of geography as is the interpretation of phenomena observed directly in the field.

THE ESSENTIAL NATURE OF THE AIR PHOTOGRAPH

The nature of the air photograph permits a rather wide range of uses. As a picture of the face of the earth, the photograph can be read, like any ordinary photograph, by direct examination of the objects it records. The trained geographer can identify the regions in which he is interested, and draw the boundaries which separate one region from another. The closer the photograph is examined the more information it yields, even with regard to areas that are not familiar to the observer from experience on the ground. Photographs can be interpreted by specialists for a great variety of purposes: for the inventory of forest resources; for the study of soils and underlying geology; for the identification of mineral outcrops; for the study of military installations; or for the analysis of the many phenomena with which geographers are especially concerned. There is no such thing as a general photo-interpreter. Professional photo-interpreters are listed as military interpreters, photo-geologists, or by other designations indicating the field of competence in which the interpreter operates [1]. In these several fields, photo keys have been developed to illustrate by examples the appearance of various kinds of objects on photographs, and to suggest how these objects can be identified.

Air photographs can be taken with the camera pointed perpendicularly toward the ground or at an oblique angle or, in certain multi-lens cameras, in combinations of verticals and obliques. Oblique photographs can be read and interpreted, but because of the perspective introduced by the tilted camera axis they have peculiar and somewhat complex geometric properties and are less map-like than vertical photographs. Although the photographic coverage of parts of the world includes oblique pictures, as explained in chapter twenty-six, vertical photographs are more common and are the ones usually sought for purposes of photo interpretation. In the United States pictures are mostly 7″ by 9″ or 9″ by 9″, and are on a nominal scale of 1/20,000 or 1/25,000.

The vertical air photograph, however, is not a map. The scale varies not only from one photograph to the next along a flight line, but also from one part to another of the same photograph. There are three chief reasons for this. Minor distortions of scale are due to the camera lens itself. Larger distortions may result from tilting of the camera so that the photograph is not truly vertical and the scale is affected by perspective. Other important variations in scale result from the relief of the surface, the fact that the earth is not a featureless plane.

The fact that points on the surface of the earth lie at differing altitudes produces, on the photograph, horizontal displacements of these points from their true map positions. This is the phenomenon of parallax, in which objects that stand above the datum plane, such as hills or chimneys, are displaced outward on the photograph and, conversely, low points are displaced inward. The parallactic displacement on a truly vertical photograph is along lines radiating from the principal point or center of the photograph. This can be located by connecting the marks on the four sides of each print. The apparent displacement is directly proportional to the relief of the terrain and to the distance from the principal point, and is inversely proportional to the camera altitude. In a tilted photograph the perspective center of the picture is displaced from the principal point, and parallax data must be referred to the true plumb point.

Despite these irregularities of scale within the photograph, the picture may be considered to have a nominal scale which is a function of the height (H) of the camera above the ground and the focal length (f) of the camera. The scale, or representative fraction (RF), can be determined by the simple formula:

$$RF = \frac{f}{H},$$

in which f and H are in the same units of measure.

The air photograph, also, is not a map for reasons other than those pertaining to variations of scale. The map is a selective representation of certain aspects of the earth's surface, aspects that are subjectively chosen and symbolically depicted. The air photograph is a pictorial likeness, mechanically achieved, or, as Clifford MacFadden says, "a nearly complete and definitive full-detail miniature-likeness record of a given landscape complex." [2: 732]. Where the map may fail in geographic expression due to its relative simplicity and the scantiness of its data, the air photograph may confound the user by its wealth of detail, by the complexity of its patterns, or the fullness of its representations. The appearance of any given object on the photograph is not recorded by a standardized symbol, but rather it is an individual phenomenon presented in accordance with the amount and angle of light reflected from it to the camera lens.

Objects that are represented by symbols on the map can usually be read from an air photograph after brief training in the recognition of the appearance of these objects when viewed from above. Norman Carls, in his *How to Read Aerial Photos for Census Work* [3], illustrates how readily one may learn to distinguish a road from a railroad. Photo interpretation begins when the objects on the photograph cannot be read directly, or cannot be identified at once from their appearance, and when, therefore, it is necessary to formulate and test hypotheses that are based on specialized, substantive knowledge.

PROCEDURES IN AIR-PHOTO INTERPRETATION

Photo interpretation requires the use of orderly procedures that will wring from the photographs the desired information and will permit the application of specialized knowledge. Kirk H. Stone [4] suggests nine steps in photo interpretation and John H. Roscoe [1] lists fourteen leading to the definition of photo keys. The first stage of the procedure is to become thoroughly familiar with the work that has been done in the area on the problem in question. All pertinent writings, maps, statistics, and other data must be assembled. The interpretation, as contrasted with the reading, of air photographs requires checking back and forth between the pictures and information derived from other sources. A second preliminary stage is the preparation of the photographs themselves. They must be plotted on base maps of suitable scale, indexed, and oriented: places and objects on the pictures must be identified and marked. Scales must be determined by formula, by comparison of measured distances on photographs and on maps of the same area, or by other kinds of evidence, such as the square-mile road pattern of parts of the United States (chapter five). As a third stage perhaps the photographs can be taken into the field for the purpose of comparing the appearance of an object in the picture and on the ground. In the fourth stage the work of detailed analysis begins, including not only the recognition of broad patterns with the unaided eyes, but also the ex-amination of details by stereoscopic methods.

STEREOSCOPY

Stereoscopy plays a major role in the procedures of photo interpretation. As early as 1922 Willis T. Lee predicted the use of stereoscopy in the drawing of contour lines on maps [5: 6]. Today the science of photo-grammetry, which utilizes air photographs for geodetic purposes and for the construction of topographic maps, employs stereoscopic instruments to transfer data on slope and elevation from photographs to the map. Nearly all modern contour maps are made from air photographs. Similarly most photo interpretation is dependent on the use of simple instruments to bring out the stereoscopic effect.

Stereoscopy enables the observer to see the photograph in three dimen-sions. The effect may be obtained by individuals without use of a stereo-scope but at the loss of freedom to move about and make notations. Procedures in the use of a stereoscope are described in detail in the *Manual of Photogrammetry* [6: chap XI], and in several text books [7; 8; 9].

The same parallax that confounds the scale is usefully employed by the stereoscope. Air photographs taken along a line of flight are snapped at intervals that permit something like a 60-percent overlap. Two contigu-ous pictures, therefore, show the same ground from somewhat different angles. By placing them side by side a two-eyed or binocular view may be obtained. If the right eye of the observer can be directed to the common area on one photograph and the left eye to the same area on the second photograph the change in parallax will be perceived as a third, or vertical, dimension. The stereoscope is a simple device that insures that the vision of each eye will be limited to "its" photograph. The lens-type stereoscope adds some magnification and has a relatively small field: the mirror-type stereoscope gives a broader view but without magnification. Stereoscopy may also be obtained through the use of complementary colors for photo-graphs and eyepieces, or by polarized light.

The stereoscopic model does not have a true-to-scale vertical dimension, although B. J. Beltman [10] believes that a stereoscope for this purpose could be designed. It has in fact a noticeable exaggeration of relief, or what C. M. Aschenbrenner [11] prefers to call "relief stretching." This is a helpful factor in depth perception but it is likely to be misleading, particularly in the study of slope. Aschenbrenner states that

$$\text{relief stretching} = \frac{\text{air base}}{\text{flying height}} \text{ x } \frac{\text{actual viewing distance}}{\text{eye base-separation}}$$

E. R. Goodale [12] presents a formula compounded of seven measurements. These two formulae, and others put forward by photogrammetrists [13; 14], use certain values that may be difficult for geographers to obtain. Stone [15] suggests another formula for use in connection with photographs taken by the United States Department of Agriculture with an 8¼″ focal length camera. To measure what he calls the "appearance ratio" (relief stretching) he uses the formula:

$$\text{appearance ratio} = \frac{\text{picture edge distance}}{\text{interpupillary distance}} \text{ x } \frac{\text{camera focal length}}{\text{stereoscopic focal length}}$$

Reference should be made to Stone's article for definitions of these measurements. Cardboard templates for different degrees of slope can be made for the appearance ratio of each individual worker. There is an obvious advantage in knowing the relief stretching factor of the individual interpreter when a photo-interpretation team is at work on a given set of prints.

Stereoscopy also makes possible the measurement of parallax differences, and therefore of terrain relief and the height of objects. This may be done by use of the parallax wedge, a transparent plastic sheet on which are marked two non-parallel rows of dots, or by any of a number of photogrammetric instruments employing a "floating dot" or dots. Useful for making spot calculations and relatively inexpensive is the parallax bar which attaches to a lens stereoscope and consists of one fixed and one movable dot, and a micrometer for reading parallax values. The parallax wedge and the parallax bar provide numerical values for parallax at two or more points when the operator causes the wedge dots or floating dots to appear to rest upon or to touch those points. The difference in parallax values then is converted to read in linear units by use of the parallax formula. These procedures are set forth by S. H. Spurr [9] and others. This work rests largely upon acuity of depth perception, and considerable accuracy can be developed with practice. It should be noted that, given some spot elevations, the "trained eye" of the experienced field geographer can "contour" air photographs by these simple means. More precise contouring passes into the realm of photogrammetry and more complex instruments.

THE PHOTO IMAGE AND ITS ELEMENTS

In addition to the quantitative results that can be obtained from stereoscopy, there are qualitative advantages to the employment of the stereoscope and its three-dimensional model. Photo interpretation basically depends on analysis of photo images. A photo image is defined by Roscoe as "a pictorial representation of a specific object or substance, irrespective of size, which is composed of the sum of such elements as tone, texture, shadow, shape, size, pattern, position, and parallax" [1: 56-57]. Tone, or the shade of grayness, ranging from black to white, is the key to shape, because the shape of a photo image is distinguishable only by the change of tone at the edge of the shape. Tone is the result of the amount of light reflected from a surface and therefore is an index to the kind of surface, as for example the dark tone of railroad ballast and the light tone of the concrete highway. However, the tone of a given surface may appear quite differently in different parts of the same photograph, as is frequently the case with water bodies according to the angular relation between the sun, the water, and the camera. On two adjoining photographs noticeable differences in tone may indicate some differences in the processes of developing and printing. Likewise tone is importantly affected by the type of film used; for example, infra-red film depicts water bodies much darker than panchromatic film. Therefore, tone is a very useful aspect of photo images; but interpretation, or photo keys, based primarily upon tone must be employed with caution.

Texture consists of the minor variations in tone which impart a mottled appearance to many images. J. A. Russell, F. W. Foster, and K. C. McMurry [16] were perhaps the first American geographers to demonstrate the importance of texture, particularly in the interpretation of natural vegetation and of crops. Texture is an important criterion in distinguishing forest trees, and the extensive literature on photo interpretation for forestry pays much heed to it. Texture permits grain to be distinguished from pasture, even in small-scale photography, and may aid in specific identification of crops at larger scales.

The photo-image element of shadow is significant in several ways. In the reading of a single photograph shadows assist in visualizing the relief dimension, particularly if the photograph is placed so that the shadows fall toward the reader, as they do in a plastic-shaded relief map which is "lighted" from the north or northwest. In some photographs a "pseudoscopic" appearance may result if shadows fall away from the observer so that a river may seem to follow a ridge rather than a valley. Shadow direction also aids in establishing the orientation of a photograph in space if the date and time of day of the picture are known, or the time of day if the date and orientation are ascertained. The vertical structure of a photo image sometimes may be more clearly shown by its shadow than by the image itself, for example the shadow cast by a bridge on the water below may provide construction data that could not be observed otherwise on the photograph. On some photographs tall thin objects, such as high-tension-line pylons, may be identified only by their shadows. Shadow dimensions may be used to calculate image dimensions and the complex geometry of shadows has received considerable attention from photogrammetrists [17; 18; 19]. Under the stereoscope, shadows may be more readily distinguished from images, and there is little need for positioning a stereopair with the shadows toward the interpreter. It must be recognized that shadows also obscure images such as roads or trails in densely wooded areas. Shadows that obscure the ground make difficult the calculation of heights, "since all known methods require an ability to see the ground at or near the base of the object" [17: 380].

The patterns of arrangement and areal associations of phenomena are matters in which geographers are interested and which can be read from air photographs. In his study of rural settlements in a Wisconsin dairy area, a North Dakota spring wheat area, and Montana and Washington irrigated areas, Clyde F. Kohn made use of both direct field observation and photo interpretation [20]. Photographs facilitate the careful and detailed observation and recognition of patterns and associations; in fact, photo interpretation may reveal facts and relationships not readily observable on the ground. Ben A. Tator notes in working from air photographs of the lower Mississippi alluvial lands that "low-relief details are normally difficult to study on the ground. Among those forms more easily seen in aerial photographs than on the ground are included meander scars, abandoned channels, bar ridges and swales, alluvial fans and ox-bow lakes. . . . Thus, even in country of low relief, features often indistinguishable in ground view are easily visible in aerial view" [21: 717]. Minor drainage lines, so difficult to observe and map in the field in some humid areas, often are well represented on photos [22]. Not only physical features but also cultural features may on occasion be virtually unobservable from the ground but evident on the photograph. Tator writes of Indian village sites on the Mississippi delta: "The flatness of the delta terrain, along with the factor of subsidence, make difficult the location of old sites, particularly on the ground. Careful study of large-scale aerial photographs, . . . with detailed analysis of contrasts of tone and shape of soil and vegetation patterns, is required" [21: 723]. The value of aerial photographs to archaeological investigation has been well established in many parts of the world [23; 24; 25; 26; 27]; their value for the analysis of modern rural and urban settlements has also been established [20; 28; 29; 30; 31].

The foregoing discussion of stereoscopy and of the photo image and its elements outlines the fourth stage in photo interpretation procedure. Ideally the fifth stage requires the geographer to enter the field once again, to devote himself to the study of processes not visible on the photographs or, like G. M. Howe [32], to the study of sample areas chosen from the photographs. The geographer-interpreter goes on this final field mission with his knowledge enlarged and his preception sharpened by his laboratory examination of the "miniature-likeness" of the field. It is worth noting, also, that the air photograph not only represents the area photographed at one instant of time but that this instant is permanently recorded and therefore the interpretation can be checked independently by different workers. Field work on the ground, on the other hand, requires days, months, or years to complete. While ground field observation is in progress, and subsequently, changes are occurring. Maps based on direct observation cannot be subjected to the same kind of rigid checking by other scholars that can be applied to photo interpretation.

Photo Keys

Photo keys are particularly helpful when systematic information that lies beyond the training and experience of the geographer is required, or when, for some good reason, field work is not possible. Photo keys provide visual examples of specified phenomena, usually in the form of stereograms (stereoscopic pairs) with explanatory text. Russell, Foster, and McMurry [16] pointed to the advantages of photo keys in a geographical inventory. During the past two decades a large number of photo keys have been prepared, based on the systematic landscape elements by topical specialists. Eighteen studies are listed in the references of this chapter in which keys are presented for various parts of the world [33 to 50 inclusive]. The military services have also worked out numerous photo keys, some of which are available for civilian use [51].

Several ways of classifying photo keys have been reviewed by Roscoe [1] and by R. N. Colwell [17; 52]. Keys are either "direct" or "associative." Direct keys deal with photo images that are recognizable in themselves for what they are, such as a petroleum storage tank, or a plowed field [1: 82]. Recognition is the end product of the direct key. On the other hand, associative keys provide collateral data not directly shown on the photos, such as the depth of the water table, salinity of the soil, or the product of a manufacturing establishment. Associative keys lie at the heart of photo interpretation while direct keys are somewhat like map symbols and provide materials for photo reading. It is obvious that the geographic value of associative keys is considerable, for they deal with many interrelationships and deductions that geographers are accustomed to seek. It might be argued that any geographical study employing photo interpretation should produce associative keys, at least as a by-product.

Photo keys may also be classified as "regional" or "analogous." The "regional" key is concerned with a range of subjects in a specified area. The analogous key is designed to aid in the photo interpretation of an area not accessible to ground observation. Keys of this kind are based on the existence of broadly similar areas occurring in different parts of the earth, such as areas of similar kinds of forest, or similar kinds of landforms, or similar kinds of settlement features. If used wisely, with due regard for the existence of local differences, such analogous keys may yield worthwhile results.

With all their manifest utility, photo keys are not a substitute for professional competence. The study of any area or topic can be done best by persons with specialized training and experience; photo keys in the hands of untrained persons can be thoroughly misleading. Furthermore, there are variations in air photography that the key cannot embrace. These variations are related to the quality of the pictures, the scales, the seasons of the year, and other variables. Nevertheless, published photo keys enrich the geographic literature with a permanent record of photo interpretation and build up a volume of experience—a worthy goal in itself.

PHOTO COVERAGE AND AVAILABILITY

The coverage of the world by air photography was greatly extended as a result of the military requirements of World War II. Since the war, air photography has been further expanded and at an increasing rate. Technical developments in airplanes and cameras make air photography better and less costly; and as more photographs become available their utility is more and more widely appreciated. A world coverage map, showing the areas photographed by June 1, 1952, published in connection with the 17th International Geographical Congress [53], was soon out of date. For example, the African dependencies of Great Britain were only partially covered in June of 1952, but by the end of that year they had been completely covered, a task which required more than 2,000,000 photographs.

In an address in April, 1953, Kirk H. Stone presented a map showing that about 22,000,000 square miles, or about 45 percent of the earth's land area, had been photographed, and he estimated that this photography covered 80 percent of the more densely settled areas [54]. Neither map includes data for the Soviet Union. The United States coverage is virtually complete and many parts of the country have been photographed vertically more than once. American photography comes from several sources, both governmental and commercial. The chief source is that of the United States Department of Agriculture.[B] Most of this coverage is approximately 1/20,000 in scale, and is organized on a county basis, with county photo-index mosaics available at scales from approximately 1/31,680 to 1/90,000, but mostly at about 1/63,360. These photo-index mosaics are reproduced primarily for the purpose of facilitating the selection of photographs, but may be found useful as a source of geographic information, and as an un-controlled base map. The purchaser of government photography must deal directly with the agency holding the coverage he desires and, since prints must be made to fill his order, a delay up to several months may occur before delivery. At the present time, Department of Agriculture photo prints in quantities of 100 or more cost fifty cents each, but copies on special paper and enlargements are more expensive. Photo-index mosaics cost $1.10 apiece and several may be needed to cover a county. Photo coverage of some foreign countries is available through the United States Air Force and Navy, but a greater delay than for domestic photos must be expected.

Listings of exemplary photographs of geomorphic and geological features have been published by the United States Air Force [55], the American Geological Institute [56], and the Geological Survey of Canada [57].

The expense of coverage, particularly stereo-coverage in which prints overlap 60 percent and side-lap about 30 percent, is sometimes a deterrent to the use of photographs. However, inasmuch as the use of air photographs may reduce field costs and enhance the quality of geographic research, the initial outlay may prove an economic investment. More logical reasons for omitting air photographs from a particular research undertaking may arise from the nature of the available coverage. As F. W. Foster [58] points out, photographs are seldom taken specifically for geographic purposes and may therefore be unsatisfactory in one or more ways; particularly as regards scale, date or season of the year, and coverage of the study area. Adjustments in scale are possible by enlargement or reduction, but with some loss of detail. Out-of-date photos may sometimes be corrected by field work and may provide useful data on changes within the area, and photographs taken at the "wrong season" may provide insights not otherwise readily obtained. Nevertheless, unsatisfactory photographs like unsatisfactory maps raise problems that may have to be solved by discarding the photographs altogether or by the revision of research objectives

B. Information about American coverage and country-wide index maps to vertical photos and mosaics may be obtained from The Map Information Office, U. S. Geological Survey, Washington 25, D. C., which serves as an informational clearing house for government agencies.

and procedures. The taking of air photographs with a 35 mm. camera by the geographer himself is a procedure discussed by Clifford MacFadden [59]. MacFadden's pictures, however, are obliques and nonstereographic (Fig. 30). This introduces additional complexities in the use of air photographs that are not discussed in this chapter [60].

THE PROSPECT

The history of air photography opened with the first recorded photograph made from a balloon in Paris in 1858. Significant developments in air photography and in photo interpretation began to take shape during World War I. Air photographs were employed in geographic studies for illustrative purposes in the early years following World War I, but the use of photo interpretation as a research technique engaged few geographers until the requirements of World War II thrust many members of the profession into its active practice. A brief history of photo interpretation, and of the roles played by geographers, both those of the United States and those of other countries, is set forth by John Roscoe [1: 70-80].

In the period since World War II some war-time geographer-interpreters have continued to develop techniques and applications. Most geographers have come to recognize the utility of air photographs, and the training of graduate students in the procedures of photo interpretation has become a part of the curricula of several of the larger centers of geographic training and research. That the position is not more advanced at mid-century is generally deplored by those familar with photo interpretation. In respect to training, John E. Kesseli points out that "only in departments sufficiently staffed to consider all parts of the geographic field can it be expected that air photography has found, or may find, its due consideration as a research field." Furthermore, the student "is inclined to neglect field and laboratory courses which provide a training in the gathering and interpretation of information, hoping that this problem will take care of itself when the time for independent research arrives" [61: 738].

It is to be expected that continued demonstration of the advantages of photo interpretation, and continued increase in the number of geographers familiar with the technique, will lead to a steady growth in the utilization of air photographs in geographic research. Two additional considerations would be most helpful in stimulating participation by geographers: a textbook in English designed for the profession; and the establishment of adequate photograph collections nationally and locally. Textbooks on aerial photography exist for a number of disciplines but none pertain primarily to geography, although a start seems to have been made by Prof. Jiri Kral [62] of Prague and the matter is under consideration by other geographers. There are a number of bibliographies [63; 64; 65; 66;

67; 68] helpful to the geographer; but for those departments in which there is no experienced photo interpreter, a textbook is a definite need. Large costs are involved in assembling a photograph collection for instructional or research use and, as a result there are only a few departments and libraries possessing significant numbers of photographs. These facts place many photographs, particularly those used for instruction, in the category of non-expendable items, thus limiting their utilization. The establishment of an American national photo depository, similar to the National Photo Library of Canada, would be advantageous to interpreters and students from all disciplines. Furthermore, a local depository program analogous to the map program of the Army Map Service would provide selected universities and libraries with collections of inestimable value, and presumably aid in the creation of trained personnel available to the government in time of national emergency. Additionally, the large quantities of photographs marked for destruction after use by the armed forces and other agencies might be parceled out among the centers of photo interpretation and reach their graveyard in the hands of students and research personnel. The need for students to examine photographs of as many different areas and subjects as possible would be met by such a program of distribution. As it is, instruction must rest in part on published studies in photo keys and on printed collections. Examples of the latter are included in the references to this chapter [5; 31; 69 to 78 inclusive]. It should be noted that under magnification half-tone reproductions that employ a screen of less than 175 lines per inch seldom do justice to the original photographs. Most of the printed collections are not suitable for detailed study.

Photo interpretation will be aided by improvements in camera, film, and instrumentation. The visual qualities of the photo image are greatly affected by the type of film and filter used. The use of color photography in the form of transparencies will become common, and offers an exciting prospect to the geographer-interpreter. According to Gomer T. McNeil [79], two new instruments will be available. These are stereomechanical devices that will eliminate computations on vertical photographs by providing for direct height finding and slope determination. An additional new instrument for oblique photographs will make semi-automatic measurements of direction, distance, area, and height, also without mathematical calculations.

As more photographs are taken, and as there is more opportunity to study the same area at different times, the value of air photography as a permanent record will come to be appreciated. David L. Linton, pointing to the advantage of having air photography and statistical data available for the same date, urges that Great Britain be photographed regularly as a part of the decennial census and of the census of production [80].

Photo interpretation is not in conflict with, nor does it supersede, the use of other methods of gathering, analyzing, and presenting geographical data. Geographical studies should never be done exclusively from the air unless this restriction is enforced by conditions beyond the geographer's control. Air photographs record only the visible aspects of the face of the earth, and there are many non-material aspects with which the geographer is concerned. But the procedures of photo interpretation offer geographic research a new and valuable tool which, methodically and artfully employed in conjunction with other techniques, will make possible a great advance in the quality and precision of geographic work.

REFERENCES

1. Roscoe, J. H. "Photogeography," *Selected Papers on Photogeology and Photo Interpretation*, Committee on Geophysics and Geography, Research and Development Board, Washington, 1953: 55-102.
2. MacFadden, C. H. "The Uses of Aerial Photographs in Geographic Research," *Photogrammetric Engineering*, 18 (1952): 732-737.
3. Carls, N. *How to Read Aerial Photos for Census Work*, United States Department of Commerce, Washington, 1947.
4. Stone, K. H. "Aerial Photographic Interpretation of Natural Vegetation in the Anchorage Area, Alaska," *Geographical Review*, 38 (1948): 465-474.
5. Lee, W. T. *The Face of the Earth as Seen from the Air*, American Geographical Society, Special Publication No. 4, New York, 1922.
6. American Society of Photogrammetry, *Manual of Photogrammetry*, Washington, 1952.
7. Eardley, A. J. *Aerial Photographs: Their Use and Interpretation*, New York, 1942.
8. Smith, H. T. U. *Aerial Photographs and their Applications*, New York, 1943.
9. Spurr, S. H. *Aerial Photographs in Forestry*, New York, 1948.
10. Beltman, B. J. "Comments on the 'Interpretation of Tri-Dimensional Forms from Stereo Pictures'," *Photogrammetric Engineering*, 18 (1952): 823-825.
11. Aschenbrenner, C. M. "A Review of the Facts and Terms Concerning the Stereoscopic Effect," *Photogrammetric Engineering*, 18 (1952): 818-823.
12. Goodale, E. R. "An Equation for Approximating the Vertical Exaggeration Ratio of a Stereoscopic View," *Photogrammetric Engineering*, 19 (1953): 607-616.
13. Thurrell, R. F. Jr. "Vertical Exaggeration in Stereoscopic Models," *Photogrammetric Engineering*, 19 (1953): 579-588.
14. Miller, V. C. "Some Factors Causing Vertical Exaggeration and Slope Distortion on Aerial Photographs," *Photogrammetric Engineering* 19 (1953): 592-607.
15. Stone, K. H. "Geographical Air-Photo Interpretation," *Photogrammetric Engineering*, 17 (1951): 754-759.
16. Russell, J. A., Foster, F. W., and McMurry, K. C. "Some Applications of Aerial Photographs to Geographic Inventory," *Papers of the Michigan Academy of Science, Arts and Letters*, 29 (1943): 315-341.
17. Colwell, R. N., and others. "Report of Commission VII (Photographic Interpretation) to the International Society of Photogrammetry," *Photogrammetric Engineering*, 18 (1952): 375-451.
18. Beltman, B. J. "Shadows on Aerial Photographs" *Photogrammetric Engineering*, 18 (1952): 831-832.

19. Spurr, S. H. "A Further Note Concerning Shadows on Aerial Photographs," *Photogrammetric Engineering*, 18 (1952): 833-834.
20. Kohn, C. F. "The Use of Aerial Photographs in the Geographical Analysis of Rural Settlements," *Photogrammetric Engineering*, 17 (1951): 759-771.
21. Tator, B. A. "Some Applications of Aerial Photographs to Geographic Studies in the Coastal Gulf Region," *Photogrammetric Engineering*, 17 (1951): 716-725.
22. Frost, R. E. "The Airphoto-Interpretation Program of Research and Instruction at Purdue University," *Photogrammetric Engineering*, 18 (1952): 701-719.
23. Crawford, O. G. S. *Air Photography for Archaeologists*, Ordnance Survey (Great Britain), Professional Paper No. 12, 1929.
24. Curwen, E. C. *Air-Photography and Economic History: The Evolution of a Cornfield*, Economic History Society, Bibliographies and Pamphlets No. 2, London, 1938.
25. St. Joseph, J. K. "Air Photography and Archaeology," *Geographical Journal*, 105 (1945): 47-61.
26. Schaedel, R. P. "The Lost Cities of Peru," *Scientific American*, 185 (1951): 18-23.
27. Horlaville, M. "Communication sur un quadrillage romain revelé par les photographies aériennes," *Rapport de la commission pour l'utilisation des photographies aériennes dans les études géographiques*, 17th International Geographical Congress, Washington, 1952, Paris: 38-40.
28. Branch, M. C. Jr. *Aerial Photography in Urban Planning and Research*, Harvard City Planning Studies No. 14, Cambridge (Mass.), 1948.
29. Pownall, L. L. "Air Photographic Interpretation of Urban Land Use in Madison, Wisconsin," *Photogrammetric Engineering*, 16 (1950): 414-426.
30. Stokes, G. A. "The Aerial Photograph: A Key to the Cultural Landscape," *Journal of Geography*, 49 (1950): 32-40.
31. Walker, F. *Geography from the Air*, London, 1953.
32. Howe, G. M. "A Note on the Application of Air Photography to the Agricultural Geography of North-West Cardiganshire," *Geography*, 36 (1951): 15-20.
33. Jenkins, D. S., Belcher, D. J., Gregg, L. E., and Woods, K. B. *The Origin, Distribution, and Airphoto Identification of United States Soils, with Special Reference to Airport and Highway Engineering*, United States Department of Commerce, Civil Aeronautics Administration, Technical Development Report No. 2 (2 vols.), Washington, 1946.
34. Frost, R. E., and Mollard, J. D. "New Glacial Features Identified by Airphotos in Soil Mapping Program," *National Research Council, Proceedings of the 26th Annual Meeting of the Highway Board*, Washington, 1946: 562-578.
35. Frost, R. E., and Woods, K. B. *Airphoto Patterns of Soils of the Western United States*, United States Department of Commerce, Technical Development Report No. 85, Washington, 1950.
36. Belcher, D. J., Hodge, R. J., Ladenheim, H. C., and staff. *Beach Accessibility and Trafficability*, Technical Report No. 1, Cornell University for Amphibious Branch, Office of Naval Research, Washington, 1949.
37. Troll, C. "Luftbildplan und Ökologische Bodenforschung," *Zeitschrift der Gesellschaft für Erdkunde zu Berlin*, 1939: 241-298.
38. Stamp, L. D. "The Aerial Survey of the Irrawaddy Delta Forest," *Journal of Ecology*, 13 (1925): 262-276.
39. Schulte, O. W. "The Use of Panchromatic, Infrared, and Color Aerial Photography in the Study of Plant Distribution," *Photogrammetric Engineering*, 17 (1951): 688-714.
40. Moessner, K. E. "Photo Interpretation in Forest Inventories," *Photogrammetric Engineering*, 19 (1953): 496-507.
41. Colwell, R. N. "Aerial Photographic Interpretation of Vegetation as an Aid to the Estimation of Terrain Conditions," *Selected Papers on Photogeology and Photo Interpretation*, Committee on Geophysics and Geography, Research and Development Board, Washington, 1953: 109-133.

42. McCurdy, P. C. *Manual of Coastal Delineation from Aerial Photographs,* United States Navy Hydrographic Office, Publication No. 592, Washington, 1947.

43. Powers, W. E. "A Key for the Photo Identification of Glacial Landforms," *Photogrammetric Engineering,* 17 (1951): 776-779.

44. Horlaville, M. *Intérêt des stéréogrammes aériens dans l'enseignement de la géographie physique,* International Geographical Union, Commission pour l'utilisation des photographies aériennes dans l'études géographiques, Paris, 1951.

45. Smith, H. T. U. "Photo Interpretation of Terrain," *Selected Papers on Photogeology and Photo Interpretation,* Committee on Geophysics and Geography, Research and Development Board, Washington, 1953: 7-53.

46. Robbins, C. R. "Northern Rhodesia: An Experiment in the Classification of Lands with the Use of Aerial Photographs," *Journal of Ecology,* 22 (1934): 88-105.

47. Sissam, J. W. B. *The Use of Aerial Survey in Forestry and Agriculture,* Imperial Agricultural Bureaux, Joint Publication No. 9, 1947.

48. Bushnell, T. M. "Use of Aerial Photography for Indiana Land Use Studies," *Photogrammetric Engineering,* 17 (1951): 725-738.

49. Troll, C. "Die wissenschaftliche Luftbildforschung als Wegbereiter in kolonialer Erschliessung," *Beitrage zur Kolonialforschung,* 1 (1942): 9-26.

50. Roscoe, J. H. *Regional Photo Interpretation Series: Antarctica,* (2 vols.) United States Air Force, Washington, 1952.

51. United States Naval Photographic Intelligence Center, *A Guide to Pacific Landforms and Vegetation for Use in Photographic Interpretation,* United States Navy, Report No. 7, Washington, 1945.

52. Colwell, R. N. "Procedures for the Construction of Photo Interpretation Keys," *Selected Papers on Photogeology and Photo Interpretation,* Committee on Geophysics and Geography, Research and Development Board, Washington, 1953: 135-154.

53. "Carte des surfaces couvertes par des photographies aériennes verticales et obliques," *Rapport de la commission pour l'utilisation des photographies aériennes dans l'études géographiques,* 17th International Geographical Congress, Washington, 1952, Paris.

54. Stone, K. H. "World Air Photo Coverage for Geographic Research," *Annals of the Association of American Geographers,* 43 (1953): 193.

55. Sager, R. C. "Index to Aerial and Ground Photographic Illustrations of Geological and Topographical Features throughout the World," *Photogrammetric Engineering,* 19 (1953): 472-473.

56. American Geological Institute, *Outstanding Aerial Photographs in North America,* Report No. 5, Washington, 1951.

57. Lang, A. H., Bostock, H. S., and Fortier, Y. O. *Interim Catalogue of the Geological Survey Collections of Outstanding Air Photographs,* Geological Survey of Canada, Paper No. 47-26, Ottawa, 1947.

58. Foster, F. W. "Some Aspects of the Field Use of Aerial Photographs by Geographers," *Photogrammetric Engineering,* 17 (1951): 771-776.

59. MacFadden, C. H. "Some Preliminary Notes on the Use of the Light Airplane and 35 mm. Camera in Geographic Field Research," *Annals of the Association of American Geographers,* 39 (1949): 188-200.

60. Katz, A. H. "Contribution to the Theory and Mechanics of Photo Interpretation from Vertical and Oblique Photographs," *Photogrammetric Engineering,* 16 (1950): 339-386.

61. Kesseli, J. E. "Use of Air Photographs by Geographers," *Photogrammetric Engineering,* 18 (1952): 737-741.

62. Kral, J. "Textbook of Aerial Geography and the Geographical Utilization of Aerial Photographs," *Rapport de la commission pour l'utilisation des photographies aériennes dans l'études géographiques,* 17th International Geographical Congress, Washington, 1952, Paris: 29-31.

63. COBB, G. C. "Bibliography on the Interpretation of Aerial Photographs and Recent Bibliographies on Aerial Photography and Related Subjects," *Bulletin of the Geological Society of America*, 54 (1943): 1195-1210.

64. PARVIS, M. *Selected Bibliography on the Interpretation of Aerial Photographs and on Aerial Photography and Related Subjects*, Purdue University, Lafayette, (Indiana), 1949, and supplements, 1950, 1951, 1952.

65. ROSCOE, J. H., WELLS, S. D., and PACE, V. W. *Abridged Bibliography of Photographic Interpretation: Selected with Emphasis upon Keys, Techniques and Research*, United States Naval Photographic Interpretation Center, Report 102A/50, Washington, 1950.

66. HART, T. *A Bibliography on the Interpretation of Vegetation from Aerial Photography*, United States Naval Photographic Interpretation Center Report 113/50, Washington, 1950.

67. GARRARD, C. W. *An Annotated Bibliography of Aerial Photographic Applications to Forestry*, State University of New York, College of Forestry, Syracuse, 1951.

68. MEYNEN, E., ed. *Geographisches Taschenbuch 1953*, Stuttgart, 1953: 222-231.

69. JOHNSON, G. R., and PLATT, R. R. *Peru from the Air*, American Geographical Society, Special Publication No. 12, New York, 1930.

70. FISCHER, E. *Lesen des Luftbildes*, Berlin, 1938.

71. FORBES, A. *Northernmost Labrador Mapped from the Air*, American Geographical Society, Special Publication No. 22, New York, 1938.

72. LIGHT, R. U. *Focus on Africa*, American Geographical Society, Special Publication No. 25, New York, 1941.

73. RICH, J. L. *The Face of South America: An Aerial Traverse*, American Geographical Society, Special Publication No. 26, New York, 1942.

74. TROLL, C. "Methoden der Luftbildforschung," *Sitzungsberichte Europäischer Geographen zu Wurzburg 1942*, Leipzig, 1943: 121-146.

75. MARTONNE, E. DE *Géographie aérienne*, Paris, 1948: 103-167.

76. CHOMBART DE LAUWE, P. H. *La découverte aérienne, du monde*, Paris, 1948.

77. ———. *Photographies aérienne*, Paris, 1951.

78. GUTKIND, E. A. *Our World from the Air*, New York, 1952.

79. MCNEIL, G. T. "Machinery for the Photo Interpreter," *Photogrammetric Engineering*, 19 (1953): 121-124.

80. LINTON, D. L. "Air Photographs as Tools for Geographical Research," *Rapport de la commission pour l'utilisation des photographies aériennes dans les études géographiques*, 17th International Geographical Congress, Washington, 1952, Paris: 17-28.

SELECTED AIR PHOTOGRAPHS

Fig. 26. Santa Maria Valley, California: low oblique photograph taken by the pilot of a light airplane with a 35mm camera held in the hand. The photograph was then annotated by the photographer, Clifford H. Mac-Fadden, to show land use.

Fig. 27. Waukesha Co., Wisconsin. Part of WW-2B-13, U.S.D.A. 10/12/41, approximately 1:20,000. North is to the top of the page. These two pictures illustrate the use of the vertical photograph as a base for field mapping, employing a code for recording data. The area is part of the Interlobate or Kettle Moraine formed by the Green Bay and Lake Michigan Lobes of the Wisconsin glaciation. The center of the photographs lies about two miles east of the city of Delafield, and a portion of the western end of Pewaukee Lake appears in the upper right. The rural economy emphasizes dairying.

The upper photograph was used to map the physical characteristics of the land. The first digit shows the degree of slope; second digit, character of drainage; third digit, type and degree of soil erosion; fourth digit, soil stoniness; and the fifth digit, soil texture.

The lower photograph of the same area was used to map rural land use. Number 1 refers to cropped land, and the lower case letter codes the specific crop. If the crop was raised for seed, fodder, or canning this fact is indicated by a number in parenthesis. Number 2 is pastured land; number 3, forested and marsh land; number 4 idle land. Each of these categories is appropriately subdivided to denote specific utilization. The numbered farmsteads are those at which interviews with the farm operator were obtained. The boundary lines separate farm operating units. (John C. Herbst and Helge E. Pearson: Northwestern University-Syracuse University Summer Field Camp in Geography, 1948).

Fig. 28. Herkimer County, New York: part of CXG-8B-110, U.S.D.A. 9/22/42, approximately 1:20,000. North is to the top of the page. This photo reveals nine transport routes associated with the Mohawk Valley corridor across the northern Appalachians:

1. Beginning at the north, New York State Highway 5 crosses Sterling Creek on a concrete bridge. Immediately east, the highway is crossed

by an overhead conveyor of the stone crushing plant whose gravel pits can be seen on each side of Sterling Creek.

2. Sterling Creek is dammed where it enters the New York State Barge Canal, and a loading dock for barges to be filled with earth materials is visible near the dam. Sterling Creek once flowed across the position occupied by the canal, as shown by a second dam in the valley on the south side of the canal. East of the mouth of Sterling Creek there is a lock in the Barge Canal. The lock is known to be 328 feet long, thus providing a check on the nominal scale of 1:20,000. The lock gates form a v-shape upstream, indicating a water movement from west to east.

3. Immediately downstream from the lock the New York Central Railroad tracks cross the canal on a steel bridge whose form is revealed by its shadow. The railroad has four tracks which are clearly distinguishable where it spans the cut off portion of Sterling Creek on a girder bridge. A steel signal tower arches over the tracks east of this point.

4. The tree-lined Mohawk River meanders on its flood plain and contrasts with the straighter Barge Canal with its white spoil banks. This portion of the river is no longer used for transportation.

5. The double-tracked West Shore Railroad (N. Y. Central System) can be seen as a gray line parallel and close to a highway.

6. The more highly light-reflective paved surface of New York State Highway 5S appears as a white line which is joined by farm driveways. These farm lanes have twisting courses on the valley margins but near their junction with the highway each has a straight segment. Stereoscopic examination shows that the straight portions of the sideroads cross a narrow ditch.

7. The ditch parallel and near to Highway 5S is the long-abandoned Erie Canal. Some parts of the ditch are wet, as revealed by dark tones. The old canal is close to the edge of the floodplain.

8. Above the valley bluffs and aligned with other routes is a white linear scar which bridges ravines leading from the wooded ridge of Dutch Hill on the south. The outlines of the white trace are not sharp due to the encroachment of vegetation upon it. Since it has no connection with farm lanes it appears to be an abandoned rail line, probably a former interurban trolley track.

9. A graded farm-to-market road lies on the southern slope of Dutch Hill. The road appears to stop near the east edge of the photograph, but to the west it bifurcates and shows evidence of greater usage.

Fig. 29. Fresno Co., California: part of ABI-11B-106, U.S.D.A. 5/30/42, approximately 1:20,000. North is to the top of the page. This photograph is a sample of a southeastern San Joaquin Valley market town and its complex rural neighborhood. The city of Reedley is located some twenty miles southeast of Fresno. The Kings River to the west is now entrenched twenty-five to fifty feet below the surface of its great alluvial fan but remnants of former channels are suggested by the longitudinal changes in soil tone. The city owes its origin to the development of a spur line railroad along the eastern side of the valley in the late 19th century. As a railroad-sponsored settlement, the streets are oriented to the tracks rather than to the cardinal points.

The rural area reflects the great agricultural diversity popularly associated with California. By analysing the tone and texture of the photograph it is possible to identify the following items numbered on the picture: 1. grapes; 2. alfalfa; and 3. fallow or idle land. A more intimate regional knowledge is required in order to make positive identification of the citrus, deciduous fruits, improved pasture, and commercial row crops also found in this area. The field boundaries pose no problems.

Several miscellaneous interpretations are worthy of note. In Reedley the large buildings adjacent to the railroad tracks are packing sheds. This association of features might lead the interpreter astray as motor transportation is of great significance in this farming region. The dark lines illustrated by number 4 are irrigation canals, distinguished from the roads mainly by tone. The oval track, number 5, is a part of the Reedley High School and Junior College. The curious pattern at number 6 is the city cemetery. (Interpretation by William G. Byron)

Fig. 30. Wilson Co., North Carolina: part of ACC-7E-161, U.S.D.A. 1/15/48, approximately 1:20,000. Northeast is to the bottom of the page.

This air photograph clearly shows the intensive agriculture typical of the Bright Tobacco Belt of the Atlantic Coastal Plain. Rectilinear contiguous fields cover most of the youthful surface with the notable exceptions of the flattest interstream divide areas and of the low floodplains and tributaries of the sluggish streams. Because the cash crop agriculture of tobacco and cotton, with corn grown for feed, is associated with high land values ($300 to $400 per acre) approximately 90% of the arable land is tilled. Most of the forested soils are poorly drained, but some occupy excessively well drained slopes too steep for agriculture.

The photo images that identify and emphasize the intense land use of this agricultural region are: numerous farmsteads (28 in this picture) which are spaced six or seven per road mile; large numbers of farm buildings per farmstead, usually about six of which three are flue-curing tobacco barns; small isolated cleared patches in the woods which are the seed beds for the tobacco plants; the multiple divisions of the fields, allowing an intimate matching of the crop to its most favorable soil; natural drainage augmented by an extensive network of man-made drainage ditches bordering almost every field on at least one side; field cultivation on the contour; and, finally, some forest management, an example being the densely planted forest between the straightened channel of Toisnot Swamp and North Carolina Highway 58. In addition to this asphalt road, a small stretch of concrete paved U. S. Highway 264 may be seen near the pecan orchard in the northeast corner of the photograph. Numerous sandy roads serve the rest of the area.

Forests of mixed hard and softwoods indicate the warm humid climate, so favorable to agriculture. The soils exhibit a catenary relationship to one another. From flat poorly drained uplands to well-drained uplands to excessively drained slopes to valley bottom lands, the soil series are respectively: Plummer, Norfolk, Ruston, and swamp. These soils, plus other less prominent series of the Norfolk family, are all of agricultural importance, especially when drained.

The intimate nature and completeness of modification of the natural environment by man emphasizes the intensiveness of agriculture in the area of this photograph. (Interpretation by James M. Jennings).

Fig. 31. Clarion Co., Pennsylvania: part of APH-5G-13, U.S.D.A. 10/21/50, approximately 1:20,000. North is to the left of the page. This photograph shows an area of changing land use in the hilly country of the Appalachian Plateau about eighty miles north of Pittsburgh. Land which was formerly cleared is being progressively abandoned and much of it has already returned to trees. The north center of the photograph shows several examples of this phenomenon. The area of woods marked by the junction of the two linear clearings made for high voltage power lines, is one of mature second growth containing various sizes of trees, which gives it a characteristic mottled appearance. The area directly to the west is even in texture and indicates immature second growth of younger trees. Note the difference in the shadow cast on the strip cleared for the power line in each of these areas, giving an indication of the height of the trees. Still further to the west is a third field which has been abandoned but is only in brush and weeds. Note again the shadow cast onto this lower vegetation by the immature second growth to the right. Several fields in an even earlier stage are to be found to the east of the above mentioned fields.

Some land in the area is being taken out of cultivation by coal mines. Strip mines are evident in the southwest corner and near the northeast corner of the photograph. They show as light toned areas following the contour of the hill with dark shadows on the shady side of the spoil dump.

The sinuous pattern of the mines should not be confused with contour cropping of grains and hay by dairy farmers.

This photograph illustrates the contrasting appearance of highway, railroad, and stream. Pennsylvania Highway 68 is the light toned strip in the east. A branch line of the New York Central Railroad follows the valley and crosses the meanders of Piney Creek. Note that it is a dark line narrowing where it bridges the stream. Piney Creek's aspect is much influenced by the presence or absence of shadows cast by trees along its banks. (Interpretation by Robert L. Layton)

Fig. 32. Schuylkill Co., Pennsylvania: a stereogram from parts of AQS-5D-52 and 53, U.S.D.A. 9/18/47, approximately 1:20,000. North is to the left of the page.

These photographs depict the small anthracite coal mining center of Branchdale crowded onto the narrow floor of a minor valley in the Appalachian Ridge and Valley Province. Except for a few "company houses" perched on the hill east of the valley, all residences are close to the main road. Immediately associated with the settlement are two large piles of mined coal, the southern one having a flat sloping surface and the northern being surmounted by a symmetrical cone of coal. At the extreme northern edge of the settlement a tipple can be seen standing below the level of the stacked coal but above the railroad cars into which the coal is fed by gravity. West of the settlement and of the coal piles the countryside is thinly wooded and without evidence of farming. Across this land run the thin white scars of unimproved roads which intersect with major white gashes of strip mines and spoil dumps. The mining cuts follow the strike of the coal bearing strata and close examination reveals that the floors of the cuts dip southward following the beds. Water stands in some of the deeper pits. This stereogram is appropriately studied with a pocket-type lens stereoscope.

Fig. 33. St. Clair Co., Illinois (and St. Louis City, Missouri): part of SK-6A-87, U.S.D.A. 7/4/40, approximately 1:20,000. North is to the top of the page. This air photograph illustrates an interesting contrast in urban land use on the two sides of the Mississippi River in its largest metropolitan area. The crowded blocks of old St. Louis press closely upon the levee, but an interval of half a mile or more separates built-up East St. Louis from the river. As Mississippi River traffic diminished in significance so did the waterfront area of St. Louis. Much of the riverine section shown in the photograph became a "blighted area" and by 1940 a considerable amount of building demolition had taken place to provide space for the Jefferson National Expansion Memorial. This land clearance may be identified by the light tone and faint outlines of building ground plans in the rectangle between the approaches to the two bridges. Two historic buildings to be retained in the plan for revitalizing and landscaping the waterfront may be recognized by their characteristic shapes: the Old Court House at the west end of the major short axis of cleared land; and the Old Cathedral near the center of the north-south axis.

The land use on the Illinois side of the Mississippi River at this point grew from the early ferry and later bridge traffic to and from St. Louis. Bridging encouraged the concentration of the railroading function here and the photograph clearly reveals the complexity of part of the East St. Louis freight yards. The bridges with which these lines are associated are the historic Eads Bridge on the north and the Douglas MacArthur Memorial Bridge on the south. Although the construction aspects of the bridges are readily perceived to be quite unlike, a study of their eastern approaches shows that they are both double-decked structures with a highway above and a railroad below. A typical railroad classification yard parallels the river between the bridges and has a spur to the dock near the southern crossing. Numerous dead-end lines leading to storage sheds and warehouses lie at right angles to the waterfront. Four railroad shops may be identified by the characteristic shapes of roundhouses, and two former engine houses may be found by careful study of abandoned trackage.

The surface configuration between built-up East St. Louis and the Mississippi River has been considerably altered to provide high and level land for tracks and structures. The numerous unfilled areas between lines are fifteen or twenty feet below the man-made surface and have dark ragged tones suggesting that they are damp. The drainage line of Cahokia Creek has been much altered by the fills but it can be traced from the northeast where it appears black, through the center of the photograph where it is gray, to its junction with the Mississippi River where it appears white. The last portion of its course roughly parallels the elevated railroad approach to MacArthur Bridge. Virtually all of the land which is not devoted to transportation appears to be idle land.

IRRIGATED VEGETABLE CROPS

Vegetable fields in foreground, Santa Maris River (dry) in background.
Direction. N 20 E
Altitude: 800 feet: Low Oblique
Weather: Clear sky. Heavy haze.
Location: Santa Maria Valley, California
Date: October 1947.

Fig. 26

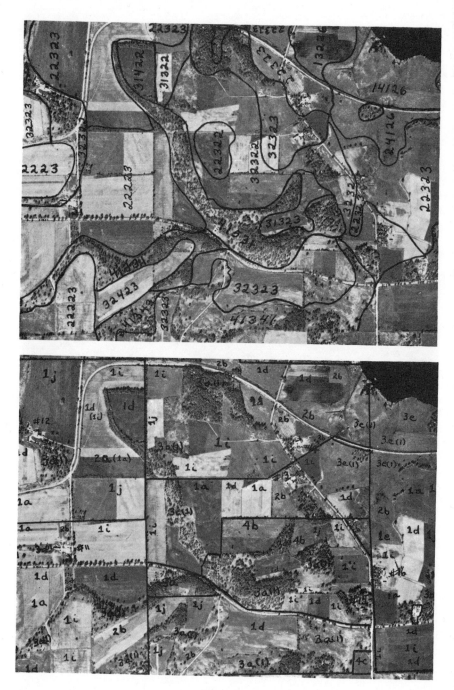

Fig. 27

Fig. 28

Fig. 29

Fig. 30

Fig. 31

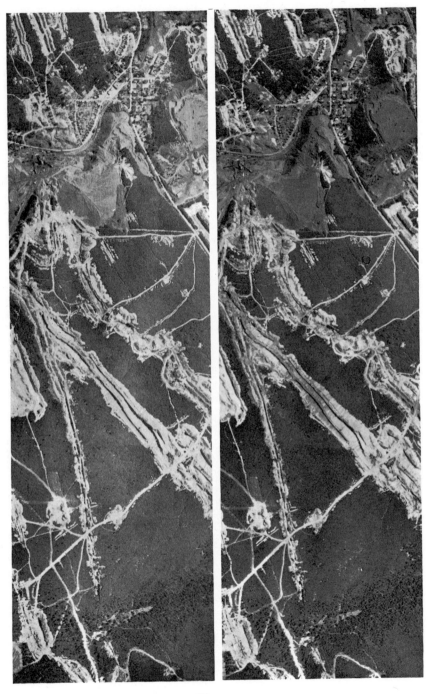

Fig. 32

Fig. 33

CHAPTER OUTLINE

GEOGRAPHIC CARTOGRAPHY[A]

THE DEVELOPMENT OF GEOGRAPHIC CARTOGRAPHY
THE LANDFORM MAP
THE MAPPING OF STATISTICAL DATA
MAP PROJECTIONS
CARTOGRAPHY IN TWO WORLD WARS

CARTOGRAPHY AT THE BEGINNING OF WORLD WAR II

CARTOGRAPHY AND MASS-PRODUCTION TECHNIQUES

THE GROWTH OF GEODETIC CONTROL

THE DEVELOPMENT OF MAPPING FROM THE AIR
TRIMETROGON PHOTOGRAPHY
THE PROSPECT FOR AIR PHOTOGRAPHY

THE DEVELOPMENT OF NEW KINDS OF MAPS
SPECIAL-PURPOSE MAPS
TERRAIN MODELS
THE STUDY OF GRAPHIC METHODS

MAP LIBRARIES AND MAP ANALYSIS

CURRENT STATUS OF CARTOGRAPHY
NEW TECHNIQUES OF MAP-MAKING AND MAP REPRODUCTION
GOVERNMENT MAPPING

(Continued on next page)

A. Original draft by Arthur H. Robinson with the cooperation of W. E. Davies, O. M. Miller, and Erwin Raisz.

553

THE PROSPECT IN GEOGRAPHIC CARTOGRAPHY
PROJECTIONS
REPRESENTATION OF TERRAIN
STATISTICAL PRESENTATION
CARTOGRAPHIC DESIGN
MAPPING ANALYSIS

CONCLUSION

CHAPTER TWENTY-SIX

GEOGRAPHIC CARTOGRAPHY

DURING THE PERIOD of recorded history, from Babylonian times to the present, cartography and geography have been intimately associated. The relationship is appropriate and necessary, for neither can the geographer deal with differences from place to place on the earth without maps, nor can the cartographer map the face of the earth without an understanding of what is significant. As a matter of fact, many of the earlier geographers might more appropriately be called cartographers; but in the long period since Ptolemy geographic investigation into the meaning of area differences on the earth has been more widely pursued than has the branch of the subject dealing with technical cartography. Today in America, cartography is still widely considered merely a means to an end, a tool; nevertheless, geographically-minded cartographers and cartographically-minded geographers are attempting to focus attention upon geographical cartography as a professional specialty.

Cartography, as an old and established professional field, is the science of mapping the earth. In modern parlance it embraces geodesy, geodetic and topographic surveying (including photogrammetry), and all the crafts and techniques connected with the compilation, drafting, and reproduction of a printed map. Thus large parts of cartography too often seem to lie outside the interests and competence of the geographer [1]. Nevertheless, if a geographer is to forge the results of original surveys and other observations into efficient tools for his analytical purposes, he must understand and even control the procedures whereby the basic map material can be used directly or summarized or generalized for these special purposes. This requires more than an ability to interpret or to make critical appraisals of published maps. It involves a working knowledge of cartographic presentation, which includes map projections and mapping of areal relationships, and also an appreciation of how the designs and scales of maps can influence the portrayal of geographical patterns and associations. The term "geographic cartography" refers to these aspects of cartography.

More than ever before in America, geographers have recently become conscious of the need for improving their cartographic methods, both in the techniques of field and office research and in the classroom presentation of geography from the elementary grades to the graduate school [2; 3]. This has come in large measure from the experience of the geographers who served in the several agencies of the government during World War II and found their training and background in cartography deficient [4].

That interest in cartography increased sharply during and after World War II is witnessed by the phenomenal growth of the American Congress on Surveying and Mapping, a professional organization representative of modern cartography as a whole. Founded in 1941, it is now affiliated with the National Research Council as a constituent society of the Division of Geology and Geography. Its Technical Division on Cartography is primarily concerned with geographic cartography and has several geographers among its officers. The Association of American Geographers has established a Committee on Cartography, and several entire numbers of *The Professional Geographer* have been devoted to cartography. The rise of cartographic interest among geographers has resulted in an increased attention to this subject in departments of geography [5]. Except in rare instances, however, research programs worthy of the name have not as yet been undertaken.

THE DEVELOPMENT OF GEOGRAPHIC CARTOGRAPHY

The development of geographic cartography in America has come about through the work of a relatively small number of geographers during the past fifty years. During the first half of the present century, geographical interest was applied at two quite different ranges of scales. The macrogeographers worked at small scales, whereas the microgeographers worked chiefly at scales greater than one inch to one mile. Those working at large scales were interested in cartography primarily as a tool and did not focus their study on cartographic methods as such. At the smaller scales, geographic cartographers were particularly concerned with such matters as the landform map, methods of plotting and presenting statistical data, and the development of new projections suitable for geographical purposes.

The Landform Map

The landform map resulted from the work of William Morris Davis and others, who in this century developed the making of block diagrams and perspective views of terrain into a high art. Although maps of this kind were prepared earlier, those of Armin K. Lobeck were the first major contributions [6]. A fresh approach was made by Erwin Raisz in 1931 with suggestions for symbolizing landforms [7]. Since then, others, notably Guy-Harold Smith, have also produced physiographic diagrams or landform maps [8]. During the thirty years that have elapsed since Lobeck's first map of the United States the entire earth has been depicted physiographically.

The variously titled landform map or physiographic diagram is possibly the only type of map that can be claimed as a wholly original contribution of American geographic cartography. Since it depends, in large part, on a thorough background of landform training, relatively few cartographers have taken part in its development.

THE MAPPING OF STATISTICAL DATA

The origins of statistical-distribution cartography are by no means ancient, and it has slowly developed over a period not much longer than two centuries, during which time cartography has utilized the isogram, the point symbol, and the graded shading in more and more useful and complex ways. In the United States geographic concern with cartographic representation of numerical distributions, beyond the well-known contour, isotherm, and others, also began early in this century with the work of O. E. Baker, Vernor C. Finch, and Wellington D. Jones. Examples of this kind of cartography are too numerous to cite, but special mention should be made of Wellington D. Jones and John K. Wright whose work with isopleths and ratios led to a widening use of statistical cartography in American geography [9; 10]. This phase of cartography has steadily progressed, and the application of the techniques today are wide indeed.

MAP PROJECTIONS

The map projection has always had a fascination for the cartographer. As might be expected in geographic cartography, because of its concern with smaller scales, interest in projections has centered primarily on those useful for world or hemispheric presentations. In 1919 J. Paul Goode published a paper concerning the interruption of the Mollweide projection which attracted considerable attention [11]. Later he combined the Sinusoidal and Mollweide projections in an interrupted manner, and the resulting graticule, which he called the homolosine, has become a favorite of geographic cartographers in America [12]. Goode's example has been followed in various other interrupted projections, such as the modification of the Aitoff projection by Finch, and the Eumorphic projection by S. W. Boggs [13]. Today the interrupted world projection is often seen in atlases and texts in America, but rarely elsewhere. It is a distinctive American contribution to cartography.

There is considerable skepticism among cartographers concerning the desirability of interruption. Although interruption reduces the inherent deformation to a considerable extent, the consequent multiple violation of the one-surface earth, and the presentation of it in pieces, may out-weigh gains achieved through this reduction [14].

American geographer-cartographers, however, did not stop with ex-periments in the interruption of projections. Brought on no doubt by the growth of interest in seeing the world as a whole, the analyzing of pro-jections and the devising of new ones for particular purposes have in-creased since World War II [15; 16; 17; 18; 19; 20; 21; 22; 23; 24]. In this connection mention should be made of the work of O. M. Miller of the American Geographical Society, whose cylindrical projection is wide-ly used at the present time.

CARTOGRAPHY IN TWO WORLD WARS

Nothing during the past fifty years has exerted so great an influence on geographic cartography as has the occurrence of two world wars (chap-ter twenty-three). A considerable amount of the work of geographers in both conflicts has been cartographic, both in federal employment while the wars were going on, and afterward in the periods of peace-making and reconstruction. Preeminent in this sort of activity during and after World War I was the American Geographical Society. During World War II no single agency or individual stood out so clearly, but the repre-sentation was large indeed. At least half of the geographic personnel pro-fessionally employed by the government during World War II were engaged in cartographic work. This exerted a tremendous influence on geographic cartography.

On the other hand, it is not correct to imagine, as some would have us do, that geographic cartography was in the doldrums until the so-called air age forced an awakening. Cartographic activity moved steadily after World War I, and in the 1930's the appearance of several volumes, in-cluding books by Erwin Raisz [25] and C. H. Deetz [26], provides evi-dence of the interest in the subject. The American Geographical Society's Map of Hispanic America at a scale of 1:1,000,000 and the Babson Insti-tute's Giant Relief Model of the United States were steadily progressing under the direction of geographer-cartographers. But, although those were evidences of a mounting interest in cartography, World War II provided an unprecedented impetus.

CARTOGRAPHY AT THE BEGINNING OF WORLD WAR II

World War II caused more cartographic activity in half a decade than had been seen in any previous century. Probably more maps were made and printed during the five years from 1941 to 1946 than had been pro-duced in the aggregate up to that time. Extensive military and naval operations over much of the earth created greater demands for new types of maps and for cartographic information than had ever arisen before. All this brought home to the cartographer the need for adequate world-wide geodetic control. The use of aerial photography and photogrammetry became a standard mapping procedure, a change that has already markedly increased world topographic coverage. Many other accomplishments took place during World War II but it is not necessary to catalog them here [27; 28]. Suffice it to say that no other event has so profoundly influenced American cartography as a whole.

The United States faced a serious manpower problem with respect to cartography in 1941. In the period after World War I military interest in mapping lagged; the General Staff of the Army maintained only a casual interest in cartography and utilized only a small staff of draftsmen for routine graphic work; and the Engineers engaged only in a continuing domestic program and gave little thought to foreign areas. Few plans for mapping in case of hostilities were drawn up, and even when the spread of World War II to global proportions was imminent the tremendous scope of military cartography to come was not envisioned.

During the partial mobilization of 1940, steps were taken to expand and coordinate existing cartographic facilities to meet the situation that was developing. The greatest problem was the production of maps for the ground forces. In late 1941 the War Department map collection was transferred from the General Staff to the Corps of Engineers and merged with the Engineer Reproduction Plant. This consolidation (after 1942 known as the Army Map Service) assumed the overall responsibility for map production for the ground forces, including the procurement of maps and map intelligence, and the compilation of maps as well as their drafting and printing. The Hydrographic Office of the Navy was more adequately staffed with experienced personnel, and the expansion of its operations was not so much of a problem [29]. The Aeronautical Chart Service did not become a producing organization in its own right until later.

When the United States entered World War II, many basic problems quickly became apparent, the most important being the development of cartographic methods that would permit the utilization, with a minimum amount of training, of large numbers of inexperienced persons. An almost complete lack of cartographic training in the United States had produced no reserve of competent cartographers. Of course, the civilian cartographic agencies of the government and of the states were staffed with specialized mappers and production personnel, but cartographers working in soils, geology, engineering, land use, forestry, and so on, did not have the general geographic competence to engage immediately in world-wide compilation work.

The larger non-military mapping agencies had grown steadily up to the time of the war. The Coast and Geodetic Survey and the Topographic Branch of the Geological Survey were early called upon to contribute to the war effort by way of technical assistance and other special services [30]. As a result their own work lagged behind and required renewed effort after the war.

CARTOGRAPHY AND MASS-PRODUCTION TECHNIQUES

To achieve success in meeting the wartime need for maps in quantity it became necessary to subdivide the work and to train each person to master only that operation to which he was assigned. The traditional and individualistic concept of cartography—wherein design of the map, evaluation of geographic data, and production of a rough draft were done by a single person—disappeared, and in its place came the specialist in specific phases of the cartographic process.

In the early stages of organization the functions were usually separated into two major divisions only. The gathering, filing, and evaluation of maps and of map information were assigned to one division, and the design and drafting of maps to another. At the time of maximum development, however, the breakdown in most cartographic institutions, large and small, was into five principal divisions: planning, design, geographic research, compilation, and drafting.

The degree to which these main divisions were further subdivided depended on the size and mission of the organization. In extreme cases there were many subdivisions each with highly specialized functions. The subdivisions within the geographic research unit were typical. The general objective of a unit of this kind was the procurement, evaluation, and preparation of geographic information for use by the compiling draftsman. Procurement was a library problem as far as ordinary maps were concerned. However, in addition to maps, a vast amount of intelligence data were being produced by a multitude of government agencies, both civilian and military, and since these data were of value in cartography they had to be located, evaluated, and procured. Of the geographers employed in research for map making, the greatest number were utilized in analyzing geographic information as collected and in maintaining it in progressive, integrated form that permitted ready use. These geographers formed the key to the mass-production technique. The scope and speed of any project depended on their ability to supply up-to-date information in a form usable when needed. To this end the geographic research unit was subdivided into smaller units, each responsible for information pertaining to a specific part of the world. This permitted each geographer to specialize in an area and build up suitable background knowledge. As personnel became available the breakdown was further subdivided until, in many cases, an individual was specializing in a country or a part of a country. It was this specialization that made possible the utilization of a vast backlog of information that had accumulated over many generations and also permitted current information to be assimilated.

Mass-production techniques were not limited to the large cartographic operation. Even a relatively small organization engaged in geographic cartography, such as the Map Division of the Office of Strategic Services, found it desirable to separate the work of the geographer-compiler from that of the draftsman. Obviously such separation of cartographic functions was not new. It was standard practice for years prior to World War II in the American Geographical Society, the United States Geological Survey, and other organizations. However, it had not been carried to such a high degree before, nor had it been reflected in the training offered in colleges and universities.

The question of whether mass-production techniques are desirable or not has plagued all who have worked with it. In monetary terms there seems to be no question that it is the best and only method where large quantities of maps are needed in a short period of time. At maximum development the quality of the finished map produced by this type of work seems to compare favorably with that obtained under the more individualistic methods of cartography.

THE GROWTH OF GEODETIC CONTROL

Artillery had long made use of local geodetic control, but, since previous military operations were confined to relatively small areas, the problem of providing integrated control was comparatively simple. The expansion of military mapping required a corresponding expansion in the use of control data. The first obstacle to be overcome was the assemblage of data. American participation in World War I, localized as it was in Western Europe, made the provision of geodetic control of significance only in that area. In 1941 the initial stages in assembly were primarily of a library nature, in which all likely publications were examined for the purpose of extracting control data. After the United States entered the war, agreements on mapping responsibilities with the British included the sharing of geodetic control, with the result that a large amount of material was made available. Processing of the data followed and in some cases paralleled their collection. The heterogeneous standards on which previous control surveys for topographic mapping had been based made it necessary to reduce the data to uniform standards before issuing them in integrated form. The collated information was furnished to users in lists keyed to specific maps.

Since enemy-held areas were inaccessible to field parties that could identify the control on photographs it was necessary to locate the points from the available descriptions. This need emphasized the importance of accurate descriptions for control points.

The impetus given to international control adjustments bore fruit after the war, when central and western European nations entered into cooperation with the International Geodetic Association for recalculation and adjustment of control in their areas [31; 32]. Other countries, recognizing the fundamental importance of adequate integrated geodetic control, became similarly internationally minded. A large program of cooperative mapping was initiated in 1946 among the nations of the Americas. The need for geodetic control and for such international cooperation is well illustrated by the following extract from an account of the work of the Inter-American Geodetic Survey [33].

In Costa Rica, a pilot obligingly flew (the author) over a peak charted at 4,500 feet which we barely cleared at 9,500.

Our geodetic reconnaissance information alone has caused several aeronautical charts to be declared "Hazardous." Pico Trujillo in the Dominican Republic is being moved over about 35 miles and raised 2,000 feet in elevation. It appears that part of the huge Xingú River in Brazil may be 30 to 40 miles from its true position on our aeronautical charts and maps.

Plotting the completed geodetic nets on existing maps reveals some amazing errors. The coastline shown on the existing Costa Rican map had to be moved several miles so the geodetic stations would not be plotted in the Pacific Ocean. In Cuba, some stations have proven to be as far as 6 kilometers from their large-scale map positions, and in one case, on the wrong side of a town, a road, and a railroad.

A recent astronomical observation indicates that the Isle of Pines, just south of Cuba, may be nearly 18 miles out of position. This is a serious discrepancy, as the island is used as a check point by aircraft approaching the Habana airport.

In eastern Peru, a large river is out of position by 10 miles.

Geographers are concerned with differences from place to place and geographer-cartographers with mapping these differences. The geodetic work that was given such an impetus by the war will help to make possible the more accurate location of many of the data with which geographers and geographer-cartographers work. Administrative divisions on cadastral maps cannot be "tied down" in any way to the physical base if the latter is unknown. Under such circumstances reconnaissance studies only can be undertaken without extensive (and expensive) personal field work. One can hardly expect to derive densities or correlations with any degree of accuracy if one cannot even locate the pertinent enumeration areas or determine their size.

THE DEVELOPMENT OF MAPPING FROM THE AIR

At the beginning of World War II, aerial photography was by no means so fully employed for cartographic purposes as it might have been. In commercial and industrial work it had been developed extensively as an efficient instrument for mapping, but mapping agencies of the federal government had been relatively slow in accepting it as a method of producing topographic maps. Only the Department of Agriculture, through the Agricultural Adjustment Administration, had completed photography approaching in scope that which was to be demanded, and in this case only a small percentage had been translated into maps.

Just before World War II, the Army had accepted multiplex methods as standard for its use in producing topographic maps from aerial photos [34]. This system was satisfactory for topographic mapping, but lacked the flexibility needed for its use in reconnaissance mapping of vast areas.

The Pacific and European theaters of war presented two distinct problems in aerial photography and mapping, the solutions of which have had a tremendous impact on cartography and geography. The European theater had fair to good maps of much of the area, and medium- and small-scale maps could be compiled from existing maps, while large-scale mapping could be revised or extended by use of multiplex and large-scale vertical photographs. In the Pacific theater conditions were the reverse; large-scale maps were generally lacking. At the start of the Pacific war inadequate maps only were available. In the Philippines, for example, only an out-of-date, inaccurate series covered Luzon. In the East Indies there were some adequate Dutch maps of Java and the larger islands. Elsewhere there were no maps, good or bad. After the limits of the Japanese conquest had been reached the road back was to be paved with maps, practically the whole way. In the case of large-scale maps, aerial photographs and multiplex methods provided the answer, but for small-scale maps and charts that must precede the large-scale work the answer was not so easy.

Trimetrogon Photography

Just before the war the Army Air Force had begun experimentation with trimetrogon aerial photography as a means of producing medium- and small-scale aeronautical charts. The trimetrogon system requires the use of three cameras arranged to take a vertical and two related oblique photographs so oriented that the three photographs form a continuous photostrip, horizon to horizon [35]. While this was not the first use of combined vertical and oblique photographs to produce a map, it was the first devised specifically for the production of small-scale maps or charts.

The capacity to cover vast areas with aerial photographs was increased tremendously by trimetrogon photography. Practically all of the theaters of operation, and the supply routes leading to them, were covered by "trimet." Obscure sections of equatorial and northern Africa, the eastern Himalayas, Alaska, parts of Latin America, and South Pacific islands were first seen in cartographic detail as a result of such photographs.

The development of methods for the cartographic utilization of trimetrogon pictures has not kept pace with the speed with which vast areas may be photographed from the air. Planimetric features, through an intricate system of lattice "lay-downs," can be plotted with accuracies commensurate with the detail necessary for medium- and small-scale maps. The radial-line or slotted-template method of extending planimetric control, one of the principal contributions of the United States to photogrammetry, permits bridging between ground-control points. This eliminates, to a large extent, the need for extensive ground-survey operations in reconnaissance mapping from areal photography. Hypsometric information, however, is more difficult to obtain by this method and in general may be portrayed far less accurately than the corresponding planimetric features [36]. Improvement can be expected with continuing research; on the other hand, advances in the techniques of taking vertical photographs from very high altitudes have reduced the costs of multiplex and similar mapping until they are comparable with those of trimetrogon photography. Such developments leave the ultimate refinement of trimetrogon photography in a questionable state.

THE PROSPECT FOR AIR PHOTOGRAPHY

There is no question that the displacement of the plane table by the air photograph as the primary tool of extensive mapping operations is a cartographic revolution in itself. While this directly concerns the mapping aspect of cartography, it also affects geographic cartography, the art of expressing the earth's patterns as collected from individual surveys the world over. Not only is the survey of the most inaccessible parts of the world possible, but the cameraman works infinitely faster and more cheaply than the men with the plane table and rod. A strip of the United States, from the Atlantic to the Pacific, for example, was crossed and photographed in a single flight, bringing in a wealth of information that no surveying party could have collected in years of work on the ground.

The increase in the amount of coverage by aerosurveying is prodigious, but for the cartographer the type of information is just as significant. Air photographs provide a richness of detail that makes their utilization for maps employing the present conventional symbols a problem. Perhaps a new type of map will be developed [37]. The use of air photography also has made it a practical procedure to enlarge the scale of topographic maps in the United States and to depict contour lines more exactly and expressively than had been possible before.

THE DEVELOPMENT OF NEW KINDS OF MAPS

The great majority of maps produced during the immediate past have been of conventional types, but this period also has seen the rise to prominence of several new kinds of maps which had hitherto hardly advanced beyond the experimental stage.

Chief among these is the aeronautical chart. Before World War II aeronautical charts were similar to the conventional topographic maps, but on a smaller scale [38]. Some experimentation had been undertaken during the 1930's and earlier, but did not immediately result in any new maps [39]. With the coming of the war the needs mushroomed, and the increase of speeds and of different flying conditions required new types of maps [40; 41]. Target charts, approach charts, special navigation charts (electronic and otherwise), and many others are required for the specialized needs of the air navigator and pilot. Civil, military, and private aviation require different kinds of charts on different scales. In order to meet these requirements in a reasonably integrated fashion there has been established an international organization (International Civil Aviation Organization) which has attempted to provide for a standardization of these types of charts.

SPECIAL-PURPOSE MAPS

The term special-purpose map has been applied broadly to a variety of chorographic maps. Many of them are interpretative in that the categories of phenomena portrayed on them are selected for the purpose of showing the significance of facts rather than the mere facts themselves. Maps portraying terrain analysis or industrial potential are typically interpretative. During the war the Military Geology Unit of the U. S. Geological Survey produced the majority of interpretative maps dealing with physical conditions, such as landing beaches or construction materials, or water supply. In most cases these maps were parts of detailed intelligence studies, although separate maps or series of maps on specific subjects or areas were common. Scales varied but tended to be in the range of 1:250,000 to 1:1,000,000 with larger scales where maps were designed for a specific operation. The physical interpretative map did not, however, originate in World War II. The Germans had made extensive use of them in World War I, and their development can be traced well back into the 19th century. However, their broad use in America was initiated during World War II.

Special-purpose maps in other fields were prepared by several agencies. The Office of Strategic Services, the Board of Economic Warfare, and the Division of Geography of the Department of State were among the more important producers. Within the military establishment various intelligence and quartermaster organizations also prepared maps of this kind.

TERRAIN MODELS

Terrain models have been used to portray earth features for almost as long as have flat maps [42: iii-ix]. However, their direct connection with cartography received but slight recognition in America before World War II. During the war both the technique of production and the utilization of terrain models advanced greatly. Until 1942, model making was a craftsman's job, devoid of any mass-production possibilities. The cost in time and money prohibited the construction of great numbers of models. During the war the demand for models for planning purposes, for use in service schools, for briefing troops in special operations, and for the recognition of bombing targets increased to such an extent that model making, as with other phases of cartography, was converted to mass-production techniques.

The Map Division of the Office of Strategic Services established a model section in 1943 that produced the bulk of the small-scale models used during the war. At the end of hostilities this unit was incorporated in the Army Map Service, where it has developed rapidly [43]. Model-making detachments, primarily concerned with operational scales, were active in the United States and in the several theaters.

Various techniques were used. At first plaster of paris or papier-maché was placed in molds that were cut and cast by hand. As the demand increased, cutting machines, special projectors, and special casting methods and materials were introduced. Among the latter was the use of sponge rubber, which permitted a large model to be rolled for ease of shipment or storage and at the same time increased its durability.

The production of plastic models, utilizing standard maps printed on plastic sheets and then molded by heat and suction around a master mold, was developed during the war. Although the method was used sparingly at first, more models are now made by this method, both for military and civilian purposes, than by any other method.

THE STUDY OF GRAPHIC METHODS

The concern with terrain representation was not limited to the three-dimensional map. The search for graphic means of representing the land surfaces on medium- and large-scale maps is an ancient struggle. With the increase of air navigation, with the increase in popular interest in maps, and with the increase in the number of map users, the development of methods of providing a graphic impression of relief have grown apace [44; 45]. Since the war, topographic maps of the United States for the first time have appeared with shaded relief expertly applied.

The development of aerial cameras, rectifiers, and half-tone screens capable of retaining detail on printed photo mosaics led to the use of photo-maps. They were employed extensively in World War II both as training aids and for operations where satisfactory topographic maps were not available. The largest project was the coverage of northern France by photo-maps at a scale of 1:125,000 for use in the Normandy invasion and subsequent operations. If good photography is available, the main problem is that of ground control and of making the necessary rectification of the photos to permit the preparation of controlled mosaics that are sufficiently accurate to permit the plotting of grids. In areas of moderate and high relief, photographs of models have been used as a terrain background for map data.

MAP LIBRARIES AND MAP ANALYSIS

When cartographers and geographers were called upon during World War II to make use of foreign maps and to evaluate them in all manner of ways, it became clear that there was a serious lack of familiarity with coverage, availability, quality, and symbolism. A knowledge of these things is obviously needed to prepare a person to analyze and utilize varied maps from diverse sources. Recognition of this led to the establishment of units specializing in what was called during the war Map Intelligence. Since the war the methods of evaluating maps and of recognizing their qualities has been continued and expanded [46]. Paralleling the growth of this cartographic counterpart of critical bibliographical work has been the expansion of map libraries and their holdings. The governmental holdings in the Army Map Service, the Library of Congress, and other collections have grown phenomenally, owing primarily to the activities of geographer-cartographers. University departments of geography have benefited tremendously from the Depository Program of the Army Map Service, in that adequate foreign map types and coverage have been made available for instruction. The increased use of governmental maps has made necessary the establishment of facilities such as the Map Information Office (located in the Geological Survey but supported by several government agencies) [47] for channeling requests and providing general analyses of coverage and quality. Publications related to this aspect of cartography since the war are numerous. The increase in the num-

ber of articles appearing in American and foreign journals concerning
map resources of various parts of the world is a welcome addition to cartographic literature [48; 49; 50]. The cataloging problem, always a difficult one, also brought forth a flood of papers.

CURRENT STATUS OF CARTOGRAPHY

The cartographic developments of the 1940's affected many geographers and sharpened their awareness of the broad relationship between
geography and cartography. Wartime experience showed that among
professionally-trained geographers are to be found the persons best equipped for many phases of cartography, especially those involving source
material, map compilation and editing, small-scale projections, landform
presentation, statistical distribution, and interpretative cartography. The
present emphasis within geography on these phases of cartography is in
marked contrast to conditions before the war when many geographers
thought of cartography merely as a research tool or a draftsman's job.
The field of geography has received considerable benefit from its enforced wartime contact with the broad field of cartography. A better
understanding of the use of maps and their specialized adaptation for portraying geographic data has been gained by geographers in connection
with all phases of the science.

New Techniques of Map-Making and Map Reproduction

Perhaps the present is the beginning of a new age in the history of
cartography, comparable to the beginning of the 16th century. At that
time the great discoveries, the revival of learning, and the invention of
engraving and printing combined to produce an unparalleled outburst of
map activity. Factors of similar importance are working today, such as
the airplane, photography, electronics, and plastics. New techniques in
the preparation and reproduction of maps, models, and globes make it
possible to create new types.

The map draftsman of the past was a highly specialized craftsman,
drawing hachure lines or lettering with infinite patience and perfection.
Once a map was drawn, it lasted; changing it was costly. At present,
nearly anyone, often with short training, may stick-up acceptable lettering and lay cellophane shading quickly and efficiently.

Machines can lay out projections, plastics with low distortion characteristics help to eliminate lack of register, and maps are produced more
cheaply and faster than ever before [51; 52]. Papers are better than in the
past, and offset machines running with incredible speed turn out copies
by the millions, some in two to four colors in the same operation.

GOVERNMENT MAPPING

Governmental agencies produce at present many times more maps than does private industry, and probably no place on earth produces more maps than does Washington, D. C. There has been an enormous increase in coverage on a variety of scales. The whole earth has been almost completely remapped on a scale of 1/1,000,000, and most parts of it on a scale of 1/500,000. There is still need for further revision, for in many areas the data these new maps are based on are sparse and unreliable. For large-scale maps the Transverse Mercator Projection seems to be becoming a favorite [53]. It is not impossible that the use of this projection will spread to air-navigation charts. Map symbolism is undergoing a change, and will change a great deal more as the methods for portraying the earth's surface are perfected; shading is now an accepted feature as an addition to contour lines; and the variety of color used on standard sheets is increasing. The 1:250,000 series of the United States has seven colors.

Washington now employs more than half of the country's cartographers and produces an ever-larger proportion of its published maps. This raises a serious problem. Not so long ago the government was concerned mostly with topographic maps and marine charts; at present it engages in every kind of map activity, requiring every type of cartographic skill. Many of the better cartographers are employed in Washington, but there they do specialized work and of a set standard; rarely can they develop their full talents, and rarely can they engage in any pure research in the field of cartography. Universities cannot compete with the government in terms of salaries; consequently there is a steady movement of cartographers to Washington, whereas the number engaged in teaching and in scholarly research remains small.

Another problem arising from the concentration of cartographic activity in Washington is the restriction of new data and new sources of information. Obviously not all the new maps used in government agencies can be made available; yet the wisdom of withholding a large proportion of the materials needed for teaching and research might be challenged. The geographer-cartographer in a university department of geography bears a heavy responsibility [54]. He must try to keep abreast of the many new technical developments; he must try to instruct students in the use of these new techniques; and, most important, he must continue to carry on research studies so that the theoretical base of his field may be adequately developed. To carry on his work without close contact with the cartographic advances in government agencies is most difficult if not impossible. If he fails to do all this geographic cartography will inevitably wither and the field will eventually be staffed only by clever draftsmen.

THE PROSPECT IN GEOGRAPHIC CARTOGRAPHY

Henceforth much of the research carried on in the field of cartography will probably be in the government or under government contract. It is not the purpose of this chapter to consider in detail the undesirable aspects of such a condition, but rather to forecast possible avenues of research that seem desirable for the profession as a whole and that may be reasonably carried on without vast technical facilities. Governmental cartographic research is dictated by current needs and, in general, will rise and fall as technical and practical requirements change. Far too large a proportion of the research being pursued is of this nature, and relatively little research is being done in universities and by non-governmental agencies. This is an unhealthy situation.

The question may be asked: when advantages in terms of technical and financial support and in terms of numbers of competent workers all lie with the government, what is left for the scholar and independent research cartographer? The answer is not in any way discouraging. The following suggestions are neither complete nor necessarily organized in such a way as to outline the field of cartographic research. They merely indicate the more obvious avenues.

PROJECTIONS

At a time when this country has awakened to the fact that it is but part of a whole earth, the study and teaching of projections is obviously important. The erroneous space concepts of the majority of Americans is living proof that our present understanding of projections is insufficient. Particularly important is the projection for large areas or the whole earth. Our standards of choice, by and large, are biased by conventions. We are not willing enough to experiment. For example, most projections can be made oblique in infinite variety with very little effort, yet the departure from the ordinary is looked upon with suspicion. Perhaps if cartographers were to devote more effort to the variation of projections, and to circulate the results in the professional journals, there would be less tendency in the geographic profession to judge projections on the basis of familiarity. Perhaps the greatest advances could be made in the employment of equal-area projections.

Much of the blame for the negative attitude toward projections among geographers (and cartographers) may be placed on inadequate teaching. Except in rare instances, the student is left with little appreciation of this fascinating subject and with no desire to construct his own projections or even to be selective. New methods of teaching and analysis of projections are necessary.

REPRESENTATION OF TERRAIN

Studies in the method of representing terrain at different scales are conspicuous by their rarity [55; 56; 57]. This is a gap that needs to be filled. There has been a considerable increase in the use of terrain drawing and shading; but the methods that have been suggested to accomplish this should be evaluated. Little attention has been paid in the United States to the actual visual efficiency of the methods of landform representation; not even of the favorite device, the contour. Although, as was pointed out earlier, American cartography can claim as an original contribution the development of the physiographic diagram or the landform drawing through the work of Davis, Lobeck, Raisz, and others, little has been done to provide a critical appraisal of these methods or to develop the teaching of the art of terrain drawing. A most interesting problem in connection with physiographic or landform drawing is the degree to which interpretation can be included in the graphic expression of the forms.

STATISTICAL PRESENTATION

Perhaps in no other aspect of cartography is the opportunity for important research greater than with regard to the problem of presenting numerical data on maps. Studies of how to map ratios and correlations effectively are sorely needed. Even the time-honored dot map is capable of considerable refinement, as J. R. Mackay has shown [58].

Probably the most valuable device for the presentation of statistical data is the line representing equal values of some kind. John K. Wright has pointed to the importance of distinguishing between those lines that connect or pass through points of equal value, and those which pass through areas in which equal ratios exist [59]. This distinction is clear, yet the full implications of it have been to some extent obscured by a confusion in terminology. Many geographers use the term isopleth to refer to all kinds of lines of equal value or ratio; others use the term isarithm for all such lines. Wright's proposal for naming the different kinds of lines has not been fully accepted. The terms adopted in this volume are tentative and provisional: as a generic term for all such lines of equal value or ratio, *isogram*; as the term for lines passing through points of equal value, *isarithm*; and as the term for lines passing through areas in which equal ratios exist: *isopleth*. Research is needed regarding the techniques of placing isarithms, and especially for the placing of the less precisely located isopleths; the problem involved in the selection of intervals in the values and the ratios for different purposes calls for further study [60; 61; 62].

There are many uses to which isogram maps can be put. All geographers and cartographers are familiar with the large amount of information concerning landforms that can be gained from the interpretation of the patterns of contours. So far as is known, however, the study of isogram patterns has been limited to the contour; yet there is no reason for believing that equally informative interpretations might not be revealed by the careful study of other families of isograms. No work of this sort has been done, but an illustration of what might be done may not be out of place. From the patterns of isohyets or rainfall lines it should be possible to identify the type of rainfall: the thunderstorm pattern, for example, would be distinguished from rainfall produced along a mountain front; or the rainfall produced along a windward coast by steady on-shore winds could be distinguished from that produced along a coast by the interaction of air masses of different origin. From the isohyets the kind of rainfall might be as well determined as the kind of landforms can be determined from contour patterns. Isarithms, because of their greater precision, will probably be the first to reveal inherent relationships. Isopleths, if their precision can be controlled, may also yield useful results when analyzed in this fashion.

CARTOGRAPHIC DESIGN

Another field of research is that connected with the techniques and media of presentation. Statistical presentations should create in the mind of the reader the correct impression that the cartographer wishes to convey. If a map cannot be read, for any reason other than the illiteracy of its reader, it is a failure. Yet the poor quality of cartographic design in many maps is well known. This requires a considerable amount of research in the visual aspects of all kinds of symbols, a field in which little has been done [63]. The studies that have been made have been aimed at the range of black and white values, but equally important is a systematic approach to the value problem in the use of color [64; 65]. Such subjects as visual acuity, apparent depth-effects of colors, simultaneous contrast, and a host of others need investigation. A valuable body of published materials has been made available from the experimental and analytic research of workers in other fields, and can be found in the optical and psychological journals, but it seems to be slow in reaching the cartographers [66]. A conspicuous exception is the research being carried on by certain government agencies in the field of map design.

Lettering is an important aspect of map design. Little research has been done on the problem of styling and the selection of lettering for cartographic use [63]. Freehand lettering that can be fitted and styled as necessary is being replaced by various forms of what cartographers call "stick-up," that is, separately printed names and letters cut out and stuck on the map in appropriate positions. So far as is known only the stick-up of the National Geographic Society has been designed specifically for cartographic use [67]. Type designed for reduction and for curved positioning is a definite requirement. Further work on legibility and on the relative visibility of various sizes of lettering on different map backgrounds and for different purposes is necessary [68]. The entire gamut of presentation techniques requires research. Relative line weights, balance, layout, movement, and the like, as applied to cartographic presentation, are capable of such infinite variation that the cartographer needs experimental evidence to aid him in his choice of media and techniques.

In view of the ever-increasing amount of cartographic production being centered in the government, it is likely that the paralysis of standardization may spread even to the small-scale map. If research is undertaken to expand our understanding and, at the same time, to point up our genuine ignorance of fundamental aspects of cartography, then cartography can continue to advance in more than the mechanical sense. Standardization within map series and in topographic scales is an obvious necessity, but it should not invade the entire field of cartographic technique. Even the standard conventions need investigation. For example, testing has shown that when groups have been asked to grade and rank graduated circles drawn in strict areal relationship with the values they represent, the readers consistently receive incorrect impressions. Clearly, grading diameters according to the square roots of the data is an incorrect technique. Some other curve is necessary. The facility with which cartographic techniques can be and are misused is ample evidence of the need for further research [69].

The tremendous strides made in the past few decades in the field of map production, particularly reproduction, makes it necessary for the cartographer continually to investigate ways and means of preparing his maps and of reducing the costs. For example, owing to the increase in the size of classes in schools the usual wall map has become too small to be read by many of the pupils. Yet larger wall maps when prepared according to current techniques are costly. The development of the newer techniques of color reproduction makes a wide range of effects possible. The application of these to cartography requires investigation.

MAPPING ANALYSIS

It is rather surprising that a tool as indispensable to the geographer as the topographic map has not received more attention from geographer-cartographers. To be sure, the shortcomings of the United States topographic coverage and quality has often been considered in a general way, and frequently overemphasized, but little real research on the quality and coverage of topographic mapping took place until World War II. As was pointed out previously, the sudden need for using foreign topographic maps of large areas made it startlingly clear that American understanding of this subject was inadequate. Since that time there has been an increasing attention to map sources and to the methods of map evaluation.

The most important facet of this complex research problem is probably the inventory aspect, that of ascertaining what mapping there is, its availability, and particularly its quality. This requires, for topographic maps, searching out the basic geodetic control, the methods of survey, the dates of production, and a host of other less important facts. Only when that is done on a regional basis will the geographer and cartographer have a proper understanding of the major source of his information. Equally important is the inventorying of published small-scale maps according to subject matter, of which some of the recent publications of the Map Division of the Library of Congress are good examples [42; 70].

Allied to such research are other, more specialized studies. For example, one of the great needs today is for an up-to-date exposition of minor political boundaries throughout the world as a basis for the mapping of census data. The problem is a continuing one because the boundaries are constantly changing. Ratios and correlations depending upon the areas of civil divisions are important geographic items. Since the areas of minor civil divisions cannot be determined except from maps, such areal data can be calculated only approximately in regions where the divisions have not been accurately mapped [71]. Similarly, analyses of slopes, and other landforms cannot be properly derived from contour maps without considerable understanding of the quality of such maps.

Historical cartography has not been developed in recent times in America to the extent that it has in Europe. This is surprising, for it has much to offer to the historical geographer. The spread of geographical ideas and the movements of peoples are reflected in the maps of a period. With the reappearance in 1948 of *Imago Mundi,* and the publication of Lloyd Brown's volume it is to be expected that historical cartography will gain in interest in America [72].

CONCLUSION

Today the need for more and better maps is fully acknowledged, not only by the geographic fraternity, by workers in the physical and social sciences, and by the military, but by all concerned with the peaceful administration of the world. The qualities of cartographic materials and the efficiency of methods of reproduction have improved considerably in the last few years. A wealth of basic information now exists concerning the visual response to color, form, and overall design. Our understanding of the methods of statistical and cartographic analysis is much greater than it used to be. Thus there are immense possibilities for introducing revolutionary improvements in maps, which should make them more legible and more truthful. It can only be hoped that the increasing calls on the services of the geographic cartographer for teaching and production will not hamper this research. The advancement of a science or technique can be seriously retarded or even brought to a standstill by subordination of fundamental research to the demands of the moment. The full possibilities cannot be explored unless there is freedom to think originally and constructively and to put new ideas to the test.

REFERENCES

1. *Modern Cartography: Base Maps for World Needs*, United Nations Secretariat, Department of Social Affairs, Lake Success, New York, 1949.
2. HARDING, G. H. "A Possible Solution to the Problems of Surveying and Mapping," *Surveying and Mapping*, 11 (1951): 104-106.
3. WILSON, L. S. "Geographic Training for the Postwar World: A Proposal," *Geographical Review*, 38 (1948): 575-589.
4. "Lessons from the War-Time Experience for Improving Graduate Training for Geographic Research," *Annals of the Association of American Geographers*, 36 (1946): 195-214.
5. KISH, G. "Teaching of Cartography in the United States and Canada," *The Professional Geographer*, 2 (1950): 20-22.
6. LOBECK, A. K. *Physiographic Diagram of the United States*, Madison (Wisconsin), 1922.
7. RAISZ, E. "The Physiographic Method of Representing Scenery on Maps," *Geographical Review*, 21 (1931): 297-304.
8. FENNEMAN, N. M. *Physiography of the Western United States*, New York, 1931; SMITH, G.-H. "Physiographic Diagram of Japan," *Geographical Review*, 24 (1934): 402; ROBINSON, A. H. "Physiographic Diagram of Tyosen (Korea)," *Geographical Review*, 31 (1941): 654.
9. JONES, W. D. "Ratios and Isopleth Maps in Regional Investigation of Agricultural Land Occupance," *Annals of the Association of American Geographers*, 20 (1930): 177-195.
10. WRIGHT, J. K. "A Method for Mapping Densities of Population with Cape Cod as an Example," *Geographical Review*, 26 (1936): 103-110.
11. GOODE, J. P. "Studies in Projections: Adapting the Homolographic Projection to the Portrayal of the Earth's Surface Entire," *Bulletin of the Geographical Society of Philadelphia*, 17 (1919): 103-113.
12. ———. "The Homolosine Projection: A New Device for Portraying the Earth's Surface Entire," *Annals of the Association of American Geographers*, 15 (1925): 119-125.
13. BOGGS, S. W. "A New Equal Area Projection for World Maps," *Geographical Journal*, 73 (1929): 241-245.
14. ROBINSON, A. H. "Interrupting a Map Projection: A Partial Analysis of its Value," *Annals of the Association of American Geographers*, 43 (1953): 216-225.
15. MARSHNER, F. J ."Structural Properties of Medium- and Small-Scale Maps," *Annals of the Association of American Geographers*, 34 (1944): 1-46.
16. FISHER, I., and MILLER, O. M. *World Maps and Globes*, New York, 1944.
17. RAISZ, E. "Map Projections and the Global War," *The Teaching Scientist*, 2 (1946): 33-39.
18. ROBINSON, A. H. "An Analytical Approach to Map Projections," *Annals of the Association of American Geographers*, 39 (1949): 283-290.

19. ———. "The Use of Deformational Data in Evaluating World Map Projections," *Annals of the Association of American Geographers*, 41 (1951): 58-74.
20. STEWART, J. Q. "The Use and Abuse of Map Projections," *Geographical Review*, 33 (1943): 589-604.
21. MILLER, O. M. "A Conformal Map Projection for the Americas," *Geographical Review*, 31 (1941): 100-104.
22. ———. "Notes on Cylindrical World Map Projections," *Geographical Review*, 32 (1942): 424-430.
23. ———. "A New Conformal Projection for Europe and Asia," *Geographical Review*, 43 (1953): 405-409.
24. CHAMBERLAIN, W. *The Round Earth on Flat Paper*, National Geographic Society, Washington, 1947.
25. RAISZ, E. *General Cartography*, New York, 1948.
26. DEETZ, C. H. *Cartography*, Washington, 1943.
27. WRIGHT, J. K. "Highlights in American Cartography, 1939-1949," in *Comptes Rendus du XVI⁰ Congrès International de Géographie*, Vol. 1, Lisbonne, 1949: 298-314.
28. ROBINSON, A. H. "Cartography," in *Ten Eventful Years*, Encyclopedia Britannica, Chicago, 1947: 547-550.
29. BRYAN, G. S. "War Charts of the U. S. Navy," *Military Engineer*, 38 (1946): 131-138.
30. WOODWARD, L. A. "Cartography at War," *Soil Conservation*, 11 (1945): 75-78, 82.
31. HOUGH, F. W. "Progress of the European Triangulation Adjustment," *Transactions of the American Geophysical Union*, 29 (1948): 915-918.
32. WHITTEN, C. A. "Progress of the European Triangulation Adjustment," *Transactions of the American Geophysical Union*, 30 (1949): 882-883.
33. STONEMAN, W. G. "International Cooperation in Mapping Latin America," *Surveying and Mapping*, 11 (1951): 149-158.
34. HARDIN, M. J. "Topographic Mapping with the Multiplex Aero Projector," *Surveying and Mapping*, 5 (1945): 39-46.
35. United States Army Air Force, *Reconnaissance Mapping with Trimetrogon Photography*, Washington, 1943.
36. RAISZ, E. "The Use of Air Photos for Landform Maps," *Annals of the Association of American Geographers*, 41 (1951): 324-330.
37. ———. "Landform, Landscape, Land-Use, and Land-Type Maps," *Journal of Geography*, 45 (1946): 85-90.
38. ROSS, R. L. "The U. S. Sectional Airway Maps," *Military Engineer*, 24 (1932): 273-276.
39. MILLER, O. M. "An Experimental Air Navigation Map," *Geographical Review*, 23 (1933): 48-60.
40. FITZGERALD, G. "Aeronautical Charts in an Air Age," *Surveying and Mapping*, 4 (1944): 13-17.
41. SMITH, P. A. "Aeronautical Chart Production," *Military Engineer*, 35 (1943): 357-361.
42. RISTOW, W. W. *Three-Dimensional Maps: An Annotated List of References Relating to the Construction and Use of Terrain Models*, Map Division, Library of Congress, Washington, 1951.
43. SPOONER, C. S. JR. "Modernization of Terrain Model Production," *Geographical Review*, 43 (1953): 60-68.
44. KINGSLEY, R. H., and HOLMES, H. C. "Terrain Representation from Aerial Photographs for Aeronautical Charts," *Photogrammetric Engineering*, 11 (1945): 267-271.
45. MUNDINE, J. E., and SHELTON, H. "Visual Topography," *Photogrammetric Engineering*, 11 (1945): 272-278.
46. WILSON, L. S. "Lessons from the Experience of the Map Information Section, O.S.S.," *Geographical Review*, 39 (1949): 298-310.

47. FUECHSEL, C. F. "The Map Information Office of the United States Geological Survey," *Surveying and Mapping*, 6 (1946): 251-253.

48. PLATT, R. R. "Official Topographic Maps: A World Index," *Geographical Review*, 35 (1945): 175-181.

49. LEITE DE CASTRO, C. "Cartography in Brazil," *Surveying and Mapping*, 5 (1945): 8-13.

50. PETERS, F. H. "Surveying and Mapping in Canada," *Surveying and Mapping*, 3 (1943): 8-11.

51. SACHS, S. G. "Map Scribing on Plastic Sheets versus Ink Drafting," *The Professional Geographer*, 4 (1952): 11-14.

52. WICKLAND, L. R. "Evolution of Paper and Plastics as Related to Mapping," *The Professional Geographer*, 4 (1952): 15-18.

53. O'KEEFE, J. A. "The Universal Transverse Mercator Grid and Projection," *The Professional Geographer*, 4 (1952): 19-24.

54. ROBINSON, A. H. "University Training for Government Cartographers," *The Professional Geographer*, 3 (1951): 4-6.

55. TANAKA, K. "The Relief Contour Method of Representing Topography on Maps," *Geographical Review*, 40 (1950): 444-456.

56. ROBINSON, A. H. "A Method for Producing Shaded Relief from Areal Slope Data," *Annals of the Association of American Geographers*, 36 (1946): 248-252.

57. BATCHELDER, R. B. "Application of Two Relative Relief Techniques to an Area of Diverse Landforms: A Comparative Study," *Surveying and Mapping*, 10 (1950): 110-118.

58. MACKAY, J. R. "Dotting the Dot Map: An Analysis of Dot Size, Number, and Visual Tone Density," *Surveying and Mapping*, 9 (1949): 3-10.

59. WRIGHT, J. K. "The Terminology of Certain Map Symbols," *Geographical Review*, 34 (1944): 653-654.

60. ALEXANDER, J. W., and ZAHORCHAK, G. A. "Population-Density Maps of the United States: Techniques and Patterns," *Geographical Review*, 33 (1943): 457-466.

61. MACKAY, J. R. "Some Problems and Techniques in Isopleth Mapping," *Economic Geography*, 27 (1951): 1-9.

62. WRIGHT, J. K. (and three other authors). *Notes on Statistical Mapping, with Special Reference to the Mapping of Population Phenomena*, American Geographical Society and the Population Association of America, New York, 1938.

63. ROBINSON, A. H. *The Look of Maps: An Examination of Cartographic Design*, Madison (Wisconsin), 1952.

64. BIRREN, F. *The Story of Color from Ancient Mysticism to Modern Science*, Westport (Connecticut), 1941.

65. EVANS, R. M. *An Introduction to Color*, New York, 1948.

66. CHANDLER, A. R., and BARNHART, E. N. *A Bibliography of Psychological and Experimental Aesthetics*, Berkeley (California), 1938.

67. RIDDIFORD, C. E. "On the Lettering of Maps," *The Professional Geographer*, 4 (1952): 7-10.

68. ROBINSON, A. H. "The Size of Lettering for Maps and Charts," *Surveying and Mapping*, 10 (1950): 37-44.

69. WRIGHT, J. K. "Map Makers are Human: Comments on the Subjective in Maps," *Geographical Review*, 32 (1942): 527-544.

70. RISTOW, W. W. *Marketing Maps of the United States: An Annotated List*, Map Division, Library of Congress, Washington, 1951.

71. PROUDFOOT, M. *Measurement of Geographic Area*, Bureau of the Census, Washington, 1947.

72. BROWN, L. A. *The Story of Maps*, Boston, 1949.

INDEX

(Numbers in italics refer to bibliographies)